ISBN 978-0-656-50345-2
PIBN 10676529

This book is a reproduction of an important historical work. Forgotten Books uses
state-of-the-art technology to digitally reconstruct the work, preserving the original format
whilst repairing imperfections present in the aged copy. In rare cases, an imperfection in
the original, such as a blemish or missing page, may be replicated in our edition. We do,
however, repair the vast majority of imperfections successfully; any imperfections that
remain are intentionally left to preserve the state of such historical works.

BULLETIN

OF THE

Harvard University.

MUSEUM OF COMPARATIVE ZOÖLOGY

AT

HARVARD COLLEGE, IN CAMBRIDGE.

VOL. XIX.

CAMBRIDGE, MASS., U. S. A.

1890.

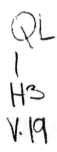
UNIVERSITY PRESS:

JOHN WILSON AND SON, CAMBRIDGE, U. S. A.

CONTENTS.

No. 1. — *Studies on Lepidosteus.* Part I. By E. L. MARK.[1]

CONTENTS.

I. Introduction.

I BECAME deeply interested in the embryology of Lepidosteus through reading the paper on that subject published by Mr. Agassiz in 1878, and determined to avail myself of the first opportunity of following up the study. I desired particularly to pursue the development of the early stages. A little later I was further incited to this by the brief account of it which Balfour gave in the second volume of his Comparative Embryology (1881).

I had already formed plans for going to Black Lake, in the vicinity of Ogdensburg, N. Y., in the spring of 1882, for the purpose of getting material for the contemplated study, when I learned from Mr. Agassiz that Balfour had in hand an extensive paper on the subject. Mr. Agassiz also informed me that he still had left a part of the material from which Balfour had been supplied, and he kindly placed this at my

[1] Contributions from the Zoölogical Laboratory of the Museum of Comparative Zoölogy at Harvard College, No. XV.

disposal. I thought it would be desirable, nevertheless, to procure an additional supply of eggs and embryos, and especially to endeavor to rear the young beyond the stages already in my possession.

On account of my duties in College it was impossible for me to leave Cambridge until nearly the middle of June, — almost a month after the usual time of spawning. Nevertheless, owing to the extreme backwardness of the season, I hoped that I might be able to procure some material, and was confirmed in this by correspondence with Mr. J. H. Perry, through whom I learned that up to a day or two before the time I had fixed upon for setting out the gar-pike had not spawned.

I arrived at Black Lake on the evening of June 13th. The weather had meantime grown warm, and the fish had already spawned, but I was able to secure some eggs which were not very far advanced in development. By a number of processes I killed and preserved at short intervals sets of embryos which presumably belonged to the same lot of spawn. The eggs which were collected from different localities were kept in separate earthen-ware dishes and supplied with fresh water every twelve hours. In this way the embryos were easily kept alive until they hatched. Then they soon attached themselves by means of their peculiar maxillary disks to the sides of the dishes, and near the surface of the water, where they clung with a tenacity truly surprising.

A number of eggs were preserved — principally by means of Kleinenberg's picro-sulphuric mixture — at intervals of a few hours, beginning on the afternoon of June 14th and extending through several succeeding days. This method was controlled at intervals by preservations made in alcohol, in chromic acid, in osmic acid followed by potassic bichromate, and in the last named reagent alone.

Besides the large number of embryos which were preserved at Black Lake, I took away with me (June 20) many more — upwards of a hundred — which had recently hatched. These living fishes were carried in a narrow-necked tin pail, to the sides of which they adhered very firmly.

Instead of returning directly to Cambridge I took the gar-pike with me to my summer residence on the south shore of Lake Erie, about forty miles southwest of Buffalo, N. Y. Although this journey extended over more than three hundred miles the fishes survived it well. The greater part of it, however, was rendered comparatively easy, since it was made by steamboat instead of railway. It is fortunate for such an undertaking that these fishes have so large a yolk-sac, since it obviates the necessity of procuring food for many days after hatching.

The question of being able to raise them beyond the stages already secured by Mr. Agassiz so evidently depended on finding a suitable food for them, that I spared no pains to accomplish this end. Many kinds of meat and fish were minced and fed to them, but none of these was acceptable. The minced liver, which Mr. Agassiz used with success at first, was likewise refused. Fragments of meat were suspended in the water by fine threads, but neither when moved about nor when left perfectly quiet did they seem to attract attention. Great numbers of water-fleas (Cladocera) were put in the water with the young fishes, but the latter made no attempts to catch them. It was not until after many fruitless trials that organisms were found which were seized with such eagerness, and so persistently, as to leave no doubt that they were the natural food of the young gar-pike. These were *the larvæ of the common mosquito*. They constituted the exclusive diet of the young fishes until the latter became large enough to catch and swallow minute "mud minnows" (Fundulus), on which they subsequently fed as long as they were kept alive.

When it was once ascertained that the young fishes would take mosquito larvæ, there was no longer any serious question about the feasibility of rearing them, nor was it doubtful that these larvæ formed their natural food, for the shallow and quiet waters at the margins of Black Lake and along the creeks which feed it abound in mosquitos. It was by means of this diet that the fishes were kept in a thriving condition during the stages immediately following the absorption of the yolk.

From the 20th of June until the 1st of July specimens were, with a few exceptions, killed every twenty-four hours; and from the 1st of July until the beginning of August, usually at intervals of about forty-eight hours.

By the 3d of August there remained besides those which had not been preserved only about a dozen living fishes. On that date I started for Newport, R. I., travelling by rail to New York. These remaining fishes were carried by hand in a tin pail suspended by a spring; but owing to the difficulty of carrying in the pail a sufficient number of mosquito larvæ, and more particularly to the impossibility of properly renewing the water, about half of them succumbed to the unfavorable conditions of railway travel and were put into alcohol. One more died on the way from New York to Newport; but the remaining ones, having been fed on larvæ after my arrival at Newport, appeared to thrive. At the end of a month they were taken to Cambridge, where they were put into a large glass jar and supplied with running hydrant water.

Here they were also kept on the diet of mosquito larvæ until from one
cause or another they had all died.

My failure to secure early stages of the eggs in the spring of 1882
made me desirous of repeating the attempt at a more seasonable time
the following year. With this object in view I left Cambridge for
Ogdensburg, May 18, 1883.

Judging from my previous experience that it would be difficult to
procure fertilized eggs in sufficient quantities without great labor, if they
were to be individually detached from the rocks, I procured several yards
of thin muslin of a color resembling the stones in the lake. I planned
sinking this and loading it with small stones in the water on some of
the "points" most frequented by the gar-pikes at spawning time. I
hoped in that way to secure a large number of eggs firmly attached to
the cloth, which I could then remove to a box suitably provided with
wire nettings to allow the necessary circulation of the water. Had it
proved successful, this device would have enabled me to have under
control the eggs thus acquired, and would have allowed me a degree of
certainty as to the age of the preserved material not otherwise easily
attainable.

Unfortunately for my plans, the weather proved to be in several ways
very unpropitious. A long period of cold and rain delayed the spawning to a time much later than common, and when at length, a few days
after the 1st of June, the weather and water became warm enough to
impel to the act of spawning, such high winds prevailed that it was
impossible to watch the movements of the fishes, and the most of them
had spawned before the water became quiet enough to allow one to discover their places of rendezvous. Some of the localities which they had
visited with the greatest constancy during the past years were apparently deserted. Moreover, the cloths, which had of necessity been anchored near the shore, were either set free by the dashing of the waves,
or rolled into ropes which presented a comparatively small surface for
the reception of ova.

The limited time in the latter part of May left at my command, after
completing some work which I was compelled to take with me, was util-
ized in studying the ovarian ova of females captured before the spawning period arrived, and in some attempts at artificial fertilization. I
then succeeded in getting a fairly satisfactory knowledge of the interesting structure of the egg membranes and of the micropyle, but I did not
learn the peculiar relation of the latter to the granulosa until a few

months later, when I had made microscopic sections of the whole ovarian egg, including membrane and granulosa.

As the outcome of this journey I secured a number of series of eggs, beginning with the early stages of segmentation, from which I prepared at intervals and by various methods a considerable quantity of material. I was also able to bring to Cambridge about two hundred young fishes just hatched. Some of these were kept alive until September 26, 1886, — nearly three and a half years.

The fishes brought to Cambridge were put into running hydrant water and fed on mosquito larvæ for several weeks, — until about the 1st of August, — when they were large enough to swallow small "mud minnows" of nearly their own size. These were gradually substituted for the larvæ, and those fishes which were large enough to avail themselves of this kind of diet grew much more rapidly than their mates.

I have also received some eggs from Mr. Perry since my last trip to Black Lake, and although his attempts at fertilization did not prove to be more successful than my own, I still hope to secure before long the early stages which are needed to fill the gaps in my material.

II. Habits of the Young Fishes.

The habits of young gar-pike have already been quite fully described by A. Agassiz ('78ᵃ) and Wilder ('76, '77), so that I shall not have much to add to what has been previously published.

When first hatched the fish is so small in comparison with the size of the yolk-sac that it swims only with the greatest difficulty, and its movements are anything but graceful. It is so disinclined to swim, that, were it left alone in water sufficiently pure to meet its requirements, I have no doubt it would not move from the point of its first attachment for many hours, or even days. When hatched in confinement the young fishes always swim nearly up to the surface of the water and attach themselves to the sides of the dish. When there are a large number of them they may attach themselves to floating objects. Frequently the superficial film of the water — aided possibly by secretions from the oral disk — serves to support an individual in the middle of the dish. Sometimes half a dozen or more individuals form in a cluster, and appear to hang suspended simply from the surface of the water. It is evident that they are not merely floating in the vertical position, because in such cases the surface of the water in their vicinity is always more or less depressed, and upon the slightest touch the fishes begin at once to sink slowly ; if the water has not been too much disturbed, they make no motion while sinking until they have nearly reached the bottom. Before they actually touch the bottom of the dish they appear to recognize their proximity to it, and then begin to make vigorous efforts to swim up to the surface again. This is apparently a very laborious undertaking, and, if they fail to attach themselves at once, they again begin to sink slowly; they seldom attach themselves at the bottom, — especially if the water has remained for some time unchanged, — but always as near the surface of the water as possible. If there are too many to be accommodated in a single row, those last to come crowd in between the tails of those already attached, thus forming a second row; but if there are still others, they usually attach themselves to other fishes rather than take a lower position on the sides of the dish. During the period of yolk absorption they hang pendent and nearly motionless, except for the respiratory movements; those which hang from the surface of the water are vertical, and any sudden motion in the water shows that their bodies are quite limp. When the absorption of the yolk is well advanced, the flexibility of the body is shown in a striking way by the snake-like motions which the animal slowly executes while remaining attached.

The disinclination to swim lasts about as long as the yolk-sac persists. With the gradual disappearance of the latter, the fishes show an increasing tendency to swim about. When at length the sac is nearly absorbed they rest in quite another way. They float near the surface, taking a horizontal position, and remaining perfectly straight and motionless until disturbed; whereupon, by vigorous strokes of the tail, they swim away with remarkable celerity. In transferring the fishes from one dish to another I was accustomed to use a small spoon, but after the absorption of the yolk-sac I found it exceedingly difficult to capture them in that way, so rapid were their movements. The stage at which the fishes begin to swim and float is reached in eight or ten days after hatching. Even at this early age locomotion is accomplished by two distinct methods. The rapid motions are executed by vigorous strokes of the whole caudal region. A slower, gliding motion is maintained by means of the very rapid vibrations of the extreme end of the tail, which are so characteristic of the caudal filament at a later stage, and by the still more rapid motions of the pectoral fins. Not only do the pectorals vibrate when the tip of the tail is motionless, and *vice versa*, but either of the pectorals may be in rapid motion while the other is at rest. This second method of locomotion is apparently very serviceable to the fish, in allowing it to approach its prey unobserved.

At a later stage of development, when the upper lobe of the tail is reduced to a caudal filament, this gliding motion is accomplished, principally at least, by the action of the pectorals. When the fish is advancing, these fins are directed obliquely backward; but when, as often happens, the motion becomes retrogressive, they are directed more nearly at right angles to the body. The motion of the fins is so rapid, that I have been unable to determine by observation if, as is probable, the direction of the *stroke* is reversed in the two cases. Not only the direction of the long axis of the fin, but also the inclination of its transverse axis to the horizon, is conspicuously changed at such times.

The vibrating movement of the caudal filament perhaps assists the forward motion of the fish, but it cannot be considered essential to it, since the filament often remains motionless while the animal is gliding by means of the pectorals. The amplitude of the vibrations made by the filament is not great in any case, — about 15°, — and the terminal half alone is vibratory. When in motion the direction of its axis is usually continuous with that of the spinal column, although it *may* droop more or less while in motion, and is quite liable to do so when at rest; it then presents an even curvature, as seen from the side, and often inclines a little either to the right or left.

When the fish is stemming a current, or, in swimming, is directing the head downward, the caudal filament is kept in rapid vibration; it then takes a dorsal turn, and the curvature is rather abrupt at its base. The whole curvature may amount to enough to make the extreme tip of the filament perpendicular to the axis of the body, but usually it is much less.

During the night of August 6–7, 1883, one of the individuals hatched in June of that year escaped from the tank, and was found in the morning lying in only sufficient water to keep the body moist. Upon being returned to the tank, though still able to swim, it showed evident signs of weakness. The body was considerably arched, just as it has been figured for somewhat younger fishes by A. Agassiz ('78[a], Plate IV. Fig. 39, and Plate V.). I think this case suggests an explanation of the peculiar curved shape exhibited by the fishes reared by Mr. Agassiz. I had already in the previous year imagined that the arched condition was not common, for all my fishes were quite straight, at least so long as they were well nourished. The curved condition of the escaped fish was apparently due to muscular weakness; the curvature was also accompanied by a slight distortion from the sagittal plane. Inasmuch as it subsequently regained its normal condition and became straight, I have no doubt that the curved condition is abnormal, and its appearance in the fishes raised by Mr. Agassiz was probably owing to insufficient nutrition. The control of the pectorals in the case of this escaped fish was not lost, although somewhat impaired. The fish was also able at once to execute vigorous though ill-planned motions with the tail, and was therefore able to swim rapidly, but in a reckless manner. The control of the caudal filament, however, was entirely lost; there was not the least trace of the rapid vibratory movement, even when the pectorals were in active motion. By the following day the use of the caudal filament had been partially regained; but its vibrations were rather feeble, and were resumed only after long intervals of repose.

The temperature of the water greatly affects the power of locomotion, and very cold water may even produce fatal results in a comparatively short time. In the summer of 1882 I placed a dozen or more fishes in the cold water of a spring; within twelve hours half of them were dead. The first signs of an abnormal condition are shown by uncorrelated movements, reckless swimming, and the inability to keep the dorso-ventral axis vertical. The appearance is as though the centre of gravity were located *above* the centre of volume, and the fish gradually became incapable of remaining in its normal position of unstable equilibrium. In swimming the body rolls from side to side. After a time the fish sinks to the bottom of the vessel, and can regain the surface only with considerable exertion; it cannot remain at the surface in a motionless condition. At length the body usually becomes curved sidewise. In this condition the fish may remain for many hours, or even several days. Restoration to fresh water of the ordinary temperature seems to have only a temporary effect, or none. The only way to afford relief is to place the fishes in direct sunlight until the water becomes warm, when, if not already too much affected, they will gradually recover.

The movements of the eyes are principally in the horizontal plane. When one eye is directed obliquely forward, thé other looks obliquely backward at about the same angle, so that the axes of the two eyes are kept approximately parallel.

The manner in which young gar-pike capture and swallow their prey is interesting, and serves to show why it is so difficult to get them to feed on anything except living objects. When very young, as previously stated, they will not feed on anything but mosquito larvæ. The fish always approaches the larva by a slow, even motion, resulting from the vibration of the pectorals and the tip of the tail, until the prey is about opposite the middle of the "bill"; then with a quick lateral motion of the head the larva is snapped up. Occasionally it is bitten through, and, whether struggling or not, is allowed to escape; but the reason of this I did not discover. Usually the larva is not snapped at until by some slight movement it shows signs of being alive, and then only a single snap is made. If unsuccessful, the fish remains quite motionless, or glides slowly away by means of the same motion of the pectoral fins. Fishes, especially when less than an inch in length and when well fed, rarely make a lunge forward toward their prey when they snap at it; the motion is usually only a sudden sidewise bend in the neck region. So cautious are the fishes not to snap at dead larvæ, that, after they have advanced so that the insect is opposite the middle of the bill, if it does not move, they begin to swim obliquely forward, pushing the larva before them without allowing it to glide backward along the bill. In this way a larva may be pushed about several inches before it makes the fatal movement which betrays its condition; or it may, if it remain entirely passive, escape altogether, since the fish, failing to discover evidences of life, leaves it and glides off in search of other food.

The number of insects caught by a single gar-pike is undoubtedly large. I was accustomed to feed the fishes twice a day. In July, 1883, I had the curiosity to watch the largest of them, then a month old and about $2\frac{1}{2}$ inches (62 mm.) long, during its feeding. In the course of ten minutes it caught twenty-one large mosquito larvæ and made nine ineffectual attempts at seizure. But the voracity of young fishes was still more forcibly exhibited when, early in August, they were given small minnows (Fundulus) for food; for then they did not hesitate to catch the minnows even when the latter were considerably larger than themselves; but they succeeded in swallowing only those which were of their own diameter or smaller. Since the minnows have much thicker heads than the gar-pikes, the total weight of the former was doubtless always somewhat less than that of the latter.

The minnows were caught in the same way as the mosquito larvæ, by a sudden sidewise motion of the head; but being too large to be swallowed at a single gulp, they were at first impaled on the sharp teeth, and then by a series of deliberate movements put in a position to be swallowed. Almost without exception, the minnows were swallowed head first, irrespective of the region of the body first seized. If caught near its tail, as usually happens, then the prey is moved between the jaws — to which it lies crosswise — by successive shiftings, until it is held near the base of its head. A few movements usually suffice to make it take a direction parallel to the jaws, and the head end of the prey is thus brought to lie in the gar-pike's throat, which is often greatly distended by it. The movements by which the shifting of the minnow is accomplished are rather complicated, and require a nice correlation to be successful. During the process of transferring it cross-wise between the jaws, the latter have to be opened quickly, and this motion is instantly followed by a quick lateral motion of the whole head in the proper direction. This lateral motion is accompanied by two others; one a forward thrust of the whole body, and the other a depression of the floor of the mouth. To accomplish the first movement there is a preparatory curving of the post-anal portion of the body, the sudden straightening of which at the instant the jaws are loosened gives the necessary forward impetus, and helps to prevent the escape of the prey; this is further guarded against by the second motion, — the depression of the floor of the mouth. The gill covers being in contact with the sides of the body, this latter motion produces a tendency to a vacuum in the mouth, which can be satisfied only by a sudden influx of water between the jaws. The current thus produced of course has the effect of carrying with it any movable object in the mouth or its immediate vicinity, and of impeding the escape of the prey until the jaws are again closed upon it. While the hold upon the prey is gradually shifted from its tail region to its head region, the part of the jaws which holds it is also not the same as at first. By the time the fish has been fully shifted laterally, it will be very near the base of the jaws, for at each loosening of the latter they have been thrust forward a little by the motion of the whole body. But that is not all, for when the prey has been shifted back to near the base of the jaws, — as one can see better in the case of the larger gar-pikes, — each subsequent movement of the latter causes the prey to take a slightly different direction, so that it is finally swung around until it is *parallel* with the jaws instead of cross-wise to them. I am not entirely certain how this is effected; but it is all accomplished when the fish is held in the jaws near their base. It is possible that the teeth of one ramus of the jaw are not loosened quite as promptly as those of the other, and that as a consequence they act for an instant as a sort of pivot for the rotation of the prey. But that both sides of the jaw are ultimately set free at each motion is probable, from the great liability of the prey to escape at this very critical step in the process. Or it may be that the lower jaw is moved slightly toward one side as the jaws are being opened, thus giving a swing to the prey which changes its axis slightly before it is again caught by the closing jaws. Whatever the means by which it is effected, it is certain that this changing of the direction of the prey is all accomplished at the time of the renewals of the grasp.

After the fish has been thus nearly oriented, its head end is introduced into the throat by a single forward lunge on the part of the gar-pike. These forward movements are continued for some time, the jaws being slightly opened at each lunge. When the head is well introduced into the gullet, the jaws are no longer suddenly opened and closed, but remain more or less gaping, while the prey slowly disappears, doubtless being drawn on by peristaltic motions of the gullet. There are usually slight motions of the jaws during this latter process, — a slow opening and partial closing of them. The movement of the lower jaw is somewhat unsteady, almost tremulous, a peculiarity which is also seen, although not so distinctly, at other times than when feeding.

The advantage of a great divaricability of the rami of the lower jaw and of the fulness of the skin connecting them is at once apparent when a comparatively large fish is half swallowed, for then the thin membranous floor of the mouth is greatly distended and the rami pushed far apart. A side view of the head of the fish then resembles somewhat the appearance of the throat of a feeding pelican.

Gar-pikes sometimes take food which they apparently discover to be objectionable only after they have partially swallowed it. In such cases, it is ejected from the throat with a sudden jet of water.

They do not hesitate to snap at each other when kept in confinement, as I have many times observed, and as the mutilated condition of the fins, and especially of the tail, makes very evident. I think it may be inferred that they are not altogether free from danger from their own kind when living in their natural haunts, for they always show a remarkable sensitiveness to being touched in the region of the tail. The caudal filament especially is so sensitive that the slightest touch from a foreign body causes the fish to dart away with utmost speed, whereas one may touch any other part of the body with comparatively little effect. The young fishes become easily accustomed to the touch of the hands, and may even be lifted altogether out of the water without offering any resistance, provided it be done gently and without any quick motion. But none of them ever become so tame as to allow the slightest contact with the caudal filament without immediate efforts to escape. It is almost invariably the tail end which is snapped at by their mates, though I have a few times seen two individuals with interlocked jaws carry on a short contest without fatal results to either. I have known of only one case in which a gar-pike swallowed one of its mates. Near the end of July, 1883, I found an individual in the process of swallow-ing a somewhat smaller mate. The bill of the victim and part of its head were still protruding from the distended jaws of the captor; so that this individual was swallowed tail first, contrary to the more usual method.

The movements of the gill covers vary in rapidity at different times, but they are always executed with considerable promptness. Their adduction is quickly followed by their abduction, but the interval that follows before another adduction is usually rather prolonged; it is the more variable element. It may be so short as to make the abduction and adduction separated by equal intervals, or it may be prolonged to several seconds. These respiratory movements seldom exceed sixty per minute, and may diminish during the torpor of winter to scarcely more than one a minute.

The emission of bubbles of gas, which begins soon after the young fishes detach themselves from their fixed supports, at first takes place through the gills of one side. It is usually preceded by a forward lurch of the body, accompanied by a slight rolling to one side and the elevation of the gill covers. The bubble usually emerges before the fish regains its normal position, and consequently comes through the gill slits of the side which happens to be uppermost. Occasionally two smaller bubbles escape from beneath the gill covers, one from each side. When the fishes have become much older, the amount of gas is so great that the bubbles often escape not only from beneath both gill covers, but also through the mouth opening, and the rolling of the body does not always accompany the escape of gas.

In the earlier stages of their growth the gar-pikes remain most of the time very near the surface of the water in a horizontal position. In such cases the only premonitory symptoms of the escape of gas are the motions just described; but as they grow older they gradually habituate themselves to lying in deeper water, and then they almost invariably ascend to the surface before emitting gas. The ascent is nearly always accomplished by a slow forward and obliquely upward motion, the body being at an angle of about 45° with the horizon. The motion is usually deliberate, and at a uniform rate. After reaching the surface the body is allowed to assume the horizontal position before any effort is made to expel the gas. This motion of ascent is so characteristic, that after studying their habits one may predict with tolerable certainty whether a given fish is about to emit gas.

I believe the slight rolling of the body to one side is for the purpose of bringing the slit in the roof of the throat, through which the gas must be forced, into a position more favorable for its escape. If the opening through which the gas is obliged to pass is directed downward, its expulsion will require greater effort than if the opening is directed sidewise. An advantage depending on the same physical properties of the gas may perhaps also explain the universal habit of coming to the surface of the water to disengage the bubble. At least, the pressure of the water to be overcome in forcing out a bubble when the fish is at a considerable depth must be greater than when it is near the surface.

The escape of gas, which may be several times repeated during the process of swallowing a large fish, shows clearly enough that the bubbles are not simply air taken in at the mouth to be immediately discharged through the gill openings. The repeated emission of gas from the same fish, without the possibility of any fresh air having been taken in through the mouth, even led me to conjecture at one time that air was never taken in through the mouth. At least, it was certain that the young fishes often discharged gas without lifting any portion of the body out of the water.

The rate at which gas escapes from the gill openings is extremely variable, depending on the temperature of the water, the recency of feeding, etc. Perhaps the following observations give a fair idea of the rate during a warm summer day. In the course of ten minutes (August 6) eight fishes together emitted forty bubbles, or an average of one in two minutes for each fish. On August 17, a single fish, 62 mm. long, came to the surface six times in ten minutes, and caused bubbles to escape from the gills.

In the case of older fishes, snapping at the prey is frequently accompanied by an escape of gas, which is apparently involuntary, the sudden motion of the head and the opening of the jaws being sufficient to cause the escape of a few bubbles.

The nature of the gas will be considered in the following section.

III. The Respiratory Function of the Air-Bladder.

Lepidosteus, as well as some other fishes, has the habit of coming to the surface of the water and emitting through the gill slits or the oral opening bubbles of gas. This habit has attracted the attention of all who have had the opportunity of examining the fish while living.

Poey ('55, p. 136) observed that, when placed in a basin of water, "every five or eight minutes he [Lepidosteus] would come to the surface to swallow a mouthful of air, returning downwards immediately. One second after, half a dozen air-bubbles, some quite large, escaped by the opening of the branchiæ. The air," he adds, " remains in the bladder one second, sometimes one and a half, and this time is probably sufficient for the absorption, digestion, and expulsion of the inspired air. Besides, it is certain that, the animal not attempting to swim, the bladder was not used in augmenting or diminishing the density of the body, as most fish do, in order to ascend and descend in water." Poey further strengthened his opinion that " some sort of pulmonary respiration existed in the Lepidosteus " by dissections and injections of the aorta, which showed the great vascularity of the bladder (pp. 134–136).

Louis Agassiz ('57), in exhibiting before the Boston Society of Natural History some young living gar-pikes, called attention to the fact that this fish is " remarkable for the large quantity of air which escapes from its mouth. The source of this Prof. Agassiz had not been able satisfactorily to determine. At certain times it approaches the surface of the water, and seems to take in air, but he could not think that so large a quantity as is seen adhering in the form of bubbles to the sides of the gills could have been swallowed, nor could he suppose that it could be secreted from the gills themselves."

More recently Wilder ('76 and '77) also has studied the young of this fish. He says ('76, pp. 151–153): " Very often these young individuals of *L. osseus*, and more frequently the adults of the smaller species (*L. platystomus*), would protrude the snouts from the water in the respiratory act; but the length of the jaws made it impossible to determine whether this was intentional, and for the purpose of inhaling as well as of exhaling the air." He inclined to the opinion that air is taken in as well as given out, because the fishes uniformly approached the surface, whereas, if exhalation were alone sought, that " could be as well accomplished at any depth." As the result of some experiments in restraining the motions of *Amia*, he says: " There seems no doubt that with *Amia* there is a true inspiration as well as expiration of the air. The same may be considered probable, though not yet proved, with *Lepidosteus.*"

So far as I know, these are the only published accounts of this peculiar habit in the case of Lepidosteus. While Agassiz maintained a conservative attitude regarding the question of the source and nature of the bubbles, Poey and Wilder expressed the conviction that it was air which had served the ordinary purposes of respiration, and both of them supported their belief with arguments.

Close attention to the movements of young fishes kept in confinement led me at first to doubt the accuracy of this conclusion. The result claimed by Moreau ('63 and '63ª, p. 820) — that, after the artificial removal of gas from the air-bladder of fishes generally, the gas is regularly renewed, *even when they are restrained from coming to the surface* — also served to confirm my doubts upon this point. Usually it was only the tip of the bill which protruded from the water, and it seemed incredible that such a movement could be sufficient for the acquisition of even a limited volume of air. In addition, it has often occurred that several times in succession a fish has been observed to come near the surface and emit bubbles of gas *without any portion of the head region breaking the surface of the water.*

This seemed conclusive upon at least one point : coming near the surface could not be *solely* for the purpose of acquiring air. I have given elsewhere (pp. 12, 13) what appeared to me to be the probable explanation of the upward movement of the fish previous to emitting gas.

The possibility that fresh air was not always — perhaps not generally — acquired, led me to reflect further about the possible functions of the air-bladder. Starting with the assumption, that no organ arises absolutely *de novo*, and that the air-bladder is the result of a differentiation in the alimentary tube, what, in a phylogenetic sense, was likely to have been its first function ? Did it arise as a hydrostatic apparatus to be ultimately diverted to the service of respiration for the higher vertebrates, or was the hydrostatic function already a superimposed one ? Might it not be that the original function was purely one of *excretion ?* Perhaps in so primitive a fish as Lepidosteus this original function was still the dominant one.

These reflections led me to think it more than ever desirable to subject the gas emitted by the gar-pike to chemical analysis, — really the only method of arriving at definite and satisfactory conclusions. I therefore determined to undertake an analysis as soon as the fishes had attained a sufficient size to make the amount of gas given off in the course of a few hours voluminous enough for easy experimentation.

Accordingly, in December, 1883, when the fishes remaining were six or eight inches in length, some preliminary experiments were begun. The analysis was attempted by using a bent tube of about 10 mm. calibre, roughly graduated to 5 mm. ; but the result showed that satisfactory conclusions were attainable only by employing more suitable apparatus. Through the kindness of Professor Cooke the mercurial bath and other apparatus for gas analysis at the Chemical Laboratory of Harvard College were placed at my disposal, and all the subsequent work of analysis was carried on in that Laboratory. I am also indebted to Professor Cooke for valuable suggestions during the progress of the work. To Prof. A. V. E. Young, then a private student in the Chemical Laboratory, I am under especial obligation, since he carried on the manipulations and measurements of the gases. From his skill and previous experience in gas analysis the results are entitled to more consideration than if the measurements had been conducted by one less familiar with such manipulations.

It was my aim in these experiments to ascertain, without sacrificing the fishes, the composition of the gas in the air-bladder, or at least of that which was emitted in the form of bubbles, presumably from the air-bladder. No attempt was made to ascertain the results of branchial respiration.

The collection of the gas, even from fishes 20 cm. long, is a tedious process. An inverted glass funnel large enough to allow the fish to swim freely inside, was first employed; but the inclined sides of the funnel not allowing the fish to assume a horizontal position except when deep in the water, appeared to interfere with the emission of the gas. After ascending to the apex of the funnel, the fish would make violent efforts, apparently for the purpose of attaining the surface of the water, but only occasionally would it emit gas under these cir-

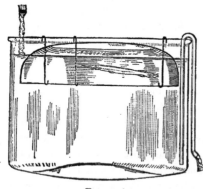

FIGURE 1.

cumstances, the emission usually taking place only after it had begun to descend. If freed from the restraint of the funnel, the fish invariably came at once to the surface outside the funnel, and emitted the customary bubbles.

Afterwards a nearly flat-bottomed glass dish, only a trifle smaller than the one containing the fish, was inverted over the latter (see Figure 1); but this apparatus was only slightly more successful than the one at first employed. The fishes came to the under surface of the inverted vessel, and after long intervals some of them emitted small portions of gas, but this was too small an amount to be satisfactorily analyzed. Another method — the one finally employed, although a tedious process — will be mentioned further on.

In my preliminary analyses I had found that a very thin stratum of air under the inverted vessel was sufficient to allow the ordinary emission of gas. The first experiment in which the gas was accurately analyzed was therefore conducted upon fishes confined under the inverted dish, beneath which was a bubble of air of known volume, which had been *deprived of all its carbon dioxide.* The details of the experiment were as follows.

Experiment 1. — Dec. 13, 1883.   At 11:45 A. M. six gar-pikes, varying in length from 13 to 21 cm., were removed from running water (+ 12° C.) to the experimental jar, containing recently distilled aerated water at about the temperature of the room (+ 20° C.), over which the glass vessel, with a bubble of air (deprived of carbon dioxide) measuring 66.71 c.c.[1] had been previously inverted. The experiment was continued until 3:15 P. M. (3½ hours) without renewing the water. The fishes were then taken out and the gas collected. In the final transfer over the mercurial bath a portion of the gas was lost, so that its total volume could not be ascertained.[2] The portion remaining was collected over mercury, and found to measure 19.27 c.c. (reduced). Treated with potassic hydrate, the volume was diminished to 19.07 c.c. Absorbed, .20 c.c. = 1.04% = carbon dioxide. This volume (19.07) transferred to water and treated with pyrogallate of potash measured over water 16.27 c.c. Absorbed, 2.80 c.c. = 14.5% = oxygen.

Experiment 2. — Dec. 14, 12:30 P. M.   Six gar-pikes (the same as in Experiment 1) were removed from water (+ 11° C.) to the experimental jar containing water at + 21° C. The inverted vessel contained 65.20 c.c. of air deprived of its carbon dioxide. The experiment was continued until 4 P. M. (3½ hours). During the last hour all the fishes remained at the top of the water, and became somewhat swollen in appearance. One of the six did not recover from the effects of the abnormal conditions. The gas, re-collected and measured, was found to have

[1] The eudiometer employed was graduated to fifths of a cubic centimeter. All the volumes given are the volumes reduced to 0° C. and 1 m. pressure.

[2] In a preliminary experiment it was found that under nearly the same conditions there had been a slight *increase* in volume. The conditions were as follows. Dec. 3, 1883, 12:30 P. M. Four fishes (17–20 cm.) were transferred from cold water into an experimental jar (Fig. 1, p. 16) containing water at + 20° C. and a bubble of ordinary air measuring 53 c.c. At 2:30 P. M. five other somewhat smaller fishes were added. At 4:30 P. M. the experiment was discontinued, and it was found that the bubble had increased to about 56.5 c.c., the increase being nearly 3.5 c.c.

diminished nearly 18.3% in volume; it measured only 53.29 c.c. This diminution, however, is evidently not to be accounted for as the result of actual absorption by the fishes, but rather as the result of their over-distention with gas swallowed, — probably in the vain endeavor to compensate themselves for the altered condition of the atmosphere. The recovered gas (53.29 c.c.) was divided into two portions (A, B), and treated separately.

A = 43.20 c.c. (reduced) measured over water. This was treated with potassic hydrate solution, but the volume remained unchanged. Therefore carbon dioxide = zero. This volume (43.20 c.c.) was then treated with pyrogallate, which reduced it to 38.88 c.c. The diminution, 4.32 cc., = 10% = oxygen.

B. The average of three readings made this volume = 10.07 c.c. Treated with pyrogallate, it was reduced to 9.12 c.c., the reduction in volume, 0.95 c.c., being equivalent to 9.4+ % = oxygen.

Experiment 3. — The gas for this experiment was secured by collecting the bubbles as they were given off by the fishes. A number of the gar-pikes were placed in an uncovered experimental jar containing recently distilled water at +20° C. It was found that by using a small glass funnel, — the one employed was about 65 mm. in diameter, — held near the surface of the water, the fishes were not prevented from the ordinary movements accompanying the emission of bubbles, as they were when confined under a larger funnel. The funnel was held so that the tip of the snout, but not the gill region, projected beyond its edge. The otherwise very slow process of collection is somewhat hastened by employing a number of fishes in the same jar. As they rise one after another, at comparatively short intervals, the collector, anticipating their movements, holds the funnel in the proper position, and seldom fails to secure the desired bubbles. But even in this manner the amount to be collected in the course of a few hours is not large. In this experiment the (reduced) volume of gas collected in about two hours was 9.54 c.c. Treated with potassic hydrate, the volume was reduced to 9.38 c.c. The difference = 0.16 c.c. = 1.7-% = carbon dioxide. Treated with pyrogallate, the volume was further reduced to 8.20 c.c. The diminution, 1.18 c.c., = 12.4-% = oxygen.

For convenience of comparison the results of measurements may be tabulated as follows : —

	Volumes in c.c. reduced to 0° C. and 1 m. pressure.			Per cent Carbon Dioxide.	Per cent Oxygen.
	Original.	After Potassic Hydrate.	After Pyrogallate.		
I. Mixed air,	19.27	19.07	16.27	1.04	14.5+
II. " "	53.29				
A,	43.20	43.20	38.88	0.	10.
B,	10.07	——	9.12	0.	9.4
III. Respired air,	9.54	9.38	8.20	1.7−	12.4−

In considering the significance of these results, it is to be kept in mind that there are two principal problems involved. I have not attempted to find out whether *nitrogen* is consumed or produced in this process ; but simply to ascertain the changes effected in the atmosphere respired as far as regards (1) the oxygen and (2) the carbon dioxide. The difficulty in drawing at once a satisfactory conclusion from the analyses rests primarily on the fact that the emitted gas could not be collected under conditions which allowed it to be assumed that at the time of analysis it was in the same condition which it presents in the air-bladder of the fish. For previous to analysis the gas had bubbled through the water in which the fishes were living, and had remained exposed to a limited portion of its surface. The reduction of the proportion of oxygen contained in the atmospheric air is so great, — between one third and one half, — that the influence of the water upon the composition of the emitted gas in this respect would certainly have been too insignificant to modify essentially the result. With the carbon dioxide, however, the case is different. The coefficient of absorption in water for the latter is immensely greater than for oxygen, and the total amount of carbon dioxide in comparison with the volume of the water to which it was exposed is so small (never having exceeded 1.7% of the volume of gas collected) as to make immediate deductions from the analyses of little value.

In a series of very carefully conducted experiments upon the intestinal respiration of *Cobitis fossilis*, Baumert ('53) arrived at the conclusion that the gas from the intestine, in bubbling through the water and during its subsequent exposure to that liquid, was *not essentially altered* either by the water or the gases contained in it (p. 48). The conditions under which his collections of gas were made [1] were so like

[1] Baumert ('53, p. 39) employed five or six fishes (Cobitis fossilis) of medium size which were kept in a vessel containing about twelve litres of water from the river Oder. The gas was collected by means of a large inverted funnel terminating in a small-necked receiver, and usually about two hours were required for its accumulation. Three analyses, made at intervals of three days without any

those which I employed, that it seems justifiable to make use of his conclusions in the present case. The ingenious device employed by Baumert ('53, p. 46), in verifying the accuracy of his previous results, to suppress the branchial respiration, and thus secure the effects of intestinal respiration alone, could not, from the nature of the difference between the modes of respiration, be made available in the case of Lepidosteus.

The means (viz. protracted boiling) which Baumert employed for extracting the absorbed gases contained in the water of experimentation were as complete as the chemical methods at his time permitted. Since then, however, Gréhant ('69) has shown that simply boiling gas-impregnated water, although sufficient to eliminate all the oxygen and nitrogen, does *not* remove all the carbon dioxide, but that a combination of the mercurial pump with the method of boiling is capable of accomplishing this result. By this process he has demonstrated that

renewal of the water, gave in 100 volumes of gas respectively 1.77, 0.47, and 0.13 volumes of carbon dioxide, and 10.46, 13.71, and 11.92 of oxygen. It will be seen, therefore, that not only the conditions under which the collections were made have been in both cases very similar, but also that the composition of the gas as determined by analysis was nearly the same in Lepidosteus that it was in Cobitis. Since the ratio of the gases in the two cases was practically the same, there is no reason to suppose that the water absorbed a greater proportion of gas in one instance than in the other.

As regards the *temperature* of the water at which the experiments were made, although not definitely stated by Baumert, it is safe to infer, from the temperature at which his numerous experiments on *branchial* respiration were made, that it was considerably *lower* than the temperature of the water in which the Lepidostei were placed, so that the tendency to an absorption of carbon dioxide in the latter case would certainly have been *less* than in Baumert's experiments.

One of the plans adopted by Baumert to test the influence of the water and its contained gases upon the emitted gas — whether the latter were either deprived of any of its oxygen or carbon dioxide, or received accessions while exposed to the water — was (1) to prepare *artificial* mixtures of atmospheric air and carbon dioxide in different proportions (4 : 1 and 6 : 1) and cause them to bubble through water in which fishes were living, and after collection to allow these gases to remain exposed to the water, as in the experiment with the *natural* gas ; (2) in a similar way, to cause atmospheric air to pass through water in which the fishes had been living for several days. In the second of these experiments, the atmosphere remained absolutely unaffected, showing not only that it did not lose oxygen, but that it also did not acquire carbon dioxide. In the first of these methods, however, while the proportion of nitrogen and oxygen remained practically the same, the carbon dioxide was diminished by absorption to the extent of about one half its original volume. To this evidence of considerable absorption, it seems to me, Baumert does not allow proper weight, when, by a more satisfactory method, he subsequently finds what he believes to be the true composition of the emitted gas.

the water of the river Seine contains *oxygen* and *nitrogen* in the proportions determined by Humboldt et Provençal ('09), but that, owing to the defective means of extracting the carbon dioxide employed by them, they secured only one fortieth of the volume of that gas which was actually present (Gréhant, '69, p. 379).

Such being the case, it is evident that reliance cannot be placed on Baumert's conclusions, for his opinion — that there was no perceptible absorption of gases under the conditions of experimentation previously detailed — rested ultimately upon his experiment with intestinal respiration in distilled water, and his ability subsequently to extract *the whole* of the gas *absorbed by the water* during the experimentation.

Being unwilling to sacrifice the limited number of young fishes in my possession for the purpose of securing the contents of the air-bladder, and not having the opportunity of getting the gas from the air-bladder of adults, I determined to employ a method which seemed likely to furnish positive evidence of the presence of carbon dioxide, if any were really eliminated, even though it would not give a rigid quantitative test.

The plan was, to place the fishes in a vessel of water having a limited air-chamber, which could be rapidly swept of its contents by a continuous flow of pure air through it, and to lead the air, thus drawn from the experimental jar, through baryta water, which would indicate the presence of even a small amount of carbon dioxide by the formation of a precipitate.

Accordingly, a glass jar (Figure 2, J) of about 20 litres' capacity, provided with a thick ground-glass cover having a central neck and orifice for a stopper, was selected. The cover was larded to secure a close joint, and in addition the edge was sealed with melted paraffine, and the jar nearly filled with recently boiled distilled water. This was siphoned into the jar with as much precaution as possible in order to prevent exposure to the atmosphere. To make the water habitable for the fishes, it was of course necessary to aerate it, and it was at the same time important not to introduce in this process any carbon dioxide. To accomplish this successfully was found to be a task much more laborious than was at first anticipated. The apparatus shown in Figure 2 was that which finally proved satisfactory.

A Bunsen pump (P) attached to the hydrant provided the suction required, and the system of tubes and chambers through which the air was drawn was arranged as follows. Four glass combustion tubes (K) each about 2 meters long and 2 cm. in diameter were loosely filled with fragments of potassic hydrate, and joined by short pieces of rubber

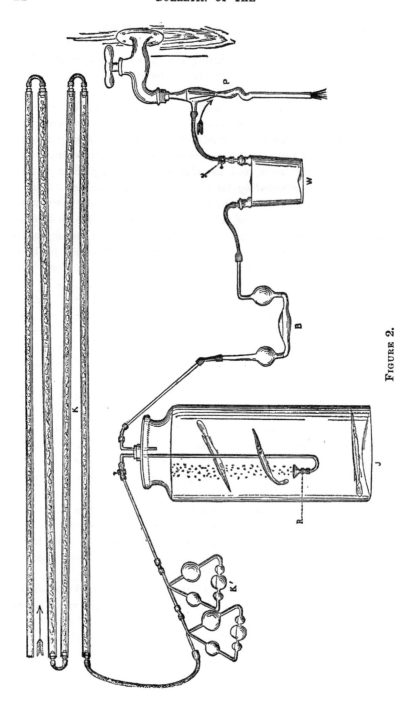

FIGURE 2.

tubing, as indicated in the figure. After traversing these tubes, the air was compelled to bubble through two sets of potash bulbs (k'), and thence by a glass tube was led through the stopper of the experimental jar (j). The glass tube was carried to within about 10 cm. of the bottom of the jar, and, after describing a U-shaped bend, terminated in a platinum "rose" (r). Thence bubbling through the water, the air was led from the air-chamber of the jar by a glass tube to the baryta bulb (b). To guard against the effect of an accidental diminution of pressure in the hydrant water, and a consequent reflow of water into the baryta bulb, the latter was not directly connected with the pump, but an ordinary Wolff's bottle (w) was interposed.

The preliminary test imposed upon the apparatus was, that it should run twenty-four hours with no fishes in the water, but otherwise under the same conditions that would be required when the fishes were introduced, without giving a perceptible precipitate in the baryta bulb. It required some time and considerable attention to details before this was attained. All the connections were made by means of close-fitting rubber tubing, which was made as limited as possible, narrow glass tubing being used wherever practicable, and all the joints were in addition sealed with melted paraffine. While the pump was in operation, the water in the jar was of course under diminished pressure,[1] the diminution being equivalent to a column of water equalling the distance between the surface of the water in the jar and the lower end of the bent glass tube to which the "rose" was attached. Evidently the removal of the stopper from the neck of the jar (through which it was proposed to introduce the fishes) under such circumstances would allow the entrance of considerable impure air, which might give a precipitate as soon as the pumping was resumed. To obviate this source of error, the tubing of the apparatus was clamped at x, and by means of a hand bellows air was injected through the potash system until equilibrium was restored in the chamber of the jar, this being indicated when the water rose inside the bent tube to the same height as outside. In the preliminary test, this was done in the same manner as subsequently when the fishes were introduced, and the stopper was also removed for the length of time which experience had shown would be necessary for the introduction of the fishes. Thus, as nearly as possible the same conditions were observed in the preliminary and the final experiments.

[1] It is to be observed that, so far as this diminution of pressure affects the absorption of gases by the water, it can only act favorably in preventing such absorption.

When the preliminary experiment had proved that the air which for twenty-four hours was being pumped through the apparatus did not contain enough carbon dioxide to cause any trace of a precipitate in the baryta bulb, then the tubing was clamped at x. Equilibrium was restored by forcing in air as before, four fishes were introduced through the neck in the cover to the jar, and the pumping was resumed. It was kept up for six hours, at the end of which time *not the slightest trace of a precipitate had been formed,* although the fishes had been emitting bubbles at the surface of the water in the ordinary manner during the whole time. The baryta water was subsequently tested to make sure that no error had been made in the matter of its sensitiveness to carbon dioxide.

The immediate inference from this experiment is, that the gas eliminated by the fishes contained no trace of carbon dioxide; but here, again, the possibility of an absorption of eliminated carbon dioxide by the water cannot be rigidly excluded.[1] It seems highly improbable, however, that with an air-chamber which required at most only a few minutes to be swept of its contents, there could have been a sufficiently prolonged exposure of the gas to allow the absorption of its carbon dioxide. It might be urged that this, being but a single experiment, can be entitled to little weight, — that, even if in this one instance there was no elimination of carbon dioxide, it is to be considered simply as an exceptional individual case. It is, however, to be borne in mind that there were *four* fishes under experimentation, and that in the case of the *analysis* which failed to show the presence of carbon dioxide there were *six.*

As the result, then, of these various experiments to determine whether carbon dioxide is eliminated, it must be concluded for the present, until opportunity is presented for a careful analysis of the gas taken from the bladder without exposure to water, that *the gas of the air-bladder of Lepidosteus is not likely to contain more than two per cent of carbon dioxide at the most, with a strong probability that in many cases it does not contain an appreciable amount of that gas.*[2]

[1] If this experiment were ever repeated, it is perhaps desirable that *control experiments* should be carried on at the same time. In the control experiments atmospheric air mixed with known volumes of carbon dioxide, varying from zero to 2% of the total volume, should be made to bubble near the surface of the water. By observing the frequency and estimating the volume of the gas emitted by the fishes in a given time during the actual experiment, practically the same conditions could be observed in the control experiments. The introduction of the carbon dioxide mixture could easily be accomplished by means of a third glass tube, piercing the rubber stopper of the jar and dipping just below the surface of the water.

[2] The recent analyses made by Jobert ('77 and '78a) on the gases taken from

But if little or no carbon dioxide is eliminated by the air-bladder, what is the purpose of this aerial respiration? The answer must be evident from the amount of oxygen which invariably disappears. The emitted gas contains only from two thirds to one half the amount of oxygen found in an equal volume of atmospheric air. I believe it is therefore definitely and satisfactorily settled by these experiments, that *the air-bladder respiration in Lepidosteus is subservient to the oxygenation of the blood.*

There arise two more questions, which I believe to be intimately related to each other. If the air-bladder does not provide for the elimination of a larger amount of carbon dioxide than is stated, how is this product disposed of? And, secondly, What is the relation between branchial and aerial respiration in Lepidosteus, and how has it been brought about?

The most probable method of elimination of the carbon dioxide is by means of the gills. To determine the question definitely would require the complete separation of the effects of branchial respiration from those accomplished by other means; but it seems to me at present extremely doubtful if any method can be devised which will enable one to do this in the case of Lepidosteus. With Cobitis, or any other fish with intestinal respiration in which the gas is emitted through the anus, the solution of the problem would be comparatively easy.

But there is already some evidence that the gills in such cases effect an increased amount of elimination. At least the comparisons which Baumert instituted between Tinca and Cyprinus on the one hand, and Cobitis on the other, indicate clearly that there is a more rapid elimination of carbon dioxide through the gills in the latter case than in the former.

If, then, it can ultimately be shown that in cases of aerial respiration the gills accomplish most of the elimination of carbon dioxide, and that the air-bladder is serviceable principally as a means of securing additional oxygen, something will have been accomplished toward appreciating the natural conditions which must have led to the imposition of this

the air-bladder and from the intestines of several Brazilian fishes gave the following results: —

I. Fishes with intestinal respiration,
   a, emitting gas from the anus, *Callichthys asper* C. & Val. : 1.5–3 8% carb. diox.
   b, emitting gas from the mouth and gill slits, *Hypostomos* : 1.5–2.8% carb. diox.
II. Fishes with air-bladder respiration, *Erythrinus tæniatus* } : 1.3–2.4% carb. diox.
   Spix, or *Erythrinus brasiliensis* Spix,

function upon the alimentary canal, or one of its dependencies. The water in which such fishes lived may have been at times incapable of furnishing the necessary amount of oxygen, but sufficiently serviceable as a means of removing, through the gills, the carbon dioxide. The solubility of carbon dioxide in water, as compared with that of oxygen, favors such an explanation.

I have not succeeded in finding many analyses of the gases held in solution by water which seem capable of throwing light on this question.

A. Morren ('41, pp. 471, 478–480, and '44, p. 12), who conducted numerous experiments to ascertain the effect of light and of green organisms on the composition of the gases dissolved in water, has recorded some interesting facts which seem to me to bear upon the problem. He ascertained that when the per cent of oxygen in the gas extracted (by boiling) from the water fell below 18, 19, or even 20, the fish contained in his experimental reservoir began to languish, and many of them died. He also deduced from his experiments these conclusions : that water which [is stagnant or] flows slowly over a slimy bottom is subject to conditions which serve to explain why it may be habitually less oxygenated than water which runs rapidly over a sandy bottom, and why it undergoes greater variations in the composition of its dissolved gases.

Morren also found that the oxygen in the gas contained in the waters of the river Marne fell as low as 18 per cent on the 18th of June, 1835, when there was a remarkable mortality among the fishes in the river, and he ascribed this mortality to the want of oxygen in the water.

If such gas is incapable of supporting the life of fishes, it might occur under certain circumstances that the proportion of oxygen would be considerably below the normal 32 per cent, and still be far from producing asphyxia. Under such conditions, which presumably happen more often in the stagnant water of shaded swamps than elsewhere, fishes which could avail themselves of the oxygen in the atmosphere would be able to survive when others could not. They might still employ their gills for the elimination of carbon dioxide into the water, — 100 volumes of which can absorb about 120 volumes of the gas, — but would have to depend largely upon the atmosphere for their supply of oxygen.

One might conclude, then, that in the transfer of the respiratory function from the gills to the prospective lungs the two components of the respiratory process were separated from each other, and that the oxygenating function was the one first to be transferred; that, so long as the animal lived in the water, the gills were, under all ordinary circumstances, the customary channel for the elimination of carbon dioxide; and that finally, during the passage from water to land, this function was also imposed on the vesicular organ of the intestinal tract, which thus became a lung in the fullest sense of the word.

IV. Embryology.

The interesting accounts of the ontogeny of Lepidosteus given by A. Agassiz and by Balfour and Parker have put us in possession of important data concerning vertebrate development. It is hoped that a more extended study in this field will serve to answer some of the questions left unsettled by them, and will afford additional results of general value in discussions concerning the phylogeny of vertebrates.

There would be some advantages in beginning the subject of the development of Lepidosteus with an account of the formation and growth of the ova. Although I have considerable material for a study of oögenesis, it has seemed to me better, on the whole, to defer a consideration of the topic until the close of that part of these studies which deals with organogeny, and to begin the description with the ovarian ovum as it exists at the time of oviposition. But I have deviated from the plan in the first part of the subject, — the egg membranes, — in order to give a more complete exposition of the structures which envelope the mature ovum.

1. Egg Membranes.

The only account of the structure of the egg membranes of Lepidosteus is that given by Balfour and Parker ('81, p. 112, and '82, p. 362 and foot-note).[1] It is as follows: "They [the ova of Lepi-

---

[1] Ryder ('85, p. 146) has since given the following brief account: "In the ova of Ganoids, *Amia* and *Lepidosteus*, the *zona* is composed, in the first instance at least, of short, parallel, elastic fibers disposed in a plane vertical to that of the membrane, these fibers being fused at their ends or just below the inner and outer surfaces of the membrane. Sections through the egg membrane of *Lepidosteus* seem to indicate the same condition of things as in *Amia*, in fact Dr. E. L. Mark of Cambridge, Mass. has kindly shown me drawings which show the fibers of the zona of the former isolated in the same condition as I have been able to separate those forming the egg membrane of the latter."

It will be seen by the following description that I do not agree with all that

dosteus] have a double investment consisting (1) of an outer covering formed of elongated, highly refractive bodies, somewhat pyriform at their outer ends (Plate 21, fig. 17, *f. e.*), which are probably metamorphosed follicular cells, and (2) of an inner membrane, divided into two zones, viz.: an outer and thicker zone, which is radially striated, and constitutes the *zona radiata* (*z. r.*), and an inner and narrow homogeneous zone (*z. r.'*)." In a foot-note the authors state, in addition, that "the ripe ova in the ovary have an investment of pyriform bodies similar to those of the just laid ova," but that their attempts to ascertain the nature of these peculiar pyriform bodies proved futile on account of the bad state of preservation of the material at their command.

A. OBSERVATIONS.

The observations which I have made on the egg membranes of Lepidosteus will be followed by an historico-critical account of what is known about the structure of egg membranes in other fishes.

I have examined the membranes in fresh eggs, as well as in those which have been treated with various reagents, and have been able to carry my investigation somewhat further than Balfour and Parker. It will appear in the course of the following account to what extent my results agree with theirs, and in what they differ.

a. *Zona Radiata and Villous Layer.*

Omitting for the present the modifications in the micropylar region of the membranes, I will consider first the structure of the envelopes in the recently laid egg which has not been subjected to the action of reagents, and subsequently will describe what is to be gained by the study of sections made from eggs that have been hardened and stained. The differences between the membranes of recently laid eggs and those of mature ovarian eggs are so unimportant that it will not be necessary to give a separate account of each. The nature of the differences, when such exist, will be pointed out in the course of the description. When the ripe eggs are artificially removed from the female by "stripping," they have at first irregular, more or less polyhedral forms, due to mutual pressure in the ovary, and the membrane investing each is in a pliable condition. This state is retained for a long time, provided the eggs do

Ryder here states. He seems to have entirely overlooked the existence of two membranes, and gives such an account of his "zona" as to make me believe that he has had in view what I have described as the villous layer.

not come in contact with water; but when immersed in water they soon exchange the flaccid for a more rigid condition, like the eggs of many other fishes. Whether the egg is in water or in air, its surface is excessively sticky, as Mr. S. Garman[1] has already accurately observed. The eggs adhere equally well to polished and to roughened surfaces. When let fall directly from the female into 95 per cent alcohol, with a view to ascertaining if there was any special superimposed layer of viscid substance such as that described by Kupffer ('78ª, p. 178) for the herring, the eggs have furnished no evidence of the existence of any such continuous film, nor of any covering additional to that which is distinguishable in the mature ovarian ovum, except small amber-colored bodies mentioned later.

When laid the egg of Lepidosteus is enclosed in a single membranous envelope about 50–60. μ thick (Plate I. Figs. 5, 11). This membrane is, however, composed of two distinct but firmly united layers. The outer layer, which embraces from one fourth to one third of the total thickness of the membrane,[2] I shall call the *villous layer* (*st. vil.*); the inner, the *zona radiata* (*z. r.*). The former is the outer covering of "elongated, highly refractive bodies," described by Balfour and Parker; the latter undoubtedly embraces both the zones — *z. r.* and *z. r.'* — described by those authors. It will be convenient to consider the two layers together rather than separately.

Examined in the fresh condition, the outer surface of the egg envelope is of a faintly yellowish or brownish tint, which is in part due to the presence of small ovoid bodies of variable size, of an amber color and a waxy appearance (Plate VIII. Fig. 5), which are scattered over the surface. I am unable to say how or where these bodies are formed, but possibly they result from the disintegration of the granulosa. Aside from these bodies the surface presents a roughened or shagreen-like appearance, which is found upon microscopic study to be due to slightly rounded prominences of nearly uniform diameter which are separated from each other by regular nearly straight lines, so that a view perpendicularly upon the surface (Plate I. Fig. 1) presents a field divided by these lines into small polygonal (four- to six-sided) nearly equal areas. The average size of the areas increases slightly as one approaches the vegetative pole of the egg. When the envelope has been removed and torn

---

[1] See A. Agassiz, '78ª, p. 66.

[2] Measurements of the fresh membrane of an ovarian egg left twelve hours in glycerine gave a total thickness of 68 μ, of which 50 μ represented the zona and 18 μ the villous layer. After the addition of weak hydrochloric acid the latter increased to twice its original thickness (36 μ).

into pieces, it is readily seen to be composed of the two layers mentioned, which for the most part remain firmly united. Along the torn edges, however, it often happens that the lines of rupture in the two layers do not coincide, so that for a considerable area there is a separation of one layer from the other (Plate I. Fig. 5). Such regions are the most satisfactory ones for the separate study of the two structures.

Both layers are translucent; to the outer belongs the brownish tint seen in surface views, while the inner is slightly opalescent. When seen from the edge or in optical section, or, better still, when cut with a razor into thin radial sections, both exhibit radial striations, which are much closer to each other in the inner layer than in the outer (Plate I. Figs. 4, 5, 11).

Aside from certain exceptions which will be considered later, the fine radiate markings of the inner layer, or zona, appear as nearly straight parallel lines, which are traceable through the whole thickness of the layer, but which become gradually less prominent toward the deep surface.

The markings of the outer or *villous layer*, on the contrary, are less uniform; they traverse the whole thickness of the outer layer, but are most clearly defined near the periphery, their deeper portions being more irregular and confused, and often exhibiting a tendency to a zigzag course. They indicate the boundaries of highly refractive prismatic bodies, of which the layer is composed, and which seen endwise produce the appearance of polygonal areas already alluded to. When viewed from its deep surface (compare Plate I. Figs. 2, 7), the villous layer has a somewhat ragged appearance; it also exhibits polygonal areas, but they are less regular and less clearly defined than those seen from the external surface. When the egg membrane has been left for some time in water, or, better still, in a mixture of water and glycerine to which a trace of hydrochloric acid has been added, the prismatic elements which compose the villous layer undergo a remarkable change, during which the cause of the peculiar zigzag appearance of their boundaries is made evident. After a time the free rounded ends of some of the prisms appear to protrude above the neighboring ones (Plate I. Fig. 4), thus giving the surface a less even contour than it had at first. On comparing the conditions before and after the application of acid, it at once becomes apparent that the layer has increased in thickness. At its free edges the prisms become more or less detached from each other, and it then is possible to appreciate their real form.

There are recognizable at least three distinct regions in each prismatic villus, and each may be roughly compared to a stalk of grain, with its head, shaft, and root. These names may be applied not inappropriately to the three regions of a villus. The peripheral portion or head (*cap.*), embracing one fourth, or sometimes as much as one third, of the original thickness of the layer, is distinctly prismatic and highly refractive ; its sides are parallel, and it is little affected by the acid, so that, although it increases very slightly in size, it still retains to some extent its angular form. Its free end is always more or less rounded (Plate I. Figs. 4, 9, *i*). Following this terminal head, and marked off from it by a slight constriction, comes the stalk (Fig. 9, *i*, *pd.*), a long, also prismatic fibre, which is less highly refractive than the head, and is so crowded as to be folded back and forth, thus giving to it the appearance of a spiral spring. In fact, many of the fibres are coiled into a tolerably regular spiral, but the majority are simply folded irregularly, evidently being accommodated to the space most available.

Through the action of acid these stalks begin to swell, and some of them — since they are affected more promptly than others — cause an earlier protrusion of the corresponding prismatic heads (Plate I. Fig. 4). The increase in the thickness of the layer — which soon reaches twice its original dimensions — is due almost entirely to the swelling of these deeper portions of the villi. When isolated, they may in some cases be elongated to ten or twelve times the length of the coils which they at first formed (*a*, *b*, Fig. 9, Plate I.). A portion of this elongation is due simply to the unfolding of the compressed stalks ; but ultimately in proportion as it elongates the stalk becomes more attenuated. It often happens in this process that different portions of the stalk are at first unequally affected. Usually it is the deeper portion which is first to uncoil and to become attenuated (*n*, Fig. 9). When fully extended the stalk is slightly tapering, being narrowest at a little distance from its basal or root end, and although generally quite uniform in calibre, it occasionally exhibits varicosities. In many cases the isolated villi (Fig. 9, *g*, *h*) appear as though temporarily prevented from straightening out because of delicate longitudinal structures of a band-like appearance. The aspect of the stalk is then remarkably similar to the pouched condition of the mammalian *colon*, to the longitudinal muscles of which these band-like structures correspond. The apparent "bagging" is usually all in one direction, namely, toward the attached end of the villus (*h*, Fig. 9). The basal end or root (*rx.*) appears to terminate regularly in a number (3–9) of tapering root-like diverging prolongations (Fig. 9, *f*, *g*, *i*, *k*), which are often apparently connected with each other by membranous expansions of the basal portion of the villus.

These roots serve to fasten the villous layer very firmly to the zona radiata, in a manner to be explained in connection with the account of that layer.

The much finer radial markings of the *zona radiata* (*z. r.*) are entirely different in character from those of the villous layer; seen from either surface with a moderately high power they appear as punctations, dark or light according to the focusing (Figs. 3, 8, Plate I.), evenly scattered over the surface, and yet so arranged as to give the whole area a very characteristic appearance. Although rather evenly distributed, they are arranged in groups or systems. One may trace over a considerable area a series of dots placed at the intersections of a system of imaginary equidistant lines crossing each other at right angles; near by may be other series, in which the systems of imaginary lines cross at angles varying widely from that of 90°; in still other series, the lines are arcs of circles; the circles may vary somewhat in size, but the arcs are never to be traced for more than a few degrees. These different systems abut upon each other in the most fortuitous manner, and the intervening spaces are filled with dots so evenly arranged as not to interfere with a fairly uniform distribution over the whole surface (compare Plate III. Fig. 5). Higher powers show that the punctations are circular in outline, of very nearly equal diameters (0.5 μ at their outer ends), and placed at intervals averaging about 1.5 μ. My notes of May 24 and 25' 1883, make the intervals between the pore-canals, as determined by measurements on the shell of an ovarian egg that had lain in glycerine over night, only $\frac{2}{3}$ μ, less than half the value given above; but I believe that the larger distance fairly represents the average condition.

Thin tangential sections show by focusing that these markings are due to minute canals (*pore-canals*), which are ordinarily hollow, or at least contain a substance that is less refractive than the common homogeneous mass of the matrix which they traverse. I have not seen any evidence of a differentiation in the optical properties of the walls of these pore canals which would allow one to speak of them as *tubules.*

Weak hydrochloric acid causes the zona to swell slightly, and ultimately renders the pore-canals less conspicuous or entirely invisible. There are certain of them which do not fade away, however, even after treatment with acid, and which at length become the only visible structures in what otherwise appears as a homogeneous layer. (Fig. 10. Compare also Figs. 4, 6.) These canals very generally have a spiral course, and are noticeably broader at the outer surface of the zona than elsewhere; they taper gradually toward the inner surface of the layer, but seldom reach it ; most of them are traceable only a short distance from the outer surface. They owe their prominence to the fact that they are filled with a highly refractive substance having the form of a corkscrew. When the villous layer has been torn from the zona, this substance appears to terminate exteriorly in a ragged, broken end, which in some instances is drawn out into a tapering appendage (Fig. 10, α). There can be no doubt — according to evidence to be gained from ovarian eggs — that this substance is continuous with that of the prismatic columns of the villous layer, of which they are in reality the roots. The roughened appearance of the inner surface of the separated villous layer is largely due both to the lacerated ends of these roots and to the fact that many of them are wholly withdrawn from the pore-canals when the two layers are torn asunder. This relationship of the layers also explains why it is so difficult to separate them over even a limited area.

Sections of stained eggs, both radial and tangential, give instructive views of the egg membranes. In radial sections the difference between the villous layer and the zona becomes at once apparent from the deeper stain which the former takes on. The pore-canals are also usually more distinct than in the fresh egg, although the effect of certain acid preservative reagents (Perenyi's fluid, picrosulphuric mixture) is such as to obscure the radial markings of the zona. In the villous layer a still more striking contrast is produced between the heads and the stalks of the villi, since the former almost invariably take a much deeper stain than the latter. Especially is this true when stained in picrocarmine, by which the heads are colored a deep carmine while the stalks and roots remain unstained or take a greenish-yellow hue from the action of the picric acid (compare Plate IV. Fig. 1). In borax carmine both portions are usually stained, and almost invariably the head much deeper than the rest of the villus ; but it has occasionally happened that the heads were less deeply colored, and presented a slightly yellowish tint (Plate IX. Fig. 2). I am unable to account for the difference, unless possibly a prolonged decoloration in hydrochloric acid is the cause of the feeble stain of the head ends. In all these stained radial sections it is to be seen that the transition from the head to the stalk, although not marked by a sharply defined line, is nevertheless abrupt. Owing to this, and the fact that the stalks, as well as the heads, are of nearly uniform lengths, radial sections of well stained specimens always exhibit the stalks in the form of a broad band or zone, sharply marked upon both edges, — more deeply stained than the zona radiata, but less deeply than the narrower well defined band which is made up of the heads of

the villi. Each of the heads has its external free surface more or less
rounded and not quite smooth, its sides nearly parallel and straight,
and its ill-defined deep face also tolerably straight. Along the last it
is distinguishable from the stalk, with which it is continuous, by its
greater refractive power as well as deeper color, and by a slight diminution in the size of the stalk. The last distinction becomes more
conspicuous the more the stalk is elongated. The differences between
head and stalk are emphasized by the fact that the villi have a greater
tendency to rupture along this line of union than elsewhere (Plate II.
Fig. 1, and Plate III. Fig. 1). The outlines of the free end and the sides
of the head are sharp, and in thin sections, especially such as cut the
heads crosswise, the margins seem to be limited by a narrow double-
bordered dark band (compare Plate III. Fig. 3), as though the head were
invested in a thick deeply staining membrane. Since I have never been
able to find evidence of the separation of any membranous structure
from the surface of the head, I am disposed to believe that the appearance simply results from a differentiation of the cortical portion of the
head, which otherwise appears perfectly homogeneous. In some cases
this cap-like cortical part seems to exert a restraining influence on the
swelling of the central portion ; at least I interpret in that sense certain
conditions of the heads frequently met with. In such cases their sides
are not strictly parallel, especially when the villi stand in an isolated
position (Plate II. Fig. 1). The head, instead of being marked off from
the stalk by a constriction or shoulder of the ordinary form, has its
outline gradually broadened or flaring as it approaches the peripheral
end of the stalk, and its cap-like sheath appears to end abruptly with
edges which are slightly everted ; the connecting portion of the stalk is
as broad as, or even broader than, the basal end of the head, so that the
direction of the resulting shoulder is just the reverse of that commonly
seen. The most natural explanation of this appearance which occurs to
me is, that the free edge of the cap-like sheath is distended, and even
sometimes everted, by the swelling which takes place in the region
where head and stalk are continuous, and that the sheath in all probability acts as a restraining investment in preventing any great distention
in the rest of the head.

The stalks, in radial sections of eggs which have been subjected for
some time to the action of water before hardening, have the appearance
of comparatively slender columns, which are often slightly sinuous, but
in general nearly parallel. They taper at first quite rapidly for a short
distance from the head, and then only very gradually toward the basal

or root end. The spaces between the stalks are much greater than those between the heads; while the latter sometimes remain — even after the prolonged action of water — in a continuous layer, the stalks often appear to stand individually isolated. It is more common, however, to find, as the result of the swelling, that both the heads and the stalks are arranged in groups or patches, — better shown in tangential sections. Even when the heads are not thus separated, the stalks may be gathered into clusters which leave in radial sections broad lenticular spaces between them (Plate III. Fig. 1).

The stalk gradually diminishes in size to near its zonal end, where it enlarges rather promptly into a sort of conical foot, which exhibits dark longitudinal or radiating markings continuous with the dark outer ends of corresponding pore-canals in the zona. In some cases the foot is split into two or three strands, between which there is then left a space that in radial sections is triangular, with its base resting on the zona and its more acute angle rising into the stalk. The roots proper embrace only the portions of the villi still occupying the pore-canals of the zona. In some cases they are to be recognized as occupying every pore-canal, in others some of the canals appear to be destitute of villous contents. The roots are highly refractive, like the stalk, and seem to stain even more deeply than the latter. They are always broadest at the outer end, and taper until they are exceedingly fine threads. They seldom reach more than a tenth or an eighth of the way through the zona, although longer and larger roots are met with at intervals. They always appear more tortuous — zigzag, or spiral — than the pore-canals which do not contain roots, and are at times so irregular in form as to have caused great distortions in the canals (Plate IX. Fig. 2). Their finest tips, however, always appear continuous with the much more faintly marked pore-canals. I cannot doubt, therefore, that they are accommodated by simple enlargements of the pore-canals. The great regularity in their distribution, too, allows no other interpretation than that the position of the roots is practically determined by that of the pore-canals.

Tangential sections of stained eggs (Plate III. Figs. 2–5) afford the most satisfactory evidence of the shape and grouping of both heads and stalks, and is the only safe means of controlling the views of the foot region gained by radial sections. The heads are at first close set, leaving only the finest narrow lines, with here and there an irregular opening where the prisms incompletely match (Plate III. Fig. 2); their cross sections are angular and range from variously proportioned triangles to six- or seven-sided polygons. After the prolonged action of water they become less angular, and begin to separate along irregular lines, so as to leave the heads arranged — as already indicated — in patches, which vary considerably in size but are for the most part of a characteristic polygonal outline, with borders which are necessarily jagged owing to the nature of the lines of separation, for the latter never split a prism, but simply separate adjacent ones. The heads may vary in diameter in the ratio of one to two.

The dark border already alluded to is best seen in thin tangential sections (Plate III. Fig. 3), and is readily distinguishable on all the heads when well stained and cut sufficiently thin. The line of separation between prisms is not always distinguishable, but whether this is due to actual contact or not it is difficult to say, since the least obliquity in the section is sufficient to obscure so faint a marking.

The stalks also are found upon cross section to be prismatic, even after the process of swelling has completely isolated them (Plate III. Fig. 4). They are also arranged in groups which correspond fairly to those of the heads, but the spaces between them are much greater. Occasionally sections of stalks are to be seen, even from the middle of the stalk-zone, the central part of which has not been stained (Plate III. Fig. 4, α, α). Careful examination shows that such stalks are really *hollow*, the boundary of the colorless area being sharply defined. I have never seen vacuoles in the middle region of any of the stalks examined in radial section; besides, these cavities can often be traced continuously on successive tangential sections toward the foot. They are, moreover, increasingly frequent as one approaches the zonal attachment of the stalks. The consideration of all these facts makes me quite sure that many of the stalks, at least in their basal halves, are really hollow prisms, although I have never been perfectly certain that I have seen this condition in radial sections. One may, however, as before stated, readily see on radial sections that the expanded foot of the stalk is often apparently split into diverging roots, and that there is an intervening unstained region. The prolongation of this space, which is triangular in side view, forms, I believe, the cavity of the stalks in question. Although the prisms appear sharply marked in cross sections, there is very generally a trace of a filmy substance projecting here and there from their edges in the form of faintly marked threads, which sometimes end indefinitely in the inter-prismatic spaces, but at other times appear loosely to connect neighboring stalks (Plate III. Fig. 4, β). This substance seems to stain less deeply than the stalks themselves, but it is exceedingly difficult to decide whether the faintness of color is due to a specific difference of substance, or is simply the result of the tenuity of the film itself. In the former case, one would perhaps be justified in concluding that there was an *inter-prismatic substance* which served the purpose of a cement to hold the stalks together. The peculiar longitudinal band-like structures noticed during the elongation of the stalks (Plate I. Fig. 9, *g*, *h*, *l*) are possibly to be referred to the same substance.[1] But on the second assumption these shreds of faintly stained substance could be hardly more than the lacerated edges of the stalks themselves. I consider the latter the more probable explanation.

Owing to the spherical form of the egg, tangential sections are circular in outline, and in a given section the centre represents the deepest part. When the centre of such a section is occupied by the superficial part of the zona radiata, the periphery is formed by a circular band of the villous layer, the deeper portions of which are nearer the centre of the section.

A segment from that portion of the band which cuts through the bases of the villi, their roots, and the superficial portion of the zona, is shown in Plate III. Fig. 5. Proceeding from the outer (in the figure upper) toward the central portion of the section, one observes that the cross sections of the villi increase somewhat in size, that the stalks which embrace cavities become more numerous, and that the outlines of the stalks become more and more star-shaped, and then irregular, and that finally they break up into detached spots, which a little farther along become smaller and smaller until they cannot be distinguished in size from the pore-canals.

Since the sections of the membrane are successively increasing in diameter, the deep face of each will pass through a broader portion of the zona than the upper face will. If for the purpose of examination the section be inverted, so that the deep face is uppermost, the relation of parts can be much more easily and satisfactorily studied than if it be viewed from the upper face only, because the zona offers less impediment to vision than the thick-set columns of the villous layer. Attentive focusing shows conclusively on such preparations that the rays or branching roots of the prismatic columns lead each to a pore-canal, and it becomes possible in many cases to note the exact number of pore-

---

[1] Dilute hydrochloric acid causes the distance between the heads of the villi to increase. This would be readily explainable as the result of the swelling of an inter-villous substance, could the existence of such a substance be satisfactorily established.

canals in which a given stalk takes root. The substance of the roots and of their rib-like extensions up the stalk appears to be more deeply stained than that of the expanded foot of the stalk; but this is perhaps only an appearance due to the fact that they are considerably thicker than the membranous portion which connects them.

As the successive sections pass through deeper and deeper portions of the zona radiata, the calibre of the pore-canals grows very gradually finer, and those which are plugged with deeply stained villous roots become less numerous, but otherwise there is no essential difference in the appearance of the sections. The characteristic arrangement of the pore-canals previously described is visible here, and may be made out more easily than on the fresh egg-shell, provided the sections are made perpendicular to the canals and are sufficiently thin.

In radial sections from eggs that have been hardened and stained, the zona is usually of a uniform faint tint (Plate III. Fig. 1), but often there is a very gradual deepening in the intensity of the color in passing from the outer to the inner boundary of the layer (Plate II. Fig. 1). In a few instances this deeper stain seems to extend toward the outer surface of the zona in flame-like jets (Plate IV. Fig. 1). The outer boundary of the zona, although appearing slightly irregular, owing to the variable lengths of the root-like prolongations of the villous layer, is in reality fairly even and sharply marked (Plate II. Figs. 1, 7, 8, Plate III. Fig. 1, and Plate IV. Fig. 1). The inner boundary is still more precisely defined, and appears as a fine continuous line, which sharply separates the zona from the peripheral layer of the yolk. Nowhere is there any evidence of a gradual transition from the yolk to the membrane. Occasionally, when the section is not exactly perpendicular to the inner surface of the zona, this boundary appears double, but careful focusing in such cases always shows this to be an optical illusion. In a few instances I have seen a similar appearance which was not thus explainable. For a considerable distance a layer of nearly uniform thickness appeared to intervene between the zona radiata and the yolk (Plate II. Fig. 7). But the line which separated this from the rest of the zona was never to be made out for more than a small portion of the circumference of a section, for it either terminated abruptly, or, gradually approaching the inner boundary of the zona, became confluent with it. Its inconstancy and its want of continuity are together sufficient to show that the layer in question is not entitled to be considered a distinct membrane, nor even a differentiated portion of the zona radiata. I may add, that I have never seen a section of this kind in which it was not possible to discover in some part of the layer evidences of pore-canals continuous with those of the remaining portion of the zona radiata. I am therefore convinced that *the zona radiata is a single homogeneous layer which is in direct contact with the surface of the yolk, and is traversed by pore-canals which reach from the yolk to its outer surface.*

When radial sections of the zona are broken, they occasionally show a tendency to rupture in lines concentric with the surface of the egg, but this is so rarely the case as hardly to be characteristic. The fracture is usually irregular, and not dependent on any structural feature; even the pore-canals do not appear to have much influence on the direction of the line of separation. The nature of these canals can be more readily studied on sections of hardened specimens than on the fresh shell. Their proximity to each other is not so readily determined from radial sections as by means of the tangential sections already described. The same general features which were mentioned in describing their appearance on the fresh egg are usually visible with even greater clearness on those which have been hardened. The distinctness of the pore-canals varies, however, considerably in different specimens, depending undoubtedly upon the refractive power of the mounting medium, which penetrates the canals, as compared with that of the matrix of the zona itself. Upon the most favorable preparations the canals can be easily traced from end to end, so straight is their general course. At the periphery of the zona they are uniformly somewhat broader than at its deep surface; but they taper so gradually as to make the difference in calibre, even at their two ends, trifling. In the case of almost every canal a slightly spiral course is noticeable near the outer end, whether it be plugged with the root of a villus or not; and throughout the whole length there is usually the faintest trace of a wavy or zigzag course. Aside from this, however, the canals are remarkably straight and parallel. There are no enlargements or irregularities in the calibre, save those which appear to result from the distention of the canal with the substance of the villous roots already described.

There still remain to be considered some peculiarities of the villous layer, which either result from particular methods of treatment, or have not been observed sufficiently often to allow one to consider them characteristic features.

Of those dyes which I have used, acetic acid carmine gives the sharpest differential staining for the heads of the villi. While the stalks and roots remain comparatively pale, the heads (Plate II. Fig. 2) take a deep rose tint, and the transition from the substance of the head to that of the stalk is rather abrupt. It happened that many of the villi from the shell of a mature egg, that was let fall into ninety per cent alcohol without contact with water, and was afterward stained for twelve hours in acetic acid carmine, exhibited a very peculiar appearance at the free surface of their heads. At or very near the middle of this surface the dark border, so characteristic of the heads of the villi, seems to be interrupted, and there projects from the free end of the head a short conical or longer finger-like process. This issues from the head, apparently through a circumscribed opening in the cortical layer, and may assume a variety of forms, several of which are shown in Plate II. Fig. 2, a–m. This peculiarity is interesting, as showing that there is a region of least resistance in the cortical layer near the apex of each head, which allows the protrusion of a part of the substance of the head when it is subjected to the swelling influence of the acetic acid; but whether this fact is capable of throwing any light on the source of this villous layer, or the method of its formation, I greatly doubt.

There is often to be seen in radial sections of the villous layer a strong tendency for the villi to fuse (Plate II. Fig. 2, l). This is especially true of the region of the stalks, although it is also to be observed among the heads. Since this tendency seems to be much greater in some cases than in others, I am induced to believe that it is due to the influence of the reagent with which the egg was hardened, and sometimes perhaps is dependent on the length of time the egg has been in the water before hardening.

In a few cases — especially in certain nearly mature ovarian eggs which were hardened in chromic acid — I have seen peculiar markings in the villi, which at first led me to think they might be traversed by spaces analogous to the pore-canals of the zona. They were first noticed on tangential sections, and appeared there like minute circular holes in the segments of the prismatic villi (Plate II. Figs. 4, 5). Focusing showed that their contents were much less refractive than the substance of the villi, and they were consequently very sharply defined. But the notion that they were optical sections of tubes, like pore-canals, was at once corrected upon finding isolated villi, which had fallen out of the layer and were seen sidewise. When the thickness of the section is about equal to the diameter of the villi, it is difficult, if not impossible, to decide whether the isolated angular blocks are seen endwise or sidewise; but by selecting the thicker sections, where the length of the villous segments is greater than their diameter, the difficulty is avoided, and it at once becomes evident that the spaces supposed to be canals are for the most part minute spheroidal cavities or vacuoles. Usually there is only a single vacuole in a villus, although occasionally two are to be seen in the same cross section (Plate II. Fig. 4). Not all of the villi contain these cavities. Taking into the account their abundance on successive sections from the villous layer, I should estimate that not more than one half or three quarters of them present this feature. The proportion to be seen upon a single section is, of course, much less than this. They are most abundant in the stalks, but occasionally one is also seen in the head. Upon the egg, where they were found most abundantly, they were rather more numerous in the micropylar region than at the opposite pole. In the latter region there were, however, sometimes as many as three or four in one villus, although the size (0.5 to 1 μ) was the same as at the micropylar pole. I have in a few cases observed that the vacuoles were elongated, and then they were always of uniform calibre and were curved. Occasionally (Plate II. Fig. 4) such a tubular vacuole appears to communicate at one end with the inter-villous spaces. Concerning the nature of the contents of these vacuoles, I can only say that they do not stain, and do not appear differently from what one would expect if they were cavities simply filled with the mounting medium.

The differences between the membranes in mature ovarian eggs and those recently deposited are principally the result of the swelling of the layers by the water, and do not require any further explanation.

The foregoing account of the zona radiata in Lepidosteus contains descriptions of two features which appear to me to bear directly on the condition of the zona radiata of fishes in general.

First. The *proof* that the striate appearance of the zona is due to *pore-canals*, although very generally assented to by the most competent observers, especially in recent years, has nevertheless hitherto rested upon comparatively slight evidence. That this evidence has been meagre depends upon the excessive minuteness of the structures in question. It is not to be overlooked, in the first place, that the tubular nature of the pore-canals in the case of the perch, as originally described by Johannes Müller ('54) for what he called the "Eikapsel," has not the slightest bearing upon the nature of the pore-canals of the zona radiata, since the egg capsule of Müller is a structure entirely different from the zona. I cannot, however, avoid the conviction that his opinion as to the tubulated condition of that capsule has had considerable influence in effecting the general acceptance of similar conclusions as to the nature of the radiate markings of the zona. Neither the evidence produced by Müller, — the possibility of pressing yolk globules through the "pore-canals" of the capsule, — nor the vacuolated condition described by Ransom ('68, p. 455), can have any direct bearing on this question.

Leuckart ('55, p. 258) appears to have been the first to assert with the utmost positiveness that the radial striations of the *zona* were due to pore-canals; and although he nowhere states the exact nature of the evidence which convinced him, we are doubtless at liberty to infer that it was, in part at least, the kind of evidence which he elsewhere ('55, p. 106, foot-note) makes use of; namely, the now well understood differences in optical effects produced by elevations and by depressions of surfaces. Kölliker ('58, p. 83) soon furnished additional evidence, derived partly from the study of *thin sections* of the zona in the trout, but more especially, as it appears to me, from the fact that maceration in fresh water causes the middle region of these supposed pore-canals to be converted into *vacuoles*. Aside from the arrangement of the dot-like appearances as seen from the surface of the zona, which has been very generally recognized, and the features emphasized by Leuckart and Kölliker, I am not aware that any additional evidence in proof of the nature of the pore-canals has yet been produced. If, then, the facts warrant the description I have given of the zona in Lepidosteus, the evidence that it is a canaliculation which produces the radial striate markings in the zona radiata of fishes' eggs has received an additional confirmation.

Secondly. Although Müller (as well as more recent observers) has shown that the pore-canals in the *outer* envelope, or capsule, in the case of the perch may have a *spiral* course, no one has hitherto observed a similar feature in the case of the canals of the true zona radiata. The natural injection of these canals in Lepidosteus with a substance continuous with that which constitutes the villous layer, renders it comparatively easy to establish the spiral course of the canals in that fish; and this makes probable the inference, that certain irregularities in the direction of these canals, shown by other observers to exist in the case of other fishes, may in reality be referable to the same spiral condition, which, from the minuteness of the canals, has not been recognized.

b. *Micropyle.*

The micropyle was apparently overlooked by Balfour and Parker, since it is not mentioned by them; nor has it been mentioned, I believe, by any one else, although it occupies a region which is so conspicuously marked that, having once seen it, one could readily find it with the aid of a simple lens. Except in eggs that have lain for some time in water, the region of the micropyle appears, when seen under a hand lens, like a minute hole in the shell; in surface views with a higher power it looks like a deep circular pit (Plate IV. Figs. 3, 4) sunk in the egg membrane. Its diameter is five or six hundredths of a millimeter. Its outline is nearly always circular, and it has a clearly cut edge. In a few cases a cross section of the pit has proved to be oval instead of circular, occasionally with one diameter of the oval more than twice as long as the other (Plate VII. Fig. 4). A similar appearance, though not so marked, is often produced, even when the pit is really circular, if the plane of the section is oblique to its axis. Sometimes the pit is partly filled by a whitish, apparently spheroidal body (Plate IV. Fig. 4). When the egg is so viewed that this depression lies in the equator, the profile of the egg in its vicinity may be slightly modified, and show a low conical elevation, at the apex of which the pit is located. This is not commonly the case, however, for usually there is nothing in the profile to denote the position of the pit. In eggs nearly mature, and in those which have been recently laid, its place can be easily found by its relation to the lighter colored animal pole of the egg. It is invariably located over some part of the germinal area, and usually precisely over its centre (Plate IV. Fig. 3).

The real nature of this pit and its relations to the two layers of the egg membrane and to the yolk can be studied on optical, but still better on actual sections. For a general survey radial sections are most instructive, but for the elucidation of some questions sections tangential to the egg at the animal pole are more valuable.

In strictly radial sections through the region of the micropyle, it is to be seen that the surface of the egg is deeply depressed. The form of the depression varies somewhat in different eggs, from that of a funnel, i. e. with sloping walls (Plate IV. Figs. 1, 5), to that in which the walls are for some distance almost parallel (Plate V. Fig. 2). This depression results from an infolding of both layers of the egg membrane; it forms, however, only an approach to the true micropyle, or *micropylar canal*, the latter being a minute passage through both layers which begins at the bottom of the depression.

The *funnel,* as I shall call that portion of the egg membrane which forms the walls of the depression, involves a modification of both the zona radiata and the villous layer. Both are affected in two ways, in thickness and in direction.

The villous layer begins to grow thinner at some distance from the edge of the funnel. Sometimes it retains its normal thickness to within a distance equal to the diameter of the funnel; at other times it begins to grow thinner at three or four times that distance from the pit. Its diminution in thickness is quite gradual and very nearly uniform until it reaches a minimum at the micropylar canal. The stalks of the villi are shortened more than the heads, in comparison with their appearance on other parts of the capsule, and the boundaries between them gradually become less distinct. The diameter of the villi also decreases considerably. Near the bottom of the funnel they become very short, but frequently it is evident that, instead of constantly diminishing in diameter, they may even increase as compared with other regions of the funnel (Plate VI. Figs. 5–8).

In all parts of the funnel the villi retain a direction perpendicular to the outer surface of the zona. In the lateral wall, and especially near the bottom of it, they are slightly wedge-shaped or conical, the head ends being narrower than the root ends. They thus accommodate themselves to the diminished space at their disposal (Plate VI. Figs. 6–8).

The zona radiata (Plate IV. Fig. 1) likewise begins to diminish in thickness at some distance from the micropylar canal, and continues to do so until it reaches the canal; but it does not, like the villous layer, grow thinner at a uniform rate. Its thickness decreases very slowly to within a short distance of the region where the membranes begin to bend inward to form the funnel, and then it suddenly narrows to one third its normal dimension, after which it again decreases more slowly until it reaches the micropylar canal. The pore-canals are not perceptibly finer nor more closely set in the vicinity of the micropyle than elsewhere. They retain in most regions a rectilinear course perpendicular to the surface of the zona, but at the region of most rapid reduction in the thickness of the latter, and for a little distance on either side of it, their course is curved, the concave side facing the micropylar canal.

The change in the direction of the two layers of the egg-shell results in the formation of an external depression, which is considerably deeper than the total thickness of the shell, so that, even with a great diminution in the thickness of the latter, its inner surface projects into the yolk as a conical elevation, which is nearly as high as the thickness of the shell. In this deflection of the membranes, the zona radiata seems to bend more abruptly than the villous layer; this, however, is due principally to the fact that the region of greatest curvature is also the region of most rapid change in the thickness of the zona. From this it results that the inner contour of the zona is much more abruptly curved than the outer, in some cases appearing almost angular. As a further consequence of this, the conical elevation appears to arise abruptly from the inner surface of the membrane; its apex is rounded, and in the ovarian egg its surface is everywhere in contact with the yolk. An inquiry as to whether this infolding is the result of a process of absorption, or is due to a peculiar local modification of the activities which produce the membrane, will best be deferred until I have given a description of the layer of cells which immediately invests the ovarian ovum.

The micropyle proper, or the *micropylar canal* (Plate I. Fig. 11ᵃ, Plate IV. Fig. 1, Plate VI. Figs. 3, 4), is straight and of uniform calibre. It begins at the centre of the bottom of the funnel, and passes through both villous and zonal layers of the egg membrane; it is about $8\,\mu$ long. Its cross section is circular and about $2\,\mu$ in diameter. There is no flare to the canal, either at the external or internal end, so far as I have been able to observe. I am unable to say whether the diameter which I have given is that which the micropyle possesses at the moment the egg is laid. From measurements of spermatozoa allowed to dry upon the slide (Plate VII. Fig. 3), one would imagine that the calibre of the micropylar canal must be at least $3\,\mu$, that being the diameter of the heads of spermatozoa thus treated; but according to measurements made upon living spermatic cells the heads are only about $1.8\,\mu$ in diameter, so that I think $2\,\mu$ is probably the normal average calibre of the canal. Still, I have sections in which its diameter is $2.5\,\mu$, and in the case of some fresh membranes it was only $1.5\,\mu$ in diameter. The narrowness in the latter case I attribute to the swelling of the zona when exposed to water and glycerine, in which the membranes were examined.

c. *Granulosa.*

Nearly mature ovarian eggs are closely enveloped by an uninterrupted cell layer, which is everywhere in contact with the outer surface of the villi. Over the greater part of the egg this layer — the follicular epithelium or *granulosa* — is composed of thin, flat polygonal cells, arranged in a sheet only one cell thick. In surface views the granulosa cells (Plate V. Fig. 4) appear of fairly uniform size, — 15–$20\,\mu$ in diameter, — are slightly granular, stain feebly, and exhibit each a single large (5–10 μ) nucleus, with an even outline and a circular or oval form. When seen in profile, — as in radial sections of the egg with its membrane and granulosa (Plate V. Fig. 3), — a majority of the cells are observed to be very thin, and their nuclei flattened; but there is occasionally a cell whose nucleus is not so much flattened, and which therefore protrudes beyond the general surface of the granulosa. Radial sections of the ovum with its granulosa are further instructive in showing the relations of the cells to the heads of the villi. *Each granulosa cell corresponds in size to from four to eight villi, but there is no constancy in the position of the cells or their nuclei in reference to the underlying villi.* Nothing intervenes, however, between the cells and the villi except occasional artificial spaces. Externally the granulosa is limited by a thin, homogeneous delicate membrane, the *membrana propria (th. fol.)* of the *theca folliculi.*

This is the condition which obtains over all parts of the egg except in the vicinity of the micropylar funnel. Elsewhere the granulosa retains great uniformity of thickness. At a considerable distance from the micropyle its cells begin to elongate so that the granulosa grows thicker; as the cells approach more and more the condition of columnar epithelium they become inclined, their outer ends being directed toward the axis of the micopyle (Plate VII. Fig. 1). They still continue to form a layer only a single cell deep until they reach the vicinity of the rapid declivity in the wall of the funnel. Here the cells, having now attained an elongated columnar form, become superposed, and *fill completely the micropylar funnel.* With a single exception the cells composing this mass are fairly similar to each other. They are considerably elongated, irregularly columnar or spindle-shaped, and contain each a single oval nucleus about 10 μ by 8 or 9 μ in diameter. The cells themselves vary from 15 μ to 40 μ in length, and are about 10 μ in diameter. When the hardened egg, with its membranes, is removed from the follicle, it often happens that this conical plug of granulosa cells is left with the rest of the granulosa in the follicle. But even when the majority of the granulosa cells of the plug are thus removed from the funnel, there is usually left behind a single one which is unlike the others. It occupies the bottom of the funnel, which it completely fills, and is much larger than any other of the granulosa cells (*m py. cl.* Plate IV. Figs. 1, 4, 5, Plate V. Fig. 2, Plate VII. Fig. 2).

When I first became aware of the existence of such a cell it was from the study of radial sections of a recently deposited egg in which a "maturation spindle" was visible near the micropylar pole (Plate IV. Fig. 1). As there were no other granulosa cells left attached to the egg, the first impulse was to regard this as one of the "polar cells" formed by the ovum during maturation. This seemed the more probable on account of the undoubted existence of a maturation spindle. A serious obstacle to this view was the great size of the cell as compared with the narrow mycropylar canal. Even the elongated condition of the cell would hardly warrant the assumption that it had passed through so narrow an orifice. The examination of suitable sections from ovarian ova (Plate IV. Fig. 5, Plate V. Fig. 2, Plate VII. Fig. 2) soon showed that this interpretation was inadmissible, and made it as certain as one could expect, without having traced it from its origin, that the cell in question was a specially modified granulosa cell. It may be appropriately called the *micropylar cell*, for, whatever may be its function, the morphological fact remains that it occupies the micropylar funnel, and lies directly over the micropylar canal. I have not been able to discover that its substance extends into the canal, but the number of favorable cases which I have examined is not enough to allow me to say that such a condition is improbable. So far as I know, nothing of this kind has been found in the case of any of the osseous fishes, unless the figure given by Hoffmann ('81, Taf. I. Fig. 20) for Leuciscus is capable of being thus interpreted.[1] Hoffmann himself has evidently not considered the condition of the granulosa in the region of the micropyle sufficiently important to give it any attention in the text, but there is not the least doubt in my mind that the accumulation of granulosa cells which he has figured is the equivalent of the granulosa plug in Lepidosteus. I am inclined to believe, moreover, that Hoffmann has overlooked a real difference between the cells in this region, and that an equivalent of the micropylar cell of Lepidosteus will be found in Leuciscus, and perhaps in many other of the osseous fishes, especially in those where there is a large micropylar funnel. In fact the three cells which in Hoffmann's figure (Plate I. Fig. 20) seem to occupy the funnel, are all slightly larger than the remaining granulosa cells, and one of them — the deepest — fairly represents in its position the micropylar cell. Since all the cells have a somewhat diagrammatic appearance, it is not too much to expect that a more careful examination would show a difference between them.

[1] Since this account was written, Owsjannikow and Cunningham have both found similar conditions in other fishes. A review of their articles will be found at the end of the historical section of the present paper, pp. 104–110.

d. *Origin of the Zona Radiata and the Villous Layer.*

The youngest ovarian eggs in which either of the egg membranes has been observed were about 430 μ in diameter, and the ovaries to which they belonged were preserved just before the period of spawning began. Sections of such an egg are shown in Plate VIII. Figs. 1 and 2. Tangential sections (Fig. 2) show that the egg is enveloped in a layer of polygonal granulosa cells whose boundaries are exceedingly faint, and whose nuclei have very irregular outlines, being lobed or deeply incised, in some cases almost to complete division. The nuclei contain one, and frequently two small nucleoli, but otherwise appear homogeneous, and are uniformly stained. Upon focusing just below this layer of granulosa cells, one sees the surface of the yolk covered with innumerable fine, close-set points, which are evenly distributed.[1]

Radial sections (Plate VIII. Fig. 1) supplement the surface views, and show that the granulosa cells are relatively thin, and easily separable from the underlying structures. Their protoplasm is finely granular, and their boundaries are not distinguishable; neither do their deep surfaces appear to be defined by any membrane. Their nuclei are considerably flattened, and irregular in outline.

Immediately beneath the granulosa the surface of the yolk exhibits fine radial, nearly parallel markings, which are close together and very short. They are so intimately joined to the yolk that they seem to form an integral part of it, and nowhere show the least tendency to become detached from it. With high powers one can recognize a very thin cortical portion of the yolk (membrane?), with which they seem to be continuous. It is very difficult to ascertain the distance between the markings, but about 21 of them may be counted in the space of 17 μ, so that the average distance is not far from 0.8 μ. The length of each is about 0.5 μ.

It would not be easy to determine from this stage alone whether the markings indicate the beginnings of the formation of the zona radiata or the villous layer. But even in this early condition the punctate markings of tangential sections appear brighter rather than darker when one focuses high, so that the inference must be that they are due to minute bodies which are *more highly refractive* than the surrounding substance.

This conclusion is abundantly confirmed by the study of somewhat larger ova. These bodies seem to increase in length with considerable

---

[1] In Fig. 2 (Plate VIII.) these punctations appear much too scattered in the middle of the area which shows them. They are better represented toward the margin of the area.

rapidity, for when the egg has attained a diameter of about 600 μ (Plate VIII. Fig. 3, Plate IX. Figs. 4, 5) they may have reached the length of 3–3.5 μ. In this stage the layer when seen from the surface presents an appearance (lower half of Fig. 3, Plate VIII.) which so closely resembles that of the zona radiata in the mature egg, that one is involuntarily led to believe that it is the zona. Even the peculiar arrangement of the markings in curved lines recalls the appearance of the zona when seen in a similar position. Notwithstanding the striking resemblance, there cannot be the slightest doubt that this layer is not the zona radiata. In radial sections it is difficult to distinguish between a layer composed of a homogeneous matrix pierced with minute parallel *canals*, and one composed of parallel *rod-like* structures, but in surface views this is much easier. Careful focusing shows the same optical properties as were observed in the earlier stage, and with much greater distinctness. The staining, too, is such as is to be observed in the villous layer rather than in the zona; for the highly refractive bodies take the deeper stain, the intervening substance having the paler color of the yolk. But the last possibility of doubt concerning the nature of this layer is dispelled by the appearance presented when the elements which compose it are separated from each other. It frequently happens in mounting thin sections that portions of the layer are detached, and even resolved into their constituent elements. In such cases clusters of two or three rod-like bodies, and even single ones, can be found in such proximity to the layer as to leave no doubt that they are elements detached from it. They have the same length and thickness as the markings of the layer; they are highly refractive and deeply stained. They can in no way correspond to anything that is observed in the zona radiata, but do resemble in several particulars the villi of older eggs.

From all this evidence I am certain that the layer which is first to make its appearance between the yolk and the follicular epithelium is the villous layer.

In this stage, too, the union of the layer with the yolk is much more intimate than its relation to the granulosa. The latter is often separated from the layer, the yolk never.

The cells of the follicular epithelium (Plate VIII. Fig. 3) have become somewhat smaller than in the previous stage, but their nuclei retain the same dimensions and the same lobed appearance which they had during the earlier stage. As a consequence, the nuclei are closer together. It will be seen that in the stage figured on Plate VIII. Fig. 3, the diameter of a single average-sized granulosa cell corresponds to the distance occupied by about a dozen of the villi. This fact will be of some interest later, when a comparison is made with the conditions in the mature egg.

In the sections figured on Plate IX. the villous layer has become still thicker and the villi are correspondingly elongated; they are also somewhat farther apart, as well as thicker.[1] The thickness of the individual villi is really greater than that of the spaces intervening between them, but the appearance as seen under the microscope is represented with tolerable accuracy in the figures. The villi of the egg shown in Fig. 3 (Plate IX.) have attained a little greater length than those of the other eggs figured, but the egg itself was probably somewhat smaller than the one shown in Plate IX. Figs. 1 and 5. I am not entirely certain of this, because the egg was incomplete, the yolk having all disappeared except a portion directly underneath the villous layer.

A more advanced condition in the development of the ovum and its membranes is to be seen in Plate VII. Fig. 5. The evidence that this egg is more advanced than those last described is found in its slightly greater size (nearly 0.7 mm.), and also in the increased size and elongated condition of the yolk bodies which already occupy all parts of the egg except a peripheral layer. The villous layer has here attained a thickness of 5.5 μ, or about one third its thickness in the mature egg, but the individual villi have not changed perceptibly from the condition in the previous stage, except in regard to length.

I have no stages between this condition and that which the eggs present at maturity, but already enough of the egg membranes has been formed to allow several conclusions as to the method of their production.

It is to be observed, that immediately before spawning there is no structure, even in the latest of the stages here described, which can be considered the zona radiata ; neither are there at this time any stages older than the one last described, except the mature ova. It seems to me, therefore, perfectly safe to infer that *the* zona radiata *is developed after a large part, if not the whole, of the villous layer has been produced, and that it is wholly formed during the twelve months immediately preceding the spawning.* From its late production and its position inside the villous layer, as well as its intimate relations to the yolk, it is further to

---

[1] In Fig. 5 (Plate IX.) they are represented a little too far apart and not quite thick enough, whereas in Fig. 3 (Plate IX.) they have been represented too close together. The granulosa cells in Fig. 5 are too sharply defined, especially on the side toward the villous layer.

be inferred that *the zona radiata is exclusively the product of the yolk.* It is also probable, from the evidence of stratification sometimes seen in the completed structure, that the zona is produced in successive layers. If such is the case, it follows that portions of the zona nearer the yolk are formed after those which have a more peripheral position.

The question as to the source of the villous layer is not so easily answered. The fundamental difference between it and the zona radiata at once suggests for it a different origin. If the latter arises from the yolk, the former might be produced by the follicular epithelium. This view would seem to receive confirmation from the peculiar way in which the roots of the villi in the mature egg penetrate the pore-canals of the zona radiata. I have no doubt that this condition would be regarded by many observers as a welcome confirmation of the theory that the pore-canals are primarily for the purpose of transmitting nutritive material to the growing egg. Such observers might look upon the villi as secretions from the granulosa, which, owing to slight physical and chemical changes, had not passed through the pore-canals as nutriment, but remained partly outside the zona to subserve other functions. This view might be further supported by the fact that during the formation of the villi the inner surfaces of the granulosa cells are not sharply marked off by membranes from the underlying structures.

Nevertheless, it seems to me that the arguments which may be adduced to support the opposite view, — that the villous layer is the product of the secretive activity of the ovum itself, — greatly outweigh these considerations.

During the early stages of their formation the villi are so intimately related to the ovum that they appear to be rods imbedded in its substance, and at no time during its formation is the villous layer separable from the yolk. If the latter is by any means removed from the membrane, there is always a superficial portion of the ovum which remains attached to its inner surface. The separation of the granulosa cells from the membrane during this period, on the contrary, is quite common. What might otherwise be a serious obstacle to considering the villi the product of the ovum, — the presence of a zona between the two, — is entirely nullified by the fact, previously established, that the villous layer is produced *before* the zona radiata.

Whatever renders improbable the formation of the villi from the follicular epithelium is, of course, favorable to the opposite view. If the villi were products of the epithelium, one would expect some constancy in the numerical relations between the two, but this is certainly wanting.

I have made some measurements and comparisons between eggs half a millimeter in diameter and those having a diameter of about two millimeters, which indicate that the *number* of the villi remains constant during the period of growth from the smaller to the larger size.

In an egg 0.5 mm. in diameter there occur about 30 villi in a space of 35 μ; i. e. the villi are about 1.15 μ from centre to centre. In an egg 2 mm. in diameter from the same ovary, treated in the same manner and cut at the same time, the villi are 4.5 μ from centre to centre (compare Plate V. Figs. 3, 4). Allowing for the growth of the smaller egg, which at the larger size would have a diameter four times as great as at first, it is evident that the interval for a villus would be four times 1.15 μ, or 4.6 μ, which agrees fairly well with the space (4.5 μ) actually occupied by a villus in the larger egg measured. There are also other reasons for believing that the villi do not increase in number after the egg has reached a diameter of half a millimeter. If new villi were interpolated, one would reasonably expect to find the younger ones shorter than the older ones; but at no stage which I have seen is there any marked difference in their lengths.[1]

In the larger eggs measured (2 mm.), the nuclei of the granulosa were on the average about 14 μ apart, from centre to centre; i. e. there were about three villi to the diameter of each cell. But in eggs about half a millimeter in diameter (compare Plate VIII. Fig. 3, and Plate IX. Fig. 5) it is to be seen that from six to fourteen villi correspond to the diameter of a single granulosa cell. If there has been no change in the number of villi, it follows that the granulosa cells must have increased in number *at least* fourfold between the half-millimeter stage and the two-millimeter stage. It is for this reason I contend that there is no constancy in the numerical relations of villi and granulosa cells, and that consequently it is improbable that the former are the product of the latter.

[1] It is evident that there has been a corresponding increase in the *diameter* of the individual villi during the growth of the ovum, for in the mature condition they form a continuous layer, with little or no intervening substance.

Ransom ('67) has claimed that the pore-canals of the zona radiata increase in number during the growth of that membrane. If one were to disagree with me, and to regard the markings which first appear at the surface of the ovum as the incipient zona instead of the villous layer, he would be compelled to adopt Ransom's view, for the intervals between the markings on eggs half a millimeter in diameter (1.15 μ) would become, unless there were interpolations, 4.6 μ apart when the eggs had increased to two millimeters in diameter. In order to reduce the intervals to the condition actually found in the zona of the mature egg (1.4 μ), the number of pore-canals would have to be increased more than threefold!

At first thought one might regard the modifications of the villous layer in the micropylar region as the direct result of an alteration in the secretive powers of the granulosa cells situated at that place; but it seems to me that the thickness of the layer ought, on this assumption, to be greater than elsewhere, since the granulosa cells are here more numerous and larger. Besides, the corresponding diminution in the thickness of zona radiata could not be thus accounted for, but must be assumed to be the result of diminished secreting activity on the part of the ovum in this region. Hence the same explanation would certainly be more reasonable in the case of the villous layer. This is a point which seems to me of considerable importance; the diminished activity of this region which is shown during the formation of the zona was already manifest during the formation of the villi.[1]

From these several considerations, I believe there can be little question that *the* VILLOUS LAYER *of the egg membranes in Lepidosteus is also the product of the ovum itself rather than of the follicular epithelium surrounding it.*

If this conclusion is established, it follows that the parts of the villi first to be produced are those which are most superficial. I believe that this is confirmed by the fact that the forming villi are readily stained in carmine. It is probable that, even in the latest stage of the immature eggs (0.7 mm.) which I have seen, not much, if anything, more than the heads of the villi have been produced. The length and the highly refractive condition of the villi at this stage, and the fact that they are not at all folded, all point to this conclusion.

There still remains much to be done in following out the exact course of the development of the membranes in Lepidosteus, — especially in determining when the formation of the zona begins in relation to the completion of the villous layer, — but I think that the main features of the process as outlined above will not be disproved by subsequent study.

I have no explanation to offer of the apparently sudden change in the nature of the secretions from the ovum which is registered in the production of structures so dissimilar as the zona and villous layer are; but it is possible that some light may be thrown on this question when the period of the transition has been carefully worked out.

---

[1] This is an evidence of the polar differentiation of ova (which exhibits itself in many other phenomena) to which attention has not hitherto been called.

B. HISTORICAL AND CRITICAL REVIEW OF THE LITERATURE ON THE PRIMARY EGG MEMBRANES [1] AND THE MICROPYLE IN FISHES. [2]

It is possible that the eggs of fishes may present as many as *four* essentially distinct kinds of enveloping membranes before separation from the ovary. The innermost of these, if it exists, may be considered a true *vitelline membrane*, the equivalent of the cell membrane in general. I have made no observations concerning it, and shall have little to say regarding the conflicting testimony as to its existence. The second, proceeding from the yolk outward, is radially striate, and I shall call it, as in the preceding description, *zona radiata*. Although this is totally different in structure from the next membrane, there are several reasons why it will be best to consider both at the same time. This third membrane I shall call, as previously, the *villous layer*. The fourth and outermost, when it exists, is formed exclusively from the granulosa cells, and may be called by the name first given to it by Johannes Müller, — *capsular membrane*. [3]

a. *Cyclostomata.*

The eggs of the myxinoids are enveloped in a "horny capsule," which was first described by Thomson ('59, pp. 50, 51) for *Myxine glutinosa*. He evidently considered it the equivalent of the egg cases of selachians. Since the latter are formed in the oviduct, they cannot be considered

---

[1] I use the expression *primary egg membranes* in the sense in which it has been employed by Ludwig ('74 p. 197), i. e. for all membranes which are the product of either the ovum itself or the follicular epithelium surrounding it.

[2] Owing to delays in publishing my studies I have been able to extend this review, and to bring it down so as to include papers which have appeared since my own account was written.

[3] I have the less hesitancy in adopting this name because Müller ('54, p. 189) — notwithstanding some misconceptions as to its real nature in the perch — gave the following concise, and, in my opinion, still perfectly applicable definition: "Eine von dem Eifollikel, Ovisac eines Wirbelthiers erzeugte Eihülle scheint von der Eischale anderer Eier unterschieden werden zu müssen als *capsulare Eihülle*, oder Eicapsel." When subsequent observers, — as for example His ('73), — ignoring the true explanation of Müller's investigations given by Leuckart ('55, pp. 257–260), transfer the name Eicapsel to the zona radiata, one is compelled to protest that that was not the structure described by Müller under the name of "Eicapsel," and that no one has yet brought forward satisfactory evidence that the zona is "produced by the egg follicle," as Müller's definition demands. It therefore seems to me that it is better, for the sake of avoiding confusion, to drop entirely the name capsule — whether egg capsule or "cartilage capsule" (His) — as a designation for the zona radiata.

as primary egg-membranes; but Steenstrup ('63) subsequently showed that the egg of Myxine glutinosa possesses this covering before it leaves the ovary, from which it follows that the "horny capsule" is really a primary membrane.

THOMSON'S ('59) account is brief: "I have found that in the Myxine glutinosa the globular yolk is enclosed in a horny capsule of similar consistence and structure [to that of the oviparous cartilaginous fishes], but of a simple elongated ellipsoidal shape, and in place of four terminal angular tubes, a number of trumpet-shaped tubular processes projecting from the middle of the two ends, which probably serve the same purposes as the differently shaped appendages of the ova of the shark and skate."

STEENSTRUP ('63, pp. 233–238, Figs. a–h) also saw the horny egg-shell and the peculiar projections from its ends. He says (p. 236): "In the last received individuals the eggs now had not only the same considerable size [as some large eggs previously described] and more oval-elliptical form, but besides they were surrounded with a somewhat firmer, almost horn-like egg-shell, which was furnished at the ends with a large number of slightly curved or S-shaped horn-threads. Each horn-thread ends in a head-shaped portion with three or four projecting spines or hooks, and has thereby some resemblance to a ship's anchor. The threads recall — even though somewhat remotely — the horn-threads projecting from the eggs of the rays and sharks, much as the shell itself recalls the firm capsule of these cartilaginous fishes. The accompanying figures exhibit both the appearance of the capsules (f, g) and the manner in which they hang in the mesovarium (h^{**} and h^{***}), together with eggs of the same appearance as c, d, e (Fig. h^*), and with a large number of only slightly developed eggs (o, o, in Fig. h)."

In the two eggs with horny shell figured by Steenstrup, the shell has been represented as though it were composed of two parts separated by a sharp continuous line; the egg appears cut through near one pole by a plane perpendicular to its long axis. The appearance recalls that seen in the egg-shells of certain trematodes, where one end serves as a lid which opens to allow the larva to escape; but whether the author regarded this as a similar provision for the escape of the young hag, or as an accidental condition, is not stated in the text.

WILHELM MÜLLER ('75, pp. 114–117, Taf. V. Figs. 14, 15) appears to regard the "Testa" of Myxine glutinosa — which I suppose to be the same as the "horny capsule" of Allen Thomson — as resulting from the secretions [metamorphosis?] of a layer of [granulosa] cells, which immediately invest the ovum. He does not expressly state this, but it seems to me he leaves one to draw such an inference. He says that the ovarian egg when 0.6 mm. in diameter is surrounded by a single layer of very flat polygonal cells, outside of which is a thick layer of fibrous connective tissue, and that when the eggs have attained a length of 18 mm. and a thickness of 6 mm. there are two connective-tissue envelopes; an outer thinner, a continuation of the mesovarium, and an inner, which at the ends of the egg is thickened (0.4 mm.) and vascular. At its inner surface the inner membrane is condensed into a lustrous membrana propria $2\,\mu$ thick, and is firmly attached to the underlying "Testa." In contact with the inner surface of this membrana propria is a layer of cells. In the middle of the egg the cells are cubical, but they become more and more cylindrical towards its poles, where the layer becomes three or four cells deep.

I believe there can be no question that this layer of cells inside the membrana propria represents the granulosa; but it seems as though Müller must have overlooked the egg membrane, if one existed at that stage, and must have taken the granulosa to be in some way the equivalent of it. Perhaps, assuming that the granulosa cells secreted the membrane, his idea was that the granulosa ought itself to be considered as a part of the "Testa," for he afterwards (p. 126) mentioned, in the case of *Petromyzon Planeri*, "a very thin folded egg membrane which exhibited a polygonal pattern when seen from the surface." Moreover, he says, with regard to two *deposited* eggs of Myxine which he examined, that there was no trace of either inner or outer connective-tissue envelope, and from this fact concludes that they must have undergone complete regressive metamorphosis, similar to that which the enamel organ of the teeth suffers after the completion of the enamel.

W. Müller is the only person who has seen anything of a *micropylar apparatus* in the myxinoids. "Exactly in the middle of the white pole of the egg," he says (p. 115), "this cell layer [granulosa] exhibits a conical infolding 0.1 mm. deep and 0.06 mm. broad, which contains a funnel-shaped opening, the micropyle, which is directed straight toward the underlying nucleus and the protoplasm surrounding it." This is the whole of his description; and from it I infer that he has seen that portion of the granulosa which occupies the micropylar *funnel*, but that the micropylar *canal* — which is *a passage through a membrane, not an involution of a cell layer* — has not been seen by him. If the condition in Myxine is at all comparable with that in Lepidosteus, it is certain

that Müller has seen the equivalent of what I have called the micropylar plug of granulosa cells, and it is therefore probable that he was
the first person to observe that peculiar structure in any fish-like animal.
If he were less positive in his assertion that the infolding contained an
opening, I should question if the cells took the form of a hollow funnel;
even as it is, I doubt if the membrana propria is infolded.[1]

The first account of the membranes in *Petromyzon Planeri* was by
MAX SCHULTZE ('56, pp. 1–5). When taken from the body, the eggs had
besides the yolk membrane a firm " Eischalenhaut," or "chorion," which
was surrounded with a scarcely discernible thin layer of gelatinous substance, which was quickly swollen, when it came in contact with water,
to a thickness of not more than a quarter of a line.   It was delicate and
fugitive, and was easily removable from the firm underlying membrane.
In the course of eight days it mostly disappeared, being dissolved in the
water; it was not an "albuminous layer," but was rather to be compared
to the gelatinous mass uniting frogs' eggs; its chemical composition was
not known.
The firm "Eischalenhaut," which closely enveloped the egg, was a
clear membrane about 0.0015$'''$ (probably should be 0.015$'''$, or about
0.03 mm.) thick, which had a tendency after being torn to roll in at
the edges.   It appeared very finely punctate when viewed from either
the inner or the outer surface.   Schultze was inclined to regard the
punctations as due to very fine canals traversing the membrane, but
on account of the delicacy of the object he could not reach a perfectly
satisfactory conclusion on this point.   For this finely punctate membrane and that found in bony fishes, the author would use the name
chorion rather than vitelline membrane, for a true vitelline membrane
(or egg-cell membrane) exists inside the punctate structure.
OWSJANNIKOW ('70^a, p. 184) says that the gelatinous layer of the outer
egg membrane is very little developed, so that the fertilized eggs are
only feebly attached to the objects on which they fall, the least current
carrying them away.
CALBERLA'S ('78, pp. 438–441) account is in some particulars more
extended than that of Schultze.   The eggs, he says, instead of being
round, are slightly ellipsoidal.   The membrane (zona) consists of two
layers, which are not, however, sharply separated from each other.   The
outer is highly refractive, rough externally owing to all sorts of elevations and tooth-like structures (Zacken); the inner is much thinner

[1] For a review of more recent work on Myxine, see pp. 91–93, 107–110.

and translucent. With low powers the outer appears as though made
up of concentric layers, but with higher powers it is seen to be a
homogeneous substance traversed by fine radial canals which are continuous with those passing through the inner layer. At the outer surface each of these canals opens out at the base of one of the elevations
(Zacken). Calberla regards this whole layer as a secretion from the
peripheral layer of the yolk. The proof of it he finds in the conditions
of the membranes in nearly ripe and in over ripe eggs. On the former,
the boundary between the two layers is sharper and the inner layer is
much thicker than on mature eggs; whereas on the latter all distinction
between inner and outer layer has disappeared.

As soon as the egg comes in contact with the water, the tooth-like projections on the surface of the egg membrane (zona) quickly swell, in consequence of which the whole egg appears as if surrounded with a delicate
area of hyaline substance. This may well be the cause, he adds, of the
stickiness of the surface of the egg.

It seems to me that there is considerable reason for believing that
these external projections described by Calberla correspond to the villi·
of Lepidosteus, both in function and in position. An examination of his
figures (Taf. XXVII.) lends support to this view. I believe also that,
when the genesis of the membrane has been studied, it will be found
that these "Zacken" are formed before the zona itself. It is true that
KUPFFER UND BENECKE ('78, pp. 9, 10) find the conditions somewhat
different from those recorded by Calberla. They claim that the envelope
of the egg consists in both P. Planeri and P. fluviatilis of a double membrane (Eihaut), and of a continuous covering of gelatinous material which
is replaced at the watch-glass-like elevation of the membrane by a structure known as A. Müller's "Flocke." The inner membrane — which
they figure as being much thicker than the outer — contains closely set
pore-canals, but these they assert positively are not continued into the
outer layer. The difference in structure between the two membranes is
demonstrable by means of 0.5 per cent hydrochloric acid. The outer
membrane swells more in water than the inner, but not quite uniformly.
It appears here and there as though it were restrained by a filament of
less easily-swelling substance. And this, they say, is probably the cause
of "Calberla's unzutreffende Angabe, dass diese Rindenschicht mit
allerlei Erhebungen und Zacken besetzt sei, an deren Basis Porencanäle mündeten."

But even if Calberla's description is not quite satisfactory, it is evident
that this outer envelope is not homogeneous, and that the toothed appear-

ance which he has figured must have had a basis in optically different portions of that envelope. According as the imbibition of water has proceeded less or more, this marking might be more or less conspicuous. From a comparison of the figures by Calberla with those by the last mentioned authors, I should think that Calberla's outer layer of the zona by no means corresponded with the outer layer of Kupffer und Benecke, and that the latter, being very thin, had been overlooked by Calberla.

The *micropyle* of Petromyzon, though sought for by Schultze ('56) and A. Müller ('64) was not found by them.

Owsjannikow ('70ª, p. 184), who discovered it, says that it is very small, but that it remains visible for several days after fertilization. In mature eggs it occupies a position over the eccentric nucleus.

Calberla ('78, pp. 439, 440) has given a careful description of the micropyle, which, he says, agrees in all essential particulars with that of osseous fishes. His account is substantially as follows. At one pole of the elongated egg its membrane is thickened, and bulges out, much as though a shallow watch-glass — with shorter radius of curvature than the rest of the egg membrane — had been set into one end of the membrane. Radial sections which pass through the centre of the elevated portion of the membrane show that in the middle of it there is a very flat saucer-shaped depression, the centre of which is further depressed into a funnel. From the narrow end of the funnel a canal is continued through the membrane, and opens on its inner surface with a slight flaring. A little below its middle the canal exhibits a spindle-shaped enlargement, which is shown in Calberla's Taf. XXVII. Figs. 2 and 3.

The views held by Kupffer und Benecke ('78, pp. 9–15) regarding the nature of the micropyle are not easily summarized. They are based on close observations of the deportment of the egg and spermatozoa at the time of fertilization, but do not appear to have been corroborated by sections of the egg membranes.

In the region of the watch-glass segment of the membrane described by Calberla, the mucilaginous envelope outside the membranes is wanting, and in its place is a hyaline dome (A. Müller's "Flocke") composed of a substance which, unlike the mucilaginous layer, is permeable for spermatozoa. Usually only one spermatozoön passes through the inner and outer egg membranes and reaches the yolk; but the place of its passage is by no means always the centre of the watch-glass area. It was such only six times out of fifty. The passage may occur even near the margin of this area. Neither is it always the spermatozoön that first reaches the outer membrane, after having traversed the "Flocke," which passes through.

The statement that the egg membrane is not alone permeable at a single spot would lead one to suppose that the authors were ready to deny the existence of a micropyle. They do not, however, directly assert its absence, although they were unable to find anything of it on the unfertilized egg. But as soon as the spermatozoön has passed through the membrane, a small circular spot may be seen from the surface; this is due to a shallow depression in the surface of the *inner* layer of the membrane, the outer layer never showing any passage through it. The authors hint at the possibility of a chemical action on the part of the spermatozoön resulting in a loosening of the two layers and a partial solution of them, and endeavor to make that view harmonize with the conclusion that the micropyle "is the remnant of an opening in the inner layer of the egg membrane, which exists during the stay of the egg in the follicle, corresponding to the condition which Herr von Jhering recently established in the case of the eggs of the mussels." The outer layer would be formed, they imagine, afterwards, and would cover over this opening, leaving a remnant of it recognizable on the inner membrane.

"The micropyle, therefore, is not an open passage, as it appeared from Calberla's description and drawings, but only a permeable place."

b. *Selachii.*

What Ludwig wrote in 1874 concerning oögenesis in the selachians, that it had been studied by only a very few investigators, was equally true of the primary egg membranes of the group. LUDWIG ('74, p. 145) himself, although he studied the development of the ova, had nothing to add to what was already known about the egg membranes, and since him there have been only two writers who have dealt with the subject, Schultz and Balfour.

LEYDIG ('52, pp. 87, 88) speaks incidentally of a vitelline membrane, and a thin albuminous layer surrounding it, in the case of Raja batis. The latter probably corresponds to one of the membranes seen by later observers.

GEGENBAUR ('61, p. 518) recognized the existence of a homogeneous egg membrane on eggs of Raja from 1''' to 2''' (2–4 mm.) in diameter; its external contour was delicate, but internally it was sharply limited. In the case of Acanthias there was only this one membrane to be observed; it attained a thickness of 0.08''' (175 μ) on eggs 4'''–5''' (9–11 mm.) in diameter.

Gegenbaur considers it probable that this membrane is produced by the follicular epithelium, but is evidently not certain of it. He says: "Es liegen hier wohl bei den Selachiern andere Verhältnisse vor als bei den Vögeln und Reptilien, und eine Dotterhaut, wie sie dort von Seite des Dotters durch Umwandlung seiner peripherischen Schichte zu Stande kam, kommt hier wohl nicht vor, sondern der Dotter bleibt auf dem früheren Stadium der Differenzirung bestehen, dagegen bildet sich eine Hülle von aussen her, wozu wahrscheinlich die Zellen des Follikelepithels das Material abscheiden, wenn man den Vorgang der Bildung jener Membran nicht auf die Oberfläche des Dotters selbst verlegen will."

Schultz and Balfour disagree in their conclusions as to the origin of the fugitive membranes which envelop the ovarian eggs of selachians. Schultz ascribes their formation to the follicular cells ; Balfour, to the ovum itself.

SCHULTZ ('75, pp. 574–576) claims that in Torpedo oculata the follicular epithelium is composed of two kinds of cells : genuine *granulosa cells*, derived from the germinal epithelium of the ovary, and, alternating with them, *lymphoid cells*, which are derived from the stroma of the sexual organ. "The cells of this follicular epithelium, especially the lymphoid cells, are merged at their deeper ends into a homogeneous cuticular layer (Fig. 8), and there form a structure having the morphological value of a chorion." This homogeneous layer at no time has a morphological relation to the egg protoplasm, but retains the closest connection with the follicular cells. On objects subjected to pressure the outer margin of the homogeneous layer appears jagged like a wood-saw,[1] the remnants of the lymphoid cells corresponding to the teeth, in the intervals between which the granulosa cells are lodged. The latter are also attached, he says, to the homogeneous layer by means of protoplasmic processes, and even appear to fuse with it, but do not show any differentiation within the substance of the layer. It is not possible even with the highest powers to demonstrate any such structural peculiarities (radial striation, pore-like perforations) as are met with in the egg membranes of most classes of animals, even in Raja batis itself.

"Finally, when the egg cell has reached maturity and the follicle approaches the stage of rupturing, the lymphoid cells together with the homogeneous layer are converted into connective issue, in the interstices of which the granulosa cells persist, although the latter finally undergo fatty degeneration. Only at a single place, corresponding to the whole extent of the germinal disk, do the follicular cells and the

[1] " Gleichsam hohlsägeformig [hohlzsägeformig ?] gezackt."

homogeneous layer persist unchanged up to the bursting of the follicle. It is from this part that those granulosa cells come which are occasionally encountered on the escaped [egg] and within the empty follicle."

On eggs of Acanthias, Scymnus, and Mustelus, Schultz found a homogeneous layer joined with the follicular layer, and inside the latter a zona radiata, the inner margin of which was sharply defined against the yolk. "The pores of this cuticular zona were traversed by protoplasmic processes, which stretched from the homogeneous layer to the egg protoplasm and fused with the latter."

The author concludes that, so far as his own observations reach, there are to be distinguished in selachians the four "following conditions of the follicular epithelium": (a) simple epithelium (embryonic stage of selachians); (b) epithelium with homogeneous basal margin (Torpedo); (c) epithelium with homogeneous perforate basal margin (Raja); (d) epithelium with broad homogeneous and narrower perforate basal margin (Squalidæ).

BALFOUR ('78[b], pp. 402, 403) has confirmed the existence and subsequent disappearance of two membranes — an outer homogeneous, and an inner striate — in one of the Squalidæ, Scyllium ; but he believes that they are produced by the ovum, not by the follicular epithelium, and that they are absorbed, not converted into connective tissue. Two similar membranes are also found in Raja, and are believed by Balfour to be common probably to all sharks. The homogeneous membrane is formed before the striate one. In Scyllium "the [homogeneous] membrane would seem indeed to be formed in some instances even before the ovum has a definite investment of follicular cells." Consequently it is called a *vitelline* membrane. In ova 0.12 mm. in diameter it is not thick enough to be accurately measured ; in those of 0.5 mm. diameter it has a thickness of 2 μ, and there may also be observed inside it faint indications of the differentiation of the outermost layer of the vitellus into the perforate or radially striate membrane of Schultz. The latter Balfour does not hesitate to call a zona radiata.

In ova 1 mm. in diameter the zona has increased in thickness (to 4 μ) and "is always very sharply separated from the vitelline membrane, but appears to be more or less continuous on its inner border with the body of the ovum, at the expense of which it no doubt grows in thickness." In larger eggs both membranes increase in thickness, especially the zona, which now becomes marked off from the yolk. "In many specimens it appears to be formed of a number of small columns as described by Gegenbaur [for the alligator] and others."

The size of the ova at the time of the maximum development of the membranes is not stated; but after this stage is reached both membranes gradually atrophy. "The zona is first to disappear, and the vitelline membrane next becomes gradually thinner. Finally, when the egg is nearly ripe, the follicular epithelium is separated from the yolk by an immeasurably thin membrane, — the remnant of the vitelline membrane," which is no longer visible when the egg becomes detached from the ovary. Both vitelline membrane and zona are found in Raja, but in a much less developed condition than in Scyllium, and the zona is developed at a much later period than in that species.

If the account given by Balfour is correct, — and his description seems to be both more complete and more accurate than that of Schultz, — then there is an interesting parallelism between the primary egg membranes in selachians and those of Lepidosteus and the bony fishes. Not that the villous layer in Lepidosteus is *structurally* comparable with that which Balfour calls in sharks a vitelline membrane, but *genetically* they are alike. *They are the membranes which are first to be produced,* — i. e. before the zona radiata, — *and they are in both instances the product of the ovum, not of the follicular epithelium.* I shall take occasion later to refer to the theoretical importance of this discovery by Balfour.

The ultimate disappearance of both membranes renders the formation of a micropyle superfluous. It would be interesting to learn, however, whether there is at any time a trace of such a structure.

c. *Ganoidei.*

The only ganoids besides Lepidosteus whose egg membranes have been described are Amia and Acipenser. I have elsewhere (p. 27) quoted Ryder's account of the membranes in the case of Amia.

From the accounts given by Kölliker, by Kowalevsky, Owsjannikow und Wagner, and by Salensky, I believe there must be considerable similarity between the conditions of the egg membranes in Acipenser and Lepidosteus.

KÖLLIKER ('57, p. 197) says that the porous membranes in the case of the sturgeon form "three layers; two inner, darker, thinner, closely porous, and an outer pale, thicker layer, apparently with fewer pores. This outer layer, which is also softer than the others, shows its outer surface divided into small polygonal areas, which appear to correspond to the epithelial cells of the egg capsule [follicle]. These cells are extremely delicate and pale, but yet seen from the surface they show a fine punctation. It as the appearance, therefore, as though these cells secreted the porous layers; however, concerning this, as well as concerning the corresponding parts of the eggs of other animals, only careful studies made on eggs of all ages can give an answer, wherefore I abstain for the present from any opinion."

According to KOWALEVSKY, OWSJANNIKOW UND WAGNER ('70[a], p. 172), the outer membrane in Acipenser is thick, shagreen-like, and possesses numerous very fine canals. When the ripe eggs fall from the oviduct, this membrane sticks to objects; with a certain amount of skill it may be rather easily detached from the egg. The inner membrane is much finer [has finer canals?], transparent, and very firm.

SALENSKY ('81, pp. 234–236) applies the name chorion to the outer of the two layers composing the "thick capsule" which envelopes the ripe egg of Acipenser; the inner he calls vitelline membrane. At first the two are so intimately joined to each other that it is difficult to separate them; but after the egg has been deposited for some time, the chorion is easily detached from the vitelline membrane, and may be removed from the whole egg. From the study of sections of stained eggs the author determined that the stickiness and the roughness of the surface were due, not to the chorion, but to two special cell layers which invest it.

"As to the origin of these two membranes, there is no doubt that they are derived from the two layers of cells which constitute the epithelial wall of the ovarian follicle."

"The chorion," he adds, "is probably a product of the secretion of the membrana granulosa of the follicle; when the latter ruptures, the epithelial cells remain adherent to the chorion and are expelled with the egg, and are again met with slightly modified at the surface of the deposited egg."

"The examination of microscopic sections of the egg shows that this envelope, which, as has been said, is divided into two after deposit, — into chorion and vitelline membrane, — presents three distinct layers. The external and the internal have about the same thickness; the middle one is thinner. In separating the chorion from the vitelline membrane one may convince himself of the fact that the chorion is composed of two layers, the vitelline membrane of only one."

From Mayzel's abstract (Salensky, '79[,] p. 220) I learn further, that the outer layer is stained deeply by hæmatoxylin, the two remaining layers not at all, and also that all three layers are radially and finely striate. In the figure of the membranes given by Salensky ('78[b],

Tab. I. Fig. 8 B) the inner membrane is only slightly thicker than the middle one, and both present a lighter appearance than the outer one.

I believe it is probable that the outer layer will be found to corre-spond closely to the villous layer of the gar-pike. There can be no doubt that the inner layer is the zona radiata, and I am inclined to regard the middle layer simply as the differentiated outer half of the zona; but the question can be answered satisfactorily only after renewed investigations which give more particular attention to this point. The principal reasons for my conclusion regarding the middle layer are, that it evidently resembles the middle layer more than the outer, especially in its capability of being stained, and that differences between the inner and outer portions of the zona have been observed in the case of other fishes. I know of no case, I admit, in which the outer half of the zona may be easily removed with the villous layer, so that it still is possible that the middle layer in Acipenser corresponds to the stalk region of the villi in Lepidosteus.

I do not clearly understand what the author means by saying that the two membranes are derived from the two layers of cells which con-stitute the epithelial wall of the follicle. It is true he claims that there are two distinct cell layers, which, according to his figures of *early stages* ('78[b], Tab. I. Figs. 5–7), are " granulosa " — next to the yolk — and "follicular epithelium " — immediately outside the latter; but he gives no figure showing both these at advanced stages of development. I doubt their existence. But even if there were two separate epithelial layers, I fail to understand how *both* could share in the production of membranes which lie wholly inside the inner cell layer.

The eggs of Acipenser are altogether unique so far as regards the con-dition of the membrane in the micropylar region.

The earliest observer of the micropyle was KÖLLIKER ('57, p. 197), who incidentally remarks that there is a *single* micropyle in the stur-geon's egg; but subsequent observers have claimed that there is a *group of several micropyles.*

KOWALEVSKY, OWSJANNIKOW UND WAGNER ('70[a], p. 172) state that at one pole of the egg there are seven micropylar openings, — one in the centre, with the other six arranged in a circle around the first.

SALENSKY ('81, pp. 235, 236, Planche XV. Fig. 1 A) gives a more com-plete account of the micropylar region, which is illustrated by figures. He says: " At the germinative pole of the egg there is found a micropyle.

The orifices of the micropyle are so small on hardened eggs that it is probable that they are narrowed by the action of the reagents, which also cause a retraction of the capsule of the egg. For this reason it is very difficult to find these orifices."

The micropylar apparatus is composed of several (5 to 13) orifices, and Salensky states that, although he has examined a great number of eggs, he has never found two which were identical either in the number or distribution of the orifices. "Each orifice consists of a quite small pit (fossette) having the form of a crater; it is surrounded by small, very slender cylinders."

Salensky has given only surface views of the micropylar region, but it has occurred to me that his "very slender cylinders" may be villi which surround the crater-like depression. They would appear to have a radial arrangement about the crater as a centre, if the villous layer were seen in optical section at a plane a little below the outer margin of the crater.

d. *Dipnoi.*

The conclusions reached by BEDDARD ('86[a]) relative to the ova of *Lepidosiren* possibly rest on too limited material to receive immediate acceptance. So far as regards the egg membranes, the story is certainly far from being satisfactorily completed. In the youngest eggs there was no trace of any membrane; but as development proceeded, a delicate homogeneous membrane encircled the ovum. This was from analogy thought to be the product of the egg protoplasm, even though it was more firmly adherent to the follicular epithelium than to the ovum, and it is called vitelline membrane. In more mature ova there was underneath this a much thicker radially striate membrane, probably corresponding to the zona radiata of other vertebrates, which in places seemed to pass gradually into the substance of the protoplasm. This membrane (zona radiata) began to disappear with the first steps in the formation of yolk. During the period of yolk formation the vitelline membrane became thicker, and also radially and coarsely striate. The author believes that there was a stage which succeeded this, during which there was no membrane of any kind, and that at this time an immense number of follicular cells migrated into the yolk. But in addition there followed still another stage, — when the ovum was entirely occupied by yolk, — in which the epithelium was separated from the contents of the ovum by an extremely delicate homogeneous membrane, which either corresponded (in some cases) to the persistent vitelline membrane, or (in other cases) was a new formation; but even in the latter case the author maintains that it is homologous with the vitelline membrane !

The fact that the layer called by Beddard *vitelline membrane* becomes radially and *coarsely* striate, suggests comparison with the villous layer of Lepidosteus. As in the ganoid, so in Lepidosiren this is the layer which is produced first.

e. *Teleostei.*

The numerous descriptions which have been given of the egg membranes in different *osseous fishes* show that there is not uniformity either in their number or structure. Besides the wall of the follicle with its epithelium, the granulosa, there is perhaps only one investment of the egg which is universally present, the zona radiata, and even this may be wanting, or at least may wholly disappear in the case of certain viviparous fishes. I believe it is certain that the homologue of the zona radiata is invariably present in oviparous fishes, and I am likewise of opinion, notwithstanding the inability of some observers to discover the presence of pore-canals in the eggs of certain fishes, that the canals are also always present.

In many cases there is a membrane intervening between the zona and the granulosa, sometimes thin and homogeneous, at other times of a more complicated nature. It may even (villous layer) resemble somewhat the zona itself, although in reality very different from it in structure. Whether the thin homogeneous layer is homologous with the more complicated villous layer cannot as yet be definitely answered, but Balfour's account of the origin of a similar structureless membrane in Elasmobranchs makes me incline to the belief that it is.

In addition, the cells of the granulosa undergo in some instances a remarkable metamorphosis, accompanied probably by a process of secretion, and thus furnish still another primary envelope to the ovum.

As to the existence of membranous structures *inside* the zona radiata, there is much less certainty. The presence of a structureless *vitelline membrane* has been maintained with more or less confidence by Vogt, Aubert, Thomson, Lereboullet, Kölliker, Eimer, and others, and more recently by Owsjannikow and Scharff. Its existence has also been denied by eminent authority.

The question as to the presence and nature of a so called *zonoid layer* seems, especially in the light of Brock's recent contributions, to demand a more extended and thorough investigation. In the case of some of the eggs studied by Eigenmann ('90), structural conditions have been observed which can hardly bear any other interpretation than that of a striate zone of substance inside the zona radiata proper; but in other cases (Esox, Amiurus) a somewhat similar though perhaps not identical appearance is due to a retraction of the vitellus from the zona, which leaves strands of vitelline substance stretching across the space thus produced.

1. *Zona Radiata and Villous Layer.*

JOHANNES MÜLLER ('54) has often been credited with having discovered the fine radial canals which traverse the zona radiata, and give to it its most characteristic appearance. This is a mistake which has arisen from Müller's misunderstanding the relation of the peculiar membranes of the perch to those of other fishes. What he described as "pore-canals" in Perca belonged to a much thicker membrane than the zona. This membrane lies outside the latter, and is a result of the activity and metamorphosis of the granulosa cells. Müller, it is true, supposed this to be the equivalent of the "shell membrane" previously described by Vogt, and therefore imagined that he had been able to demonstrate on a more favorable object what Vogt had claimed on grounds of analogy rather than on satisfactory proof.

In VOGT's studies on *Coregonus palæa* he ('42, pp. 1, 8–10, 27, 28) claimed the presence of two membranes. The inner one — being thin, transparent, and without apparent texture — he called *vitelline membrane;* the outer one he called a *shell membrane,* and homologized it with the "membrane coquillière" of birds' eggs. This outer membrane presented the appearance of shagreen, which seemed to result from a quantity of small opaque points uniformly distributed over its surface. Treated with hydrochloric acid, the points became more transparent, and then resembled minute warts. Valentine called Vogt's attention to the resemblance between this structure and that of the carapace of the cray-fish, where he had found that a similar effect was due to perpendicular tubes, filled with lime, traversing a membrane composed of regularly polyhedral cells. Vogt, admitting that the "shell membrane" was too thin to allow the attainment of exact results relative to the nature of the "points," nevertheless claimed that the position, behavior, and reticulate appearance of the latter warranted one in supposing that the structure was analogous to that of the carapace of the cray-fish. Thus it appears, he continues, that the shell membrane is formed by the union of flattened cells, which are arranged around the primitive egg only toward the epoch of its maturity; the presence of these minute tubes, which traverse the membrane, would in his opinion sufficiently explain the absorption of the water into the interior of the shell membrane. (Compare also Vogt, '42, pp. 27, 28.)

From this summary it is evident that Vogt had under consideration appearances which were due to the pore-canals of the zona radiata, and that he moreover believed in the presence of canals, but it cannot be claimed that he demonstrated their existence. Unlike his predecessors, he rightly claims that the shell membrane (in Salmo umbla) originates in the ovary.

◆ In a paper written in 1845, but not published till many years later, VOGT ET PAPPENHEIM ('59, pp. 357, 361, 362) also maintain that the shell membrane of the eggs of fishes, which is uniform and elastic, is constituted by the fusion of a cell layer formed *in the ovary,* and therefore not to be compared with shell membranes which are produced in the oviduct. They made the mistake of insisting that "this cell layer is not to be confounded with another epithelial layer which one finds in the ovisacs of the youngest ovules, and which is composed of large extremely pale cells which subsequently disappear and give place to this second layer."

MECKEL VON HEMSBACH ('52, p. 421, Taf. XV. Fig. 1) saw and figured a *radial structure* of the "zona pellucida" in the case of Cyprinus auratus, after treating the membrane with acetic acid and crushing it, but he expresses no opinion as to the real nature of the striation.

LEUCKART ('53, pp. 796, 797), who probably had not yet seen Meckel's paper, was evidently not impressed with the explanation which Vogt adopted to explain the appearance of the outer membrane in Coregonus ; for he says simply, "That which characterizes the eggs of teleosts is the possession of a special firm egg-shell (*chorion*), which is already formed in the follicle around the primitive yolk membrane, and generally presents a delicate marking resulting from regularly grouped granules or points."

During the five years beginning with 1853 there appeared a large number of papers dealing with the egg membranes of fishes, and the subject was brought to a temporary conclusion by the thorough work of Kölliker.

AUBERT ('53, pp. 94, 95, Taf. VI. Fig. 1) was the first to figure well the appearances due to the pore-canals of the zona radiata. In addition to "a very finely granular, but otherwise structureless skin, which envelops the yolk," and which I believe must have been in his opinion the vitelline membrane,[1] Aubert describes the "shell" of the egg of Esox as a transparent thin membrane, furnished with fine points, which closely envelops the yolk.   These points exhibit a great regularity of arrangement, being placed at the intersection of symmetrical curved lines. When the shell has lain some time in water, it separates in many places into two membranes, of which the outer is very thin, finely granular, and irregularly elevated, while the inner is thicker, uniform, and upon sections exhibits fine radially placed streaks.

I believe it is certain that the radial streaks described were due to the pore-canals, although the author does not fully commit himself to that view. "The spermatozoa are so large," he says, "that it would be difficult for them to pass through the 'points' of the shell, in case the latter are regarded as the lumina of fine canals."

Also LEREBOULLET ('54, pp. 240, 242, 245, 249) wrote concerning the pike : "The ripe egg is surrounded by two membranes : the external is pierced by microscopic tubes, which serve for the absorption of water, and consequently for the respiration of the egg ; the internal, applied to the vitellus, is a simple, extremely thin and amorphous protecting envelope."   He also saw the pore-canals in the perch, and argued that the expulsion of albuminous globules from the fertilized egg proved the absence of a vitelline membrane at that time, and that it went to confirm the opinion of those who regarded the chorion as produced by the primitive vitelline membrane, which was itself detached from the vitellus.

Although J. MÜLLER ('54) contributed much to the knowledge of the egg membranes, especially of the perch, and was also the first to appreciate the importance of the difference in origin between the egg-shell in birds and what he called the "egg capsule" in fishes, he did not fully comprehend the structure of what he called the "Dotterhaut" (zona).

[1] I cannot agree with His ('73, p. 2, foot-note, compare also p. 14) in his criticism of Aubert when he says: "Der Name Dotterhaut, welchen die früheren Schriftsteller für eine besondere den Dotter unmittelbar umhüllende structurlose Membran gebraucht haben, wird von H. Aubert (Beiträge, etc., 94) auf die Eikapsel angewendet.   Er spricht nämlich bein Hecht-Ei von einer Trennung der Dotterhaut [!] in zwei Schichten, eine äussere dünne, fein granulirte und eine innere, dicke, mit radiären Streifen.   Eine Begründung seiner abweichenden Bezeichnungsweise giebt er nicht."

Aubert says distinctly enough that it is the "Schale" which is divided into two membranes; and although he nowhere employs the word "Dotterhaut," there seems to me no doubt that he has Dotterhaut in mind when he says : "Der Dotter wird von einer sehr fein körnigen, sonst structurlosen Haut überzogen."   His may have been misled by the statement that "Die Schale etc. den Dotter eng umgibt."

He claimed for Cyprinus erythrophthalmus, Perca fluviatilis, and Acerina vulgaris a velvety appearance of the external surface of this membrane "as if beset with tufts" (Zotten [1]), which he ascribed to "very small cylindrical projections or rods with rounded ends," — prolongations of the vitelline membrane itself. Here again Müller unfortunately confounded the unlike conditions of different eggs. While his conclusions have been confirmed in the case of Cyprinus erythrophthalmus, I know of only one author whose observations give any evidence of the presence of such a "pile" or velvety structure outside the zona in either of the other fishes mentioned. Hoffmann ('81, pp. 19, 20, Taf. I. Figs. 9, 10), it is true, not only figures an external layer of the zona in a nearly ripe egg of Perca, which is thinner and *much more sparsely striate* than its inner layer, but he also describes and figures an October egg as possessing outside the still thin zona a layer of *minute, close-set tubercular projections* which "fully correspond to the Zöttchen of the Cyprinoids." I believe that Hoffmann has in some unaccountable way fallen into an error in this matter. At least, no other observer has seen any trace of the structure which he describes, and an examination of the eggs of our American perch in October reveals nothing of the kind. I think the condition in Acerina is probably similar to that in Perca, — certainly no one has shown the presence of a "Zöttchen" layer. That being the case, what was it that Müller mistook in Perca and Acerina for "Zöttchen"? A statement by Owsjannikow ('85, p. 18) makes me believe that Müller may have had under view the branching deep ends of the tubular structures which according to Owsjannikow traverse the gelatinous envelope, and which are left sticking in the pore-canals of the zona when the two layers are artificially separated. It may be, however, that Müller saw the more conspicuously — *but not*, as Hoffmann makes it, *more sparsely* — striated outer portion of the zona itself, in which event it might be

---

[1] Müller was not the first to see the appearance presented by these "Zotten," nor even to suggest the name. Von Baer ('35, p. 7) had, twenty years before, seen the same thing in species of carp. "Die äussere Eihaut ist aber nicht ganz formlos und gleichartig in sich. Sie enthält in den Karpfenarten, die ich zu untersuchen Gelegenheit hatte, keine [should be *kleine*] *dunkleren Vorragungen*, die ihr bei starker Vergrösserung ein *zottiges* Ansehen gaben." It is evident that Von Baer confounded the radial "tubes" in the outer envelope of the perch egg with these villi of the carps, for he adds: "Im Barsche ist diese Hülle noch sehr viel dicker und man sieht, dass die *dunklern Flecken*, die hier lang und schmal sind, in der äussersten Schicht sich befinden." One will find little occasion for surprise at this parallelism, in view of the fact that some of the most recent observers, with the best modern appliances at command, have arrived at a similar, though I believe erroneous conclusion.

fairly claimed that he had seen the pore-canals; but even in that case
it remains perfectly evident that he did not understand at all the
structure of the zona radiata.

When, a few months later, Remak ('54) announced the discovery of
radial striations traversing the whole thickness of the zona pellucida in
the case of the rabbit's egg, and attempted to determine the cause of
the appearance by a comparison with the conditions found in the egg
membrane of a fish (presumably Gobio fluviatilis), Müller ('54ª, p. 256)
was unable to agree with him in the conclusion that the appearance was
probably due to "an alternation of canals and cylinders," but endeavored to show that it was "merely an optical expression of the sum-
mation and partial superposition of the images of the rods [Zapfen] as
seen when viewing the vitelline membrane in profile." The images of
the overlying rods which fall in one line would, in his opinion, cause
the striations to appear much longer than the individual rods really
were, and thus make the lines *appear* to traverse the whole thickness
of the membrane.

Ransom ('56) maintained that there was present at an early stage
in the growth of the ovum of Gasterosteus a very thin membrane, having a finely and regularly dotted structure; but he does not appear to
have realized as yet that the dots were evidence of pore-canals. He also
discovered that in older eggs the part of the membrane immediately
surrounding the micropylar depression exhibited a number of cup-
shaped pediculated bodies scattered over its surface. These have since
been claimed by Kölliker ('58, p. 81) and subsequent authors to be the
localized equivalents of the "Zapfen" layer discovered by Müller.

In the same category must also be placed the remarkable filamentous
structures discovered by Haeckel ('55) on the eggs of several of the
Scomberesocidæ. Although Haeckel described the filaments as lying
inside the finely punctate vitelline membrane (zona radiata) and having
no connection either with it or the yolk, there can be no doubt, as
Kölliker ('58, p. 81) first showed, that they are really outside the zona.
I have not yet had the opportunity of examining the eggs of any of the
Scomberesocidæ, but conjecture that the sheath which envelops the bases
of their filaments may be only a part of a membrane external to the
true zona radiata, and comparable with that which Eigenmann ('90)
has found in Fundulus.

Leuckart ('55, pp. 257–264), who in an appendix to his celebrated
paper on the micropyle of insects' eggs deals, at least incidentally, with
the structure of the egg membranes in fishes, was the first to perceive

the true significance of Müller's discoveries in the perch. He retained the name chorion for the zona radiata, and from a study of the trout was fully convinced that the punctate appearance is due to " delicate tubules or canals which traverse the membrane perpendicularly, without opening, however, at its inner surface." The latter part of this statement has not been confirmed by subsequent observers. Leuckart, however, gave an excellent description of the structure of the zona radiata in the perch, for he not only recognized that it was composed of an outer thinner, firmer membrane, and an inner thick layer of viscid sarcode-like substance, but he also saw that the two layers were so intimately joined to each other that the canals were continued through both. While I prefer to regard these two layers as substantially a unit, basing my conclusions on a variability in the apparent independence of the outer layer and on the continuity of the pore-canals through both, I recognize that this is a minor point, and that already Leuckart was in possession of the important facts of structure. It was the presence of this " chorion " in addition to Müller's capsule with its coarser pore-canals which convinced Leuckart that the latter could not be considered the equivalent of the radially striate membrane in other fishes, such as the salmon and trout. I believe that Leuckart was less fortunate when he concluded that there was in Esox a layer immediately outside the " chorion " which was homologous with the Müllerian " capsule" of the perch ; for in my opinion there can be no doubt that the layer in question is the same as that in which Eigenmann has found the pore-canals to be continuous with those of the deeper portion of the zona. Even Leuckart describes the canals as straight, not spiral as in Perca. In my judgment, therefore, this layer corresponds to the thin outer layer of the zona seen by Leuckart in the perch, rather than to the capsular layer described by Müller.

Although Volume V. (Supplementary Volume) of Todd's Cyclopædia of Anatomy and Physiology was not issued until 1859, the article " Ovum " by ALLEN THOMSON was published much earlier, and in two parts, according to Gegenbaur ('61, p. 495). The first part (pp. 1–80) appeared in 1852, and the second part (pp. [81]–[142]), which contains the portion devoted to osseous fishes, in 1855.

THOMSON ('59, pp. [99], [100], [103]), besides giving a very brief summary of previous work on the subject, treated at some length the structure of the zona radiata, basing his conclusions partly on the work of Ransom and partly on his own studies. On the strength of Ransom's work he claimed that " the structure [Eikapsel] described by Müller in the perch was peculiar to that fish, and belonged only to an outer cover-

ing superadded to the surface of the dotted membrane, which last resembles in all respects that of other fishes." "This outer covering," he
adds, "appears to be of cellular origin; and Dr. Ransom thinks it may
be due to the separation of the tunica granulosa along with the ovum."
Thomson was able "to perceive the circle or lumen of the tubes" in
the dotted membrane by using a high magnifying power, and also
thought he could distinguish a hexagonal marking of the intervals between the pores (which he figures) in the salmon; he also pointed out
that the size of these pores was only about one third of that of the tubes
in the perch as described by Müller. The author was less successful,
when, in explaining Fig. 67*, he said: "A granular or dotted appearance of these [granulosa] cells seems to indicate their conversion into
the dotted membrane, which is probably formed in successive layers from
the exterior. . . . The ovum (p. [103]) receives its firm porous membrane [zona] while within the ovarian capsule, but only in the latter
part of the time of its formation." The origin of the membrane he
is inclined to connect with the interior of the ovarian follicle; "but
whether by exudation from it, or by amalgamation of the innermost layer
of epithelial cells of the follicle," he has not been able to determine.
The latter he believes the more probable, and that the membrane is the
true vitelline membrane. I am entirely unable to comprehend how
Thomson could have reconciled the two statements in the last sentence,
for surely the vitelline membrane was not, even at that time, regarded
as the product of anything but the ovum itself.

On ripe fish-eggs REICHERT ('56, p. 89) was able to distinguish two
membranes, both of which were formed within the follicle. The inner
was the punctate membrane, but owing to the fineness of the markings
it was impossible to determine whether they were the result of elevations
or depressions of the surface. Reichert was unable to adopt without
reserve the conclusion that the dotted appearance is due to radial canals,
even though such an explanation was suggested by his finding the markings on the inner as well as the outer surface of the membrane in the
case of Cyprinus carpio. He evidently reposes great confidence in
Müller's explanation of the striation as an optical illusion, which in his
opinion accounts for the appearances figured by Aubert. He is also
uncertain in regard to the existence of a vitelline membrane, so that a
positive conclusion as to the nature of the zona radiata was not reached.
The smallest eggs possess a transparent homogeneous membrane without
punctations, and too thin to be measured, which may be regarded as a
vitelline membrane. With an increase in the size of the egg this mem-

brane becomes thicker, and in Acerina punctate markings become visible on its outer surface when it is only about 2.8 μ thick. There is never any appearance which allows the supposition that the thickening is produced by secretions from the epithelium of the follicle. Since the punctate appearance becomes visible only after the thickening of the original membrane, it is to be concluded, says Reichert, that the punctate membrane of the ripe egg is not the original vitelline membrane, but a secondary egg-membrane, which, however, has been formed by the deposition of thickened layers ("Verdickungschichten") of the egg outside the vitelline membrane. But what has meantime become of the vitelline membrane is not stated.

Reichert distinguished on the majority of the eggs which he studied a second membrane outside the punctate one. In Esox it was clear, homogeneous, and viscid. In cyprinoids it had the velvety appearance already described by Müller, from whom the author differs in regarding this membrane as not a part of the punctate one. His reasons for this conclusion are, that it is as sharply marked from the punctate membrane as is the capsular membrane in the perch and may even be detached after treatment with chromic or nitric acids, that the rod-layer is not so firm as the punctate, and finally that the rods are much fewer than the punctations of the same egg. The rods are set in a clear, homogeneous layer, only their rounded ends protruding.

Reichert was evidently influenced by his belief in the probability that the capsular membrane in the perch owed its origin to the membrana granulosa, and consequently he left unsettled the question of the origin of this villous layer, although its intimate adhesion to the punctate membrane indicated a common origin with the latter.

KÖLLIKER ('57, p. 197) found in 1856 that the porous membrane in the case of sturgeons' eggs presented conditions which favored the view that it was formed by the cells of the follicle, but he discreetly abstained from forming an ultimate judgment before eggs in all stages of development had been studied. Such studies he found the opportunity of beginning during a sojourn at Nice a few months later, and he continued them on fresh-water fishes during the beginning of the following year; the results were published in 1858.

KÖLLIKER'S ('58, pp. 80–93) observations embraced a large number of fishes, and — what is of more importance — he also studied eggs in several stages of development. With Reichert, he recognized two capsular egg membranes, which he called "Dotterhaut" (zona radiata) and "Gallerthülle" (the latter in Perca), but he dissented from Reichert's explanation of the striate appearance of the former, and demonstrated by means of *thin sections* that it was due to the presence of poré-canals. Kölliker also claimed that there was an outer thin, resistant layer of the porous vitelline membrane in the case of all fishes, — such as Leuckart alone had recognized in Perca, — and that this layer might retain the striate appearance even when the rest of the membrane had been made pale and apparently homogeneous by the use of reagents. It is in this outer layer that the viscid and fatty-looking "Zöttchen" are rooted with their slightly enlarged bases. Kölliker therefore regards the "Zöttchen" layer as belonging to the "Dotterhaut," rather than as a separate layer, and compares its elements to the peculiar appendages discovered by Haeckel in the Scomberesocidæ. To these he adds a description of peculiar mushroom-shaped appendages of the vitelline membrane in Gasterosteus and Cottus gobio. There is, however, one difference between the "Zöttchen" and the mushroom-shaped bodies; in caustic potash the former are greatly swollen and become pale, whereas the latter are made to shrink somewhat and to become darker.

In reference to the origin of this membrane and its appendages, Kölliker gives the first positive information which we have, for Reichert's conclusions were at best only theoretical and tentative. Both in the case of Gasterosteus and Cobitus barbatula he showed that the villous structures made their appearance before the zona radiata. In the case of the first-mentioned species, the mushroom-shaped bodies were distinguishable as minute wart-like points resting on the outer surface of a membrane so thin that it presented only a single contour. As these wart-like structures continue to occupy the outer surface of this vitelline membrane while the latter increases in thickness, it is not to be doubted, he says, that the increase is due to deposits upon the inner surface of the membrane. The warts continue to increase in size, while the membrane becomes still thicker and shows radial markings.

The first appearance of the villi in Cobitis is to be seen in eggs 0.08″ [0.08‴ ? $= 175\,\mu$] in diameter, where they appear as deposits or outgrowths on the external surface of a thin membrane (Reichert's primitive vitelline membrane); at first they are low and narrow, but they gradually increase in length, and also, though more slowly, in breadth. It is only when the villi have attained their full length that the porous layer begins to be formed by deposits on the *inner* surface of the thin membrane, but this proceeds with such energy that the porous layer soon exceeds in thickness the villous. At the same time the villi increase in breadth, though not in length, and the thin membrane persists as the outer layer of the porous membrane, which in the ripe egg bears the villi.   Upon the first formation of the villi the appearance of a sur- face view of the membrane so closely resembles that of a porous vitelline membrane (zona radiata) with fine close-set pores, that one must follow the whole course of development, and convince himself of the late ap- pearance of the porous layer before he can be certain that the fine points are due to the villi.

Kölliker believed that the formation of this peculiar " Dotterhaut" (zona radiata) of fishes could be easily understood as one of the so called secondary cell secretions, if there were on the inner side of it another membrane, which latter would then be regarded as the origi- nal cell membrane of the egg.   Although he found some evidence of the existence of such a thin structureless membrane in Cobitis, he was unwilling to give much importance to that fact, but inclined to a belief in its existence, rather upon the theoretical ground that it would offer a satisfactory explanation of the radial pores as being the equiva- lents of pores in cuticular structures.

This assumption (p. 104) that the fine pore-canals are to be explained as resulting from the presence of fine openings in the cell membrane may be satisfactory enough for those cases in which a definite cell membrane is demonstrable previous to the appearance of the cuticular secretion; but it seems to me superfluous to assume the universal exist- ence of a cell membrane in order to explain the conditions.   Where no cell membrane exists, the same phenomenon may take place, and is no more difficult of explanation than in the case where there is a cell membrane with the supposed structure; for the latter must in its turn be explained as the result of localized activity on the part of the cell protoplasm in secreting its membrane.   So, in the end, it comes to one and the same thing, whether we assume the presence of a cell membrane or not: the explanation must rest on the ability of protoplasm to localize its activities; but further than that we are at present unable to advance.   Why or just how protoplasm is able to effect such a histo- logical division of labor is still unexplained.

The important paper by GEGENBAUR ('61) on the structure and devel- opment of the vertebrate ovum adds nothing to our knowledge of the zona radiata in bony fishes, but is valuable for the way in which it illuminates the subject of the vitelline membrane.

The only articles of much importance during this decade were one by Buchholz and two by Ransom.

RATHKE'S ('61) posthumous work on the development of vertebrates evidently treats the subject from the standpoint of thirty years before, when little was known about the matter ; and

LEREBOULLET ('61, pp. 120, 121, 123) does not greatly add to the knowledge of the zona radiata when he says that the chorion of the trout egg is thin and very soft at the moment the egg is laid, and does not present the resistance and elasticity which it acquires after it has remained for some time in the water. For the absorption of water and the passage of gases necessary for the respiration of the egg and embryo, the chorion is pierced, he says, by an infinite number of excessively narrow, short parallel tubes, which give a striate appearance to perpendicular sections of the shell.

OWEN ('66, pp. 593–595), although following the accounts by Ransom and Thomson, fails to recognize one of the important points established by the latter author, for he does not distinguish between the " ectosac " in the perch, and that of salmon and other fishes. Moreover, this " ectosac " (evidently the zona radiata) " is composed of close-set series of hollow columns." (!) As Ransom ('67, p. 3) has since pointed out, Owen also erroneously states, possibly under the influence of Rathke's exposition, that the villi are formed after the ova escape into the cavity of the ovary.

BUCHHOLZ ('63, pp. 71–81, and 63a, pp. 367–372) was the first to describe a very peculiar appearance of the egg membranes in Osmerus eperlanus. In addition to the porous membrane (zona radiata), which continues to invest the egg after it is laid, there is a second one *external* to the first, and like it traversed by similar pore-canals. Buchholz states that these canals are much more readily recognized to be pore-canals in the case of this fish than in that of other fishes (p. 73). When the egg has lain for some time in water, the outer envelope, or at least a portion of it, is found to be attached to the inner membrane around the circumference of the micropylar canal, whence it depends as a loose funnel-shaped frill with its originally inner surface now directed outward. The pore-canals which traverse these two membranes, instead of being cylindrical, are funnel-shaped, the wider end being directed outward. By treating the fresh ovarian egg with acetic acid the *outer* membrane, which at first lies closely in contact with the inner, is made to swell up with irregular foldings until it becomes entirely separated from the inner ; but a striate appearance, which is visible for a moment, becomes quickly obliterated by the action of the acid. If, however, the eggs are first treated for twenty-four hours with very dilute chromic acid, and then with acetic acid, the radiate structure remains easily distinguishable in both membranes.

With regard to the formation of the two membranes, Buchholz argues that the outer is the older, since one often finds, in the earlier stages of their formation, that the inner is thinner than the outer, whereas subsequently they are both of equal thickness. The increase in the thickness of the inner membrane was observed between the middle of February and the middle of April, — the spawning time, — and meanwhile the outer membrane was found to be thinner over about one third of the surface. It is to be assumed, according to the author, that this attenuated portion finally disappears altogether, since the persistent portion which remains attached in the region of the micropyle is too small to have completely enveloped the egg. Even nearly up to the time of maturity there is no fusion between the two membranes, which must, therefore, take place rather late.

The homology of the outer membrane in Osmerus is not at once evident from this account by Buchholz. If one were to accept unquestioned his description, it would be most natural to regard it simply as a detached portion of the zona radiata, for he maintains that the two are identical in structure. There are, however, two other possibilities; it may be homologous either with the villous layer of Lepidosteus, or with the capsular envelope of the perch. I regret that I have not yet found the opportunity to acquire from personal study additional evidence in favor of one or the other of these explanations; but there are two or three things connected with the account given which incline me to believe that the outer membrane is the equivalent of the villous layer. The very fact of its becoming detached from the deeper layer and thrown into folds after the egg has lain in the water suggests a similar though less striking feature of the villous layer in Lepidosteus; and although there is no evidence that any such eversion of the membrane takes place in the latter case, or that it even becomes regularly attenuated on one side, as in the case of Osmerus, still I can imagine that a similar condition might be artificially produced in Lepidosteus, so far at least as regards the peeling off and eversion of the outer covering, and it is possible that a slightly different physical condition of the villi would cause them to adhere to each other so persistently as to allow the attenuation of the whole membrane on one side of the egg without separating the individual elements. Since Buchholz asserts that the pore-canals are more readily distinguished as such in this fish than in other instances, I infer that the markings which he observed must have been coarser than is usual. If his attention was mainly directed to the outer layer, and if, as I imagine, this is a villous layer, the reason of his statement would be obvious. Besides, he mentions that the canals are not cylindrical, but funnel-shaped, the wider end outermost; this, too, though suggestive of the capsular membrane, would be entirely compatible with the idea that he had under view club-shaped villi. And finally, the argument to show that the inner layer is produced after the formation of the outer is exactly applicable in the case of the villous layer, as the observations of Kölliker and my own conclusively show.

RANSOM's ('67) account of the "yelk-sac" (zona radiata) in Gasterosteus is principally interesting to me from his asserting that the pore-canals as well as the villi increase in number during the growth of the egg, and from his consequent conclusions as to the method in which membranes are formed. The "yelk-sac," he says, is formed in very young ova ($\frac{1}{200}''$, or 125 μ, in diameter), in which it is easily recognized by the button-shaped villi attached to the outer surface surrounding the micropyle. The finely dotted structure is first discoverable in eggs $\frac{1}{140}''$ (180 μ) in diameter, and it is the same in character in these as in the ripest eggs. The membrane is composed of very fine concentrically arranged laminæ, each of which is marked by dots of equal size, so arranged as to mark (in surface view) the angles of equal-sized lozenge-shaped spaces, and corresponding in position in the successive laminæ so as to form vertically placed lines or striæ. In eggs .01 inch in diameter there were about 24,000 dots to the [linear] inch, and when the egg had attained .06 inch there were 11,000 to the inch; the distance between dots being scarcely more than *doubled*, while the diameter of the egg had been multiplied about *six* times. From this the author argues that there must have been an increase in the number of the dots during the growth of the sac, and therefore that the membrane does not increase by apposition of layers either from the inside or outside, either by the hardening of an exudation or by the conversion of the substance of the yolk into that of the yolk-sac. "It grows in some way by interstitial molecular deposit." A similar increase in the number of the button-shaped villi was also observed to occur during the development of the ova.

I do not recall that any one has corroborated or disproved these observations, or the deductions made from them; but I have shown that in the case of Lepidosteus there does not seem to be sufficient evidence to prove that there is any increase in the *number* of the *villi*. I believe that a careful investigation of the question in the case of the pore-canals would well repay one who should undertake it, for it could not be without influence upon theories on the method of the growth of membranes.

In his more extended paper RANSOM ('68) gives an account of the egg membranes of a number of fishes, but more particularly of Gasterosteus, Esox, and Perca. To what he had already stated about the "yelk-sac" of Gasterosteus he adds (pp. 440, 444, 448) that it is difficult, if not impossible, to determine the precise period at which it is formed. In eggs $\frac{1}{800}''$ (31.5 μ) in diameter it is not found, but is probably indicated by the smooth, hard outline which the yolk shows on its surface. "It is separable in eggs $\frac{1}{200}''$ in diameter, and may be seen in the fluids on the slide as a homogeneous-looking collapsed sac." With a power of 500 diameters the dots appear round, and with one of 3,000 they are but obscurely hexagonal; they are the same distance apart on the inner surface as on the outer. Besides these minute regular dots, there are larger and darker ones of a stellate form, which the author suggests may in some way be connected with the interstitial growth of the membrane. They occur irregularly at intervals of about $\frac{1}{3000}''$, and act like bodies of low refractive power; "at the cut edge they may be seen to pass radially about two thirds into the substance of the yelk-sac, gradually coming to a point and ceasing." I am not aware that any other observer has confirmed this appearance, which I imagine may be due to the presence of protoplasmic prolongations of the yolk into some of the pore-canals, just as in Lepidosteus the substance of the villi is traceable in many cases for some distance into the zona, although from the opposite direction.

"There are no facts known to me," says Ransom, "to point out whether the pabulum for the growth of this membrane is derived directly from the currents passing inwards, or from the material elaborated in the egg and passing out of it, or from both sources indifferently."

WALDEYER ('70, pp. 80, 81, 83), however, did not experience any such uncertainty concerning the source of the zona radiata, for while he continued to call it the "Dotterhaut" and could find no vitelline membrane inside it, he was very explicit in stating that it was a *cuticular formation produced by the follicular epithelium,* and that the pore-canals were occupied by delicate protoplasmic filaments which were in direct connection with the epithelial cells of the follicle on the one side, and with the finely granular yolk substance on the other. In his general conclusions concerning the eggs of vertebrates he says: "The complete homology of the zona pellucida [mammals] with the vitelline membrane of other vertebrates can . . . no longer be denied. The vitelline membrane is certainly a structure which does not belong to the primordial egg, but *is deposited upon it from without.*"

Like Waldeyer, EIMER ('72ᵃ, pp. 417–428, Taf. XVIII. Figs. 9–13) regards the zona as a *cuticular* product, but, unlike him, he maintains that it is produced (precisely as in reptiles) by the egg, not by the granulosa. But the homology between the egg membranes in these two groups goes, in his opinion, still further. The delicate membrane described by Aubert and others as external to the zona in Esox and other fishes is to be regarded as a chorion, — which in reptiles is produced by the follicular epithelium. The trumpet-shaped structures of the outer membrane (Eikapsel) in the perch are formed from granulosa cells. They are the homologues of the trumpet-like structures which the author described for Coluber and placed in the category of "beaker cells." The villi which repose on the outer surface of the zona are, as held by Reichert, to be regarded as belonging to the second membrane, and not in accordance with Kölliker's views as the outer layer of the zona itself. In Eimer's opinion they are simply yolk substance which has emerged from the egg through the pores of the zona. Eimer even maintains that he has observed the protrusion and subsequent disappearance of such villi.

I have already (p. 54) given my reasons for dissenting from the use which HIS ('73, pp. 1–3) makes of the word "Eikapsel" to signify the zona radiata. According to His, the "capsule" in Salmo salar is from 33 to 35 μ in thickness, the pore-canals are straight, only 1.5 to 2 μ apart, and not funnel-shaped at either end. In Esox it is 16 to 17 μ thick, and has fine parallel (concentric) as well as radial markings (p. 13).

His ('73, pp. 17, 35, 36, *et passim*) also discovered that in numerous cases the young eggs possessed a peripheral layer of clear substance which exhibited fine radial markings after treatment with certain acids. This did not appear to be constantly present, nor necessarily of uniform thickness. He called it the *zonoid layer,* and thought there was probably a physiological connection between it and the porous egg capsule, but what the nature of that connection was, remained to be ascertained. Nothing in either figures or text allows one to draw conclusions as to the relative fineness of the striations in the zonoid layer and the "capsule."

In 1878 appeared three papers which dealt with the structure and development of the egg membranes of bony fishes.

A pupil of Waldeyer, KOLESSNIKOW ('78, pp. 402, 403, 407–409), states that in both Perca and Gobio the "Dotterhaut" consists of two membranes, which are not easily separable either in the fresh state or after treatment with osmic acid; but on thin sections of eggs hardened in the latter reagent or potassic bichromate the two are sharply marked. This condition is held to indicate a gradual growth of the striate Dotter-haut, " *und zwar* ihrer äusseren Stäbchenschicht," which is the youngest formation, the *inner* layer being the older. Both layers consist of rods radially placed side by side, those of the inner layer being much finer, longer, and nearer together than the outer. It is evident, I believe, that the inner layer referred to must be the zona radiata, and the outer the "Eikapsel" of Müller. Fine granules, which are colored black by the osmic acid and are regarded as yolk granules, are found in the fine processes between the rods, as well as in the follicular epithelium itself. When the ovary hardened in alcohol is stained in hæmatoxylin, the inner surface of the follicular epithelium resembles ciliated epithelium, since the rods rest like cilia on the epithelium. On sections of a very young follicle ($4.65\,\mu$) the membrane, composed of rods, can be seen on the inner surface of the follicular epithelium. It is still thin and not easily detached, and the rods are fine and close together. When the follicle has reached a diameter of $465\,\mu$, the membrane is much thicker, and composed of two layers. It is then to be seen that in some places the rods of the outer layer are continued into the inner layer, being consequently longer. In this case the rods gradually diminish in thick-ness as one passes from the follicular epithelium toward the yolk sub-stance, but they are always sharply limited from the latter. Both layers, the author concludes, are cuticular formations of the follicular epithelium, and in no case is the inner layer to be regarded as a special membrane produced by the egg itself.

One of the best contributions recently made to this subject is that of BROCK ('78, pp. 547–559), who gives, besides a condensed summary of previous observations, his own valuable results. Aside from the capsular membrane of the perch, which he calls "Gallertkapsel," and the related structures of other fishes, the author is able to recognize only one egg membrane, which, to avoid prejudging its genetic relation, he calls, from its most evident morphological peculiarity, the zona radiata, reserving the term *membrana vitellina* for egg envelopes, which are the equivalent of other *cell membranes* and therefore differentiations of the [egg] proto-plasm. The outer lamella of the zona described by Reichert and by Kölliker he finds with varying distinctness in different cases. Being in some instances unable to discover it at all, he doubts its constancy. It may often be demonstrated by the use of acetic acid when not visible

in the fresh condition. In the perch it has a striate appearance, and
is much more coarsely marked than the true zona,[1] but in Serranus
hepatus it is homogeneous.

The elongated club-shaped villi resting upon the outer surface of the
zona were found in many cyprinoids, and also in Osmerus. They are
not, as asserted by Eimer, expressed droplets of yolk substance; they
are secondary appendages of the zona, which "have nothing whatever
to do with either follicular epithelium or yolk." Brock finds the zonoid
layer of His well developed in Alburnus lucidus, Salmo fario, and Perca
fluviatilis. He is inclined to regard it as a general structural condition,
often overlooked because distinctly shown only at certain periods in the
development of the egg. It is often divisible into two layers, of which
the inner remains homogeneous. When the zona radiata is removed
from the yolk, the "zonoid" remains attached to the latter, to which it
must therefore belong. Its striations are intermediate in fineness between those of the villous layer and those of the zona. Notwithstanding
certain objections, Brock regards the follicular epithelium as the principal, if not the exclusive, source of nutrition and growth for the yolk,
which are accomplished by means of cell processes sent by the granulosa
through the zona into the yolk. The evidence, aside from that which
Waldeyer produced for other groups, is to be found, says Brock, in the
fact that, when the granulosa is separated from the zona by a secondary
membrane (Perca, Serranus), it sends processes through the latter which
are traceable up to the zona.

According to the opinion which I have formed, however, in regard to
these so called processes, they are not outgrowths from the granulosa
cells; on the contrary, the cells, retaining their original contact with
the zona, are by the accumulation of the capsular secretion greatly
attenuated.

In regard to the order in which the different membranes make their
appearance, Brock comes to views diametrically opposite to those expressed by Kölliker. Leaving out of consideration the capsular membrane, which, as Brock rightly states, is late in being formed, and the
outer lamella of the zona, concerning the origin of which he could dis-

---

[1] It is not possible to say with certainty from his figure (Fig. 7, f) whether
Brock regarded the striations of the outer lamella as less *numerous* than those of
the rest of the zona. They are represented as *broader* than the latter, and as
thickest at their peripheral ends, which agrees with Eigenmann's observations;
but they certainly are not represented as *continuous* with the striations of the inner
portion, and in this I believe that Brock is in error.

cover nothing, he asserts that he saw in all cases with the greatest distinctness that the zona radiata first appeared, and that when it had attained a certain thickness then for the first time the villi and the zonoid layer made their appearance, almost simultaneously. Unfortunately, Brock has given no details concerning the proof of this assertion, either in figures or description. I am tolerably certain, notwithstanding the very positive way in which he maintains his conclusion, that Kölliker was right, and that he is wrong, for I cannot believe that in so fundamental a matter there is such a difference between fishes as would be implied by admitting both views to be correct.

That the differences of opinion which the egg membranes have given rise to are not exclusively due to the study of different fishes, is clearly seen from the results reached by Kupffer, and soon after by Hoffmann, in the study of the herring's egg. KUPFFER ('78^a, pp. 177, 178) found the yolk to be closely invested by an egg capsule 6 to 8 μ in thickness, and the latter to be covered by a layer of viscid semifluid substance, which was found to be of nearly uniform thickness if the eggs were dropped into alcohol without contact with water. In water it soon becomes solid. The capsule consists of two firmly united layers, the inner one being finely striate radially, and alone equivalent to the porous capsule (zona radiata) of other eggs. The striation is due to pore-canals. These do not traverse the outer layer, which has concentric striations. A boiling ten per cent solution of potash dissolves the porous layer quickly, but does not affect the outer layer of the capsule, which is believed to be the same as that described for Esox by Aubert, and for other fishes by Kölliker. This difference in the two layers Kupffer regards as favoring Eimer's view, that one is produced from the egg and the other from the follicular cells.

On the other hand, HOFFMANN ('81, pp. 15–33), who has given the subject of egg membranes in fishes the most extensive treatment of any recent writer, differs materially from Kupffer in his account of the herring, although he offers a partial explanation of their differences in saying that Kupffer examined only fully mature eggs, and such as had been in contact with sea-water. Hoffmann makes the total thickness of the membranes in not quite mature eggs to be 32.5 μ. The outer layer, 10 μ thick, is not separable from the inner, although a dark line marks the boundary. Both layers are traversed by numerous pore-canals, which, to judge from his figures, appear to be finer (not necessarily nearer together) in the inner than in the outer layer; but whether they are continuous it is impossible to say on account of the sharp line separating the layers. The inner layer is composed of two parts, the outer part being often striated concentrically. The membrane of the fully ripe egg is different. Before contact with water the pore-canals of the outer layer are not so easily distinguishable as in immature eggs, and in this layer a great number of small lustrous spherules are now visible. In eggs which have been in contact with sea-water the outer layer is raised up from the inner, forming a viscid sheet 10–12 μ thick, which causes the adhesion of the eggs. The outer portion of the inner layer now exhibits a tendency to split into concentric layers, which obscures the radial pores, although they are still visible in the deeper part of this portion.

It appears to me probable that the two portions of the inner layer represent the whole of Kupffer's capsule, and that Hoffmann's outer layer is the equivalent of Kupffer's viscid semifluid substance. In view of the striate appearance of the outer layer described by Hoffmann, and the greater coarseness of the striations as compared with the inner layer, and also in view of its viscid nature, I am strongly inclined to believe the outer layer will ultimately be shown to be equivalent to the villous layer in other bony fishes.

Two layers resembling those of the herring were also found in Crenilabrus. Greater interest attaches, however, to the account of Leuciscus rutilus, in which, as the author says, one again finds the two layers of the zona radiata. But in the next sentence he shows that he does not distinguish sharply between zona and villous layer: "The outer layer here forms the well known *Zöttchenschicht.*" The villi are club-shaped, close set, and clothe the zona as a uniform layer, with the exception of the place where the micropyle is situated. In the extent of their distribution, therefore, the villi in Leuciscus differ from those of Lepidosteus, — which, though shorter, are not wanting in the periphery of the micropyle, — and also present a condition which is the complement of that exhibited by Heliasis, Gobius, etc., in which villous structures are restricted to the region of the micropyle.

Hoffmann arrives at these general conclusions : In adhesive eggs the zona radiata consists of two layers, of which the outer effects the adhesion. The latter may form a part of the whole zona, or may exist in the form of villi over the whole surface, or of long filaments limited to the micropylar region. "But whatever form this outer layer may assume, it always has a like origin with the rest of the zona ; it is nothing else than a part of the zona itself, which sooner or later undergoes peculiar metamorphoses." Hoffmann recognizes the difficulty of determining how the zona radiata arises, whether as a product of the yolk or of the granulosa cells. An examination of maturing eggs shows him that its peripheral layers are always the most distinct. It is as though new layers were being deposited from within, and this leads to the conclusion that the zona radiata represents a true vitelline membrane.

Although Hoffmann seems to me to come very near the true solution of the problem, this presentation of the matter appears altogether unsatisfactory, because he insists that what he calls the zona is practically a unit in structure, and fails to recognize a fundamental difference between the outer villous layer and the "*true* zona radiata," as he terms the inner portion of his zona. From this last conclusion one would be led to infer that the zona (including both layers) was the result of a continuous process, and that there could not be any radical structural difference between villous and porous layer. But to my mind a common origin from the yolk by no means implies identity of structure, nor the continuous operation of the same formative process.

RYDER ('81^c, '82^c, '83) has contributed a good deal to our knowledge of the occurrence and functions of those modified forms of villi first described by Haeckel, but I think he cannot have given their structure very close attention or he would not have said ('83, p. 195) that "they are apparently composed of the same tough material as that which enters into the formation of the egg-membrane itself." He ('81^e, pp. 137, 138) regards it as probable that the egg membrane (zona radiata) is secreted from the cellular walls of the follicle.

Since writing the above summary of and comments on the observations of authors there have appeared a number of papers, some of them of considerable importance, which I was unable to utilize in forming my opinions of the nature of the membranes in Lepidosteus and other fishes.

STOCKMAN'S ('83) account of the egg capsule in Salmo describes the appearances of the pore-canals under a Reichert $\frac{1}{20}$ homogeneous immersion lens. In sections the limits of the pore-canals appear toothed, owing to the presence of minute folds which have for the most part a direction tangential to the capsule. These extend to the two ends of the canal, and consequently its mouth appears angular rather than circular. The substance of the capsule is beset with minute spaces which communicate with the pore-canals between the tooth-like projections, and are believed to have a function in the transmission of nutritive material to the egg.

RYDER ('84, pp. 3, 4, 11, Plate I. Fig. 5) states that the cod's egg is

"covered by a vitelline membrane which is not porous or enveloped in
adhesive material. It is thin, very transparent, and laminated, as has
been stated by Sars, and at one point is perforated by a single minute
opening, the micropyle." Ryder was unable to discover the lamination
until after the action of osmic acid, and was uncertain whether it was
a natural condition or the result of the action of the acid. "The cod's
egg," he says, "is without the zona radiata found enclosing the egg
proper of the shad, whitefish, and sculpin, and, inasmuch as it is unquestionably true that a micropyle perforates the zona in a number of these
cases, it does not appear that sufficient grounds exist for the declaration
that a micropyle perforates the zona radiata alone, in the face of the fact
that the vitelline membrane only is perforated in this one instance." I
have no doubt that the membrane in question possesses pore-canals, and
that it is therefore a true zona radiata. I can confirm Eigenmann's
observations in this particular, and believe that Ryder himself would
have come to the same conclusion had he observed the membrane under
the same favorable conditions.

In an extensive paper on the eggs of bony fishes OWSJANNIKOW ('85)
describes the egg membranes of a number of the more common fresh-
water forms. The most important part of his paper deals with the capsular membrane in Perca and the equivalent structure in other fishes;
but the consideration of that part will best be deferred until I come to
review the other papers which deal with that subject. I may here add,
however, that he does not recognize the existence of a villous layer
outside the zona, but regards the structure which immediately envelops
the zona in Osmerus and other fishes as the equivalent of the capsular
membrane of the perch. His description of a thin transparent membrane (membrana vitellina) inside the zona radiata in Salmo trutta is
materially affected by the subsequent statement that it is not found in
other cases (Lota, e. g.), and by the admission that it may have been
an artificial product.

The pore-canals of the zona are often more deeply stained than the
substance of the matrix, and by treatment with certain reagents minute
points can be seen in the canals when highly magnified. In Lota vulgaris the zona is very thin, and the pore-canals in patches do not penetrate to its inner surface. It is generally stratified, the strata being
laid down successively and all being perforated. The zona might, in his
opinion, better be called *perforata* than radiata. Concerning its development in Gasterosteus the author says that the first trace of it is
seen to be a very thin membrane without any pores. These appear when

the zona has become thicker, but they are much finer than in the mature egg. After describing the condition in Acerina vulgaris (p. 18), the following statement is made as embodying the author's idea of what takes place in the formation of the zona. The granulosa cells secrete a substance (Zwischensubstanz), which surrounds the egg, one layer upon another. The pores in this substance arise by the growth into it of the points of the granulosa cells, or plasmatic processes from them. The way in which the author describes the mushroom-shaped villi which surround the micropyle in Gasterosteus shakes one's confidence in this part of his work. It is very probable, he says, that they are derivatives of the granulosa cells. After admitting that they have been described in a masterly way by Kölliker, he gives an interpretation of them that must appear to that author very remarkable. They are nothing less than individual cells (!), each with a nucleus that is stained red in carmine while the cell protoplasm resists for a time the action of the dye. From the base of the cell emerges a thread which can be traced into the zona. Very young eggs possess these appendages, as Kölliker maintained, but they are much smaller, and consist essentially of a nucleus which is attached by a thread and surrounded by a thin film of protoplasm.

If everything that is stainable is to be called either nucleus or cell protoplasm, then the current notions of what constitutes a cell will have to be abandoned; that will give room, it is true, for the admission into this category not only of the villi and the *yolk cells* of His, but also, I fear, of many other structures as well.

The author's views concerning the villous layer in other cases appear from his account of Coregonus. The granulosa cells send processes into the zona radiata. When they are removed from the immature egg, the zona appears to be covered with small more or less pointed " Zöttchen," which are therefore to be regarded as processes of the granulosa. Outside the zona radiata, Owsjannikow finds in many cases a thin viscid layer, which he suggests is derived from the oviduct in the case of S. trutta, but looks in Lota as though it resulted from the fusion of endothelial [granulosa?] cells.

SOLGER ('85, pp. 330–332) is evidently inclined to bring the villous layer into connection with the presence of intracapsular corpuscles found in the perivitelline fluid. Without committing himself unreservedly to the views of Eimer, — that the villi are simply exuded drops of vitelline substance, — he confirms the statement that they are not of uniform length or shape in the case of Leuciscus rutilus, and says that Eimer must certainly have the credit of having especially emphasized the fact that after a certain epoch there exists a contrivance which prevents the further entrance of water into the intracapsular space.

I believe a comparison of the conditions in Lepidosteus with the early account by Kölliker will convince the author that the villi have no connection whatever with the interesting conditions of the perivitelline space which he has discussed.

RYDER ('86, pp. 18, 23, 30, 35, 36, and '87) has recently noted the existence of a zona radiata in several species of fishes, but without having given the structure special attention. In some cases the eggs when laid are covered with an adhesive material, the source of which is not alluded to. In the case of Ictalurus albidus (white catfish), there is an interesting condition of the egg membranes which he ('86, p. 47, and '87, p. 535) describes as follows : " The egg membrane is double, that is, there is a thin inner membrane representing the zona radiata, external to the latter and supported on columnar processes of itself which rest upon the inner membrane ; there is a second one composed entirely of a highly elastic adhesive substance. The columns supporting the outer elastic layer rest on the zona and cause the outer layer to be separated very distinctly from the inner one. . . . This peculiar double egg membrane, with a well defined space between its inner and outer layers, is highly characteristic, and bears no resemblance to the thick, simple zona investing the egg of Ælurichthys, nor has anything resembling it ever been described, as far as I am aware, in the ova of any other Teleostean." Eigenmann ('90) has attempted to show by comparisons that the whole of this double membrane is probably a true zona radiata, and that the columns are protoplasmic substance which occupied the pore-canals before the separation of the two portions of the zona ; but it seems to me more probable, from the " highly elastic adhesive " condition of the substance of the outer membrane, that it corresponds to the villous layer in other fishes. There can be no certainty, however, as to the real homology of the outer membrane until it has been subjected to a more careful study with especial reference to the time and manner of its production.

CUNNINGHAM ('86) has recently rediscovered what Buchholz found out upwards of twenty years ago about the peculiar egg membranes of Osmerus. It is unfortunate that Cunningham overlooked the valuable work of Buchholz, and the more surprising since he refers to Owsjannikow's paper, — which I should suppose he must have consulted, — in which Buchholz is cited. Cunningham gives figures and an account which practically confirms that of Buchholz as to the eversion of the outer membrane. According to Cunningham it serves as the so called suspensory filament by which the deposited eggs are attached. He calls the outer membrane the external zona, and agrees with Owsjannikow that it is traversed by pores which are larger and (according to his figure, about four times) farther apart than those of the internal zona. He makes no mention of the outer membrane differing in any other way from the inner, except that it is somewhat thinner.

SCHARFF ('87, '87ᵃ) regards Beddard's zona radiata[1] in Lepidosiren as a "zonoid" layer, which he also finds present in Trigla gurnardus, where it has a thickness of 25 μ, while the true zona is only about 8 μ thick. "Both layers are striped, i. e. provided with minute radial pores," which are apparently continuous through both layers. The zona is firm, granular, and stains deeply; the zonoid is semifluid, usually devoid of granules, and stains only slightly. In ripe ova the latter disappears entirely. The zona is formed before the zonoid layer. In Blennius pholis there is no zonoid layer. Scharff "has no doubt that the egg membranes originate from the yolk," although, as far as I can understand, he advances no new arguments to prove the fact.

HENNEGUY ('88, p. 419) says that, notwithstanding the use of high powers, he has been unable to find in the trout the three layers of Owsjannikow or the denticulations described by Stockman.

A recent paper by CUNNINGHAM ('86ᵃ) deals with the interesting question of the structure and origin of the egg membrane in *Myxine glutinosa*. Even when the eggs have become 11 mm. long there is no trace of any membrane between the yolk and the single layer of granulosa cells, which latter project irregularly into the surface of the former. Sections of an egg 16 mm. in length showed that the granulosa had become several cells deep, though not arranged in regular layers, and that there was beneath this, and in direct contact with the yolk, a thin homogeneous membrane.

It certainly is an interesting fact, to which Cunningham calls attention, that the epithelium is thicker at the poles than at the equator of the elongated ovum, and that the thickness of the membrane varies [directly] with that of the epithelium. W. Müller also observed the same fact.

In eggs 20 mm. long the follicular cells are elongated, and form practically a single-cell layer, with the nuclei at the ends nearest the yolk.

[1] Compare Beddard ('86ᵃ), and the summary of his paper given on pp. 66, 67.

The membrane at the poles presents thickenings which are shown by the later stages to be the beginnings of the peculiar polar filaments of the egg membrane. These thickenings project into corresponding depressions of the granulosa, the cells of which are so thinned at these places as to be barely discernible. When the egg has increased slightly in size (21 mm. long, 7 mm. in breadth), these thickenings affect the external appearance of the follicle. They have the form of finger-shaped processes covered with a single layer of granulosa cells, and project into the connective-tissue wall of the follicle. The statement that in the thinnest places (i. e. over the tips of the projections) the granulosa is only .02 mm. thick, while its thickness at the exact pole of the egg is .5 mm., is not borne out by the figure, and I therefore suppose that it is a typographical error for .2 mm. In either event, however, the granulosa does not regain its normal thickness over the ends of the rapidly growing filamentous projections of the membrane, and this fact may have significance in determining the source of the membrane. The author says that the membrane, if prepared in chromic acid, appears homogeneous even when highly magnified; but, if hardened in Perenyi's fluid, striæ perpendicular to the surface are to be seen with a high power. The striæ are not represented as rigidly straight and parallel; they may even branch, and are often moniliform. They have been traced to the outer surface of the membrane, where the author believes they are continuous with fibrils of the epithelium. He is also convinced that the membrane is homologous with the zona radiata of teleosts. I cannot agree with him, however, in the statement, "When there are two zonæ radiatæ, as in Perca fluviatilis and Osmerus eperlanus, according to Owsjannikow, these seem to be simply parts of one membrane differentiated in physical properties, but essentially similar in structure."

Cunningham believes it more probable that the zona is produced by the follicular epithelium than by the outermost layer of the yolk, and in the following manner: "The deeper part of the elongated epithelium cells is gradually changed into the zona radiata, the substance of the cells being partly transformed into the substance of the membrane, while threads of protoplasm, at more or less regular intervals, remain unchanged, and thus give rise to the pores of the membrane."

Against this view I would urge that a metamorphosis of the epithelial cells, especially if prolongations into the membrane occur at intervals, would be likely to result in the closest union between the membrane already formed and its generatrix, whereas it is exactly along this line that the artificial separation, which the author notes in all his eggs, takes place. Moreover, a diminished thickness in the epithelial cells corresponding to the most rapidly increasing part of the membrane (the filaments), *not at the close of the period of their formation, but at its beginning,* is the reverse of favorable to his "hypothesis."[1] And, finally, I would suggest that the zona is everywhere in contact with the substance of the ovum, and that the increased thickness of the membrane at the poles may be due to the accumulation there of a greater proportion of the active protoplasm than is found at the surface elsewhere. Perhaps it might be urged against this view that the explanation would be only partial, — that, while it might account for an increased thickness of the membrane at the formative pole, it would leave the condition at the opposite pole unaccounted for, and therefore could not fully satisfy the needs of the case.

I have recently sectioned the eggs of a rather poorly preserved specimen of *Myxine australis,* with a view to getting additional light on this question. Although the eggs had attained a considerable size, — 22 mm. long by 6 mm. in diameter, — still there was as yet no indication of the filamentous projections; in fact, I could not trace a membrane continuously around the egg. At the formative pole there was unquestionably a membrane about 3.5 μ thick; it was faintly marked like the zona radiata of teleosts, and it presented a deep micropylar infolding, with a cellular epithelial plug. Nevertheless this membrane gradually grew thinner in passing from the formative pole, until it could no longer be recognized. It had about the same extent as the protoplasmic cap. At this stage there was no more accumulation of protoplasm at the opposite pole than at the sides of the egg; *but there also was no more evidence of a zona radiata at this pole than at the sides of the egg.* That is all I can offer at present in reply to the possible objection which I have suggested. If it could be shown that the zona is developed at the nutritive pole of the egg without the presence of an accumulation of protoplasm there, and that the granulosa is more highly developed there than on the sides of the egg, I should admit that a strong case would be made against the view I defend.

[1] It is true the author has offered an explanation of this; viz. that the filaments are formed from the cells *at the sides* of the process, where the epithelium is very thick, and that they are pushed up by the growth at the base. But I should imagine it would be difficult to explain how secretions from *lateral* cells could do anything more than increase the diameter of the process.

2.   *Capsular Membrane.*

As I have already pointed out, the capsular membrane, since it was first described by Müller in the perch under the erroneous supposition that it was the same as the zona radiata of the salmon, has often been confounded with that membrane.   In looking over the literature on the egg membranes of fishes, after I had worked out the structure of the villous layer in Lepidosteus, I was forcibly impressed by the resemblance of that layer to the descriptions that had been given of the outer membrane in the perch, and at first thought they might be homologous structures.   It was particularly the account given by Ransom ('68, p. 455, Plate XVI. Figs. 30, 31) of the root-like prolongations of the tubules in the capsular membrane which suggested comparison.   It therefore seemed necessary to examine carefully all that had been written on the egg capsule of the perch.   The result has not confirmed my first supposition.

MÜLLER ('54) himself gave an excellent account of the structure of the capsule, and accurately formulated the most interesting question concerning its morphological significance.   He described the egg envelope (capsular membrane) as about 0.11 mm. thick; its outer surface as covered with six-sided facets, which average 19 μ in diameter.   Each facet contained in its centre an open funnel, which was continued into a vertical tubule as long as the thickness of the capsule, and from 2.2 μ to 4.7 μ in diameter.   In fineness these were comparable to dentinal tubules.   They terminated on the inner surface of the capsule in funnel-shaped enlargements, just as they did on the outer surface.   Upon eggs that had been boiled or hardened in chromic acid, it was possible to see that the tubules had a spiral course, but they also appeared narrower (1.1 μ) than in the fresh state.   The tubules were filled with a thickish (albuminous?) mass, which in the fresh egg was clear, without deposits, and under pressure projected from the funnel like a rounded stopper or cylinder, but appeared to be coagulated by boiling and treatment with chromic acid.   When one compressed the fresh egg to bursting, the oily substance of the yolk might be pressed into and through the tubules; thus was effected a delicate injection which might greatly distend them.   Between the tubules, however, there was nothing pressed out, which proved that on its deeper surface between the tubules the capsule was closed.   In the inter-tubular portions of the membrane, after the eggs are hardened, there were to be recognized, besides a gelatinous nearly invisible material, exceedingly delicate projections or

filaments placed alternately and running across between adjacent tubules. These were thickest next to the tubule, and rapidly tapered to
very fine threads. In his opinion, the method of the formation of the
tubules might be made out during the winter. " *The question is, whether
each of the tubes arises from a single cell, which becomes open, or whether
the tubes are originally inter-cellular, and whether their walls result from
the remnants of several cells in contact with each other.*"

A similar condition is maintained by Müller for Acerina vulgaris, but
the membrane in this case was much thinner and the tubules consequently shorter.

LEREBOULLET ('54, pp. 242, 246) also discovered independently, perhaps even before Müller,[1] that there were in the perch what he called
hollow closely interlaced piliform appendages (also called stiff curved
filaments), which traversed the whole thickness of the shell, and to which
he attributed the agglutination of the eggs into a network. He also saw
besides these the much finer pore-canals.

In regard to the chemical nature of the capsular membrane, it was
maintained by VON BAER ('35) and by LEUCKART ('55, p. 260) that it
was an albuminoid substance. KÖLLIKER ('58) called it gelatinous, but
HIS ('73, p. 15) proved that it at least closely resembled chondrin, and
consequently claimed the right to call it a cartilage capsule.

REICHERT ('56, p. 93) was not able to add much to Müller's account of
the structure of the capsular membrane. Concerning its origin he was
at first inclined to believe that it resulted from [a metamorphosis of?]
cells (" aus Zellen hervorgegangen "), and therefore to regard it as a product of the membrana granulosa. This conclusion was strengthened by
finding the granulosa composed of *cylindrical* cells in the case of Esox,
and that when this membrane appeared in the perch the granulosa cells
had disappeared ; but subsequently, finding that the follicular cells in
the perch were round, and not finding any transitional stages from the
epithelium to the membrane, he was compelled to leave the question
unsettled. Reichert was certainly looking in the right direction, and
evidently very near to a fair settlement of the question.

It remained for KÖLLIKER ('58, p. 90) to confirm this supposition of
Reichert. He found that in February the capsular membrane had a
thickness of 45 μ to 75 μ. The tubules, he says, are formed by the
outgrowth of the epithelial cells of the follicle, so that the jelly which
joins them can only be a substance secreted by these cells. These so called
tubules were after all not hollow structures ("noch keine deutlichen

---

[1] See Kölliker ('58), p. 81, foot-note.

Hohlgebilde "), but apparently solid pale processes of the epithelial cells, on which the anastomosing filaments found by Müller were visible. Kölliker did not doubt, however, that they were from the beginning hollow cell processes, but they still contained at the time of their formation cell contents, and only subsequently became clear. Their independent nature was shown by the fact that in chromic preparations they could be drawn out from the jelly without losing their union with the [rest of the] epithelial cells. As long as the eggs remained in the follicle the epithelial cells probably continued in union with the tubules; but at the liberation of the eggs the cell bodies probably fell off, with the exception of the walls, which were continuous with the tubules, and then constituted the hexagonal facets of Müller. Kölliker was able to produce a similar effect by artificially separating the cells from the capsule.

RANSOM ('68, p. 455, Plate XVI.), who does not seem to have been acquainted with the papers of either Reichert or Kölliker, compares the capsular membrane in consistence with fresh fibrine. "The striæ look like tubes, have a distinct double contour for each wall (Fig. 28), but are filled with a vacuolating material, and do not seem to convey anything into or out of the egg." The outer surface was thrown into folds which radiated from the ends of the "tubes," but the hexagonal markings seen by Müller could not be made out. The tubes, instead of being funnel-shaped, at their inner terminations divided into root-like processes, and were in some way intimately adherent to the outside of the dotted yolk-sac (zona). The clear matrix was elastic and concentrically laminated. "The appearance described by Müller, of oil granules passing through the tubes, may possibly have been due to vacuolation in them." Experiments with colored fluids to ascertain if there were any absorption of fluids along the "tubes" always gave negative results: the cleavage went on, the yolk-sac was dyed throughout, the clear matrix more than the tubes, the germinal mass not at all. Either, therefore, the tubes did not subserve imbibition at all, he contends, or in a much smaller degree than the clear matrix.

WALDEYER'S ('70, p. 81) conclusion about the origin of the zona from the granulosa appears to me to have resulted, in part at least, from the fact that he was unable to discover any essential difference between it and the capsular membrane of the perch. He says the latter does not differ from the former in the principle of its structure. He rightly adds, that "here [capsule] it is to be seen with the greatest distinctness that the filaments are connected with the subsequently somewhat degenerated remnants of the follicular cells." He further adds, that occasionally it appeared as though there were between these two membranes a thin flat expanse of granular protoplasm in which the filaments terminated.

HIS ('73, pp. 14, 15), who examined perch eggs in April, also confirms the opinion of Kölliker, and gives a figure to show the relation of the radial processes to follicle cells. The radial streaks consist, he says, of a turbid substance, which stains in osmic acid, and is continuous with conical nucleated bodies which form a continuous layer between the follicle wall and the outer surface of the capsule. Kölliker is therefore right in considering the layer as "granulosa," and the capsule as its product.

BROCK ('78, p. 556) gives the following clear, and I believe correct, account of the capsular membrane of the perch. The follicular cells, which at first are in close contact with the young egg, are raised up from the zona radiata by the developing gelatinous layer, and with the advancing growth of that layer are drawn out on the side toward the egg into long processes which can be followed up to the zona. In older eggs these follicular cells, separated by considerable intervals, (an indication that their multiplication soon stops,) lie in shallow depressions of the gelatinous capsule, and with their lower pointed ends continuous with the processes. These appear to end at the zona with conical enlargements, but the author will not affirm that this is a constant feature. Brock also maintains that Serranus hepatus has a very similar gelatinous capsule. The follicular cells, however, are very peculiar. They form a network of thin flat cells, which are in contact with each other only by means of lateral processes, while perpendicular processes, which are sometimes branched and exceedingly fine, can be traced through the jelly to the zona radiata. Concerning the development of the capsule, nothing is known.

So far as regards the capsular membrane of the perch, HOFFMANN ('81, pp. 19, 20, 27–29) comes to totally different conclusions from Brock, and expresses views which seem to me untenable. In October, ovarian eggs from 600 to 700 μ in diameter possess a membrane 5 μ thick, which is composed of two layers of nearly equal thickness. "The inner is the true zona radiata; the outer is composed of very numerous, small knob-like projections, which stand very close together and *correspond exactly to the villi of the cyprinoids.* On the free surface of the conical villi lie the granulosa cells."

It cannot be denied that the figure cited (*l. c.*, Taf. I. Fig. 9) corroborates the description given. But there is one fact which I should imagine would have caused the author more concern than it seems to have done. If his drawing accurately reproduces the conditions, there must have been about *four times as many villi as there were granulosa cells.* That in itself alone might not be of any significance to the author, especially as he disclaims any genetic connection between the granulosa and the underlying capsular membrane, but it does seem as though it should have received some explanation in view of the ultimate relation (Taf. I. Fig. 10) which Hoffmann admits to exist between the radial fibres of the capsular membrane and the cells of the granulosa. This is what he says about the later (February) stage of the egg : The zona itself is seen to consist of two layers, the inner much thicker than the outer. From the latter there arise with small triangular bases long peculiar fibres only 1 μ in thickness, which are stained in osmic acid precisely like the outer layer of the zona from which they arise. The outer ends of these fibres are thickened, even more than their inner ends, and form a continuous layer, between which and the zona radiata the fibres themselves are stretched like so many columns. "Over the proximal [1] ends of these fibres the granulosa cells are arranged *in such a way that a granulosa cell fits into* EACH *thickened end* (Taf. I. Fig. 10)."

If after the [ovarian] eggs have lain in water a short time they are transferred to osmic acid, it becomes very easy, says Hoffmann, to isolate both this layer (formed by the expanded ends of the fibres) and the granulosa in the form of large shreds ("Lappen"). From such preparations one can readily convince himself, he says, that the expanded ends of the fibres form a continuous sheet, and *are not processes of the granulosa cells,* as one is at first inclined to assume, and that *the fibres are, as the examination of the early stages proves, only the greatly grownout conical villi.*

That which seems to me to need explanation is, Why is the numerical relation between the villi and the granulosa cells so different at different stages in the growth of the egg, and why does this relation become such an invariable one in the later stages of development ?

It might be answered, in reply to the first question, that the granulosa cells undergo rapid multiplication, and that cell division occurring twice for all granulosa cells between October and February would explain the altered relations ; but is it not more reasonable to suppose that, through some unexplainable accident, Hoffmann has been led to attribute eggs to a perch which were taken in October from some other

---

[1] It is not clear to me in what sense the author can use the word "proximal" of the ends of the fibres which are directed *away* from the centre of the egg.

fish possessing a villous layer, than to ignore the evident constancy between villi and granulosa cells, and to assume an extensive multiplication of the latter in eggs during the period of their growth from a diameter of .7 mm. to .75 mm.?

The latest paper dealing with the capsular membrane of Perca is that of OWSJANNIKOW ('85, pp. 3–8), who reaffirms Müller's claim that the "tubules" are hollow structures, and corroborates Ransom's discovery that their inner ends are divided into branches which penetrate the pore-canals of the zona radiata.

On gold or silver preparations of ovarian eggs, one finds the granulosa cells [1] bounded by broad lines of a precipitate, so that there must be present a large amount of intercellular substance. The cells themselves lie, as previous authors have shown, over the beginnings of the corkscrew-shaped canals. These "beginnings," in the fully developed egg, are not at all cells, and have no nuclei; they are little funnels, with the narrow end sunk as it were into the jelly. The finely granular substance lies more compactly in the bottoms of the funnels; it is scantier on the margin, and in many places extends beyond the rim, of the funnel. This tissue (granulosa cells) often presents the appearance of stellate cells joined together by numerous processes, and separated from one another by abundant intercellular substance. The more closely packed molecules at the bottom of the funnel had given occasion, he says, to the assumption that there was a nucleus there.

---

[1] In his account of the Graafian follicle, not always readily harmonized with his figures, and sometimes obscure, he claims the presence of a greater number of epithelial (or endothelial, for he recognizes no difference between the two) cell layers than have usually been admitted. Thus, if I rightly understand the explanation of Fig. 4, Taf. I., there are in Perca, e. g., two epithelial cell layers between the vascular layer and the capsular membrane, — an outer layer of flat endothelium and a deeper layer of cylindrical granulosa cells; but in the text (pp. 4–8) he speaks of only two cell layers, an inner and an outer granulosa (!), which are separated by the vascular layer. From his description of the latter as the source of the new eggs, there can be no doubt that it is the "germinal epithelium" of authors. I can reconcile this apparent contradiction between text and figures only on the assumption that Figure 4 *and its explanation* belong to a period in Owsjannikow's studies when he was not as yet convinced of the error of taking "the granular accumulation in the bottom of the funnels of the 'tortuous canals' for a cell nucleus." That he ultimately supposed that to be an error appears from his description, and the statement (p. 5), "Die am Grunde des Kelches dichter an einander gelagerten Moleceln gaben die Veranlassung dort einen Kern anzunehmen."

In the case of Osmerus, moreover, he recognizes the existence of *two* layers of large flat endothelial cells in addition to the granulosa. (Compare his explanation of Fig. 8, Taf. I.)

From the bottom of the funnel there proceeds to the zona radiata a strongly and evenly twisted canal, which breaks off easily at the bottom of the funnel. Owsjannikow cannot agree with Kölliker, who maintained that in February these were solid fibres, because already in November and December he finds them hollow. The superficial layer of the gelatinous mass, as well as that part which immediately surrounds the canals, appears to be more compact than the rest of it, and the vicinage of the funnel is more deeply stained by aniline red or gold chloride than the other parts. The granules which occupy the canals or the funnels never enter into the gelatinous substance when the canals are ruptured, but escape into spaces which surround the canals. The inner end of the canal does not terminate in a pointed manner, as figured by His, but is often enlarged into a funnel, and sometimes divided into two or three fibres, — in one case into so many that it looked like a brush. On one occasion these branches were traced through the zona. These processes are to be seen only in stained specimens (gold chloride followed by aniline blue), because, having the same refractive power as the substance of the zona, they are otherwise undistinguishable. The fine molecules which lie on the inner surface of the zona were found to be deeply stained, and the author concludes that the dye must have penetrated through the spiral canals. The function of these canals must consist in the transportation of nutritive material to the yolk. They arise out of the granulosa cells, are similar to those seen by Eimer in the adder, and are not processes of the zona radiata, as affirmed by Hoffmann. The lateral processes from the canals were also seen by Owsjannikow, but he has for them another interpretation. The matrix (Zwischensubstanz) appears to lie in layers parallel to the surface. Upon its being swollen by any fluid, narrow *fissures* are formed between these layers, which join the canals and appear as processes from them.

Besides the difficulty of trying to comprehend how fissures could arise as a result of the swelling of a gelatinous mass, — it would seem that the reverse process ought to be more favorable to their appearance, — the sufficient answer to this last claim is, that the transverse processes are more deeply stained than the remaining portions of the matrix, which could hardly be the case if they were simply fissures.

It was in the hope of ascertaining something more about the interesting capsular membrane in the perch, that I advised Mr. Eigenmann ('90) to include that fish in his studies on the development of the micropyle. I believe it will be seen from his results that there is still very good reason for maintaining that the tubular or columnar structures of the capsular membrane, which have been the objects of so much study, are derived from the granulosa cells, one from each cell, and that the process by which the capsular membrane is formed is neither simply a cell secretion nor exclusively a cell metamorphosis. Although Eigenmann has not succeeded in getting stages which show clearly all the steps in the formation of the capsule in Perca, he has shown that there exist conditions in the later stages of the development of the egg in Esox (Eigenmann, '90, Plate III. Fig. 37) which seem to me of considerable importance in interpreting the conditions in Perca. In Esox the cells become elongated, and the central (axial) portion retains the granular and stainable properties of unmodified cell protoplasm. This axial portion is not cylindrical, but conical; its base is directed outward and contains the nucleus. The peripheral portion — which is more and more abundant as one approaches the zona — is more homogeneous than the axial part, and reacts with dyes in a different way. The cell boundaries have been previously lost. The boundary between these two constituents of the cell is not at first sharp, so that this phase of the process may perhaps be regarded as one of metamorphosis rather than of secretion.

I believe that Perca must pass through some such stage as this during the earlier part of the process which produces the capsular membrane. I imagine that the distinction between the axial and the peripheral portions of the cell becomes more and more sharply defined as the thickening of the capsule goes on. Meanwhile the axial portion does not long retain the indifferent condition, but is metamorphosed, especially at its periphery, into a highly refracting substance, so that there is reason for regarding the structure as tubular. This metamorphosis advances till it has practically obliterated the cell, even though a nucleus with a small amount of enveloping protoplasm may still be made out at its distal end in very late stages of ovarian growth. At any time before this, and after the distinction between a funnel part and a tubular part has arisen, the less modified distal portion of the cell may doubtless be easily separated from the secreted gelatinous substance and also from the metamorphosed cell process. Such at least is the view which I have formed, after a comparison of the granulosa in Perca and in Esox.

It would certainly be remarkable if the perch were the only representative of bony fishes in which such a process took place. I believe that there are a few cases already known which may prove upon renewed inquiry to be essentially the same as the perch. The similarity to the perch shown by Acerina vulgaris was recognized by Müller ('54, p. 189). "In Acerina," he states, "the egg membrane has the same structure [as in Perca], only it is much thinner, and consequently the tubules are only short, not longer than the breadth of the areas." Ransom ('68, pp. 453, 454) also says of this species that its "yelk-sac has an outer layer or 'Eikapsel.'" But he adds, that "the outer layer appears to be continuous with, and similar in structure to, the yelk-sac proper." However, Owsjannikow ('85, pp. 17, 18, Taf. I. Fig. 13) has given an account of it which points still more strongly to the resemblance claimed by Müller. That part of his account is especially significant in which he states that in some preparations the follicle cells have the form of very narrow cylindrical epithelium; the broad end of the cell is directed outward, the pointed end inward toward the zona. The cells, he adds, lie in a transparent non-staining layer, similar to that in which the spiral canals (in the perch) are located. Finally, a third condition is described in which the cell form is lost. The structure begins with a broad short funnel, and passes at once into a narrow, straight hollow fibre which imbeds itself in the zona radiata.

The peculiar follicular layer described by Scharff ('88, p. 69, Plate V. Fig. 15) in the interesting egg of the shanny (Blennius pholis) also appears to have begun to undergo a modification in the same direction that leads in the perch to the formation of a capsular membrane. The substance which Scharff calls "interstitial" is, I believe, *morphologically* the same as the gelatinous secretion of the follicular cells in Perca.

3. *Micropyle and Micropylar Plug.*

Since Doyère ('50) discovered, in 1850, an aperture leading through the membrane of the egg of Syngnathus, and gave to it the name of micropyle, there has been a good deal of attention given to that structure.

Independently of each other, and probably without knowledge of Doyère's discovery, Ransom ('56) in England and Bruch ('55ᵃ) in Germany rediscovered, in 1854, this structure in fish eggs, and both applied the name which had meantime become current through Müller's ('51 and '54ᵇ) discovery of a similar canal in the egg membrane of Holothuria, to which he also gave the name of micropyle, borrowed from the usage of botanists.

Ransom in fact succeeded in observing in the egg of the stickleback the passage of spermatozoa through this opening; Bruch was less fortunate with Salmo fario and S. salar, although he made special effort to discover if such a passage took place,. and particularly emphasized the fact that the orifice of the micropyle is of exactly the same size as the spermatozoön.

These discoveries were soon (1855) confirmed by LEUCKART ('55, pp. 257–264) and REICHERT ('56, pp. 83, 84, 98–104, Taf. IV. Figs. 1–4) on the Continent, and by THOMSON ('59, p. [100]–[104]) in England. Their observations established the fact of the existence of the micropyle in numerous fishes, and under several modifications of form. Ransom had given a fairly accurate account of the structure of the micropylar region, but Reichert especially insisted upon the differences between an invaginated portion of the membrane and a passage through the latter in the case of cyprinoids. He distinguished three regions, — an approach (Eingang), a fundus, and a neck or cylindrical canal, the length of which was diminished from what it would otherwise have been by a reduction in the thickness of the membrane to one third its normal dimension.

It was, however, the interest in fertilization stimulated by Newport's researches on the impregnation of the ovum in Amphibia, and by Keber's paper, " De introitu spermatozoorum in ovula," etc., Königsberg, 1853, that gave paramount importance to these discoveries, and attracted general attention to them.

Perhaps it is not surprising, in view of this fact, that the *ovarian* egg was less studied, and that the relation of the micropyle to the granulosa cells in its vicinity was not especially examined; and yet those observers who concerned themselves with the questions relating to the origin of the different egg envelopes must have been very near to inquiring what share the granulosa had in producing so remarkable a modification in the egg membranes. KÖLLIKER ('58, p. 92), although only incidentally making observations on the micropyle, established the fact of its existence in a large number of cases. He regarded it simply as an enlarged radial pore of the secondary vitelline membrane (zona), which might be produced, he thought, by a process of resorption.

Several subsequent writers have concerned themselves only with questions relating to the form and position of the micropyle, and its probable function. Thus BUCHHOLZ ('63, p. 72) compared the micropyle in Osmerus eperlanus to a crater the floor of which is closed except for a minute canal in the middle of it, which traverses the thickness of the wall; RANSOM ('68), besides adding some unimportant details to his earlier description of Gasterosteus, described briefly the micropyle in a large number of other (fresh-water) fishes, claiming that it always terminated

in a small elevation lying directly over the germ, and concluded, as the
result of experiments (pp. 459–462) made on Gasterosteus pungitius,
that "the function of the micropyle is to admit the spermatozoids to the
surface of the yolk"; and His ('73, pp. 3, 4) described with some detail
the structure of the micropyle in Salmo salar and S. fario, in both of
which he recognized a shallow depression ("Mulde") surrounding the
crater, which in S. fario terminated in a deep funnel, and this in the
canal.   He also showed that only one spermatozoön at a time can pass
the micropylar canal, which terminates somewhat eccentrically over the
germinal disk.

HOFFMANN ('81, pp. 33–36) has confirmed for a large number of fishes
the observations that the inner end of the micropyle terminates in a
papillary elevation of the zona radiata, and that in the ovarian egg it
lies directly over the germ.   From a comparison of the dimensions of
the spermatozoa and the calibre of the canal, he also draws the conclusion that not more than one spermatozoön can traverse the micropyle
at a time.   In nearly all the micropyles figured by Hoffmann the canal
is a tubular passage without any special enlargement; but in the case
of the herring's egg — which is the one most carefully described — the
outer half of the passage is enlarged into a conspicuous bulbous cavity
(Taf. I. Fig. 19), which, so far as I recall, has been seen in only one
other instance, that of Petromyzon as figured by Calberla.   But the
greatest interest attaches to the conditions figured for Leuciscus rutilus
(Taf. I. Fig. 20).   In this case there exists a distinct plug of granulosa
cells occupying the depression in the egg membrane at the micropylar
region.   Since the author does not mention the fact in the text, it is
probable that he attached to it no importance.   I think, however, it is
the first clear proof published of the existence of a specialization in the
granulosa of the micropylar region in any teleost.   On account of the
rather diagrammatic rendering of the granulosa cells, it is not possible
to be very confident about the existence of a specialized micropylar cell,
but the fact that a single cell forms the apex of the plug favors that
view, and I shall be surprised if such a structure is not hereafter demonstrated in this European fish.

Of the more recent writers on the micropyle, OWSJANNIKOW ('85,
pp. 11–13, Taf. I. Figs. 5–7) describes for Osmerus eperlanus a micropylar apparatus composed of two portions, an external and an internal, corresponding respectively to the two membranes which envelop
the egg, — the "external zona radiata" (which corresponds, in his
opinion, to the outer [capsular] layer in Perca) and the "internal zona

radiata." The apparatus has the form of a crater-like depression, about as described by Buchholz and Ransom. Where the membranes take the direction of the crater, they form folds with the pointed ends directed inward. But of more interest are his statements, that the zona radiata externa takes the greater share in the formation of the crater, and that "*other tissues, especially the endothelium and granulosa cells, participate in the same.*" Owsjannikow was thus, I believe, the first person after W. Müller ('75) to call attention to an intimate relation between granulosa and micropyle.[1] But there are some elements of uncertainty about his descriptions and figures that seem to baffle every attempt to reduce them to harmony. The most perplexing thing about his description is the use of the term "endothelium," which is at first used for Osmerus in the following connection (p. 10): "Die Graafschen Follikel der Osmeruseier bestehen aus Endothel, Gefässen, Bindegewebe und Follikelzellen." In the description of other eggs (Perca) the word "Endothel" is also used as though applied to cells which lie *outside* the vascular layer, and even as though including the germinal epithelium of other authors (p. 4). Unless an endothelium having a very different position from that previously described by him is meant, when he says that it participates (as well as the granulosa) in the micropylar structure, I believe that the author has fallen into some error; for I am of opinion that neither the connective-tissue layer with its blood-vessels nor the germinal epithelium shares directly in the formation of the micropylar apparatus. Neither do I believe that there exists *inside* the connective tissue any layer of cells except the granulosa. Moreover, I do not think that any layer of endothelium, either inside or outside the vascular layer, has been figured as sharing in the formation of the micropyle. It does not help matters in the least to add that the author discountenances (p. 4) any attempt to draw a distinction between *epithelium* and *endothelium;* for after saying that "Endothel *und* Granulosazellen" share in this formation, he proceeds with a description which certainly allows the assumption that there is only *a single layer of cells involved*, to which, however, he gives successively the names of granulosa, endothelium, and epithelium.

[1] Although published several years ago, Owsjannikow's studies were not made until some time after I had demonstrated the conditions in Lepidosteus which have been described above. His paper, as well as the more recent one of Cunningham, has therefore had no influence in determining the course or the results of my studies on Lepidosteus, nor did it influence me to suggest a comparative study on the eggs of bony fishes, such as Mr. Eigenmann has undertaken; for I had already proposed that question to one of my students before either paper was published.

On " teased " preparations Owsjannikow often finds the zona deprived of its cellular covering, but ordinarily the detached cells are to be found in the form of a continuous membrane, on which a conical projection is to be seen. The form of the projection corresponds exactly to that of the external micropyle; it is hollow; the cells at the entrance to the micropyle have the form of a crown, and become smaller and smaller toward the bottom of the crater. I am not sure that I fully understand the figure (Taf. I. Fig. 6) which the author refers to in this connection, but it appears to me to be a view of an egg from the animal pole ; the granulosa cells of the crater, having been detached, are seen partly in side view, but somewhat obliquely, as a conical structure, and the pore-canals of the *external* zona of the crater are visible where the granulosa cells have been lifted. In the middle of the figure is an optical section of that portion of the zona which forms the internal projection, and in its centre the micropylar canal. There is apparently a single layer of cells, and this I take to be the granulosa. It is to be regretted that the author has not furnished us with a strictly radial section through the micropyle and the accompanying structures on a sufficiently large scale to enable one to determine what becomes of the membrana propria of the theca folliculi in the region of the micropyle. One would infer, from the statement that this granulosa cone was hollow, that the theca must follow the course of the crater; but if it does, it must be different from all other known cases. Not even in Perca does the membrana propria suffer any deflection or infolding due to the participation of the granulosa cells in the formation of a micropylar structure. I have also been considerably perplexed by Owsjannikow's Figure 5 (Taf. I.). At first I took it to represent a strictly diametric section of the egg and its membranes. With that understanding of it, I imagined that the large oval body just above the micropyle might possibly represent a single micropylar cell in some way loosened from its natural position; but more careful study leads me to believe that this is a figure representing in part an optical section, in part a surface view, and that the oval structure presents an oblique view of the external entrance to the funnel-shaped cavity of the crater, still lined with granulosa cells, while all the rest of the figure represents a view of the egg as it would be seen in optical section. Owsjannikow states that the inner micropyle can be regarded as a somewhat enlarged canal of the zona, and claims that it subserves the nutrition and growth of the egg; for he has traced from the inner end of the canal a row of granular bodies which were continued in the yolk as a fine thread, which at last disappeared. This row

of granules and its thread-like continuation the author regards as a product of the granulosa cells. But at present there are no data, he says,
either teleological or phylogenetic, which can explain the remarkable
structure of the external micropyle.

RYDER ('86, p. 30, Plate VII. Fig. 35), who has demonstrated the
existence of the micropyle in the eggs of several fishes, speaks of the
passage through the capsular membrane in Perca Americana as "a
wider pore-canal which leads to the micropyle." It is evident from
Eigenmann's ('90) account and from his figures that this statement is
inaccurate.

CUNNINGHAM ('86ª, pp. 59, 61–63, etc., Figs. 2–4, 12) has shown much
more satisfactorily than W. Müller ('75) did, that in Myxine glutinosa
the follicular epithelium plays an important part in the formation of the
micropyle. Of an egg 16 mm. in length he says : " At the exact pole of
the egg there is a differentiated portion of epithelium, where a proliferation of the latter has taken place. This portion is composed of polygonal cells, which are little or not at all elongated, and towards the egg
it runs out into a thin cylindrical process which penetrates the next layer
[zona radiata] as shown at e. p. [Fig. 2]. . . . This process penetrates
the vitelline membrane [zona radiata], occupying a tubular cavity in the
latter, and passing through it to form a hemispherical projection on its inner surface. . . . This cellular projection is covered by a thin membrane
continuous with the vitelline membrane, and is not in immediate contact with the germinal disk. . . . There is thus at one pole of the nearly
ripe ovum a tubular canal extending through the chorion [zona radiata],
but not open internally, filled up by a cylinder of cells projecting from
the follicular epithelium. . . . It is evident," the author adds, " that this
.aperture is to form the *micropyle* in the ripe ovum. . . . It is probable
that careful investigation would show that in all Teleosteans whose ova
possess a micropyle that structure is produced by a projection of cells
from the follicular epithelium." Cunningham also believes it " at least
possible that in all vertebrates the micropyle will be found on investigation to be produced in the same way as in Myxine, namely, by the
growth of a cellular process from the follicular epithelium towards the
vitellus while the vitelline membrane is being formed."

In an egg 20 mm. in length " the proliferation and differentiation of
cells at the pole in the follicular epithelium have disappeared, but the
cylinder of cells, though reduced in size, still remains in the micropyle,
and is evidently destined to keep the latter open until the maturation
of the ovum is complete." This egg was from material obtained in

December. Older eggs, obtained at the end of January, although only slightly larger than the December egg (21 mm. long), presented a very different appearance at the micropylar region. Cunningham says of the latter: " The micropyle is somewhat narrower, and the cells present in it at previous stages have disappeared almost completely, only a little débris remaining. The micropyle seemed also in these ova to be open internally, though of this point I am not absolutely certain. If there is a membrane closing the inner end, it is an extremely thin one."

Cunningham has shown conclusively, I believe, that the granulosa has much to do with the modifications of the egg membrane in the micropylar region, but there are several particulars concerning which his description and figures leave me in a rather unsettled state of mind. In the first place, the author does not seem to distinguish with sufficient sharpness between a funnel-like region, which may be partly the result of an infolding of the membrane, and a *passage through* the membrane, which I have called the micropylar canal. It seems to me possible that his uncertainty as to whether the micropyle is closed at its inner end at a *late* stage may be due to this fact. The disproportion between the calibre of the " micropyle " and the size of the spermatozoa[1] is not alluded to, but at once suggests to me that the structure in question may be the equivalent of the micropylar funnel only. I should be quite certain that it was so, were it not that the drawing of the latest stage (Fig. 12) — which is not sufficiently explained — admitted two interpretations. In this figure the whole passage is divided into two portions of about equal length, but of very unlike calibre. The inner half is a narrow canal with parallel walls, about one third of a millimeter in diameter (actual size about 10μ); the outer half is *abruptly* widened to 6 or 8 mm. in diameter, and gradually increases toward the granulosa to 10 mm. (nearly 300μ actual size). There are no granulosa cells, however, in either portion of the passage. The almost flat-bottomed outer half of the passage would appear to be the equivalent of the micropylar funnel in bony fishes, and I should certainly have so regarded it if Cunningham had not evidently considered the inner narrower portion as a part of the " micropyle " of previous stages from which the cells had disappeared, leaving " only a little débris." If the author is right in this assumption, that the *narrow* part of the ap-

---

[1] In its narrowest place the " micropyle " of the author's Figure 3 is represented as 5 mm. in diameter in the drawing, which, being magnified 280 diameters, makes its actual diameter about 18μ, whereas the actual diameter of the head of a spermatozoön (Fig. 14) is not over 7μ.

paratus was once occupied by granulosa cells, then either the true micropylar *canal* has not been seen in Myxine, or it is formed in a different manner from that of bony fishes, for in no case is it occupied in the latter by a plug of cells from the follicular epithelium. The former alternative is, in my opinion, the more probable. This conclusion I base partly on the appearance of Cunningham's Figure 3, and partly on the conditions presented by sections of an ovarian ovum of Myxine australis, which I have studied since reading Cunningham's paper. In his Figure 3, the plug of granulosa cells which is sunk into the yolk is *completely* enveloped in a uniform layer not more than 2 μ thick, which not only separates its deep end from the yolk, but also its sides from the membrane called by him vitelline membrane, or zona radiata. What the significance of the part of this thin membrane lining the chimney-like elevation of the zona radiata may be, I cannot say, unless it is reflected at the upper edge of the chimney to form the outermost layer of the zona; but the portion which separates the yolk from the granulosa plug I regard as the equivalent of the first formed portion of the zona radiata, and believe that the true micropylar canal will be found in the form of a minute passage through that membrane. My principal reason for this opinion rests on the condition of this region in the egg of M. australis. This egg was about 22 mm. long and 6 mm. in diameter. It was enveloped by a follicular epithelium composed at the animal pole of a single layer of cells, averaging about 10 μ thick. Over the region of the flattened germinative vesicle the granulosa was gradually thickened to about 25 μ, and from the middle of the thickening a solid plug of cells about 35 μ long and 25 μ in diameter projected into the yolk. The membrana propria of the follicular theca passed over the micropylar region without being at all infolded, so that the total thickness of the granulosa, measured from the apex of the plug, was about 60 μ. The cells of the plug were not well preserved, but appeared to be of about the same relative size as in Cunningham's Figure 3, — i. e. of one half or one third the diameter of the plug. There was no enlargement at the apex of the plug as seen in his figure. Between the granulosa and the yolk, and in contact with both, was a highly refractive thin (3.5 μ) membrane, which at first appeared homogeneous, but in which I believe I have detected at intervals radial markings. This membrane became thinner at some distance from the pole.

The whole apparatus had such a striking resemblance to the micropylar apparatus in Lepidosteus that I cannot doubt that the granulosa plug represents the same structure in Lepidosteus, and that conse-quently the micropylar canal is to be found at the bottom of the funnel-shaped infolding produced by the plug.

4. *Micropylar Cell.*

Although no other cases are yet known in which a single cell of the follicular micropyle-plug is so evidently differentiated from its neighbors, as in Lepidosteus, still it is clear from the results of Eigen-mann's ('90) studies that the existence of such a specialized cell is not uncommon.

The most interesting question relative to the micropylar cell is that of its physiological signification.

That it sustains some intimate and constant relation to the micropyle itself can scarcely be doubted. Perhaps its primary function is fulfilled by serving as a source of passive resistance to the forming membranes in the region of the micropyle — a kind of mould for them — during the process of their formation, therefore a mechanical device for pro-ducing a micropylar funnel. In that case it would doubtless often be more than the single micropylar cell on which devolved the function ; it would be rather that plug-like accumulation of granulosa cells, with the micropylar cell at its apex, which attains such an extensive development in Lepidosteus and some other cases.

But the very fact that one cell is generally, if not always, differen-tiated more than the rest, suggests that the function referred to may not be the only one, — perhaps, indeed, not the primary one. There are two other possible functions which are naturally suggested in this connection.

Concerning the first of these it may be said that it still remains to be shown to what extent the micropylar canal and its funnel are the result of an exclusively progressive process of development ; whether, in other words, any part of this structure is produced by a process of *resorption.* Such a process would not be without a parallel. At least Leuckart ('55, pp. 108, 216, 247) has asserted in most positive terms that the micropyle of certain insects' eggs is not to be found in the chorion from the very beginning of its formation, but that it arises after the production of the chorion by means of resorption.[1]

---

[1] More recently, it is true, Korschelt ('84, pp. 421, 422) has shown that the micropyle in Meconema varians is formed by a single cell, and he apparently leaves no room for a process of resorption, since he says : "Die Entstehung des Canals ist wohl so zu denken, dass die Zellen schon frühzeitig einen Fortsatz ausstrecken, der

The fact that Eigenmann ('90) has been unable to discover a micropyle in the earlier stages of the formation of the egg membranes in fishes may also point in the same direction. I would not wish, however, to place too much weight upon such negative evidence; it requires extensive and indeed the most exhaustive examinations to make such testimony satisfactory. Especially am I compelled to this reserve, in view of the fact that Cunningham ('86[a]) has found the micropylar apparatus well developed at a relatively early stage in the formation of the zona in the case of Myxine glutinosa; but it will be observed that he says (p. 61) : "This cellular projection *is covered by a thin membrane continuous with the vitelline membrane, and is not in immediate contact with the germinal disk.*" An actual opening does not exist, therefore, at the time of which he speaks. Although I have found in M. australis at an apparently early stage in the development of the zona a deep infolding of that membrane, as described above, still I have not satisfied myself in the single specimen sectioned that there is an *orifice* through the membrane at this stage of development. But on this observation I would not care to speculate were it not confirmed by Cunningham's studies on more extensive and I presume histologically more favorable material, for I know how easily one may be deceived as to the existence of so minute a structure as that with which we have to deal.

The facts which I have given above do have a certain weight with me as rendering it possible that in Myxine at least the micropylar canal, in the strict sense of the word, is not present until near the maturity of the ovum, and that consequently it may be the function of the cell nearest to the bottom of the funnel — the micropylar cell — to absorb so much of the already formed membrane as is necessary to allow the

anfangs nur kurz ist, später mit dem Dickerwerden des Chorions und dem entsprechenden Zurückweichen der Epithelschicht aber länger und länger wird."

In his final paper, Korschelt ('87, p. [43] 223) suggests a method of reconciling Leuckart's views with his own observations: "Die Beobachtung Leuckart's, nach welcher die Mikropylkanäle sich nicht von Anfang an am Chorion finden sollen, liesse sich vielleicht mit unseren Befunden in Uebersinstimmung bringen. Wir erkannten in mehreren Fällen, dass die Masse des in der Bildung begriffenen Chorions eine durchaus weiche und plastische sein muss. Deshalb wäre es leicht denkbar, dass die Zellfortsätze, weiche die Kanäle entstehen lassen, nicht von Anfang der Bildung an vorhanden wären, sondern erst später in die weiche Masse des jungen Chorions hinein entsendet würden. Dann würde es ein Stadium geben, in welchem das Chorion eine ununterbrochene Oberfläche besässe. Seine Ansicht, das die Mikropylkanäle an dem bereits gebildeten Chorion durch Resorption entstehen sollen, dürfte Leuckart wohl aufgegeben haben, sobald er die Entstehungsweise der Porenkanäle der Eischale kennen lernte."

passage of the spermatozoön.[1] I realize that this is a speculation on very narrow foundations ; for even if it could be shown to result from absorption, it might be the protoplasm of the ovum, not the granulosa cell, which accomplished the work. There are, however, still more serious objections to this view, which, though not disproving it, render it very doubtful. The fact that in general pore-canals and orifices in cuticular secretions are the results, not of resorption, but of the previous existence of protoplasmic projections, makes it probable, without positive proof to the contrary, that the same would be true in this case.

The serious, and indeed, as it seems to me, insurmountable objection to considering the whole micropylar funnel as the result of absorption, is the condition which it exhibits in Lepidosteus, where there is a very gradual diminution in the thickness, not only of the zona, but also of the villous layer. It is not probable that any process of absorption could result in diminishing the thickness of *both* layers so evenly without affecting their mutual relations, unless perchance it should be imagined that the zona was absorbed through the agency of the yolk, and the villous layer by means of the granulosa cells. But even that assumption would not help the case very much, for it would still have to be explained why the shorter villi retained all the parts of the longer ones, *and in nearly the same proportions.*

While, then, the conditions clearly preclude the possibility of looking upon the micropylar apparatus in general as resulting from a process of absorption, it by no means follows that the micropylar *canal* may not be produced by such activity.

The other purpose which it has occurred to me the micropylar cell may subserve, is to facilitate the penetration of spermatozoa. Not precisely that a minute drop of slimy substance, resulting from the degeneration of this cell and covering the orifice of the narrow canal, would offer less resistance than water, but that its presence might prevent the occlusion of the orifice by the accidental introduction of impenetrable substances without itself offering any serious obstacle to the free entrance of the fertilizing element.

If one were to attempt a phylogenetic explanation of the *micropylar funnel and canal* in bony fishes, he might reason somewhat as follows.

[1] In Cunningham's Figure 3, the cells of the granulosa plug which form the layer nearest the yolk — four of them being cut in the section figured — are all larger than the remaining cells of the plug, but I am unable to say that any one of them is the largest of all.

The development of a persistent egg membrane impervious to sperma tozoa would evidently be possible only with the concomitant production of one or more orifices; for without such provision no egg could be fertilized, and the transmission of such tendencies would clearly be impossible. That would necessitate the development of the micropyle by what I should call an exclusively *progressive* method. It would not imply any *regressive* or resorbent process. How, then, could one find reason for claiming any such process? I believe it would only be necessary to assume that the zona radiata in its original development subserved some useful end *during the development of the ovum*, in order to form an idea of the possible course of events which has led to the present condition. Imagine eggs at oviposition provided with a zona radiata which remained or had become penetrable by spermatozoa; such eggs would be in the most favorable condition for fertilization, but on account of the condition of the zona they would be poorly *protected* against external agencies. If, under these conditions, a *portion* of the zona radiata in some eggs should become more resistant, even to the point of preventing the entrance of spermatozoa, the eggs thus modified would be better protected from injurious external influences than those which remained in the original condition, and yet they would be almost as readily fertilized as the latter, *provided* some portion of the zona remained, as at first, penetrable. In short, the advantages of such a changed condition would be greater than the disadvantages, and consequently in the long run the more favorable condition would predominate. Evidently the *optimum protection* would have been reached when a region no larger than that absolutely necessary to admit a single spermatozoön had been left for that purpose. But this process of restriction in the area accessible to the spermatozoa may easily have been accompanied by another process, which may have begun as early as the former. The zona was assumed to remain, or *become* at the time of oviposition, penetrable to spermatozoa. It seems to me entirely reasonable that a process tending to modify a portion of the zona and make it more readily penetrable should be set up in the ovary, and that such eggs in the matter of fertilization alone would have some slight advantage over eggs less easily penetrable. If now these two tendencies were operative at the same time, — the one serving to soften a part of the zona, in order to make it the more readily penetrable, the other to harden *another portion*, to make the egg less subject to adverse environment, — the former would become localized by the encroachment of the latter until at length there would be only a limited area in which the process of softening went on; but this might be — as we must believe often happens in the animal economy — correspondingly intensified, until an activity which in the beginning resulted in only a feeble modification in the condition of the zona ultimately terminated in its complete liquefaction and absorption.

A single argument, which it seems to me may have some value, in support of this hypothesis, is to be drawn from the condition of the zona in Elasmobranch fishes. So far as I now see, the complete disappearance of the zona at the maturity of the egg would be entirely in harmony with the hypothesis. The condition there at any rate seems to me to favor the assumption previously made, that the zona originally had a function distinct from that which it now appears to possess in protecting the embryo after fertilization. For if not, it must be an inheritance from ancestors which, like bony fishes, had their embryos thus protected. There are probably few who would defend the idea that the Elasmobranchs are descended from bony fishes, and the evidence of common ancestors with eggs thus protected still remains to be found.

If, as the Russian naturalists assert, there are several micropylar openings in the egg-shell of sturgeons, it may be that those fishes present a condition which is intermediate between an extensive region of penetrability and the extreme restriction which now prevails in bony fishes.

The funnel portion of the micropylar region is certainly the less essential and least constant part of the structure. It may reasonably be considered, I believe, a secondary condition, and the explanation of its development might lie either in the fact that it served, in a passive way, to direct the motion of a greater number of spermatozoa toward the actual orifice, or, possibly, that it served to preserve from accidental removal the protective products of the degenerated micropylar cell.

Thus the micropylar *canal* might be regarded as the result of two to a certain extent antagonistic tendencies, — the fittest solution of a problem requiring the fulfilment of two conditions. The micropylar *funnel* could obviously be regarded as a partial compensation for the diminished opportunities for fertilization caused by a restriction of the area available for that purpose, and might have arisen simultaneously with the restriction, or only after the latter had attained its present maximum.

Summary.

1. The first food of young gar-pikes after the absorption of the yolk is mosquito larvæ ; later, they feed on small fishes.

2. Very cold water is injurious to young gars.

3. The arched form which the body sometimes exhibits is probably always the result of an abnormal condition.

4. The acts of catching and swallowing the prey are complicated. Fishes are usually swallowed head first.

5. The young gar-pike lives at the surface of the water after the absorption of the yolk-sac.

6. The emission of bubbles of gas through the gill slits is accompanied by a rolling of the body to one side and a forward movement.

----

7. Analyses of the bubbles of gas emitted by young fishes, and also of a limited amount of atmospheric air which had been used by the fishes in respiration, showed a reduction of the oxygen to 10–15 per cent, and the presence of only a small amount of carbon dioxide, 0–1.7 per cent.

8. Air which had been previously deprived of its carbon dioxide gave no evidence of containing even a trace of that gas, after having been respired. Consult the text for the conditions of experimentation.

9. It is probable from these experiments that the air-bladder respiration serves to oxygenate the blood, but that the elimination of carbon dioxide is effected in some other manner. It is possible that the two elements of the respiratory function of higher vertebrates were *successively* transferred to the air-bladder, that of oxygenating the blood being the first to be transferred.

----

10. There are two egg membranes in Lepidosteus, — a zona radiata and a villous layer, — and they are intimately joined together. Both are radially striate, the zona finely, the villous layer coarsely.

11. Balfour and Parker were mistaken in stating that there was a homogeneous non-striate membrane inside the striate zona, and also in supposing that the pyriform bodies (villi) of the outer covering were metamorphosed follicular cells.

12. The outer layer is not as thick as the zona, and is made up of radially compressed and folded columnar structures, the villi.

13. Each villus is composed of three parts, — head, stalk, and roots. The roots project into the pore-canals of the zona.

14. The zona presents the usual structure of that membrane in bony fishes. The pore-canals are slightly enlarged and spiral at their peripheral ends.

15. The egg of Lepidosteus has a single micropyle, but it has been overlooked by previous observers.

16. The micropylar apparatus embraces a funnel and a canal. The funnel results from an infolding and a reduction in thickness of both the villous layer and the zona radiata. The micropylar canal is a narrow passage of uniform calibre ($2\,\mu$) and circular cross section, through the thinnest part of the two membranes, namely, at the bottom of the funnel.

17. The granulosa of the mature ovarian egg consists of a single layer of polygonal cells, except in the region of the micropylar funnel, where it forms a plug of cells completely filling the funnel.

18. The membrana propria of the theca folliculi is not infolded in the region of the micropylar funnel.

19. A single large granulosa cell, the "micropylar cell," forms the apex of the plug, and occupies the bottom of the funnel.

20. The villous layer is formed before the zona radiata. It first appears at the surface of the yolk in eggs about 0.4 mm. in diameter, and reaches one third its maximum thickness a year before the egg matures. It is produced by the ovum, not by the granulosa cells.

21. The number of the villi is not increased during the growth of the villous layer.

22. The zona radiata is likewise the product of the ovum, and its formation requires less than twelve months.

23. The name "capsule" should not be used for the zona radiata; it ought to be restricted to membranes outside the zona, which, like that of the perch, are the product of the granulosa.

24. An egg membrane comparable structurally and genetically with the *zona radiata* of bony fishes is to be found in representatives of all the groups of fishes except Amphioxus. It is fugitive in selachians and Lepidosiren, and probably in viviparous teleosts. *The zona is produced by the ovum, not by the follicular cells,* and is traversed in all cases by

pore-canals, which rarely (Myxine?) branch.   J. Müller is wrongly credited with their discovery.

25. An egg membrane genetically, but not always structurally, comparable with the *villous layer* of Lepidosteus, is found in several other cases: *possibly* in Petromyzon, probably in selachians and Lepidosiren, and certainly in several teleosts.   *This membrane is also produced by the ovum, and earlier than the zona.*

There is some reason to believe that it exists in the herring and the smelt (Osmerus). Hoffmann is probably in error in attributing the presence of villi to the eggs of Perca in October, and Owsjannikow certainly is in asserting that the mushroom-shaped villi in Gasterosteus are individual cells.

26. The *capsular membrane* is produced, as originally defined by J. Müller, by the follicle or follicular cells.   It has often been confounded with the zona, and also with the villous layer.

27. Although the " tubules " in Perca have been described as possessing root-like prolongations which penetrate the pore-canals of the zona, they are genetically unlike the villi found on other eggs, being produced by the granulosa cells alone.   Hoffmann's statement to the contrary rests on insufficient evidence.

28. A comparison of the condition of the granulosa in Blennius pholis and Esox with that in Perca, is believed to shed some light on the probable steps by which the capsular membrane is produced.

29. The most important paper on the *origin* of the zona and the villous layer I believe to be that of Kölliker, whose conclusions I have confirmed in the case of Lepidosteus.   All authors who have maintained that either zona radiata or villous layer is the product of the granulosa I believe to be in error; and in particular I maintain that the reasons assigned by Cunningham to prove that the zona in Myxine is produced by follicular epithelium are not adequate to establish his proposition.

---

30. It is doubtful whether the micropylar *canal* has been seen in Myxine.   What W. Müller called the micropyle was the micropylar funnel, and possibly that which Cunningham describes as the micropyle does not embrace the canal.

31. What I have called the *micropylar plug* of granulosa cells was first seen by W. Müller in Myxine ; later it was figured, but apparently not appreciated, by Hoffmann in Leuciscus ; it was described as hollow by Owsjannikow in the case of Osmerus, who recognized its relation to

the micropylar funnel ; and finally, it was described more fully for Myxine by Cunningham. Eigenmann has found the same structure in a number of bony fishes.

32. The *micropylar cell* has never before been recognized. Eigenmann has now found it in a number of osseous fishes, — Perca, Pygosteus, Esox, etc. I believe it may fairly be assumed to exist in the greater number of those fish eggs which possess a micropyle, and that it has an important function in connection with the formation of the micropyle or the fertilization of the egg.

33. I have made the following suggestions as to the possible functions and history of the micropylar apparatus : The *micropyle*, being evidently a provision for the fertilization of the ovum, may have its present structure as the result of two to some extent conflicting tendencies ; one induced by the advantages of protection to the egg, the other by the necessity of some provision for the penetration of the fertilizing element. But the best protection is not compatible with penetrability of the membrane at all points. Any reduction in the extent of the penetrable surface would be favorable to protection. An optimum condition would be reached when the penetrable area is reduced to a minimum, and that is the diameter of the head of a spermatozoön.

The *funnel* may be a partial compensation for such reduction.

The *micropylar plug* may *mechanically* determine the presence and form of a funnel.

The *micropylar cell* may serve to form the *canal* by resorption, *or* to prevent the occlusion of the canal by less penetrable matter at the time of oviposition.

CAMBRIDGE, April 7, 1889.

Postscript.

Since the completion of the present paper, I have received from the author, Dr. J. Beard, a copy of his "Preliminary Notice, — On the Early Development of *Lepidosteus osseus.*" This paper was presented to the Royal Society of London, April 20, 1889, and was printed in the Proceedings of the Society, Vol. XLVI. pp. 108–118, May 16, 1889.

Dr. Beard's material was procured by him in the spring of 1888 from the same place as that which supplied Mr. Agassiz and myself, — Black Lake, New York.

In this preliminary notice the author does not devote much attention to the egg membranes. What he says about *the inner egg membrane* coincides with the views which I have expressed. He says that it "is not composed of two layers either in Lepidosteus or in the sturgeon: It is a simple zona radiata, the striæ reaching to the innermost portions of the membrane. The division into two layers, sometimes seen, is the optical effect of thick sections."

I cannot agree with his conclusions regarding the external layer, and am confident that his final paper will not contain proof of the accuracy of his statements. He says: "The *pyriform bodies* are certainly modified cells, each with the remains of a nucleus at its outer end. These modified cells have degenerated into a sort of glue, which causes the excessive stickiness of the newly laid eggs. . . . In the ovarian egg these 'pyriform bodies' are probably nutritive cells to the ovum, for their outer ends near the nuclei contain a number of minute yolk particles."

As far as regards the external layer, it is difficult to conceive how our views, whether morphological or physiological, could have been more divergent.

BIBLIOGRAPHY.

Agassiz, Alexander.

'78[a]. The Development of Lepidosteus. (Presented Oct. 8, 1878.) Proceed. Amer. Acad. Arts and Sci., Vol. XIV. No. 4, pp. 65–76, Pls. I.–V.
1879.

Agassiz, Louis.

'42. *See* Vogt, Carl, '42.

'57. Living Specimens of young Gar-pikes from Lake Ontario. Proceed.
Bost. Soc. Nat. Hist., Vol. VI. (1859), pp. 47, 48. Jan., 1857.

Aubert, Hermann.

'53. Beiträge zur Entwickelungsgeschichte der Fische. Zeitschr. f. wiss.
Zool., Bd. V. Heft 1, pp. 94–102, Taf. VI. 16 Aug., 1853.

Baer, Karl Ernst von.

'35. Untersuchungen über die Entwickelungsgeschichte der Fische, nebst
einem Anhange über die Schwimmblase. 52 pp., 1 Taf. u. mehreren
Holzschn. im Texte. Leipzig: Vogel. 1835.

Balfour, Francis M.

'78[b]. On the Structure and the Development of the Vertebrate Ovary.
Quart. Jour. Micr. Sci., Vol. XVIII. No. 72, pp. 383–438, Pls. XVII.–
XIX. Oct., 1878.

'81. A Treatise on Comparative Embryology. Vol. II., xi + 655 + xxii pp.
429 figs. London: Macmillan & Co. 1881.

Balfour, F. M., and W. N. Parker.

'81. On the Structure and Development of Lepidosteus. (Rec'd Nov. 24,
1881.—*Abstract.*) Proceed. Roy. Soc. London, Vol. XXXIII. No. 217,
pp. 112–119. 8 Dec., 1881.

'82. On the Structure and Development of Lepidosteus. (Rec'd Nov. 24.
—Read Dec. 8, 1881.) Philos. Trans. Roy. Soc. London, 1882, Pt. II.
pp. 359–442, Pls. XXI.–XXIX.

Baumert, Friedr. Moritz.

'53. Chemische Untersuchungen über die Respiration des Schlammpeizgers
(*Cobitis fossilis*). Annalen der Chemie u. Pharmacie (Liebig), Bd.
LXXXVIII. pp. 1–56. 1853.

Beddard, Frank E.

'86[a]. Observations on the Ovarian Ovum of Lepidosiren (Protopterus).
Proceed. Zoöl. Soc. London, for 1886, Pt. III. pp. 272–292, Pls. XXVIII.,
XXIX. 1 Oct., 1886.

Brock, J.
'78· Beiträge zur Anatomie und Histologie der Geschlechtsorgane der Knochenfische. Morph. Jahrb., Bd. IV. Heft 4, pp. 505–572, Taf. XXVIII.,
XXIX. 1878.

Bruch, C.
'55· Ueber die Befruchtung des thierischen Eies u. über die histologische
Deutung desselben. Mainz. 1855.
'55ᵃ· Ueber die Mikropyle der Fische. Zeitschr. f. wiss. Zool., Bd. VII.
Hefte 1 u. 2, pp. 172–175, Taf. IX B. 20 May, 1855.

Buchholz, Reinhold.
'63· Ueber die Mikropyle von Osmerus eperlanus. Arch. f. Anat., Physiol.
u. wiss. Med., Jahrg. 1863, pp. 71–81, Taf. III A., Figs. 1–4. 1863.
'63ᵃ· Nachträgliche Bemerkungen über die Mikropyle von Osmerus eperlanus. Arch. f. Anat., Physiol. u. wiss. Med., Jahrg. 1863, pp. 367–372,
Taf. VIII A. 1863.

Calberla, Ernst.
'78· Der Befruchtungsvorgang beim Ei von Petromyzon Planeri. Zeitschr.
f. wiss. Zool., Bd. XXX. Heft 3, pp. 437–486, Taf. XXVII.–XXIX.
7 March, 1878.

Cunningham, J. T.
'86· On the Mode of Attachment of the Ovum of Osmerus eperlanus. Proceed. Zoöl. Soc. London, for 1886, Pt. III. pp. 292–295, Pl. XXX. (Read
May 4.) 1886.
'86ᵃ· On the Structure and Development of the Reproductive Elements in
Myxine glutinosa, L. Quart. Jour. Micr. Sci., Vol. XXVII. Pt. I. pp.
49–76, Pls. VI., VII. Aug., 1886.

Doyère, M. P. L. N.
'50· Note sur l'Œuf du Loligo media et sur celui du Synguathe. L'Institut, Tom. XVIII. p. 12. 1850.

Eigenmann, Carl H.
'90· On the Egg Membranes and Micropyle of some Osseous Fishes. Bull.
Mus. Comp. Zoöl., Vol. XIX. No. 2, pp. 133–154, Pls. I.–III. 1890.

Eimer, Th.
'72ᵃ· Untersuchungen über die Eier der Reptilien. Arch. f. mikr. Anat.,
Bd. VIII. pp. 216–243, 397–434, Taf. XI., XII., XVIII. 1872.

Gegenbaur, Carl.
'61· Ueber den Bau und die Entwickelung der Wirbelthiereier mit partielle
Dottertheilung. Arch. f. Anat., Physiol. u. wiss. Med., Jahrg. 1861, pp.
491–529, Taf. XI. 1861.

Gréhant, [Nestor].
'69· Recherches physiologiques sur la Respiration des Poissons. Ann. Sci.
Nat., sér. 5, Zool., Tom. XII. pp. 371–382. 1869.

Haeckel, Ernst.
'55· Ueber die Eier der Scomberesoces. Arch. f. Anat., Physiol. u. wiss.
Med., Jahrg. 1855, pp. 23–31, Taf. IV., V. 1855.

Henneguy, Felix.
'88·   Recherches sur le Développement des Poissons osseux. Embryoge-
nie de la Truite. Jour. de l'Anat. Physiol. norm. et path., Tom. XXIV.
No. 5, pp. 413–502; No. 6, pp. 525–617, Pls. XVIII.–XXI. 16 Dec.,
1888, and 20 Feb., 1889.

His, Wilhelm.
'73·   Untersuchungen über das Ei und die Entwickelung bei Knochenfischen.
Leipzig: F. C. W. Vogel. 1873. 4 + 54 pp., 4 Taf., 4to.

Hoffmann, C. K.
'81·   Zur Ontogenie der Knochenfische. Verhandl. d. koninkl. Akad. v.
Wetenschappen, Amsterdam, Deel XXI., 164 pp., 7 Taf. 1881.

Humboldt, A. von, et Provençal.
See PROVENÇAL UND HUMBOLDT.

Jobert.
'77·   Recherches pour servir à l'Histoire de la Respiration chez les Poissons.
Ann. Sci. Nat., sér. 6, Zool., Tom. V. No. 8, 4 pp. 1877.
'78ª.   Recherches Anatomiques et Physiologiques pour servir à l'Histoire de
la Respiration chez les Poissons. Ann. Sci. Nat., sér. 6, Zool., Tom. VII.
No. 5, 7 pp. 5 Aug., 1878.

Kölliker, Albert von.
'57·   Nachweis von Porenkanälchen in den Epidermiszellen von Ammocetes
durch Professor Leuckart in Giessen nebst allgemeinen Bemerkungen über
Porenkanäle in Zellmembranen. Verhandl. physical.-med. Gesellschaft in
Würzburg, Bd. VII. pp. 193–198. 1857.
'58·   Untersuchungen zur vergleichenden Gewebelehre, angestellt in Nizza im
Herbste 1856. Verhandl. physical.-med. Gesellschaft in Würzburg, Bd.
VIII. pp. 1–128, Taf. I.–III. 1858.

Kolessnikow, N.
'78·   Ueber die Eientwickelung bei Batrachiern und Knochenfischen. Arch.
f. mikr. Anat., Bd. XV. Heft 3, pp. 382–414, Taf. XXV. 23 Aug., 1878.

Korschelt, Eugen.
'84·   Ueber die Bildung des Chorions und der Micropylen bei den Insecteneiern. Zool. Anzeiger, Jahrg. VII. Nos. 172, 173, pp. 394–398, 420–425.
21 July, 4 Aug., 1884.
'87·   Zur Bildung der Eihüllen, der Mikropylen und Chorionanhänge bei den
Insekten. (Eing. 10 Aug., 1885.) Nova Acta Leop.-Carol. Deutsch.
Akad. der Naturforscher, Bd. LI. No. 3, pp. 181–252, Taf. XXXV.-
XXXIX. Halle. 1887.

Kowalevsky, A., Ph. Owsjannikow und N. Wagner.
'70·   Die Entwickelungsgeschichte der Störe. Vorläufige Mittheilung. Bull.
Acad. Imp. Sci. St. Pétersbourg, Tom. XIV. col. 317–325, 7 figg. 1870.
'70ª.   *Also in* Mélanges Biol. tirés du Bull. Acad. Imp. Sci. St. Pétersbourg,
Tom. VII. (1869–70), pp. 171–183. 1870.

Kupffer, Carl.
'78ª· Die Entwicklung des Herings im Ei. Jahresber. der Commission zur wissensch. Untersuchung der deutsch. Meere in Kiel für die Jahre 1874 –76, Jahrg. 4–6, pp. 175–226, 4 Taf. Berlin. 1878.

Kupffer, C., und B. Benecke.
'78· Der Vorgang der Befruchtung am Ei der Neunaugen. 24 pp., 1 Taf. Königsberg: Hartung. [1878.]

Lereboullet, Auguste.
'54· Résumé d'un Travail d'Embryologie comparée sur le Développement du Brochet, de la Perche et de l'Écrevisse. Ann. Sci. Nat., sér. 4, Zool., Tom. I. pp. 237–289. 1854.
'61· Recherches d'Embryologie comparée sur le Développement de la Truite, du Lézard, et du Limnée. Première Partie. Embryologie de la Truite commune (Salar Ausoni, Val.; S. fario, L. Bl.). Ann. Sci. Nat., sér. 4, Zool., Tom. XVI. pp. 113–196, Pl. II., III. 1861.

Leuckart, Rudolph.
'53· [Article] Zeugung. Wagner's Handwörterbuch der Physiologie u. s. w., Bd. IV. pp. 707–1000. 1853.
'55· Ueber die Mikropyle und den feinern Bau der Schalenhaut bei den Insecteneiern. Zugleich ein Beitrag zur Lehre von der Befruchtung. Arch. f. Anat., Physiol. u. wiss. Med., Jahrg. 1855, pp. 90–264, Taf. VII.–XI. 1855.

Leydig, Franz.
'52· Beiträge zur mikroskopischen Anatomie und Entwicklungsgeschichte der Rochen und Haie. Leipzig: Engelmann. 1852. iv. + 127 pp., 4 Taf.

Ludwig, Hubert.
'74· Ueber die Eibildung im Thierreiche. Arbeiten aus d. zoolog.-zoot. Institut in Würzburg, Bd. I. pp. 287–510, Taf. XIII.–XV. 1874.

Meckel von Hemsbach, Heinrich.
'52· Die Bildung der für partielle Furchung bestimmten Eier der Vögel, im Vergleich mit dem Graafschen Follikel der Decidua des Menschen. Zeitschr. f. wiss. Zool., Bd. III. Heft 4, pp. 420–434, Taf. XV. 1852.

Moreau, Armand.
'63· Sur l'Air de la Vessie natatoire des Poissons. Comptes Rendus de Paris, Tom. LVII. No. 1, pp. 37–39. 6 July, 1863.
'63ª· Sur l'Air de la Vessie natatoire des Poissons. Comptes Rendus de Paris, Tom. LVII. No. 20, pp. 816–820. 16 Nov., 1863.

Morren, Auguste.
'41· Recherches sur l'Influence qu'exercent et la Lumière et la Substance organique de Couleur verte souvent contenue dans l'Eau stagnante, sur le Qualité et la Quantité des Gaz que celle-ci peut contenir. Ann. de Chim. et de Phys., sér. 3, Tom. I. pp. 456–489. 1841.
'44· Recherches sur les Gaz que l'Eau de Mer peut tenir en Dissolution en

différente Moments de la Journée, et dans les Saisons diverses de l'Année. Ann. de Chim. et de Phys., sér. 3, Tom. XII. pp. 1–56. 1844.

Morris, Charles.

'85. On the Air-bladder of Fishes. Proceed. Acad. Nat. Sci., Philadelphia, 1885, Part II. pp. 124–135. 1885.

Müller, August.

'64. Ueber die Befruchtungs-Erscheinungen im Eie der Neunaugen. Schriften d. k. phys.-ökonom. Gesellsch. zu Königsberg, Jahrg. V. pp. 109–119, Taf. IV. 1864.

Müller, Johannes.

'51. [*See also* JOHANNES MÜLLER, '54b.] Monatsberichte d. Akad. d. Wissenschaften Berlin. 28 April, 1851.

'54. Ueber zahlreiche Porencanäle in der Eicapsel der Fische. Arch. f. Anat., Physiol. u. wiss. Med., Jahrg. 1854, pp. 186–190, Taf. VIII. Figs. 4–7. 1854.

'54ª. Anmerkung des Herausgabers [*on* R. Remak, Ueber Eihüllen und Spermatozoen]. Arch. f. Anat., Physiol. u. wiss. Med., Jahrg. 1854, p. 256.

'54b. Ueber den Canal in den Eiern der Holothurien. Arch. f. Anat., Physiol. u. wiss. Med., Jahrg. 1854, pp. 60–68.

Müller, Wilhelm.

'75. Ueber das Urogenitalsystem des Amphioxus und der Cyclostomen. Jena. Zeitschr., Bd. IX. Heft 1, pp. 94–129, Taf. IV., V. 30 June, 1875.

Owen, Richard.

'66. On the Anatomy of Vertebrates. Vol. I. Fishes and Reptiles. London: Longmans, Green, & Co. 1866. 42 + 650 pp., 452 figures.

Owsjannikow, Ph.

'70ª. Die Entwickelungsgeschichte der Flussneunaugen (Petromyzon fluviatilis). Vorläufige Mittheilung. Mélang. Biolog. tirés du Bull. de l'Acad. Imp. Sci. St. Pétersbourg, Tom. VII. (1869–71), pp. 184–189. 1870.

'85. Studien über das Ei, hauptsächlich bei Knochenfischen. Mém. Acad. Imp. Sci. St. Pétersbourg, sér. 7, Tom. XXXIII. No. 4, 54 pp., 3 Taf. 1885.

Poey, Felipe.

'55. Observations on different Points of the Natural History of the Island of Cuba, with reference to the Ichthyology of the United States. Ann. Lyc. Nat. Hist. New York, Vol. VI., 1858, pp. 133–137. Oct., 1855.

Provençal et A. von Humboldt.

'09. Recherches sur la Respiration des Poissons. Mém. de Phys. et de Chim. de la Soc. d'Arcueil, Tom. II., Paris, pp. 359–404. 1809.

'09ª. *Also in* Journ. de Phys., Tom. LXIX. pp. 261–286. 1809.

Provençal und A. von Humboldt.

'11. Untersuchungen über die Respiration der Fische. Jour. für Chemie u. Physik (Schweigger), Bd. I. Heft 1, pp. 86–121. 6 Feb., 1811. *German translation of* PROVENÇAL ET A. VON HUMBOLDT, '09, *by* Dr. Sigwart.

Ransom, W. H.
'56· On the Impregnation of the Ovum of the Stickleback. Proceed. Roy. Soc. London, Vol. VII. pp. 168–172. 1856.
'67· On the Structure and Growth of the ovarian Ovum in Gasterosteus leiurus. Quart. Jour. Micr. Sci., n. ser., Vol. VII. pp. 1-4, Pl. I. Jan., 1867.
'68· Observations on the Ovum of Osseous Fishes. Philos. Trans. Roy. Soc. London, Vol. CLVII. Pt. II. pp. 431–502, Pls. XV.–XVIII. 1868.

Rathke, Heinrich.
'61· Entwickelungsgeschichte der Wirbelthiere. Mit einem Vorwort von A. Kölliker. Leipzig: W. Engelmann. 1861. 9 + 201 pp., 8vo.

Reichert, Karl Bogislaus.
'56· Ueber die Micropyle der Fischeier und über einen bisher unbekannten, eigenthümlichen Bau des Nahrungsdotters reifer und unbefruchteter Fischeier (Hecht). Arch. f. Anat., Physiol. u. wiss. Med., Jahrg. 1856, pp. 83–124, 141, 142, Taf. II., III., und IV. Figg. 1–4. 1856.

Remak, Robert.
'54· Ueber Eihüllen und Spermatozoen. Arch. f. Anat., Physiol. u. wiss. Med., Jahrg. 1854, pp. 252–255. 1854.

Ryder, John A.
'81ᶜ. Notes on the Development, Spinning Habits, and Structure of the Four-spined Stickleback (Apeltes quadracus). Bull. U. S. Fish Commiss., Vol. I. pp. 24–29. [1881] 1882.
'81ᵉ· Development of the Spanish Mackerel (Cybium maculatum). Bull. U. S. Fish Commiss., Vol. I. pp. 135–172, 4 pls. [1881] 1882.
'82ᶜ. Development of the Silver Gar (Belone longirostris), with Observations on the Genesis of the Blood in Embryo Fishes, and a Comparison of Fish Ova with those of other Vertebrates. Bull. U. S. Fish. Commiss., Vol. I. pp. 283–301, Pls. XIX.–XXI. May 2 and 19, 1882.
'83· On the Thread-bearing Eggs of the Silversides (Menidia). Bull. U. S. Fish Commiss., Vol. III. pp. 193–196. 1883.
'84· A Contribution to the Embryography of Osseous Fishes, with special Reference to the Development of the Cod (Gadus morrhua). Ann. Report U. S. Commissioner of Fish and Fisheries for 1882, XVII. pp. 455–605, Pls. I.–XII.
'84ᵃ. Also separate, with title-page and cover. 149 pp., 12 pls. Washington: Government Printing Office. 1884.
'85· On the Development of Viviparous Osseous Fishes. Proceed. U. S. National Museum, Vol. VIII. Nos. 8–10, pp. 128–155. Pls. VI.–XI. 25 May, 1885.
'86· On the Development of Osseous Fishes, including Marine and Fresh-Water Forms. Extracted from Ann. Report U. S. Commissioner of Fish and Fisheries for 1885 pp. [1]–[116], Pls. I.–XXX. 1886.
'87· [Same as RYDER, '86.] Ann. Report U. S. Commissioner of Fish and Fisheries for 1885, pp. 484–604, Pls. I.–XXX. 1887.

Salensky, W.

'77· Entwickelungsgeschichte des Sterlets. Vorläufige Mittheilung. (Russian.) *In den* Beilagen zu den Protokollen d. 84 u. 89 Sitzung. d. Gesellsch. d. Naturforscher an d. Kaiserl. Universität in Kasan. 34 pp. 1877. (Cited from Hofmann u. Schwalbe, Jahresb. f. 1878, Bd. VII. Entwickelungsgeschichte, p. 213.)

'78· Entwickelungsgeschichte des Sterlets. Vorläufige Mittheilung. (Continuation of W. Salensky, '77.) 5. Postembryonale Entwickelung. (Russian.) *In den* Beilagen zu den Protokollen d. 97 Sitzung d. Gesellsch. d Naturforscher an d. Kaiserl. Universität in Kasan. 21 pp. 1878. (Cited from Hofmann u. Schwalbe, Jahresb. f. 1878, Bd. VII. Entwickelungsgeschichte, p. 213.)

'78ª. Zur Embryologie der Ganoiden. I. Befruchtung u. Furchung des Sterlet-Eies. II. Entwickelungsgeschichte des Skelets beim Sterlet. Zool. Anzeiger, Jahrg. I. Nos. 11–13, pp. 243–245, 266–269, 288–291. Nov. 4, 11, 25, 1878.

'78ᵇ. History of the Development of the Sterlet (Acipenser ruthenus). Pt. I. Embryonic Development. (Russian.) Mém. Soc. Naturalists Imp. Univ. Kasan, Tom. VII. No. 3, pp., 1–226, Pls. I.–IX. 1878.

'79· *Abstr.* of W. Salensky, '77, '78, and '78ª, *by* Mayzel *in* Hofmann u. Schwalbe, Jahresb. f. 1878, Bd. VII. Entwickelungsgeschichte, pp. 213, 218–225. 1879.

'80· History of the Development of the Sterlet (Acipenser ruthenus). Part. II. Post-embryonic Development and Development of the Organs. (Russian.) Mém. Soc. Naturalists Imp. Univ. Kasan, Tom. X. Part II. pp. 227–545, Pls. X.–XIX. Kasan, 1880. (Continuation of W. Salensky, '78ᵇ.)

'81· Recherches sur le Développement du Sterlet (Acipenser ruthenus). Arch. de Biol., Tom. II. fasc. 2, pp. 233–341, Pls. XV.–XVIII. 1881.

Scharff, Robert.

'87· On the Intra-ovarian Egg of some Osseous Fishes. (Rec'd Nov. 17, 1886. — *Abstract.*) Proceed. Roy. Soc. London, Vol. XIV. No. 249, pp. 447–449. 1887.

'87ª. On the Intra-ovarian Egg of some Osseous Fishes· Quart. Jour. Micr. Sci., Vol. XXVIII. pp. 53–74, Pl. V. Aug., 1887.

Schultz, Alexander.

'75· Zur Entwickelungsgeschichte des Selachiereis. Arch. f. mikr. Anat., Bd. XI. pp. 569–582, Taf. XXXV. 1875.

Schultze, Max. Sigmund.

'56· Die Entwickelungsgeschichte von Petromyzon Planeri. Eine von der holländischen Societät der Wissenschaften zu Haarlem i. J. 1856 gekrönte Preisschrift. 50 pp., 8 Taf. 4to. [1856?]

Solger, Bernhard.

'85· Dottertropfen in der intracapsulären Flüssigkeit von Fischeiern. Arch. f. mikr. Anat., Bd. XXVI. Heft 2, pp. 321–334, Taf. XII. Dec. 14, 1885.

Steenstrup, J.

'63· En Iagttagelse af Æg med hornagtige Æghylstre hos Slimaalen (*Myxine glutinosa*, Linn.) med Hensyn til det om denne Fisk udsatte Prissporgsmaal. (An Observation on Eggs with horn-like Egg-case, in the Slime-Eel, Myxine, etc.) Oversigt o. d. kgl. danske Videnskabernes Selskabs Forhandlinger i Aaret 1863, pp. 233–239. [1864?]

Stockman, Ralph.

'83· Die äussere Eikapsel der Forelle. Mittheil. a. d. Embryol. Institut Wien, Bd. II. Heft 3, pp. 195–199. 1883.

Thomson, Allen.

'59· [Article] Ovum *in* The Cyclopædia of Anat. and Physiol., edited by Robert B. Todd, Vol. V. (Suppl. Vol.), 1859, pp. 1–80 and [81]–[142].

Note.—Part I., pp. 1–80, was issued in 1852; Part II., pp. [81]–[142], in 1855.

Vogt, Carl.

'42· Embryologie des Salmones. Neuchatel. 1842. 6 + 328 pp., 8vo. Avec Atlas, fol. obl. de 7 pls.

Being Tome I. of L. Agassiz, Histoire Naturelle des Poissons d'Eau douce de l'Europe Centrale.

Vogt, Carl, et S. Pappenheim.

'59· Recherches sur l'Anatomie comparée des Organes de la Génération chez les Animaux Vertèbres. (Déposé dans les Archives de l'Acad. le 30 Dec., 1845.) Ann. Sci. Nat., sér. 4, Zool, Tom. XI. pp. 331–369, Pl. XIII.; Tom. XII. pp. 100–131, Pls. II., III. 1859.

Waldeyer, Wilhelm.

'70· Eierstock und Ei. Ein Beitrag zur Anatomie u. Entwickelungsgeschichte der Sexualorgane. Leipzig: W. Engelmann. 1870. 8 + 174 pp., 6 Taf. 8vo.

Wilder, Burt G.

'76· Notes on the North American Ganoids, Amia, Lepidosteus, Acipenser, and Polyodon. Proceed. Amer. Assoc. Adv. Sci., Vol. XXIV B, Detroit Meeting, pp. 151–196, Pls. I.–III. 1876.

'77· Gar-Pikes, Old and Young. Popular Sci. Monthly, Vol. XI. Nos. 61, 62, pp. 1–12, 186–195, 10 figures. May and June, 1877.

EXPLANATION OF FIGURES.

All the figures were drawn with the aid of the camera lucida, and were made from preparations of *Lepidosteus osseus.*

ABBREVIATIONS.

cap.	Head of villus.	*pd.*	Stalk of villus.
c.-t. cp.	Connective-tissue corpuscle.	*rx.*	Root of villus.
fus. mat.	Maturation spindle.	*st. vil.*	Villous layer of egg membrane.
gran.	Granulosa.	*th. fol.*	Membrana propria of theca folliculi.
m py. can.	Micropylar canal.	*vac.*	Vacuole.
m py. cl.	Micropylar cell.	*vit.*	Vitellus.
nl.	Nucleus.	*vs. g.*	Germinative vesicle.
nl. gran.	Nucleus of granulosa cell.	*z. r.*	Zona radiata.

PLATE I.

All the figures of this plate were made from material that had not been hardened, and all the figures except Figs. 7 and 11 are magnified 472 diameters.

Fig. 1. A surface view of a small portion of the villous layer of egg membrane.

" 2. The appearance presented by the same layer when the region near the boundary between it and the zona radiata is in focus. Some of the roots of the villi are seen between the stalks.

" 3. The zona radiata when the focusing is a little below its outer surface. A few pore-canals are occupied by roots of villi and appear darker.

" 4. A portion of a radial section after being treated with weak hydrochloric acid. Two of the villi much more elongated than the others.

" 5. A radial section of a fresh egg-shell, showing the relative thickness of the zona radiata and the villous layer.

" 6. A portion of the same with the villous layer removed, but leaving its roots in the spiral pore-canals. Examined in glycerine.

" 7. Portions of the villous layer removed from the zona radiata and much swollen in water. The roots appear like a fine fringe. × 145.

" 8. The appearance of the pore-canals after treatment with hydrochloric acid. The most of them, especially toward the margin of the figure, should have been drawn larger but faint. A few are conspicuous from the presence of roots of villi.

" 9. *a* to *h* and *j* to *n*, isolated villi in various stages of elongation after imbibition of very dilute hydrochloric acid; *i*, after soaking in water only.

" 10. A fragment of the zona radiata deprived of the villous layer and treated with weak hydrochloric acid until all the pore-canals except those containing villous roots had disappeared. The zona, having become soft, was partly crushed, so that the roots were seen obliquely, the ends toward the top of the plate being the ones torn from the stalk.

" 11. Optical radial section of the micropylar region of a fresh egg, the wall of the membrane beyond the micropyle being projected on the same plane. × 145.

" 11ᵃ. Optical cross-section of the same, at plane *a* of Fig. 11.

" 11ᵇ. Optical cross-section of the same at plane *b* of Fig. 11.

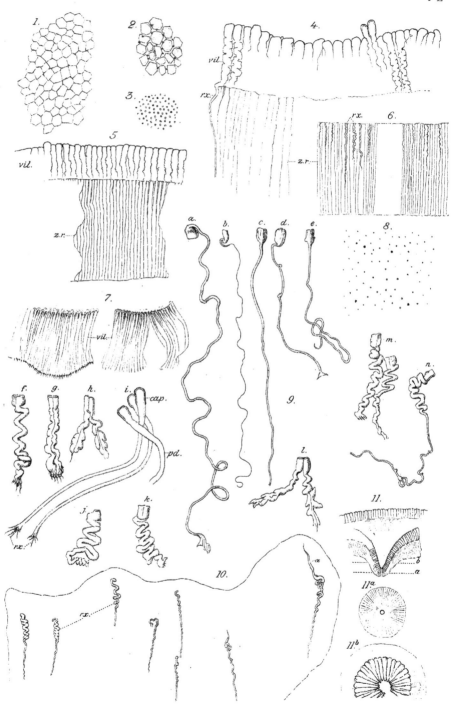

ABBREVIATIONS.

cap.	Head of villus.	*pd.*	Stalk of villus.
c.-t. cp.	Connective-tissue corpuscle.	*rx.*	Root of villus.
fus. mat.	Maturation spindle.	*st. vil.*	Villous layer of egg membrane.
gran.	Granulosa.	*th. fol.*	Membrana propria of theca folliculi.
m py. can.	Micropylar canal.	*vac.*	Vacuole.
m py. cl.	Micropylar cell.	*vit.*	Vitellus.
nl.	Nucleus.	*vs. g.*	Germinative vesicle.
nl. gran.	Nucleus of granulosa cell.	*z. r.*	Zona radiata.

PLATE II.

All the figures of this plate are magnified 515 diameters.

Fig. 1. Portion of a radial section through the shell of a deposited egg preserved in 5 per cent potassic bichromate. Stained in carminic acid dissolved in 80 per cent alcohol. Mounted in benzole-damar.

" 2, 3. Isolation preparations of villi from a mature egg pressed from female. The egg was let fall into 90 per cent alcohol, afterwards soaked in water, and then stained in acetic-acid carmine. Examined in glycerine. The beads were of a much brighter rose-color than is shown in the lithographic print. In Fig. 3, *c*, the villi are seen edgewise; in *a*, *b*, and *d*, flatwise.

" 4–6. Sections through villi of a mature ovarian egg which was preserved in 0.25 per cent chromic acid forty-eight hours, washed in water six hours, and further hardened in grades of alcohol. Stained in alcoholic borax-carmine (Grenacher). In Figs. 4, 5, the ends of the villi are seen; in Fig. 6, the sides.

" 4 Sections of stalks from the pole opposite the micropyle.

" 5, 6. Sections of stalks from near the micropyle.

" 7. Radial section of an egg preserved in Perenyi's fluid (4½ hours) followed by alcohol and stained in alcoholic borax-carmine, showing an inner portion of the zona radiata partly detached from the outer portion. It is not a separate membrane.

" 8. Radial section through zona and villous layer of an egg preserved in 5 per cent potassic bichromate, stained in carminic acid in 80 per cent alcohol, and mounted in benzole-damar.

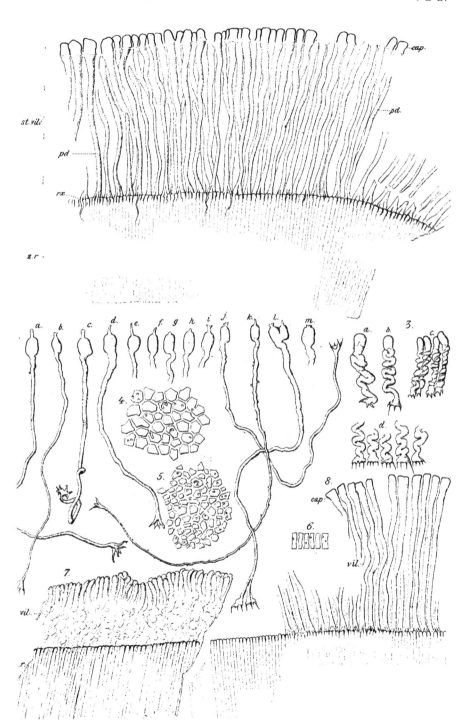

ABBREVIATIONS.

cap.	Head of villus.	*pd.*	Stalk of villus.
c.-t. cp.	Connective-tissue corpuscle.	*rx.*	Root of villus.
fus. mat.	Maturation spindle.	*st. vil.*	Villous layer of egg membrane.
gran.	Granulosa.	*th. fol.*	Membrana propria of theca folliculi.
m py. can.	Micropylar canal.	*vac.*	Vacuole.
m py. cl.	Micropylar cell.	*vit.*	Vitellus.
nl.	Nucleus.	*vs. g.*	Germinative vesicle.
nl. gran.	Nucleus of granulosa cell.	*z. r.*	Zona radiata.

PLATE III.

All the figures of this plate were drawn from the shell of an egg preserved in cold corrosive sublimate (4 hours) followed by alcohol, stained in Kleinenberg's hæmatoxylin, sectioned in paraffine, and mounted in benzole-damar. All except Fig. 3 magnified 515 diameters.

Fig. 1. Radial section, the heads of some of the villi broken off.

" 2. Tangential section through the heads, at A of Fig. 1.

" 3. Similar section of four heads more highly magnified to show the deeply stained peripheral portion. × 750.

" 4. Tangential section through the middle region (B of Fig. 1) of the stalk.

" 5. Section parallel to preceding through the region of the roots of the villi (C of Fig. 1). The lower portion of the figure cuts through a deeper part of the membrane (zona) than the upper portion does. The middle portion shows the branching roots of the villi as they enter the pore-canals.

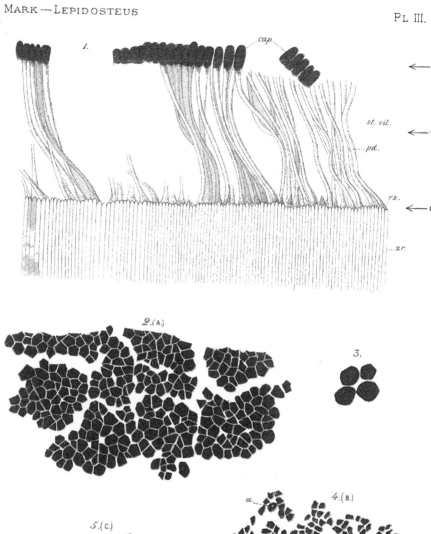

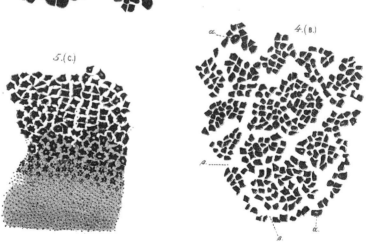

ABBREVIATIONS.

cap.	Head of villus.	*pd.*	Stalk of villus.
c.-t. cp.	Connective-tissue corpuscle.	*rx.*	Root of villus.
fus. mat.	Maturation spindle.	*st. vil.*	Villous layer of egg membrane.
gran.	Granulosa.	*th. fol.*	Membrana propria of theca folliculi.
m py. can.	Micropylar canal.	*vac.*	Vacuole.
m py. cl.	Micropylar cell.	*vit.*	Vitellus.
nl.	Nucleus.	*vs. g.*	Germinative vesicle.
nl. gran.	Nucleus of granulosa cell.	*z. r.*	Zona radiata.

PLATE IV.

Fig. 1. Radial section through the micropyle and micropylar funnel, showing the micropylar cell and a portion of the maturation spindle of an egg "stripped" from the fish, preserved in 0.5 per cent chromic acid (5 hours) followed by washing in water, and hardened in alcohol. Stained in picrocarmine. × 515.

" 2. The second section preceding that shown in Figure 1, and passing nearly through the middle of the maturation spindle. × 515.

" 3. View of the animal pole of an egg preserved in Merkel's fluid. The germinal disk was rather more than half as broad as the diameter of the egg, and its outline should have been represented more distinctly by the lithographer; it was of a yellowish color, but much lighter than the rest of the egg. The micropylar funnel is seen exactly over the centre of the disk. × about 10.

" 4. View of the micropylar funnel and contained micropylar cell of the egg, a section of which is shown in Figure 1. × 158.

" 5. Radial section through the micropylar region and germinative vesicle of an ovarian egg preserved in alcohol. A portion of the granulosa still adheres to the outer surface of the villous layer. Stained in alcoholic borax-carmine. × 158.

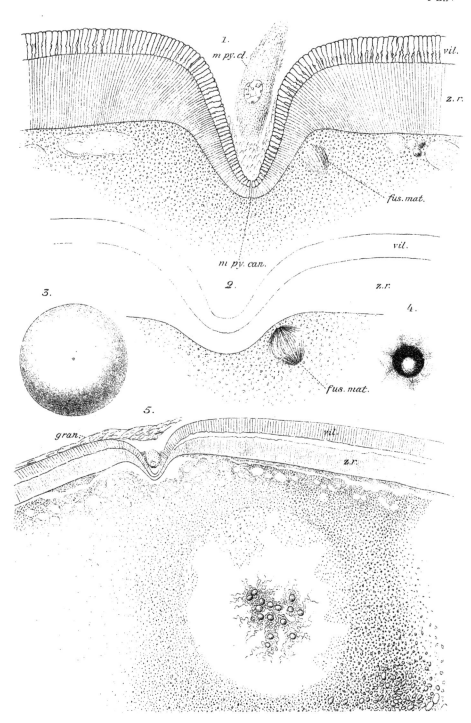

ABBREVIATIONS.

cap.	Head of villus.	*pd.*	Stalk of villus.
c.-t. cp.	Connective-tissue corpuscle.	*rx.*	Root of villus.
fus. mat.	Maturation spindle.	*st. vil.*	Villous layer of egg membrane.
gran.	Granulosa.	*th. fol.*	Membrana propria of theca folliculi.
m py. can.	Micropylar canal.	*vac.*	Vacuole.
m py. cl.	Micropylar cell.	*vit.*	Vitellus.
nl.	Nucleus.	*vs. g.*	Germinative vesicle.
nl. gran.	Nucleus of granulosa cell.	*z. r.*	Zona radiata.

PLATE V.

Fig. 1. Radial section through germinative vesicle and micropylar funnel. Owing to a distortion of the section, the curvature of the part of the membrane shown is less than it should be. The finely granular and vacuolated portion of the yolk (*vac.*) beneath the germinative vesicle is in the centre of the egg. The egg was from an ovary which was hardened in 0.25 per cent chromic acid and stained in alcoholic borax-carmine. × 72.

" 2. The section following that shown in Figure 1, more highly magnified. × 335.

" 3. Radial section through villi and granulosa of an ovarian egg preserved in alcohol and stained in alcoholic borax-carmine. × 515.

" 4. Surface view of a portion of the granulosa from the same egg as that of Figuré 3. × 515.

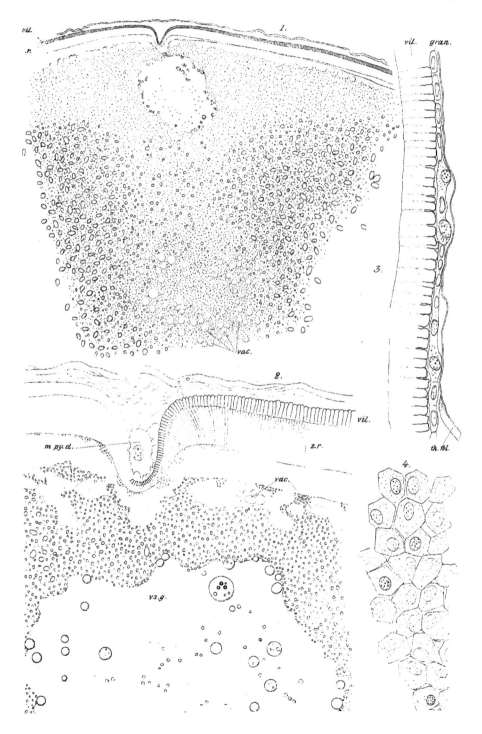

ABBREVIATIONS.

cap.	Head of villus.	pd.	Stalk of villus.
c.-t. cp.	Connective-tissue corpuscle.	rx.	Root of villus.
fus. mat.	Maturation spindle.	st. vil.	Villous layer of egg membrane.
gran.	Granulosa.	th. fol.	Membrana propria of theca folliculi.
m py. can.	Micropylar canal.	vac.	Vacuole.
m py. cl.	Micropylar cell.	vit.	Vitellus.
nl.	Nucleus.	vs. g.	Germinative vesicle.
nl. gran.	Nucleus of granulosa cell.	z. r.	Zona radiata.

PLATE VI.

All figures on this plate are magnified 515 diameters.

Fig. 1. Radial section through the micropylar canal, somewhat oblique to its axis. The egg was preserved in 5 per cent potassic bichromate and stained in carminic acid dissolved in 80 per cent alcohol.

" 2-4. Three successive tangential sections through the bottom of the micropylar funnel and the micropylar canal of an egg stripped from the fish preserved in 90 per cent alcohol, and stained in alcoholic borax-carmine. The zona radiata is closely enveloped by the yolk.

" 5-8. Tangential sections through the deeper portions of the micropylar funnel of an ovarian egg hardened in 0.25 per cent chromic acid and stained in alcoholic borax-carmine. In Figures 5 and 6 the sections pass through the deep portion of the zona radiata which is not infolded to form the funnel, but in Figures 7 and 8 only that portion of the zona is cut which projects as a conical elevation into the substance of the yolk. Only alternate sections were drawn.

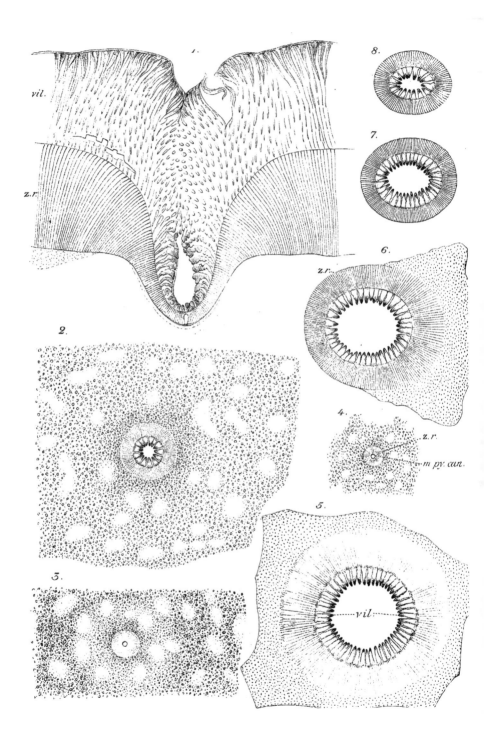

ABBREVIATIONS.

cap.	Head of villus.	*pd.*	Stalk of villus.
c.-t.cp.	Connective-tissue corpuscle.	*rx.*	Root of villus.
fus. mat.	Maturation spindle.	*st. vil.*	Villous layer of egg membrane.
gran.	Granulosa.	*th.fol.*	Membrana propria of theca folliculi.
m py. can.	Micropylar canal.	*vac.*	Vacuole.
m py. cl.	Micropylar cell.	*vit.*	Vitellus.
nl.	Nucleus.	*vs. g.*	Germinative vesicle.
nl. gran.	Nucleus of granulosa cell.	*z. r.*	Zona radiata.

PLATE VII.

Fig. 1. Radial section through the granulosa plug which fills the micropylar funnel. From an ovarian egg preserved in 0.25 per cent chromic acid, and stained in picrocarminate of ammonia. × 515.

" 2. Micropylar cell and outlines of the egg membranes in the region of the micropylar funnel, from an ovarian egg preserved in alcohol. Radial section. × 515.

" 3. Four spermatozoa, dried on the slide. × 472.

" 4. Micropylar funnel; *optical* cross-section as seen from the yolk side of the egg membranes; showing the oval form of the funnel which is sometimes met with. × 515.

" 5. Section of an ovarian egg through the germinative vesicle. Only one membrane besides the granulosa present; it is the villous layer. Preserved in 0.25 per cent chromic acid (48 hours). Stained in alcoholic borax-carmine. × 158.

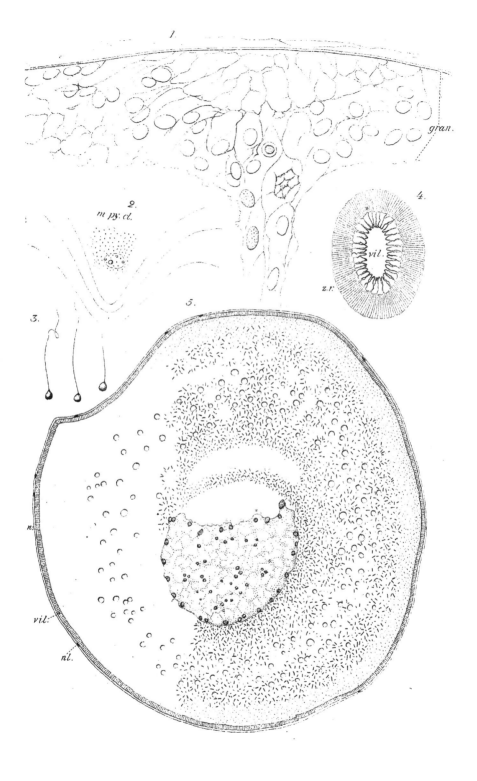

ABBREVIATIONS.

cap.	Head of villus.	*pd.*	Stalk of villus.
c.-t. cp.	Connective-tissue corpuscle.	*rx.*	Root of villus.
fus. mat.	Maturation spindle.	*st. vil.*	Villous layer of egg membrane.
gran.	Granulosa.	*th. fol.*	Membrana propria of theca folliculi.
m py. can.	Micropylar canal.	*vac.*	Vacuity.
m py. cl.	Micropylar cell.	*vit.*	Vitellus.
nl.	Nucleus.	*vs. g.*	Germinative vesicle.
nl. gran.	Nucleus of granulosa cell.	*z. r.*	Zona radiata.

PLATE VIII.

Figures 1 and 2 are from sections of an ovarian egg about 0.4 mm. in diameter which was hardened in chromic acid. × 510.

Fig. 1. Part of the peripheral portion of a radial section in which the earliest observed trace of the villous layer has made its appearance. The membrana propria of the theca and the follicular epithelium are artificially separated from the yolk and villous projections.

" 2. Tangential section from the same egg. The section embraces connective-tissue cells of the stroma, as well as follicular epithelium, and has also cut off a portion of the periphery of the yolk, with its villous projections, which last give it a dotted appearance. The nuclei of the epithelium are often lobed. *vac.* indicates vacuities evidently due to depressions in the surface of the yolk, not to vacuoles in its substance.

" 3. Portion of a section which, owing to the wrinkled condition of the surface of the egg, affords a surface view of the granulosa, as well as a radial section and surface view of the villous layer. Some of the detached villi are seen at one side. The nuclei of the granulosa cells still have irregular lobed forms. Chromic acid preparation of an egg about 0.6 mm. in diameter. × 510.

" 4. View of the villi as seen from the surface of an egg after it has lain for some time in water. × 472.

" 5. Amber-colored bodies found at the outer surface of the villous layer of the egg membrane. × 472.

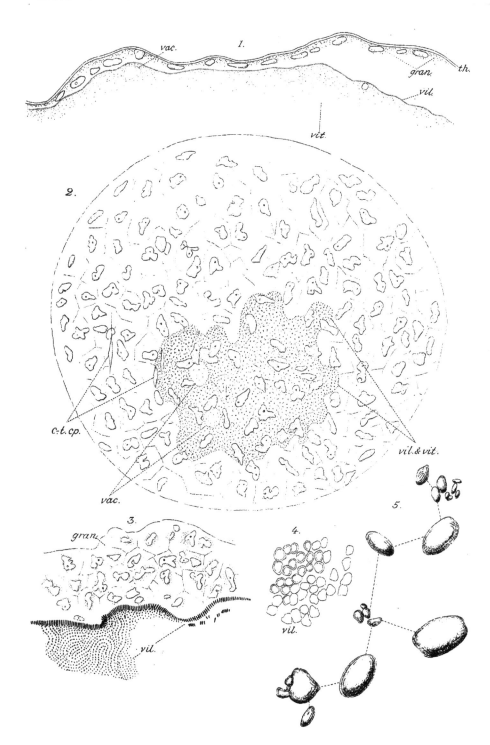

ABBREVIATIONS.

cap.	Head of villus.	*pd.*	Stalk of villus.
c.-t. cp.	Connective-tissue corpuscle.	*rx.*	Root of villus.
fus. mat.	Maturation spindle.	*st. vil.*	Villous layer of egg membrane.
gran.	Granulosa.	*th. fol.*	Membrana propria of theca folliculi.
m py. can.	Micropylar canal.	*vac.*	Vacuole.
m py. cl.	Micropylar cell.	*vit.*	Vitellus.
nl.	Nucleus.	*vs. g.*	Germinative vesicle.
nl. gran.	Nucleus of granulosa cell.	*z. r.*	Zona radiata.

PLATE IX.

Fig. 1. Section of an ovarian egg about 0.6 mm. in diameter through the germinative vesicle. The villous layer is at all points in contact with the yolk; but it is separated from the granulosa at intervals. The egg was hardened in 0.25 per cent chromic acid and stained in alcoholic borax-carmine. × 158.

" 2. Portion of a radial section through a mature ovarian ovum, hardened in 0.25 per cent chromic acid, showing the penetration of the roots of the villi into the pore-canals of the zona radiata. × 515.

" 3, 4. Radial sections of ovarian eggs preserved in alcohol, showing stages in the formation of the villous layer. The eggs were somewhat more than 0.5 mm. in diameter, and were stained in alcoholic borax-carmine. × 510.

" 5. A portion of Figure 1 enlarged. The outlines of the granulosa cells, especially on the side toward the villi, are much too sharp. × 515.

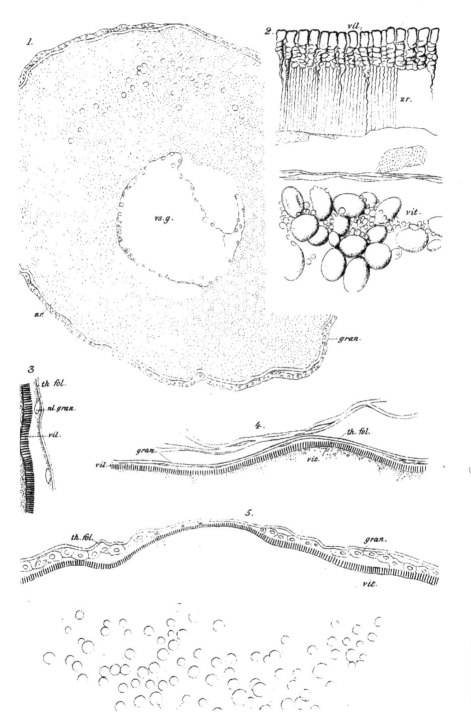

No. 2. — On the Egg Membranes and Micropyle of some Osseous Fishes. By CARL H. EIGENMANN.[1]

At the suggestion of Dr. E. L. Mark, I undertook the study of the development of the micropyle and egg membranes in some of the bony fishes.

The eggs of the following species were examined : Amiurus catus, Tachisurus sp. (?), Catostomus teres, Notemigonus chrysoleucus, Carassius auratus, Clupea vernalis, Alosa sapidissima, Fundulus heteroclitus, F. diaphanus, Apeltes quadracus, Pygosteus pungitius, Lepomis megalotis, Morone americana, Esox reticulatus, Anguilla anguilla rostrata, Cyclogaster [2] lineatus, Gadus morrhua, and Hippoglossoides platessoides. In many of these species the eggs were not in a condition favorable for tracing the development of the micropyle or even the membranes. My account will be confined to the eggs of Amiurus catus, Notemigonus chrysoleucus, Clupea vernalis, Fundulus heteroclitus, Pygosteus pungitius, Perca americana, Morone americana, Esox reticulatus, and Cyclogaster lineatus.

I am indebted to Dr. Mark for the use of his manuscript abstracts of the papers on egg membranes published before 1881.

It has long been known that fish ova are provided with a membrane, the zona radiata. The eggs of certain fishes have, in addition to and outside of the zona radiata, a second membrane which bears in some cases long filaments, in others short processes which serve to attach the eggs to foreign bodies.

Fundulus heteroclitus and F. diaphanus.

The fact that the eggs of some fishes are provided with long filaments was first noted by Haeckel ('55). He found them on the eggs of many species of Scomberesocidæ, but mistook their position, describing them as thin fibres lying inside the egg membrane (zona radiata). A connection of the fibres with cells could not be traced.

[1] Contributions from the Zoölogical Laboratory of the Museum of Comparative Zoölogy at Harvard College, under the Direction of E. L. Mark. — No. XVI.

[2] Liparis of authors.

Kölliker ('58) corrected the mistake made by Haeckel as to the position of the filaments.

Hoffmann ('81) found filaments on the eggs of Heliasis, Gobius, and Blennius.

Ryder ('82ᶜ) described the filaments of the eggs of Belone longirostris, and, in passing, mentioned the probability of their existence in the eggs of Mugil albula. He afterwards ('86ᵃ) found them on the eggs of Fundulus heteroclitus, and has also ('83) shown that the eggs of Menidia (Chirostoma) notata are provided with four of them.

I have examined eggs of Fundulus heteroclitus at intervals of about two weeks from October, 1887, till June, 1888. The eggs undergo scarcely any change between October and April. A series taken between April 1st and June 1st shows all the stages covered by the longer period. The filaments can best be studied in fresh material. They appear in the form of hyaline threads, which are more highly refractive than any other part of the egg membrane. In an ovary of October 27 there were filament-bearing eggs in three stages of development.

In the smallest eggs — about 0.16 mm. in diameter — in which filaments can be seen (Plate I. Fig. 1) they appear as hyaline dots, or as conical bodies with rounded bases, uniformly distributed over the entire surface. They either lie wholly below the granulosa, or the tips of the longer ones may lie in between the granulosa cells (Fig. 8). In this stage the diameter of the threads is much greater than the thickness of the membrane, which can scarcely be distinguished in sections. I was not able to discover sheaths enveloping the filaments such as Haeckel describes for the Scomberesocidæ. In other slightly larger eggs belonging to this same stage of development (Figs. 2–4, 6) the filaments are no longer conical, but appear in the fresh condition as short, curved threads equally blunt at both ends.

In the second stage, the eggs being intermediate in size between those just mentioned and the largest, the threads (Fig. 5) are of about the same thickness as those of the earlier stage, but they are much longer, and taper near the free end. They do not seem to be closer together than in the smaller eggs. The filaments are bent in a more or less regular manner, first to one side and then to another. On stained sections it was to be seen that the threads usually follow the margins of the granulosa cells, and that they are correspondingly curved (Fig. 6).

On the largest eggs — about 0.4 mm. in diameter — the filaments are much longer, and cover about as much of the surface of the egg as they leave exposed. They are so long and so tortuous that it is almost impossible to follow a single filament throughout its whole length. It often happens (Fig. 10) that several filaments are parallel to each other for a considerable distance. In sections the filaments are found to lie in between the bases of the granulosa cells, and also to rise between these cells nearly to their outer surfaces.

In the ripe ovarian eggs the basal ends of the filaments pass directly through the granulosa layer, and the greater part of the filament thus comes to lie between the outer portions of the granulosa cells, or even quite outside of them (Fig. 9). The regularity of their windings cannot be seen as well as in eggs of the third size. The filaments are of varying lengths, but most of them are several times as long as the diameter of the egg. The distances between filaments are not materially altered during the growth of the egg; but since the surface of the egg increases during its development to mány times the size which it had when the filaments first appeared, the total number of the latter must also be greatly increased. The earliest stages in the formation of new filaments would be difficult to find after the egg has reached its second stage, because they would be hidden by the larger filaments.

In ripe eggs forced from the ovary, the filaments extend out from the egg for some distance, and then form a network, several filaments deep, over the whole surface (Fig. 11).

Concerning the origin of the filaments it may be said that they do not have any connection with the granulosa cells at any stage of their growth (Figs. 3, 4, 6, 8). In tangential sections it is seen that they arise at places corresponding to the boundaries between two or three cells. In a ripe egg examined in the fresh state under pressure (Fig. 12) indistinct processes are seen to radiate from the base of each filament, forming a stellate figure. In no case, either in fresh specimens or sections, could the filaments be traced into the substance of the zona radiata. They are outgrowths from a thin membrane which lies outside of the zona and is formed before the latter, not processes of the zona itself.

When the filaments first make their appearance, the egg membrane, as stated above, is much thinner than the diameter of a filament, and the granulosa cells are lens-shaped, barely touching by their margins (Fig. 8). In the largest eggs found in the ovary of April 2d, the granulosa was about 8 μ thick, but the egg membrane had only reached the thickness of 2 μ. That it is radially striate is rather to be inferred than directly seen. In places the outer surface shows slight elevations at regular distances, which I believe to be prolongations of granulosa cells sunk into the radial canals (Fig. 13). While the largest eggs of April 2d were only about 0.4 mm. in diameter, and therefore scarcely exceeded in size those of October and November, the largest ovarian eggs of May 2d measured over 0.8 mm.   Between May 1 and June 1, — by which time the eggs have reached their full size, — the growth is still more rapid.   The egg membrane of early May eggs measures about 6.5 μ in thickness, and has distinct pore-canals.

There exists an exceedingly thin outer membrane overlying the zona radiata.   It was discovered in the examination of fresh ripe eggs, in which the striation of the zona itself could be seen much better than in sections of hardened eggs.   In one instance, in which the zona of a fresh ripe egg was ruptured, this overlying membrane was left intact. It is with this membrane that the bases of the filaments are continuous.

In view of this condition in Fundulus, and of the fact that other process-bearing eggs (Cyprinidæ and Gasterosteidæ) possess a thin outer membrane, it would be interesting to re-examine the eggs of the Scomberesocidæ, of Menidia, and of Mugil to find whether they do not also possess this structure.

The outer surface of the fresh ripe egg of Fundulus heteroclitus shows a network of lines (Fig. 7).   This appearance is doubtless due to the presence of superficial ridges, which in radial sections have the appearance of minute projections fitting in between the bases of the granulosa cells (Fig. 9).   Where two or more lines meet, there is a thickening. The whole arrangement bears a superficial resemblance to the appearance presented near the surface of the zona in the perch (Fig. 31).   In the case of the latter, however, the thickenings correspond in position to filaments, each of which corresponds to the middle of a granulosa cell, whereas in Fundulus heteroclitus the thickenings correspond in position to the boundaries between granulosa cells.   From the position of the filaments in Fundulus it is probable that, like the ridges, they are outgrowths of the outer structureless egg membrane.   It is evident from what has been said that there is a fundamental difference between the filaments found in Perca and those in Fundulus.   In Perca they owe their origin to the granulosa, and are formed after the zona has nearly reached its full growth ; in Fundulus, on the contrary, they owe their origin to the activity of the egg itself, and they begin to be formed before the zona.

Pygosteus pungitius.

After Haeckel had described the long filaments peculiar to the eggs of the Gobiesocidæ, Kölliker ('58) described external appendages in the eggs of Abramis brama, Chondrosteus nasus, Squalinus argenteus, Cobitis barbatula, Gobio fluviatilis, Cyprinus rufus, and Gasterosteus pungitius. In all these species he found the appendages inserted in a very thin membrane, which ultimately lies just outside the zona radiata and which makes its appearance before the latter.

The most important paper on Pygosteus is that of Ransom ('67). He studied Gasterosteus pungitius and G. leiurus, and found that the eggs of the two species do not differ greatly. He says that in the oviduct the eggs are surrounded by a viscid layer, and that the zona radiata lies below this layer. The zona is in contact with the yolk except in ripe eggs, in which a thin homogeneous membrane covers the yolk and follows the constrictions at the time of cleavage. The micropyle and the dotted appearance of the egg membrane were first made out in eggs $\frac{1}{140}''$ thick, and in eggs $\frac{1}{200}''$ in diameter the membrane could be separated from the yolk. The button-shaped processes can be made out in eggs somewhat less than $\frac{1}{140}''$ (0.17 mm.) in diameter. They are attached to the outer surface of the yolk-sac by a bright, highly refractive point. In the case of the smallest ova there are on an average seventy buttons, in that of the largest two hundred and seven. They serve to attach the egg to foreign substances. Ransom describes and figures the micropyle.

Owsjannikow ('85) found that in ovarian eggs the granulosa cells cover the micropyle. In fully grown eggs only a single membrane is present, while in the younger ones the zona seems differentiated into two layers, owing to the fact that the zona is laid down by successive additions. The pores do not appear till the membrane has attained considerable thickness, and they are then much finer than in the ripe egg. The mushroom-shaped processes are maintained by him to be cells that possess nuclei which are colored red with carmine. From the base of the process a thread can be traced into the zona radiata. In young eggs the processes consist principally of a nucleus attached to a filament. He does not believe that they are derived from the zona, but thinks they come from the granulosa ; why he thinks so is not stated. Inside the zona he has found the zonoid layer of His.

I have examined ovarian eggs of fishes taken in November, December, and April. A few days after the spawning, in early April, the ovaries contain a considerable number of eggs (about 0.55 mm. in diameter) in which the formation of the yolk is well advanced. These are evidently destined to be laid before the recurrence of the next annual spawning season, for they are much larger than any of the ovarian eggs found in December. These eggs show no signs of degeneration, and their presence can therefore hardly be explained as due to their failure to pass off with the first lot of eggs laid; nor can they be eggs which properly belong to the first set of spawn, as their size in comparison with that of the mature eggs (1.1 mm.) sufficiently proves. Therefore I believe that, as Ransom has inferred, these fishes deposit eggs more than once during the season.

The ovaries are most available for study after the first set of eggs are deposited. As in the case of Fundulus heteroclitus, all stages of growth are shown in ovaries taken during a period extending from one or two months before till a short time after spawning, — the months of March and April.

In eggs 0.15 mm. (Plate I. Fig. 14) or more in diameter there are two membranes, — an outer more highly refractive, and an inner striated one. In many sections the two are artificially separated (Fig. 15^a). In ripe eggs the outer membrane had either entirely disappeared, or its structure had become so much like that of the true zona that the two could not be distinguished from each other. Their total thickness is from 15 to 18 μ. In many sections of ripe eggs an outer layer, much thicker than the outer layer seen in the earlier stages of development, was in places separated from the rest of the zona. If it represents the outer membrane of the earlier stage, then the latter must undergo a great change in its later development, for it is now much thicker, and is traversed by the same pore-canals as the deeper portion.

The rivet-shaped processes which are found in the region of the micropyle are inserted, as Kölliker says, in the thin membrane which lies outside the zona, and which is formed before the latter makes its appearance. They take a much deeper stain than the thin membrane, but I have seen nothing which would warrant one in claiming that they contain each a nucleus. The smallest egg in which these processes could be seen had a diameter of about 0.14 mm.; only a single thin, structureless membrane was to be made out in this stage. The largest eggs examined had a diameter of about 1 mm.

When the processes make their appearance, the granulosa is so thin that it is difficult to determine from surface views whether they lie above or below it; but radial sections show that they lie below. There is no such constant relation between the processes and individual cells of the granulosa as to suggest the origin of the former from the latter; but at a later stage the heads of the rivets occupy nearly the same plane as the nuclei of the granulosa cells (Fig. 16), and therefore appear to have an intimate connection with the granulosa cells. When the granulosa is torn from the egg membranes, as, owing to the shrinkage of the egg, it frequently is, the processes no longer show the same sharp outer margins. Their edges are often frayed, and are not stained as deeply as when the granulosa and the membranes are in their normal relations to each other. With the separation of the granulosa the thin outer membrane is sometimes torn (Fig. 15ᵃ); and whether torn or not, it is often separated from the inner membrane. This may be due to the fact that the processes are from the beginning adhesive, and have thus acquired an intimate secondary relation to the cells of the granulosa. In such sections it can be clearly seen that the rivet-shaped processes are joined to the outer membrane and not to the zona, though their bases have projected into the zona for a greater or less distance. When the granulosa is torn from the egg membranes, the processes always, even in the smallest eggs, remain attached to the membranes rather than to the granulosa. I have been able to find neither the nuclear structure within nor the prolongations from these processes which Owsjannikow has described.

I have not succeeded in finding the micropyle in eggs that were much less than 0.4 mm. in diameter; in such the zona has an average thickness of about 5 μ. The portion immediately surrounding the micropyle shows a considerahle local thickening. Owing to the variation in the thickness of the zona in different regions of the same egg, and to the inconstancy of the position of the micropyle in relation to this variation, it sometimes happens that the zona at the micropylar region has already reached a thickness of 10 or 11 μ.

It is a noticeable fact, that at this earlier stage the micropyles of nearly all the eggs were cut radially when the sections were made in planes perpendicular to the axis of the ovary. Furthermore, the micropyles uniformly lie in the half of the egg opposite the side of attachment.

In the vicinity of the micropyle the zona becomes thickened by the elevation of its *outer* surface, the deeper surface undergoing no change of direction. At a distance of about 10 μ on either side of the micropylar canal it attains its greatest thickness, and then its outer contour curves inward until it becomes continuous with the wall of the micropylar canal. The inner end of the canal is sometimes slightly enlarged (Figs. 19–22).

At this stage the pore canals of the zona radiata do not seem to be modified in direction in the region of the micropyle; they are all radially arranged. The outer membrane could not be distinguished in this region ; it probably is entirely wanting in the area immediately surrounding the micropyle. The granulosa cells are two or three layers deep in the vicinity of the micropyle, and a single cell larger than the others is always to be found directly above the cana$_1$. It usually sends a prolongation into the canal itself (Figs. 18, 21).

In eggs about to be laid, the greatest thickness of the zona in the vicinity of the micropyle is approximately 24 μ, and the thickening in this region is not so conspicuous as at the earlier stage. The zona bends inward slightly, so that its inner surface no longer forms a simple curve. The micropylar passage through the zona presents three regions : a shallow funnel-shaped depression, which occupies the outer third of the layer ; a narrower tubular portion, which is a prolongation of the bottom of the funnel, and is rounded at its lower end ; and finally a very narrow canal, which traverses the inner sixth or eighth of the zona, and opens at the apex of the low elevation of the inner surface (Fig. 18).

The outer or funnel-shaped portion is wholly filled even at this advanced stage by the single large micropylar cell which was seen at the earlier stage (Figs. 18, 21).

Perca.

The egg of the perch has been a favorite subject for study. Almost every writer on teleostean ova has examined it. Von Baer ('35, pp. 6, 7) first described it as having a double membrane, the outer portion being traversed by long narrow dark spots (" dunklern Flecken ").

Müller ('54) gives a fuller account. He separates the membrane into an inner, the zona radiata, and an outer, the capsule. The outer surface of the zona is described as being covered with exceedingly small cylindrical projections. These are doubtless nothing but the elevations between the pore-canals, which are rather wide on the outer half of the zona. The capsule is radially traversed by small spiral tubes, which are enlarged and funnel-shaped at both ends. Transverse filaments are sometimes seen between these radial tubes. On applying pressure, yolk granules were forced into the spiral tubes, but in no case was any yolk matter forced between the tubules ; from which he concludes that the capsule must be closed between them.

Kölliker ('58) discusses the origin of the " tubules." He considers them to be outgrowths from the follicular cells, and the substance between them as a secretion from those cells. He denies the statement made by Müller, that they are hollow, but has seen the anastomosing filaments described by him. The tubules are independent of their jelly matrix, and in chromic acid preparations they can be separated from the latter. When the eggs are deposited, the granulosa cells probably fall off, leaving shallow depressions having polygonal outlines, from the centres of which "tubules" arose.

Ransom ('68) described the canals passing through the outer portion as having a double contour for each wall, and as filled with material containing vacuoles; but they do not seem to him to convey anything either fluid or solid into or out of the egg. This outer layer is separable only by tearing it from the yolk-sac (zona), and does not leave a distinct outline. The tubes divide at their inner terminations into branch-like roots, and adhere closely to the zona radiata. The internal ends are not expanded as Müller described, and it is rarely that filaments pass from one to the other. He supposes that the granules seen by Müller were only vacuoles. The eggs when deposited are arranged in the form of hollow tubes with the micropyles all turned to the inside.

His ('73) mentions having seen the micropyle, but neither figures nor describes it.

Brock ('78) describes the zonoid layer, and finds its striations intermediate in fineness between those of the villous layer and those of the zona. Judging by his drawing of Alburnus lucidus there are about three striations in the zonoid layer to four in the zona. The latter, he says, makes its appearance before the villous layer.

Hoffmann ('81) finds that in October the zona and the villous layer are of equal thickness. The latter is said to be composed of numerous small projections which correspond exactly to the villi of the Cyprinoids. At the free ends of the villi lie the granulosa cells. In February the zona is differentiated into two layers, of which the inner is four times as thick as the outer. There arise from the outer layer long fibres with triangular bases and with their distal ends expanded to form a continuous layer on which the granulosa cells rest. Each filament corresponds to, but is not a process of, a granulosa cell.

Owsjannikow ('85) recognizes the usual divisions of the egg membrane. The contents of the distal ends of the filaments are granular, which has given rise to the belief that they are nuclei. The filaments end externally in funnel-shaped enlargements described by Müller. He succeeded in forcing granular matter from the yolk into their deep ends. The latter divide and enter the pores of the zona, through the whole thickness of which they can be traced. He states (p. 7) that, contrary to Hoffmann's belief, the filaments are derived from the granulosa. In a subsequent part of his paper (pp. 29–31), where he gives an account of the development of the ovarian egg, his statements seem to be conflicting as to the relation of the spiral canals to the granulosa cells, but at the end he repeats that the canals are outgrowths of cells as stated by Kölliker. The interstitial matter (Zwischensubstanz) is arranged in lamellæ which are parallel to the surface of the egg. By the swelling of the lamellæ fissures arise which have the appearance of processes from the canals.

I have studied the ovarian eggs of Perca killed in October, February, and May. It is probable that the formation of the egg membranes is less advanced in the American species of this latitude than in the European species at a corresponding season.

Contrary to Hoffmann's statement that in October the capsular layer and zona are of equal thickness, not a trace of the capsular layer, distinct from the granulosa, could be found at this time of the year. The zona is well developed, and is differentiated into two layers of about equal thickness. The outer layer is radially striate, while the inner appears to be structureless. The granulosa cells lie immediately in contact with the zona radiata (Fig. 23, Plate III.). I have not been able to find the micropyle in October eggs.

In February the zona remains practically as it was in October, but vacuoles — which may be caused by the method of treatment — are to be seen in the inner portion (Fig. 25, Plate III.). They are much flattened radially, and thus suggest an approach to a stratified condition of this portion of the zona. The radial striations of the outer half of the zona are more strongly marked than at the earlier stage, and much fainter striations may also be seen traversing the inner half. The latter, though less distinct, are just as numerous as, and continuous with, those of the outer half. At this date the capsular layer is already well developed, but it has attained only half the thickness which it has in May.

Up to the month of May the thickness of the zona radiata has not changed, but the pore-canals can now be more readily traced passing entirely through it. They still remain much more evident in the outer than in the inner half of the zona. This is due to the greater calibre of the canals, not to their being farther apart in the outer half.

The different descriptions of the capsular layer are in part due to the fact that it presents different conditions according to varying circumstances. The radially arranged spiral structures traversing this layer arise as funnel-shaped tubules, one beneath each cell of the granulosa.

In the early stages of their development the tubules have a more or less spiral course, while in the later stages they become more nearly straight. In February eggs (Fig. 25, Plate III.) their inner ends are slightly expanded, and terminate in a thin structureless film overlying the zona. In radial sections of eggs taken in May, they often appear triangular at the base, and their contents divide into branches which enter the pores of the zona. The "filaments" connecting the canals are sometimes much more abundant than at others. In the vicinity of the micropyle one finds on tangential sections (Fig. 31, Plate II.) that the tubules at or near their bases are joined to each other by what appear like slender filaments, but these may be the cut edges of nearly perpendicular membranes. This results in the production of an irregular network with meshes of variable size and shape, at the angles of which the spiral tubules are located.

The micropyle was seen in eggs taken in February and in May. Immediately surrounding it, the zona radiata is thickened by a slight elevation of its internal surface (Fig. 24, Plate III.). The micropyle consists of a funnel-shaped opening in the zona with the wide end directed outward. In some cases the inner end of the canal also flares slightly. In a February egg in which the micropylar region was somewhat distorted (Fig. 26, Plate II.) the micropyle seems to have been composed of two regions, separated from each other by a distinct shoulder, the inner end of the outer portion being much wider than the outer end of the inner portion. The granulosa cells and their tubules are greatly crowded above this region (Fig. 24, Plate III.). At some distance on either side of the micropyle it is to be seen that the outer funnel-shaped ends of the canals begin to be more elongated than in other parts of the egg, and continue to increase in length up to the micropyle. The nuclei of the granulosa cells, which are situated near the bottom of the funnel-shaped expansions, also become more and more elongated as one approaches the centre of the micropylar region, and at the same time they come closer to the zona radiata. The effect of this is to produce in radial sections through the micropyle the appearance of an immense funnel-shaped depression in the whole capsular layer (Fig. 24). But the appearance is misleading; there is no such broad depression; the granulosa cells of this region extend outward beyond their nuclei until they reach the theca folliculi at the same level that the neighboring cells do. The thickness of the capsular layer is therefore not changed in the vicinity of the micropyle, and the theca folliculi does not bend inward, but stretches over this region with a uniform curvature. The granulosa cells stain more deeply than the inter-tubular substance of the capsular layer. This peculiarity is very serviceable when one is searching for the micropyle. Notwithstanding the absence of a broad depression, there is a narrow irregular canal left in the centre between the modified granulosa cells, which can best be seen upon sections tangential to this part of the egg. (Figs. 27–32, Plate II. Compare Explanation of Figures.) The appearance is similar to what one might imagine would result if the central cell of this region had dropped out of its original place. That such a cell has not wholly disappeared, but has simply lost its peripheral connection with the wall of the theca, is rendered probable by the presence of a peculiar cell at the bottom of this canal. Directly over the micropyle, in contact with the zona and filling more or less completely its micropylar depression, lies a single cell of large size. Its nucleus is more nearly spherical than the nuclei of the other cells, and it is not stained as deeply as they are. (Fig. 24, Plate III.; and Figs. 26, 31, Plate II.) There can be no doubt that it is a peculiarly modified granulosa cell.

Morone americana.

The egg membrane of the white perch has never been described, but Ryder ('82) has described the micropyle.

There is only a single membrane, the zona radiata, but it is composed of two distinct layers, both of which are traversed by pore canals. The eggs examined were taken from fishes caught in February, April, and May. In February the ovary contained eggs in four stages of development; in the older stages there are well developed membranes. Eggs of 0.16 mm. in diameter have a single homogeneous membrane 1.2 μ thick. When they have reached a diameter of 0.28 mm. the zona is composed of two layers (Fig. 33, Plate II.), a very thin inner and a thicker outer one; together they measure 39 μ in thickness. By the time the eggs have reached a diameter of 0.40 mm. (Fig. 34, Plate II.) the total thickness of the membrane is more than doubled; that of the outer layer is 49 μ and that of the inner 39 μ. The outer layer is formed first and takes a deeper stain. It does not increase much in thickness after the appearance of the inner layer, and in the older eggs it contains vacuoles. The inner is at first apparently homogeneous, but with its great increase in thickness there appear in it the radial striations characteristic of the zona. The granulosa cells are small and low, and have flattened nuclei situated in the middle of the cell.

Esox reticulatus.

The egg membrane of Esox was first described by Aubert ('53). He says the shell of the egg is a thin, transparent punctate membrane, which closely envelops the yolk and in sections exhibits radially placed streaks. After lying in water some time, an outer very thin granular membrane makes its appearance.

Lereboullet ('54) describes two membranes, the outer of which is pierced by microscopic tubes. The inner is a simple extremely thin and amorphous envelope, which has no homologue in the perch.

Reichert ('56, p. 94) states that the membrane discovered by Aubert surrounding the zona radiata is to be found on all eggs of this species, but that it is in the fresh condition entirely homogeneous.

Kölliker ('58, pp. 84 and 85) maintains the existence of a thin outer, resistant layer in all fish eggs, and was able to isolate it in fresh eggs of Esox.

Ransom ('68) says that in Esox the egg membrane is similar to that of Gasterosteus; he also, as I think erroneously, supposes the thin outer membrane to be homologous with the "Eikapsel" of the perch. He figures the micropyle.

Finally, His ('73) described for the zona radiata concentric as well as radial striæ.

The eggs examined by me were taken from the ovary in February. Leaving out of consideration the smallest eggs, 0.063 mm. and less in diameter, which have no membrane except the granulosa, the ovary contained eggs in three stages of development, respectively about 0.50, 1.00, and 1.50 mm. in diameter. In eggs of the first stage the zona radiata is about 3 μ thick and very faintly striate. There is no evidence of its being differentiated into concentric layers. At the micropyle (Fig. 35, Plate III.) it reaches a thickness of 7 μ. Very generally the yolk is more or less retracted from the zona by the action of the hardening reagents, so that a narrow space, which varies a good deal in thickness over different parts of the egg, is left between the two structures. Spanning this interval are numerous fine threads, which have the appearance of being prolongations of the substance of the yolk continued into pore-canals of the zona. This is a condition which remains at subsequent stages, and will therefore be discussed further on. The granulosa cells are still thin, and their nuclei much flattened.

In the second stage (Fig. 36, Plate III.) the zona has a total thickness of 11 or 12 μ, and is distinctly differentiated into two layers, the outer of which is only about one fifth as thick as the inner. The latter is faintly stained, and distinctly striate radially ; the outer is deeply stained, and striations are usually not to be seen in it, but on favorable sections, especially such as are very thin, the striations may frequently be made out to pass continuously through the whole thickness of both layers. Upon this point there is not the least doubt, so that it is certain the outer layer in question is truly a part of the zona, and I have been unable to find in ovarian eggs any membrane intervening between this and the granulosa cells. In sections of the micropylar region, the inner portion of the zona radiata exhibits vacuoles elongated in the direction of the pore-canals. In this region the latter are not strictly radial, but converge towards the outer end of the micropylar canal. Inside the zona there is a region to be seen which bears some resemblance to a membrane with coarser (more distant) striations than those of the zona. It varies in thickness on different parts of the egg, and corresponds, I believe, to the sub-zonal space seen in the eggs of the first stage ; but it may represent the zonoid layer of His.

The membranes of eggs of the third or oldest stage (Fig. 37) differ somewhat from the conditions just described. The vacuoles of the zona radiata, found in the second stage near the micropyle only, are here found over all portions of the egg ; they are always most numerous near the inner surface, and are not found at all in the outer fifth of the membrane. They are more or less regularly arranged in series parallel with the surface of the zona. Kölliker ('58, p. 84) attributed the presence of such vacuoles in the pore-canals to the effect of fresh water on the zona.

The granulosa cells in the second and third stages have nearly spherical nuclei, which lie at their distal ends (Fig. 37, Plate III.). Below the nuclei, tapering columns of granular protoplasm extend to the zona radiata. These columns are separated by less deeply stained tracts of substance, but the boundaries of the columns are not sharply marked. The appearance is as though the columnar cells were being gradually metamorphosed into an intercellular substance. This condition is evidently an approach to that found in Perca.

The micropyle was found in eggs of both the first and second stages. In the first stage (Fig. 35, Plate III.) the zona is twice as thick around the micropyle as in other regions. This thickening results in a considerable elevation of the inner surface of the zona, the outer surface being only very slightly changed. The micropyle is a wide canal, the outer third of which tapers rapidly and is continuous with the inner two thirds, which taper only slightly from without inward. The micropylar canal is partially filled with a plug of substance which appears to be continuous with the yolk. The granulosa cells overlying the micropyle do not appear different in size from those which envelop the rest of the egg, but a single cell is sometimes seen to overlie the micropyle in addition to the regular layer of granulosa cells. In the second stage (Fig. 36, Plate III.) the micropylar canal is narrower than in the first; it no longer tapers gradually from the outside inward, but is slightly narrowed at two points, one near the outside and one at its deep end. By the retraction of the yolk from immediate contact with the zona near the micropylar canal in the case of one of the eggs, a space was formed through which could be traced a cord of substance continuous with that which occupied the canal itself. The portion of the substance which traversed this space was funnel-shaped, with the wide end next the yolk. The thickness of the zona does not now differ so greatly in different regions as at the first stage. At some distance from the micropyle in the egg last mentioned (Fig. 36), the inner surface of the zona was raised rather abruptly; nearer the micropyle it was slightly depressed, but the margin of the canal was raised in the form of a low cone, which thus occupied the centre of a very shallow inverted crater, the rim of which was formed by the outer circular elevation. Above the micropyle in the granulosa was a large spheroidal space nearly filled with a granular mass somewhat denser than the yolk. The mass was slightly contracted, leaving a narrow space at its periphery. I am in doubt whether to regard it as a cell or not, since no nucleus could be detected. On both sides of this granular mass there were several highly refractive homogeneous bodies (Fig. 36, $x\,x$). It is however doubtful if they have any significance in relation to the micropyle. The granulosa cells at this stage are tall and have elongated nuclei, which are broad at the exterior end, and taper towards the egg membranes.

Notemigonus chrysoleucus.

The ovary of this species contained ova in four stages of development on May 9th. In all but the smallest eggs the zona radiata was present. The largest had a diameter of 0.6 mm., and the zona varied gradually from a thickness of 2 μ on one side of the egg to that of 4 μ on the opposite side. The pore-canals are very fine, being almost invisible in balsam preparations.

The micropyle was observed in only a single case; it was found in the middle of the thickest portion of the membrane, which is exactly in the middle of the attached side of the egg. The direction of the inner surface of the zona was not altered in the vicinity of the micropyle (Fig. 38, Plate II.), but its outer surface exhibited a broad circular depression, by which the thickness of the zona was diminished about one half. The micropyle proper at the centre of the depression appeared as a narrow canal of uniform calibre. Between the zona and the yolk there was a narrow space, probably formed by the contraction of the yolk; beneath the micropylar region, this space was abruptly enlarged into a hemispherical depression. Across this space the radial strands of protoplasm characteristic of almost-all the spaces between the zona and the yolk were plainly visible.

The granulosa, which over all other parts of the egg is composed, as usual, of a single layer of cells, is thickened in the region of the micropyle. As the direction of the long axes of their oval nuclei show, the cells near the margin of the micropylar depression in the zona have their peripheral ends inclined toward the axis of the micropyle. The cells which fill up the depression have larger and more elongated nuclei, and the obliquity of the latter has become so great that the depression appears to be filled with a granulosa layer two or three cells deep. It would seem that in this case the single micropylar cell found in other eggs was represented by a number of enlarged granulosa cells.

Clupea vernalis.

The chief interest in the egg membranes of this species centres in the presence of a thin, highly refractive, structureless membrane overlying the zona radiata of eggs in an advanced stage of development (Fig. 39, Plate III.). This outer membrane is intimately connected with the granulosa cells, so that it usually retains its connection with the granulosa when the latter is artificially separated from the zona. In all such cases slender striations extend from it to the zona radiata. The appearance of these markings is such as to show clearly that they are prolongations of the substance of the outer membrane, and there can be little doubt that the projections penetrate the pore canals of the zona radiata, from which they are partially withdrawn by the artificial separation of the two membranes. This structural condition suggests an explanation for similar appearances *below* the zona radiata in Esox and other fishes, and between the zona radiata and an *inner* layer in Amiurus (Fig. 45) and Ictalurus. It will be more fully discussed later.

Cyclogaster lineatus (*Liparis lineatus* Auct.).

The ovaries of this species contain ripe eggs in May, the time at which I examined them. The largest eggs were about 0.63 mm. in diameter, the membrane averaged about 0.043 mm. in thickness. The zona radiata seems to be filled with small spaces connected by the much finer radial canals (Fig. 40, Plate III.), the spaces causing the latter to appear moniliform. Near the inner and outer margins of the zona the canals are simple tubes, as in most other fishes.

The eggs next in size are 0.25 to 0.30 mm. in diameter; their zona is always only half as thick on one side as it is on the opposite, the change in thickness being nowhere abrupt. In eggs of this stage the zona is traversed by simple pore-canals, which are indistinct near its outer surface. In some cases (Fig. 41, Plate III.) the transition from the inner to the outer portion is so abrupt that the zona appears to be composed of two layers of unequal thickness, — an outer, thinner, more nearly homogeneous and unstained, and an inner which is thicker, more distinctly striate, and usually faintly stained.

The micropyle (Figs. 42–44, Plate III.) was observed only in eggs about 0.16 mm. in diameter. As in the case of Pygosteus it seems to lie in a plane perpendicular to the long axis of the ovary. The micropyle is a long narrow tube, with parallel sides, in a local thickening of the zona. This increase in the thickness of the zona affects the outline of the inner surface more than that of the outer, and is entirely independent of the above mentioned gradual change in thickness between opposite poles of the egg. It is produced principally by additions to the inner surface of the zona. The outer surface is slightly elevated at a little distance from the micropyle, but is abruptly depressed immediately over it. The regularity in the arrangement of the nuclei of the granulosa cells is disturbed in the immediate vicinity of the micropyle, where the whole layer is slightly thickened. Usually an enlarged single nucleus lies immediately above the micropylar canal (Fig. 44).

On the Number of Egg Membranes.

The views held concerning the number of egg membranes in teleosts have been many and various. Authors have generally been agreed about the presence of a membrane perforate with radial canals, the zona radiata; but doubts have been raised by Ryder whether this membrane is always present. He ('82^c) found no striations in the egg membrane of

Belone longirostris or ('84, p. 457) the cod, and states ('85, p. 145) that the eggs of Gambusia patruelis do not possess any membrane. I have found striations in the membrane of the fresh cod egg. It may be stated here that the striations of the zona sometimes show plainest in fresh eggs, sometimes not until reagents have been applied. Haeckel says, for forms related to Belone longirostris, that the membrane is structureless, but that it is covered with minute black dots. These dots were doubtless pore-canals seen from the surface. The zona radiata of Osmerus eperlanus was found by Buchholz ('63) to consist of an inner and outer portion, joined together in the micropylar region only. On deposition of the eggs the outer membrane is turned wrong side out, and serves to attach the eggs to foreign substances. These conditions have been redescribed by Cunningham ('86). Hoffmann ('81) found that the zona is differentiated into two layers in all adhesive eggs, the outer portion being ultimately transformed into a viscid mass.

Ryder ('86) describes a peculiar arrangement of the egg membranes of Ictalurus albidus. He says: "The egg-membrane is double, that is, there is a thin inner membrane representing the zona radiata, external to the latter and supported on columnar processes of itself, which rest upon the inner membrane; there is a second one composed entirely of a highly elastic adhesive substance. The columns supporting the outer elastic layer rest on the zona, and cause the outer layer to separate very distinctly from the inner one." I have found similar conditions in Amiurus catus (Plate II. Fig. 45), but am inclined to think that the two membranes represent the outer and inner portions of the zona radiata; for the outer shows the striations peculiar to the zona, and the columnar layer is of varying thickness. The inner membrane, being closely associated with the yolk, would cling to it when the yolk contracts; the protoplasm in the pore-canals being partially withdrawn would give rise to these columnar processes. Where the two membranes were separated for a considerable distance, the columnar structure was destroyed. Similar conditions obtain in the eggs of Clupea vernalis, but in this case the columnar structures lie between the zona radiata and a thin *outer* homogeneous layer which is in contact with the granulosa. There cannot be the least doubt concerning the meaning of the columnar layer in Clupea vernalis, for the two membranes lie directly in contact in some parts of the egg. The peculiar structures in Ictalurus and Amiurus doubtless have an origin and meaning similar to that of Clupea.

The eggs of all the species of fishes examined by me possess a perforate zona radiata. The radial striæ could never be made out on the first appearance of the egg membrane. The absence of striæ in these younger eggs may be accounted for by assuming, as Reichert has suggested, that the zona radiata is a later growth, and that the imperforate membrane of younger eggs is a different structure, or that during the earlier stages the material composing the membrane is less dense, allowing the food material to have ready access to the yolk. Granting for the moment that the zona grows by apposition of layers from within, the latter view is the more probable, because in the perch the inner portion of the zona is not perforate even after the outer is distinctly so, and in most cases the pore-canals are much more distinct and wider in the outer than in the inner portion of the zona. The meaning of the pore-canals, in the intra-ovarian egg at least, needs little discussion. In most of the sections prepared, where the granulosa cells are slightly raised from the zona radiata, processes of the granulosa cells can be seen to enter the pore-canals.

Various membranes have been described for different fishes as overlying the zona radiata. The peculiar capsular layer of the perch has been seen by all authors who have examined the eggs of this fish. It was first described by Von Baer ('35).

Rusconi ('36) describes a thin membrane overlying the ovarian eggs of Cyprinus. Aubert ('53) saw an outer membrane on eggs of Esox which had lain in water some time. Kölliker ('58) succeeded in isolating this membrane in the case of Esox.

Reichert ('56) discovered that whenever processes are present, as in many cyprinoids, they are set in a thin outer membrane. Kölliker ('58) confirmed this statement, and added that this outer membrane is developed before the zona radiata. Reichert also found that the membrane of the smallest membrane-bearing ovarian eggs is not striate, and concluded that the zona radiata must be a secondary formation.

Vogt ('42) was the first to describe a membrane within the zona radiata. He found that in the eggs of Coregonus palea and Salmo umbla this membrane cannot be readily seen until after the eggs have been in water for some time, and that it passes (p. 29) gradually into the germ. Ransom ('56) found a similar structure in eggs of Gasterosteus pungitius, in which this inner membrane takes part in cleavage. Eimer ('72[a]) claims to have isolated this vitelline membrane, which he saw in trout, pike, white-fish, and perch. Oellacher ('72) also succeeded in separating it in the brook trout. I believe that the structures described by Vogt, Ransom, Eimer, and Oellacher are, as others have maintained, not to be considered as vitelline membranes, but as the superficial part of the protoplasm of the egg.

His ('73) found that the cortical layer of the yolk in many ovarian eggs is more finely granular than the rest of the yolk, and that it is radially striate. This outer portion of the yolk he called the *zonoid layer*. Many others have seen similar structures. According to the accounts of some authors the zonoid layer is found only in eggs which are not mature, and even then it is not always present.

The condition of the egg membranes in Amiurus and in certain stages of Esox has suggested the idea that similar appearances may in some cases have given rise to a belief in the existence of a zonoid layer when there really was none. A partial withdrawal of the egg protoplasm occupying the pore-canals produces an appearance which at first sight suggests the presence of a striate membrane internal to the zona; in fact, I at first supposed it to be a distinct membrane, and was the more easily misled because in some cases it seems to be of nearly uniform thickness. However, more careful study showed that it was not a membrane, and that the appearance was due to fine threads of highly refractive substance stretching across a space between the inner surface of the zona and the yolk. There are two things especially which make it impossible for me to believe that this is a normal condition : the great variability in the thickness of the supposed membrane in different parts of the same egg, and the fact that the radial striations are due to a substance which is more highly refractive than the substance, if any, filling the intervening spaces. If, on applying reagents, there is great contraction of the yolk, either it is torn from the protoplasm in the pore-canals, or the protoplasm contained in the pore-canals is suddenly withdrawn from them and distorted; in either case, there would be no appearance of a zonoid layer. If, however, the protoplasm should not be withdrawn from *all* the pores, but should in the case of many remain stretched across the space between the zona and the yolk, as might no doubt frequently happen, we should find the supposed zonoid layer more coarsely striate than the zona, a condition described by recent authors. Such an origin of the zonoid layer might also explain its absence in ripe eggs. After the egg has attained its full size, the connection of the yolk substance with the canals would naturally be less intimate than at an earlier stage, and then a contraction of the yolk would not be accompanied by the stretching of any filaments across the space thus produced.

Scharff ('87 and '87ª) has recently described, within the zona radiata in young eggs of Trigla, a zonoid layer, which subsequently disappears.

The eggs examined by me may be divided as follows : —

I. Eggs with a single membrane, the zona radiata.
   a. Zona radiata a single layer of uniform structure. Notemigonus chrysoleucus, Carassius auratus.
   aa. Zona radiata differentiated into an inner and outer layer. Morone americana, Esox reticulatus, Cyclogaster lineatus, Amiurus catus.

II. Eggs with a zona radiata and a thin homogeneous outer layer.
   b. Outer membrane without appendages. Clupea vernalis.
   bb. Outer membrane bearing filiform appendages. Fundulus heteroclitus, F. diaphanus.
   bbb. Outer membrane with short appendages. Pygosteus pungitius.

III. Eggs with a zona and a thick outer layer produced by a secretion from and metamorphosis of the granulosa cells. Perca americana.

Origin of the Egg Membranes.

Concerning the origin of the different egg membranes of fishes several views have been held.

Vogt ('42) and Vogt and Pappenheim ('59) maintained that the zona radiata is formed by the compression of a layer of cells surrounding the egg; Reichert ('56), Kölliker ('58), Gegenbaur ('61), and Eimer ('72^a), that it is derived from the yolk; Thomson ('59) and Waldeyer ('70), that it is derived from the follicular epithelium; Ransom ('67) argued that it cannot grow by apposition of layers from within or without, and that it must grow by interstitial deposition of material. Whether this material comes from ingoing or outgoing currents, he was unable to determine.

I think that the zona is undoubtedly derived from the yolk. Kölliker found that in all the filament-bearing eggs studied by him the zona radiata was formed after the filament-bearing membrane. I have found the same to be true in Fundulus. In the case of Morone the outer layer of the zona does not become much thicker after the inner layer has begun to be formed, whereas the latter continues to grow rapidly. In the case of Cyclogaster lineatus, where the outer layer of the zona shows columnar structures, these do not bear any definite numerical relation to the overlying cells of the granulosa. The outer portion of the zona is almost always more uniform in its structure, and stains deeper, than the inner portion.

Reichert ('56) and Kölliker ('58) are inclined to believe that the capsular layer of the perch is derived from the granulosa cells, an opinion with which Hoffmann ('81) does not agree. It certainly does not make its appearance till after the zona is well developed; if it were derived from the yolk, its substance would first have to traverse the zona radiata. How the nourishment for the egg could pass into the latter through the pore-canals, and the formative substance of the villous layer at the same time pass out through them, is scarcely conceivable. Moreover, at the distal end of each of the villi lies the nucleus of a granulosa cell, there being as many villi as there are cells, a fact which proves beyond a doubt the intimate relation of the two structures.

The membrane just external to the zona in Clupea vernalis may be considered homologous to that in Gasterosteus, Fundulus, and many Cyprinoids, even though it does not in Clupea bear appendages as it does in Gasterosteus. From the development of the appendages in Fundulus and Gasterosteus it is evident that this membrane has no connection with the granulosa cells. In these cases each of the appendages does not correspond to a single cell as in the perch, nor to any definite number of cells. If Reichert is correct in saying that the homogeneous membrane found in young eggs is a different structure from the zona radiata, the membrane under consideration may perhaps be looked upon as the primitive membrane described by him. It is certain that it appears before the zona, and I am inclined to think that it is derived from the yolk.

CAMBRIDGE, December, 1888.

BIBLIOGRAPHY.

Aubert, Hermann.
    '53. Beiträge zur Entwickelungsgeschichte der Fische. Zeitschr. f. wiss. Zool., Bd. V. Heft 1, pp. 94–102, Taf. VI. 16 Aug., 1853.

Baer, Karl Ernst von.
    '35. Untersuchungen über die Entwickelungsgeschichte der Fische, nebst einem Anhange über die Schwimmblase. 52 pp., 1 Taf. u. mehreren Holzschn. im Texte. Leipzig: Vogel. 1835.

Brock, J.
    '78. Beiträge zur Anatomie und Histologie der Geschlechtsorgane der Knochenfische. Morph. Jahrb., Bd. IV. Heft 4, pp. 505–572, Taf. XXVIII., XXIX. 1878.

Buchholz, Reinhold.
    '63. Ueber die Mikropyle von Osmerus eperlanus. Arch. f. Anat., Physiol. u. wiss. Med., Jahrg. 1863, pp. 71–81, Taf. III A., Figs. 1–4. 1863.
    '63ª. Nachträgliche Bemerkungen über die Mikropyle von Osmerus eperlanus. Arch. f. Anat., Physiol. u. wiss. Med., Jahrg. 1863, pp. 367–372, Taf. VIII A. 1863.

Cunningham, J. T.
    '86. On the Mode of Attachment of the Ovum of Osmerus eperlanus. Proceed. Zoöl. Soc. London, for 1886, Pt. III. pp. 292–295, Pl. XXX. (Read May 4.) 1886.

Eimer, Th.
    '72ª. Untersuchungen über die Eier der Reptilien. Arch. f. mikr. Anat., Bd. VIII. pp. 216–243, 397–434, Taf. XI., XII., XVIII. 1872.

Gegenbaur, Carl.
    '61. Ueber den Bau und die Entwickelung der Wirbelthiereier mit partielle Dottertheilung. Arch. f. Anat., Physiol. u. wiss. Med., Jahrg. 1861, pp. 491–529, Taf. XI. 1861.

Haeckel, Ernst.
    '55. Ueber die Eier der Scomberesoces. Arch. f. Anat., Physiol. u. wiss. Med., Jahrg. 1855, pp. 23–31, Taf. IV., V. 1855.

His, Wilhelm.
'73· Untersuchungen über das Ei und die Entwickelung bei Knochenfischen.
Leipzig: F. C. W. Vogel. 1873. 4 + 54 pp., 4 Taf., 4to.

Hoffmann, C. K.
'81· Zur Ontogenie der Knochenfische. Verhandl. d. koninkl. Akad. v.
Wetenschappen, Amsterdam, Deel XXI., 164 pp., 7 Taf. 1881.

Kölliker, Albert von.
'58· Untersuchungen zur vergleichenden Gewebelehre, angestellt in Nizza im
Herbste 1856. Verhandl. physical.-med. Gesellschaft in Würzburg, Bd.
VIII. pp. 1–128, Taf. I.–III. 1858.

Lereboullet, Auguste.
'54· Résumé d'un Travail d'Embryologie comparée sur le Développement du
Brochet, de la Perche et de l'Écrevisse. Ann. Sci. Nat., sér. 4, Zool.,
Tom. I. pp. 237–289. 1854.

Müller, Johannes.
'54· Ueber zahlreiche Porencanäle in der Eicapsel der Fische. Arch. f. Anat.,
Physiol. u. wiss. Med., Jahrg. 1854, pp. 186–190, Taf. VIII. Figs. 4–7.
1854.

Oellacher, J.
'72· Beiträge zur Entwickelungsgeschichte der Knochenfische nach Beobachtungen am Bachforelleneie. Zeitschr. f. wiss. Zool., Bd. XXII. Heft 4,
pp. 373–421, Taf. XXXII., XXXIII. 20 Sept., 1872.

Owsjannikow, Ph.
'85· Studien über das Ei, hauptsächlich bei Knochenfischen. Mém. Acad.
Imp. Sci. St. Pétersbourg, sér. 7, Tom. XXXIII. No. 4, 54 pp., 3 Taf.
1885.

Ransom, W. H.
'56· On the Impregnation of the Ovum of the Stickleback. Proceed. Roy.
Soc. London, Vol. VII. pp. 168–172. 1856.
'67· On the Structure and Growth of the ovarian Ovum in Gasterosteus leiurus. Quart. Jour. Micr. Sci., n. ser., Vol. VII. pp. 1–4, Pl. I. Jan.,
1867.
'68· Observations on the Ovum of Osseous Fishes. Philos. Trans. Roy. Soc.
London, Vol. CLVII. Pt. II. pp. 431–502, Pls. XV.–XVIII. 1868.

Reichert, Karl Bogislaus.
'56· Ueber die Micropyle der Fischeier und über einen bisher unbekannten,
eigenthümlichen Bau des Nahrungsdotters reifer und unbefruchteter
Fischeier (Hecht). Arch. f. Anat., Physiol. u. wiss. Med., Jahrg. 1856,
pp. 83–124, 141, 142, Taf. II., III., und IV. Figg. 1–4. 1856.

Rusconi, R.
'36· Ueber die Metamorphosen des Eies der Fische vor der Bildung des
Embryos. Arch. f. Anat., Physiol. u. wiss. Med., Jahrg. 1836, pp. 278–
288.

Ryder, John A.

'81e. Development of the Spanish Mackerel (Cybium maculatum). Bull. U. S. Fish Commiss., Vol. I. pp. 135–172, 4 pls. [1881] 1882.

'82. The Micropyle of the Egg of the White Perch. Bull. U. S. Fish Commiss., Vol. I., p. 282. May 2, 1882.

'82ᵃ. Development of the Silver Gar (Belone longirostris), with Observations on the Genesis of the Blood in Embryo Fishes, and a Comparison of Fish Ova with those of other Vertebrates. Bull. U. S. Fish. Commiss., Vol. I· pp. 283–301, Pls. XIX.–XXI. May 2 and 19, 1882.

'83. On the Thread-bearing Eggs of the Silversides (Menidia). Bull. U. S. Fish Commiss., Vol. III. pp. 193–196. 1883.

'84. A Contribution to the Embryography of Osseous Fishes, with special Reference to the Development of the Cod (Gadus morrhua). Ann. Report U. S. Commissioner of Fish and Fisheries for 1882, XVII. pp. 455–605, Pls. I.–XII.

'84ᵃ. *Also separate,* with title-page and cover. 149 pp., 12 pls. Washington: Government Printing Office. 1884.

'85. On the Development of Viviparous Osseous Fishes. Proceed. U. S. National Museum, Vol. VIII. Nos. 8–10, pp. 128–155. Pls. VI.–XI. 25 May, 1885.

'86. On the Development of Osseous Fishes, including Marine and Fresh-Water Forms. Extracted from Ann. Report U. S. Commissioner of Fish and Fisheries for 1885. pp. [1]–[116], Pls. I.–XXX. 1886.

'86ᵃ. The Development of Fundulus heteroclitus. American Naturalist, Vol. XX. p. 824. Sept., 1886.

'87. [*Same as* RYDER, '86.] Ann. Report U. S. Commissioner of Fish and Fisheries for 1885, pp. 484–604, Pls. I.–XXX. 1887.

Scharff, Robert.

'87. On the Intra-ovarian Egg of some Osseous Fishes. (Rec'd Nov. 17, 1886. — *Abstract.*) Proceed. Roy. Soc. London, Vol. XIV. No. 249, pp. 447–449. 1887.

'87ᵃ. On the Intra-ovarian Egg of some Osseous Fishes. Quart. Jour. Micr. Sci., Vol. XXVIII. pp. 53–74, Pl. V. Aug., 1887.

Thomson, Allen.

'59. [Article] Ovum *in* The Cyclopædia of Anat. and Physiol., edited by Robert B. Todd, Vol. V. (Suppl. Vol.), 1859, pp. 1–80 and [81]–[142].

Note. — Part I., pp. 1–80, was issued in 1852; Part II., pp. [81]–[142], in 1855.

Vogt, Carl.

'42. Embryologie des Salmones. Neuchatel. 1842. 6 + 328 pp., 8vo. Avec Atlás, fol. obl. de 7 pls.

Being Tome I. of L. Agassiz, Histoire Naturelle des Poissons d'Eau douce de l'Europe Centrale.

Vogt, Carl, et S. Pappenheim.

'59· Recherches sur l'Anatomie comparée des Organes de la Génération chez les Animaux Vertèbres. (Déposé dans les Archives de l'Acad. le 30 Dec., 1845.) Ann. Sci. Nat., sér. 4, Zool., Tom. XI. pp. 331–369, Pl. XIII.; Tom. XII. pp. 100–131, Pls. II., III. 1859.

Waldeyer, Wilhelm.

'70· Eierstock und Ei. Ein Beitrag zur Anatomie u. Entwickelungsgeschichte der Sexualorgane. Leipzig: W. Engelmann. 1870. 8 + 174 pp., 6 Taf. 8vo.

EXPLANATION OF FIGURES.

ABBREVIATIONS.

cp.	Blood corpuscles.	*pr j. i cl.*	Intercellular ridges.
fil.	Filaments of Fundulus.	*spa.*	Space below micropyle.
fil. vt.	Filaments of vitellus.	*tbl.*	Tubules of the capsular membrane in Perca.
gran.	Granulosa.		
m py.	Micropyle.	*thc. fol.*	Theca folliculi.
m py. cl.	Micropylar cell.	*vac.*	Vacuole.
nl. gran.	Nucleus of granulosa cell.	*yk.*	Yolk.
nl. m py.	Nucleus of micropylar cell.	*z. r.*	Zona radiata.
po. can.	Pore-canals of the zona radiata.	*z. r.'*	Zona radiata externa.
pr c.	Rivet-shaped processes of zona.	*z. r."*	Zona radiata interna.

All the figures were made with the aid of the camera lucida, and all except Figs. 1, 2, 5, 7, 11, and 12 from preparations mounted in benzole-balsam. Figs. 89–41 were drawn by Dr. Mark, and the others by the author.

PLATE I.

Figures 1–13 are of Fundulus heteroclitus.

Fig. 1. Surface view of one of the smallest filament-bearing eggs of October 27. Diameter of egg, 0.16 mm. Examined fresh. × 750.

" 2. Surface view of another egg of the same size and date, with somewhat larger filaments. Examined fresh. × 425.

" 3. Tangential section of an ovarian egg 0.15 mm. in diameter. The ovary was hardened, December 23, in Flemming's chromic-osmic-acetic mixture, and subsequently stained with hæmatoxylin. × 425. The section is seen from its inner surface.

" 4. Tangential section of an egg from the same ovary with longer filaments. × 425. This section is also seen from its inner surface.

" 5. Surface view of an egg of October 27, about 0.23 mm. in diameter. Examined fresh. × 425.

" 6. Tangential section of an egg 0.25 mm. in diameter, from the same ovary as Fig. 3. × 750.

" 7. Surface view of a ripe (June) egg from which the granulosa cells had been removed, showing the network of ridges between their bases. Examined fresh. × 750.

" 8. Radial section of the egg represented in Fig. 3. Transsections of filaments are seen at *fil*. × 425.

" 9. Radial section of an egg of May 2, about 0.8 mm. in diameter. Preserved in Perenyi's fluid, and stained with picrocarminate of lithium. × 750.

" 10. Tangential section of an egg of December 23, about 0.4 mm. in diameter. From the ovary mentioned under Fig. 3. × 425.

" 11. Radial optical section of a ripe egg shortly after being forced from the ovary (June 1). Examined fresh. × 50.

" 12. Base of one of the filaments of the ripe egg. Examined fresh under pressure. × 750.

" 13. Radial section of an ovarian egg of November 23. Preserved in Perenyi's fluid, and stained with Grenacher's alcoholic borax-carmine. × 750.

Figures 14–22 are of Pygosteus pungitius, all except Figure 20 being of eggs from a single ovary, which was cut transversely.

" 14. Radial section through an ovarian egg 0.15 mm. in diameter. The ovary was preserved in Perenyi's fluid, April 18, shortly after spawning, and subsequently stained in picrocarminate of lithium. × 750.

" 15. Tangential section near the micropyle. × 112.

" 15ᵃ. Radial section of an egg 0.37 mm. in diameter. × 750.

" 16. Radial section of an egg 0.33 mm. in diameter. × 750.

" 17. Radial section of an egg 0.37 mm. in diameter. × 750.

" 18–22. Radial sections through the micropyles of eggs, about 0.4 mm. in diameter. In Figs. 20 and 21 the micropyle is cut obliquely. × 750.

" 20 is from an ovary hardened April 4, i. e. some time before spawning.

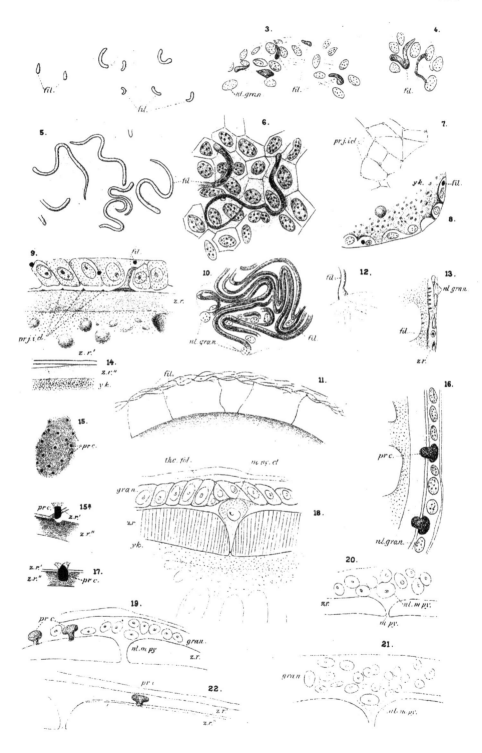

⸲ ABBREVIATIONS.

cp.	Blood corpuscles.	*pr j. i cl.*	Intercellular ridges.
fil.	Filaments of Fundulus.	*spa.*	Space below micropyle.
fil. vt.	Filaments of vitellus.	*tbl.*	Tubules of the capsular mem-
gran.	Granulosa.		brane in Perca.
m py.	Micropyle.	*thc. fol.*	Theca folliculi.
m py. cl.	Micropylar cell.	*vac.*	Vacuole.
nl. gran.	Nucleus of granulosa cell.	*yk.*	Yolk.
nl. m py.	Nucleus of micropylar cell.	*z. r.*	Zona radiata.
po. can.	Pore-canals of the zona radiata.	*z. r.'*	Zona radiata externa.
pr. c.	Rivet-shaped processes of zona.	*z. r.''*	Zona radiata interna.

PLATE II.

Figures 23, 24, and 25 are on Plate III.; Figures 26–32 are from Perca.

Fig. 26. Radial section through the micropyle of an ovarian egg from *Perca* killed in February. × 750.

" 26ᵃ. Section through the micropyle (?) of an egg of *Perca*. × 750.

" 27–32. A series of sections tangential to the surface of an egg of *Perca* 1 mm. in diameter at a point somewhat to the right of the micropyle. The portion of the sections to the right of the micropylar region lies deeper than the portion to the left. In

" 27, the cells lying to the right of the region of the micropyle are crowded, and have a curved band-like arrangement. The cells which contain dark points and lie farther to the right are from the central portion of the section, and are therefore cut across deeper than the others. In the second section,

" 28, only the deeper, filamentous prolongations of these cells are seen. In

" 28, 29, a median pit can be traced through the centre of the column of cells which occupies the micropylar region. In

" 30 the nucleus of the enlarged micropylar cell (*nl. m py.*) is seen. In

" 31 is seen the enlarged mouth of the mycropyle (*m py.*) and a few cells which lie somewhat higher and to the left of it.

" 32 is a section through the zona radiata and micropylar canal. This egg was preserved February 15 in Perenyi's fluid, and was stained with Czoker's alum-cochineal. × 425.

" 33. Radial section of an ovarian egg of *Morone americana* 0.28 mm. in diameter. Hardened in chromic acid February 25, and stained with picro-carminate of lithium. × 750.

" 34. Radial section of a larger egg (0.4 mm. in diameter) of *Morone americana* from the same series of sections as that of Fig. 33. × 750.

" 38. Radial section through the micropyle of an egg of *Notemigonus chrysoleucus* 0.63 mm. in diameter. Ovary of May 5, killed in Perenyi's fluid, and stained with picrocarminate of lithium. × 750.

" 45. Radial section through the egg of *Amiurus catus*. Ovary of May 9, killed in Perenyi's fluid, and stained with picrocarminate of lithium. × 750. The radial markings have been accidentally omitted and *fil. vl.* placed for *fil. vt.*

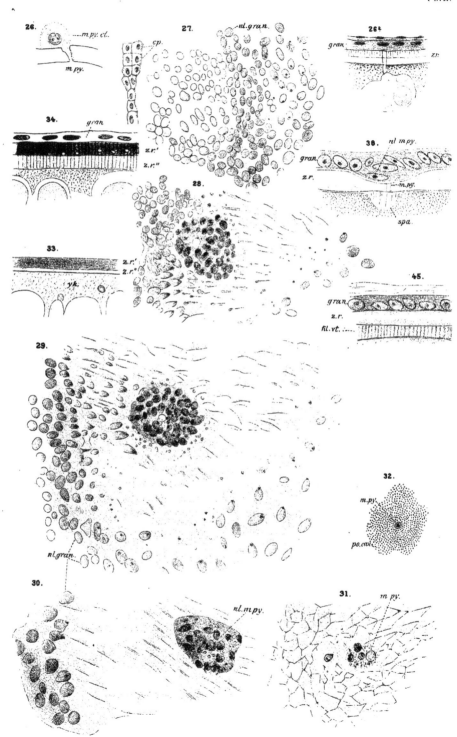

ABBREVIATIONS.

cp.	Blood corpuscles.	*pr j. i cl.*	Intercellular ridges.
fil.	Filaments of Fundulus.	*spa.*	Space below micropyle.
fil. vt.	Filaments of vitellus.	*tbl.*	Tubules of the capsular mem.
gran.	Granulosa.		brane in Perca.
m py.	Micropyle.	*thc. fol.*	Theca folliculi.
m py. cl.	Micropylar cell.	*vac.*	Vacuole.
nl. gran.	Nucleus of granulosa cell.	*yk.*	Yolk.
nl. m py.	Nucleus of micropylar cell.	*z. r.*	Zona radiata.
po. can.	Pore-canals of the zona radiata.	*z. r.'*	Zona radiata externa.
pr c.	Rivet-shaped processes of zona.	*z. r.''*	Zona radiata interna.

PLATE III.

Fig. 23. Radial section of an egg of *Perca* in October, 0 5 mm. in diameter. The ovary was hardened in 0.25 per cent chromic acid, and subsequently stained with Czoker's alum-cochineal. × 750.

" 24. Radial section through the micropyle of an egg of *Perca*. The ovary was preserved in Perenyi's fluid, May 9, and stained in carminate of lithium. × 425. The definite line at the outer margin of the zona radiata should have been omitted.

" 25. Radial section of an egg of *Perca*, 0.9 mm. in diameter. From an ovary hardened in February. × 425.

Figures 26–34 are on Plate II.

" 35. Radial section through the micropyle of an egg of *Esox*, 0.47 mm. in diameter. Ovary of February 23 killed in chromic-osmic-acetic mixture, and stained with picrocarminate of lithium. × 750.

" 36. Radial section through the micropyle of an egg of *Esox*, 0.94 mm. in diameter, from the same series represented in Fig. 35. × 750.

" 37. Radial section of an egg of *Esox*, 1.5 mm. in diameter, from the same series. × 750.

" 39. Radial section of an egg of *Clupea vernalis*, 0.54 mm. in diameter. Preserved in Perenyi's fluid, and stained with picrocarminate of lithium. × 515.

" 40. Radial section through the egg of *Cyclogaster liparis*, 0.7 mm. in diameter. Ovary of April 26 preserved in Perenyi's fluid, and stained with picrocarminate of lithium. × 515.

" 41. Radial section through the egg of *Cyclogaster liparis*, from an ovary of May 7 preserved in Perenyi's fluid, and stained with picrocarminate of lithium. × 515.

" 42, 43, and 44. Radial sections through the micropyles of three eggs of *Cyclogaster*, about 0.25 mm. in diameter. Ovary of May 7 preserved in Perenyi's fluid. × 515.

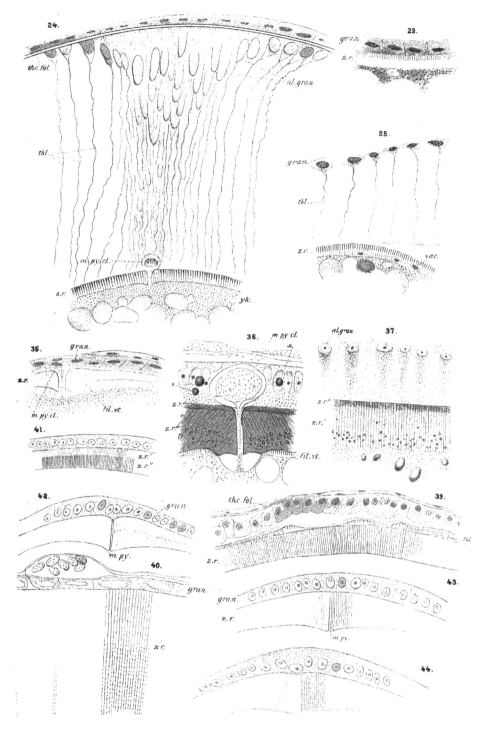

No. 3. — *Report on the Results of Dredging, under the Supervision of* ALEXANDER AGASSIZ, *in the Gulf of Mexico* (1877–78), *and in the Caribbean Sea* (1879–80), *by the U. S. Coast Survey Steamer "Blake,"* LIEUT.-COMMANDER C. D. SIGSBEE, U. S. N., *and* COMMANDER J. R. BARTLETT, U. S. N., Commanding.

[Published by Permission of CARLILE P. PATERSON and J. E. HILGARD, Superintendents of the U. S. Coast and Geodetic Survey.]

XXXII.

Report on the Nudibranchs. By RUD. BERGH.

Während dieser Expedition wurden nur ganz wenige Formen von Nudibranchien gefischt, aber fast alle neu und darunter noch dazu ein Paar ziemlich interessante. Diese Formen waren die folgenden : —

1. *Tethys leporina,* **L.** var.
2. *Chromodoris scabriuscula,* Bgh., n. sp.
3. *Chromodoris punctilucens,* Bgh., n. sp.
4. *Chromodoris sycilla,* Bgh., n. sp.
5. *Phlegmodoris? anceps,* Bgh., n. sp.
6. *Nembrotha gratiosa,* Bgh., n. sp.
7. *Phyllidiopsis papilligera,* Bgh., n. sp.

Fam. TETHYMELIBIDÆ.

TETHYS, L.

R. Bergh, Malacolog Untersuch. (Semper, Philipp. II. ii.), Heft IX., 1875, pp. 346–362, Taf. XLV. Fig. 19-26, Taf. XLVI. Fig. 1-22, Taf. XLVII. Fig. 1, 2.

H. v. Hering, Tethys. Ein Beitrag zur Phylogenie der Gastropoden. Morpholog. Jahrb., II., 1876, pp. 27-62, Taf. II.

R. Bergh, Notizen üb. Tethys leporina. Jahrb. d. deutschen Malakolog. Ges., IV. 4, 1877, pp. 335-339.

R. Bergh, Beitr. zur Kenntn. d. Aeolidiaden. VII. Verh. d. k. k. zool. bot. Ges. in Wien, XXXII., 1882, pp. 67-68.

H. de Lacaze-Duthiers, Sur le Phænicurus. Comptes Rend., Cl. I., 1885, pp. 30-35.
R. Bergh, Sur la Nature du Phænicure. Arch. de Zool., 2 ser., III., 1886, pp. 73-76.
H. de Lacaze-Duthiers, Contrib. à l'Hist. du Phænicure. Arch. de zool., 2 ser., IV., 1887, pp. 77-108, Pl. IV.
List, Zur Kenntn. d. Drüsen im Fuss von Tethys fimbriata, L. Arb. aus dem zoolog. Institut zu Graz, I. 6, 1887, pp. 287-305.

Diese merkliche aberrante Nudibranchien-Gruppe ist erst durch die zwei ersten der obengenannten Arbeiten näher bekannt worden und diese Kenntniss ist nicht ohne wesentlichen Einfluss auf das Studium der ganzen Gruppe gewesen.

Die untenstehende Untersuchung hat wesentlich nur dadurch Interesse, dass sie das Vorkommen von einer *Tethys*, der altbekannten oder einer neuen Form, *im westlichen Theile des atlantischen Meeres* nachweisst.

T. leporına, L. var.

Tafel I. Fig. 1.3.

Hab. M. atlant. occ. (Dominica).

Von dieser Form wurde ein Individüum in der Nähe von Dominica aus einer Tiefe von 138 Faden hinaufgefischt.

Das in Alkohol ganz schlecht bewahrte, verdrehte, theilweise erhärtete und Papillenlose Individuum hatte eine *Länge* von 4.3 cm., von welchen die volle Hälfte auf dem Segel kamen, der Querdurchmesser des letzteren 3 cm.; die Höhe der Rhinophorscheide 7 mm., der Keule 2 mm.; die Länge der Randfäden bis 10 mm.; die Länge des Mundrohres 4 bei einem Durchmesser am Grunde von 3.5 mm.; die Breite des Rückens bis 13 mm.; die Höhe des Körpers bis 10 mm.; die Länge des Fusses 2.5 bei einer Breite bis fast 2 cm., der Vorderrand 7 mm. frei vortretend. — Die *Farbe* der Aussenseite des colossalen Segels ist gelblichweiss wegen dichtgedrängter ganz feiner gelblichweisser Püncktchen, die gegen den Rand hin zu unregelmässigen Fleckchen fast zusammenfliessen. Die Unterseite des Segels ist hinten kohlenschwarz so wie auch das grosse Mundrohr (aussen und innen), wird dann in der mittleren Strecke mehr braungrau, gegen den Rand hin schwärzlich und (theilweise fleckig) schwarz; die Randfäden des Segels meistens gelblichweiss, der Boden, auf dem sie sitzen, aber schwarz. Die Scheide der Rhinophorien schwarz, mit grossen gelblichweissen Flecken; die Keule am Grunde schwarz, sonst weisslichgelb. Der Rücken und die Körperseiten fast von der Farbe der Oberseite des Segels, aber mehr gelblich und im Genicke so wie in der Gegend des Rückenrandes starke, grosse, kohlenschwarze Flecken. Die Rückenpapillen fehlten ganz; die Kiemen (am Grunde der Papillenfacetten) weisslich, Die obere Seite des Fusses ringsum wie die Körperseiten gefärbt; der Fussrand weisslichgelb; die Fusssohle braungrau.

Der grosse *Segel* wesentlich wie bei der typischen Tethys des Mittelmeeres; an der Innenseite der Randparthie stehen die *Randfäden* in meistens 4–6 (8) sehr undeutlich geschiedenen Reihen; die äussersten sind ganz klein, die innersten von bedeutender Länge; die *Dorsalen Cirrhen* des Segelrandes kamen in gewöhnlicher etwas sparsamer Menge vor und von einer Höhe bis 2 mm. Das starke *Mundrohr* am Vorderende (Fig. 1 a) in gewöhnlicher Weise geklüftet; der gähnende Aussenmund bis an den Rand und bis in die Tiefe, bis an die schnürlochartige Pharynxöffnung mit starken Höckerchen, nur ausnahmsweise reihegeordnet, besetzt.   Im Genick, dicht an der Gegend des hintersten Theils des Segels, ziemlich weit von einander stehend, die zusammengedrückten, oben etwas breiteren *Rhinophorien*, deren vorderer Theil oben eine Vertiefung mit umgeschlagenem Rande trägt, in welcher sich die zurückgebogene Keule fand; diese letztere etwas abgeplattet, mit 11 breiten Blättern. — Die Körperform wie in der typischen Tethys, der *Rücken* nur vielleicht etwas breiter.   Am gerundeten Rückenrande, wie es schien, 7 rundliche *Papillenfacetten* gewöhnlicher Art, die Papillen selbst aber fehlend (wie so oft bei Exemplaren von Tethys) ; dicht neben jeder Facette zwei Kiemenbüschel, ein vorderer kleinerer, ein hinterer grösserer; die Kiemenbüschel wie gewöhnlich. Vor der zweiten rechten Papillenfacette die etwas hervorragende *Anal-Protuberanz*, neben derselben die *Nierenpore*.   Die *Körperseiten* vorne ziemlich hoch ; aus der Genitalöffnung ragte ein Theil des Penis etwa 2 mm. hervor. Der grosse *Fuss* ganz wie bei der typischen Tethys; eine mediane Längsfurche fehlte nicht hinten an der Sohle.

Die Eingeweidemasse an die Körperwände durch Bindesubstanz geheftet.

Das weisslichgelbe *Centralnervensystem* zeigte die Hauptganglien von einander viel deutlicher geschieden, als ich es sonst bei Tethyden gesehen habe, nur zwischen den beiden pleuralen Ganglien war die Grenze undeutlich.   Die buccalen (vorderen Eingeweide-) Ganglien zwischen dem hinteren Theile der Speicheldrüsen liegend (Fig. 1 d), oval, durch eine Commissur verbunden, die länger als der Querdurchmesser des Ganglions war ; oberhalb der Wurzel des nach vorne gehenden Nerven fanden sich mehrere Nervencellen eingelagert (Ganglion gastro-oesophagale).   Der Riechknoten am Grunde der Keule des Rhinophors.   Kleine (sympathische) Ganglien kamen an und zwischen den Eingeweiden zerstreut vor, besonders im Gebiete des Genitalsystems.

Die kleinen schwarzen *Augen* an der Oberfläche der Gehirnknoten nach aussen fast sessil, oval, von 0.12 mm. grösstem Diam., mit gelber Linse, reichlichem schwarzem Pigmente und ziemlich grossen Retinazellen.   Die *Ohrblasen* als kalkweisse Punkten aussen an der oberen Seite der cerebralen Ganglien neben den pleuralen gelagert, kugelförmig, ganz kurzgestielt (Fig. 3), etwas kleiner als die grossen Nervenzellen, von 0.16 mm. Diam., mit zahlreichen runden und ovalen Otokonien von einem grössten Durchmesser von 0.016–0.02 mm.   Die *Haut* mit Drüsencellen und Drüschen überall reichlichst ausgestattet.

Die Pharynxöffnung unten am Grunde der Mundröhre in die *Speiseröhre* übergehend ; die etwas länger als die Mundröhre war; das vordere Ende (Fig. 1 b) derselben aussen schwärzlich, dann ringartig gelblichweiss (Fig. 1 c), dann wieder und in der übrigen Strecke schwärzlich. Die Innenseite vorne schwarz, mit etwa 15 starken Längsfalten, die sich vorne in den Pharynx hinein' fort-setzen, hinten an dem erwähnten, nicht ganz schmalen, fast farblosen Ringe plötzlich anhalten ; im vorderen Theile der folgenden Strecke kamen wieder etwa 15 starke Falten vor ; diese Falten waren von einer schwach gelblichen Cuticula überzogen, die ganz fein und zierlich gefaltet war. In dem hinteren Theile' der den Schlundkopf repräsentirenden (Fig. 1 b) vorderen Strecke der Speiseröhre mündet jederseits die langgestreckte, feinknotige (Fig. 1 dd, 2) gelblichweisse *Speicheldrüse* ein; der Ausführungsgang ganz (Fig. 2 a) kurz.— Der *eigentliche* aussen schmutzig schwarzblaue *Magen* 5.5 mm. lang, oval, von 3.5 mm. Durchmesser, von den gelblichen vorderen Lebern mit Ausnahme der Mitte der Rückenseite (und des Hinterendes) bedeckt (Fig. 1 c). Ge-öffnet zeigt der Magen feine Längsfalten der Innenseite ; etwas nach vorne findet sich rechts die Oeffnung des Gallenganges der rechten Nebenleber ; schräg gegenüber die Oeffnung für die mit einander verbundenen linke Neben- und Hauptleber. Hinten und rechts setzt sich der Magen in den Darm fort (Fig. 1 g); die schwarze Farbe hört plötzlich und scharf am Pylorus auf. Der Pylorustheil des Darmes ist gelblichweiss, und hier öffnet sich, dicht neben dem Pylorus, wie durch ein Schnürloch der sogenannte zweite Magen. Dieses ziemlich grosse *Diverticulum* (Fig. 1f) ist gelblichweiss, fast kugelförmig, von 3 mm. Durchmesser ; die Innenseite mit einem feinen pennaten Faltensystem. Der (Fig. 1 g) *Darm* erst nach unten und hinten, dann hinaufsteigend, kurz, ziemlich weit, nur in der letzten Strecke enger; aussen mit Ausnahme der ersten Strecke schwärzlich ; die Innenseite schwarz, mit feinerer Längsfalten und einer stärkeren, die von der Oeffnung des Diverticu-lums anzufangen scheint. — Der Magen und der Darm von Nahrung vollge-stopft ; dieselbe bestand aus Massen von kleinen niederen Crustaceen (Cope-poden, Ostracoden) und Stücken von kleinen Decapoden, mit Bruchstücken von kleineren Gasteropod-Schalen und Sandkörnern vermischt.

Die *rechte Nebenleber*, wie erwähnt (Fig. 1 hi), den rechten Theil des Magens mit einem dicken gelblichen Lager einhüllend ; von derselben geht (wenigstens) ein Zweig an die erste rechte Papille (und wahrscheinlich an den (Fig. 1 h) Rhinophorstiel) ab ; diese Lebermasse öffnet sich durch einen ganz kurzen Gallengang in den Magen. Die *linke*, der vorigen ganz ähnliche, *Nebenleber*, den linken Theil des Magens (Fig. 1 kl) einhüllend, sich nach hinten etwas verlängernd und sich mit (Fig. 1 m) dem Ausführungsgange der Hauptleber vereinigend; auch von dieser Leber geht ein Zweig an die Gegend der Facette der ersten Papille ab; diese Leber öffnet sich links in den Magen. Die *Hauptleber* viel grösser als die vorigen, an Länge etwa 1.8 cm. betragend bei einer Breite vorne von 11 und einer Höhe von fast 9.5 mm.; das Vorderende schief nach rechts- hinten- unten abgestutzt und (wegen der vordern Genitalmasse) vertieft; das Hinterende gerundet ; nur central am Vorderende trat die graubraune Farbe der Leber hervor, sonst war sie von

der gelblichen Zwitterdrüse gedeckt; das Organ bestand aus Lappen von ver-
zweigten Läppchen, deren Ausführungsgänge sich allmählich traubenartig
vereinigen und nach und nach den central verlaufenden Hauptgallengang
bilden, welche links am Vorderende frei hervortritt (Fig. 1 *m*) und sich mit
der linken Nebenleber vereinigt. An den Seitentheilen des Ruckens der
hinteren Eingeweidenmasse durchbrechen mehrere Leberzweige das Zwitter-
drüsenlager und steigen an die Papillenfacetten auf.

Das Pericardium und das Herz wie gewöhnlich. — Die *Niere* mit ihrer baum-
artigen Veranstelung von schönen Kolben und Röhren den grössten Theil
der hinteren Eingeweidemasse überziehend und die Längsfurche derselben aus-
kleidend; in der Auskleidung von jenen viele horngelbe und braungelbe rundliche Concremente von einem Durchmesser von meistens 0.025–0.035 mm.
Der Ureter wie gewöhnlich; in denselben öffnet sich der *Pericardialtrichter*,
der kurz- birnförmig war, von 1 mm. Länge, gelblichweiss, mit stark durchschimmernden Längsfalten; der Gang kurz, fast ohne Vegetationen der
Innenseite.

Die gelbliche, die Leber mit Ausnahme des grössten Theils ihres Vorderen-
des überziehende *Zwitterdrüse* wie gewöhnlich; in den Läppchen entwickelte
Zoospermien. Der rechts am Vorderende der hinteren Eingeweidemasse ent-
springende *Zwitterdrüsengang* an die Hinterseite der *vorderen Genitalmasse* über-
tretend. Diese letztere 9 mm. lang bei einer Höhe von 7 und einer Dicke
von 5 mm.; am oberen Rande vorne die Prostata, hinter derselben der Knäuel
der Windungen der Ampulle des Zwitterdrüssenganges, unter dem letzteren
die Samenblase; die Hauptmasse ist von der Schleimdrüse gebildet. Die gelbliche Ampulle durchgehends von beiläufig 0.5 mm. Diam.; aufgerollt hinter
der Prostata einen Knäuel bildend, der ein wenig kleiner als die Prostata war;
ausgerollt mass dieselbe 2.5 cm. Der aus der Theilungsstelle der Ampulle
ausgehende *Samenleiter* etwa doppelt so lang wie der Durchmesser der Prostata. Diese letztere gelblich, fast kugelförmig, von 4 mm. Diam., mit einem
kleinen Nabel der hinteren und einer tiefen Kluft der Vorderseite, aus welcher
die Fortsetzung des Samenleiters hervortritt; die Oberfläche fein körnig, der
Bau ganz der gewöhnliche. Die aus der tiefen Kluft vortretende Fortsetzung des Samenleiters gräulich, ziemlich dünn, etwa doppelt so lang wie der
Durchmesser der Prostata, sich durch den Penis bis an seine Spitze windend.
Der halb hervorgestreckte *Penis* gelblich, lang, kegelförmig; der gewöhnliche Nebensack konnte nicht gefunden werden. Der *Eiergang* geschlängelt
an den Schleimdrüsengang gehend, ausgestreckt beiläufig 1.5 cm. messend,
etwa so dick wie die Ampulle. Die gelbliche, sich in das Vestibulum genitale neben dem Schleimdrüsengange öffnende *Samenblase* birnförmig, von
etwa 5 mm. Länge bei einem Durchmesser von etwa 2.3 mm., von Samen
strotzend; der Ausführungsgang fast ebenso lang, mit starken Längefalten
der Innenseite. Die *Schleimdrüse* gross, kalkweisslich; die Eiweissdrüse
gelblich; der Schleimdrüsengang mit der gewöhnlichen starken Doppel-
falte.

Ob diese Form nun eine (locale) Varietät der bisher nur im Mittelmeere und bei den canarischen Inseln gefundenen Tethys leporina darstellt oder eine eigene Art, muss vorläufig hingestellt werden. Das Erste ist wohl das wahrscheinlichste, obgleich die schwarze Farbe der Verdauungshöle und Abweichungen im Genitalsystem wohl auch die letzte Annahme ermöglichten. Die so träge und nie schwimmende Staurodoris verrucosa kommt doch auch im westlichsten Theile des atlantischen Meeres (unweit von Rio Janeiro) vor (Staurod. Januarii, Bgh.).[1]

Fam. DORIDIDÆ CRYPTOBRANCHIATÆ.

CHROMODORIS, Ald. et Hauc.

Vgl. R. Bergh, Report on the Nudibranchiata. Challenger Exped., Zoöl., X., 1884, pp. 64–78.
Vgl. R. Bergh, Malakolog. Unters., Heft XV. 2, 1884, pp. 64–78, pp. 347–350; Heft XVI. 2, 1889, pp. 831–837.

Die fast immer schlanken und meistens lebhaft gefärbten Chromodoriden haben *einfach gefiederte Kiemenblätter, starke Lippenplatten,* und *die Rhachisparthie der Radula* trägt höchstens nur Verdickungen, aber *keine Zahnplatten.*[2] Die Aphelodoriden,[3] die sonst sehr ähnlich sind, unterscheiden sich durch mehrfach gefiederte Kiemenblätter und durch Fehlen von Lippenplatten.

Die Gattung ist bisher nur aus den wärmeren (Mittelmeere) und den tropicalen Meeresgegenden bekannt. Sie scheint die Artenreichste Gruppe von Doriden zu sein; sie wird hier wieder durch mehrere neue Arten bereichert.

1. Chr. scabriuscula, Bgh., n. sp.

Tafel I. Fig. 11-19.

Hab. M. atlant. occidentale.
Von dieser form wurden am 24° 44′ Lon. und 83° 26′ Lat. (d. h. in der Nähe von Straits of Florida) aus einer Tiefe von 37 Faden 3 Individuen gefischt, die fast vollständig von derselben Grösse und Formverhältnissen waren.

[1] Vgl. Ihering, Zur Kenntn. d. Nudibranchien d. brasilianischen Küste. Jahrb. f. d. Malacolog. Ges., XIII., 1886, pp. 230-233.
[2] Nur die Chr. Scabriuscula, B. macht hier eine Ausnahme.
[3] R. Bergh, Neue Chromodoriden. Malakolog. Bl. N. F. I., 1879, pp. 107-113.
R. Bergh, On the Nudibr. Gaster. Molls. of the North Pacific Oc. (Dall, Explor. of Alaska), II., 1880, Pl. VIII. (XVI.), Figs. 12-18.

Die in Alkohol gut bewahrten Individuum hatten eine *Länge* von 12 bei einer Breite bis 6 und einer Höhe bis 3.5 mm., die Länge des Fusses 10 bei einer Breite bis 2.5 mm.; die Breite des Mantelgebrämes 1.5 mm.; die Länge der Tentakel 0.6 mm.; die Höhe der (zurückgezogenen) Rhinophorien 1.8, der (zurückgezogenen) Kieme 1.5 mm. — Die *Farbe* war durchgehends gelblich-weiss, die Keule der Rhinophorien und die Kieme mehr gelblich.

Die *Form* war länglich-oval, etwas niedergedrückt; die Rückenseite etwas gewölbt, überall bis an den Rand mit ziemlich zahlreichen kleinsten conischen Höckerchen besetzt, die am Mantelgebräme zahlreicher waren; die weit nach vorn stehenden Rhinophorlöcher, und die weit nach hinten stehende Kiemen-spalte schnürlochartig zusammengezogen; die Keule der Rhinophorien stark, mit beiläufig 20 nicht dünnen Blättern; die Kieme aus 9, einem vorderen und jederseits 4, nach hinten an Grösse allmählich abnehmenden, einfach-pinnaten Blättern gebildet; die Analpapille niedrig. Der Kopf klein; die Tentakel kurz-cylindrisch, am Ende gleichsam eingestülpt. Die Unterseite des Man-telgebrämes eben, mit durchschimmernden, gegen den Rand senkrecht gehen-den Spikelzügen. Der Fuss langgestrekt, mit parallelen Seitenrändern; der Vorderrand mit Furche und gerundeten Ecken; der Schwanz 2.2 mm. lang, etwas zugespitzt. Die Genitalpapille mit zwei Oeffnungen neben einander.

Die Eingeweide schimmerten nirgends durch, waren an der Körperwand angeheftet.

Das *Centralnervensystem* zeigte die cerebro-pleuralen Ganglien kurz-nieren-förmig, die nach unten stehenden pedalen Ganglien grösser als die pleuralen; die gemeinschaftliche Commissur ziemlich kurz, kaum so lang wie der Quer-durchmesser des Fussknoten. Die ganz kurzstieligen Ganglia olfactoria un-gewöhnlich gross, fast halb so gross wie die Ganglia cerebralia; die buccalen und gastro-oesophagalen Ganglien wie gewöhnlich. ·

Die ganz kurzstieligen *Augen* ziemlich gross, mit schwarzem Pigment. Die *Ohrblasen* kleiner als die Augen; mit Otokonien gewöhnlicher Art prall ge-füllt, unter denen ein kugelförmiger, der doppelt so gross wie die anderen war. In den Blättern der *Rhinophorien* zahlreiche, auf den Rand senkrecht stehende, gelbliche, harte Spikeln von einem Durchmesser bis 0.03 mm. Die Rückenhaut im Ganzen und besonders die Höckerchen derselben mit ähnlichen Spikeln stark ausgestattet.

Die *Mundröhre* stark, 1.5 mm. lang, wie gewöhnlich. Der kurze *Schlund-kopf* 1.6 mm. lang; hinten an der Unterseite trat die Raspelscheide hervor. Die horngelbe ringartige Bewaffnung der Lippenscheibe unten viel breiter als oben, aus den gewöhnlichen, bis beiläufig 0.027 mm. langen, an der Spitze geklufteten (Fig. 11, 12), dicht zusammengedrängten Häkchen zusammenge-setzt. Die *Zunge* von gewöhnlicher breiter Form mit tiefer Kluft; in der hellgelben Raspel 58 Zahnplattenreihen, weiter nach hinten in der starken Scheide 46 entwickelte und etwa 4 jüngere Reihen; die Gesammtzahl dersel-ben somit 108. In der Raspel jederseits 25 Platten, und die Anzahl weiter nach hinten kaum 30 übersteigend. Die Zahnplatten schwach gelblich; die Breite der medianen Platten 0.01, die ersten lateralen 0.016 mm.; die Höhe der äussersten Platten meistens 0.028, die Höhe der Seitenplatten bis 0.04 mm. Es kamen wirkliche mediane, am Rande gezähnte Platten vor (Fig. 13 *a*, 14). Die innerste laterale Platte (Fig. 13) mit 8–10 Dentikeln des äusseren und 4–5 des inneren Hakenrandes ; an den übrigen Seitenplatten fanden sich, wie gewöhnlich, nur Zähnchen am Aussenrande, aber in sehr variabler Menge, mitunter 5–6, mitunter nur 2–3 (Fig. 15–17) ; an den äussersten (Fig. 18) Platten war der Grundtheil kürzer, und unterhalb der Hakenspitze fanden sich nur 2–3 Zähnchen. — Die langen und weisslichen *Speicheldrüsen* wie gewöhnlich.

Die Speiseröhre etwa so lang wie der Schlundkopf; der Magen wie gewöhnlich; der Darm vor der Mitte der hinteren Eingeweidemasse hervortretend, sein Knie in gewöhnlicher Weise bildend, und in gewöhnlicher Weise verlaufend, gelb. — Die hintere Eingeweidemasse (*Leber*) 6.5 mm. lang bei einer Höhe und Breite von 3.2 und 3.5 mm., vorne sehr schief abgestutzt und hinten gerundet, (gelblich-) weiss. Die *Gallenblase* langgestreckt-birnförmig, weisslich, links am Pylorus erscheinend.

Das Pericardium mit dem Herzen, die weisslichen Blutdrüsen, die Niere und der Pericardialtrichter wie gewöhnlich.

In den gelblichen Lappen der *Zwitterdrüse* grosse Eierzellen. — Die *vordere Genitalmasse* gross, etwa 4 mm. lang, von ovaler Form, planconvex, gelblich ; am Vorderende die starken Windungen des Samenleiters. Die Ampulle des Zwitterdrüsenganges weisslich, geschlängelt. Der Samenleiter lang ; der weissliche prostatische Theil kürzer als der gelbliche muskelöse (Fig. 19 *a*); die kurzkegelförmige, ziemlich dicke glans penis am Boden des (Fig. 19 *b*) räumigen Praeputiums kaum vortretend. Die Spermatotheke kurz-birnförmig, die Spermatocyste wurstförmig und kleiner; der vaginale Gang lang, nach vorne weiter, mit einer starken gelben Cuticula ausgefuttert ; die Vagina fast so lang wie das Praeputium, doppelt so dick wie der vaginale Gang, von einer dünneren Cuticula ausgekleidet. Die Schleimdrüse gross; die Eiweissdrüse etwas mehr gelblich.

Diese Form unterscheidet sich von den allermeisten Chromodoriden durch die harten Höckerchen des Rückens und damit durch die ziemlich starke Entwickelung der cutanen Spikeln, so wie besonders durch wirkliche mediane (rhachidiale) Zahnplatten. Auch die Auskleidung des vaginalen Ganges ist eigenthümlich.

2. Chr. punctilucens, Bgh., n. sp.

Tafel I. Fig. 4–10.

Hab. M. atlant. occid.

Ein einziges Individuum wurde aus einer Tiefe von 37 Faden auf 24° 44′ Lon. und 83° 26′ Lat. (d. h. in der Nähe der Straits of Florida) gefischt.

Das in Alkohol bewahrte Individuum hatte eine *Länge* von 3.5 bei einer Breite von 1.6 und einer Höhe von 1.5 cm.; die Breite des Mantelgebrämes 2 (vorne) bis 4.5 mm. ; die Höhe der (zurückgezogenen) Rhinophorien 3, der (zurückgezogenen) Kieme 5 mm.; die Länge des Fusses 3 bei einer Breite bis 1 cm. — Die *Farbe* der obern Seite war durchgehends olivenbraungrau mit ziemlich zahlreich zerstreuten gelben und weissen Punkten, die oft eine schwarze oder schwärzliche Areola zeigten ; am Mantelrande ein schmales, schwarzes, seiner Länge nach durch eine weisslichgelbe oder gelbe Linie getheiltes Band ; die Unterseite des Mantelgebrämes von der Grundfarbe des Rückens oder mehr grau, hier und da schwarzfleckig ; der Rand der Rhinophorlöcher so wie der Kiemenspalte schwarz mit gelben Punkten und Bruchstücken von gelben Linien ; die Keule der Rhinophorien schwarz, am Vorderrande und gegen die Spitze gelblich ; die Kiemenblätter schwärzlich, die Rhachisparthien, die Spitze und theilweise die Ränder der Blätter gelbfleckig ; die Analpapille schwarz mit gelblichem Rande. Die Körperseiten von der Farbe des Rückens, die gelben und weissen Punkte kommen aber sehr sparsam vor. Die Tentakel mit gelber Spitze ; der Aussenmund schwarz. Die Fusssohle graulich ; das Fussgebräme oben von der Farbe der Körperseiten, aber mit starken schwarzen Flecken ; der Fussrand gelb, hier und da mit schwarzen Fleckchen ; am Schwanzrücken zerstreute gelbe Punkte.

Im Aeusseren *simulirte* diese Form (in Alkohol bewahrt) ganz *eine Doriopse*, nur war der Mund wie bei den Doriden, und die Tentakel kurz kegelförmig (an der Spitze, wie bei so vielen Chromodoriden, gleichsam halb eingestülpt). Die Form länglich, die Consistenz weich. Der Rücken etwas gewölbt, das Mantelgebräme ziemlich breit, wellenförmig gebogen, an der Unterseite wie der ganze Rücken eben. Die Rhinophorlócher fast glatrandig ; die Keule der Rhinophorien kräftig, mit beiläufig 30 breiten Blättern. Die Kiemenspalte queroval, fein rundzackig. Die Kieme jederseits aus 7 einfach pinnaten Blättern gebildet, denen sich hinten eine Spirale von 13 Blättern anschliest ; diese letzteren etwas schmächtiger und unbedeutend niedriger als die andern, die alle fast von gleicher Grösse waren. Hinten zwischen den Spiralen die cylindrische, oben abgestutzte, etwa 3 mm. hohe Analpapille ; rechts und vorne neben derselben die Nierenpore. Die Körperseiten ziemlich hoch ; die Genitalpapille wie gewöhnlich. Der Fuss vorne gerundet-abgestutzt, mit feiner Randfurche ; das Fussgebräme nicht schmal ; der Schwanz stark, nicht kurz.

Das gelbe *Centralnervensystem* von den Blutdrüsen bedeckt, in reichliche, fest anhängende Bindesubstanz gehüllt ; die Ganglien ziemlich dick. Die zwei Abtheilungen der cerebro-pleuralen Ganglien sehr ausgeprägt ; die pedalen ausserhalb und unterhalb der vorigen liegend ; die pleuralen grösser als die cerebralen, die pedalen wieder grösser als die pleuralen ; die gemeinschaftliche Commissur weit, doppelt so lang wie der Querdurchmesser des Centralnervensystems. Die Riechknoten, die buccalen und die gastro-oesophagalen Ganglien wie gewöhnlich.

Die *Augen* fast sessil, mit schwarzem Pigment. Die *Ohrblasen* so gross wie die Augen, mit Otokonien gewöhnlicher Art prall gefüllt. In den Blättern der Keule der *Rhinophorien* keine Spikel. In der *Haut* des Rückens kamen erhärtete Zellen sparsam vor.

Die *Mundröhre* sehr stark, etwa 6 mm. lang bei einem Diam. hinten von 6 mm.; aussen gelblich, innen vorne schwarz und hinten gelblich; die starken Retractoren wie gewöhnlich. — Der sehr kräftige *Schlundkopf* 5.5 mm. lang bei einer Breite von 4.5 und einer Höhe von 4.5 mm.; das abgeplattete Hinterende stark schräge; von der Unterseite ragt die starke (1.1 mm. in Diam. haltende) Raspelscheide 3 mm. nach oben und links empor. Die runde Lippenscheibe von 4 mm. Diam., von der schön dunkel ambergelben *Lippenplatte* (Fig. 4) überzogen, welche oben schmäler, unten (von vorn nach hinten) viel breiter (bis 2.5 mm.) war, unten continuirlich, oben durch einen ganz schmalen Zwischenraum in zwei Hälften geschieden. Die Lippenplatte in gewöhnlicher Weise von dicht zusammengedrängten gelblichen Stäbchen gebildet, welche (in gerader Linie gemessen) eine Länge bis zu fast 0.06 mm. erreichten, gebogen und in der Spitze gekluftet (Fig. 5, 6) waren. Die *Zunge* breit, abgeplattet, mit breiter Kluft; in der gelblichen Raspel 60 Zahnplatten, weiter nach hinten und in der ziemlich langen Raspelscheide 98 entwickelte und 12 jüngere Reihen; die Gesammtzahl derselben somit 170. Die vordersten 16–18 Reihen sehr incomplet. In den hintersten Reihen der Zunge fanden sich jederseits bis 53 Seitenzahnplatten, und die Anzahl stieg kaum wesentlich weiter nach hinten. Die Zahnplatten gelblich; die Höhe der äussersten Platten 0.04–0.05 mm. betragend, allmählig stieg die Höhe der Platten bis zu etwa 0.1 mm.; die Länge der medianen (Fig. 7 *a*) Verdickungen meistens 0.035 mm. Die Zahnplatten von der gewöhnlichsten Hakenform; an den äussersten ist der Körper in gewöhnlicher Weise reducirt, und die Platten mehr aufrecht. Die innerste (Fig. 7 *bb*) Zahnplatte an beiden Rändern des Hakens gezähnelt; alle die übrigen (Fig. 8, 9) nur am äusseren Rande mit 6–10 feinen Dentikeln; die 5–7 äussersten (Fig. 10) ohne Dentikel.

Die weissen *Speicheldrüsen* sehr langgestreckt, vorne etwas dicker, sich bis an die Unterseite der hinteren Eingeweidemasse hinab erstreckend.

Die *Speiseröhre* kaum länger wie der Schlundkopf bei einem Durchmesser von beiläufig 1 mm. Der in die hintere Eingeweidemasse eingeschlossene *Magen* rundlich, nicht klein. Der *Darm* vor der Mitte der oberen Seite die hintere Eingeweidemasse durchbrechend, in gewöhnlicher Weise verlaufend und sein Knie bildend; im ganzen 7 cm. lang bei einem Durchmesser von 1.5–2 mm. — Der Inhalt der Verdauungshöhle war ganz unbestimmbare thierische Masse, worin Stücke von Zahnplattenreihen des Thieres selbst.

Die hintere Eingeweidemasse (*Leber*) 2 cm. lang, bei einer Höhe und Breite von 1.2 cm.; nach hinten zugespitzt; die vordere Hälfte der rechten Seite (durch die vordere Genitalmasse) stark abgeplattet; die Substanz gelb. Die *Gallenblase* links neben dem Pylorus, sackförmig, von 4 mm. Länge, gräulich.

Das *Pericardium* gross, queroval, von 8 mm. kurzestem Diam. Die gelbe Herzkammer von 3.5 mm. Länge. Die *Blutdrüsen* in den Rändern etwas lappig, graugelb, abgeplattet; die vordere gestreckt-herzförmig mit der Spitze nach vorn, 6 mm. lang; die hintere breit, querliegend, 7 mm. breit. — Die Verbreitung der *Niere* über die hintere Eingeweidemasse sehr schön, die͵Urinkammer weit; der *Pericardialtrichter* stark, birnförmig, 2 mm. lang.

Die *Zwitterdrüse* mit einem 0.5–1 mm. dicken, mehr gelben Lager den grössten Theil der Leber überziehend; in ihren Läppchen grosse Eizellen. — Die (sehr stark erhärtete) *vordere Genitalmasse* gross, planconvex, 14 mm. lang bei einer Breite von 7 und einer Höhe von 11 mm. Am Vorderende die ziemlich dicke, geschlängelte, opak-weissliche Ampulle des Zwitterdrüsenganges. Der Samenleiter lang, gewunden, der gelbliche prostatische Theil kürzer als der muskulöse; die glans penis kegelförmig. Die Spermatotheke von ovaler Form, von 3.5 mm. Länge; die Spermatocyste wurstförmig, gebogen, ein wenig länger. Die Schleimdrüse gräulichgelb und kalkweiss, die Eiweissdrüse bräunlich; der äusserste Theil des Schleimdrüsenganges schwarz.

Dieses Thier repräsentiert gewiss eine neue Art. Unter den wenigen bisher [1] bekannten Arten aus dem westlichen atlantischen Ocean (Chr. Moerchii, B.; Chr. gonatophora, B.) giebt es keine zu welcher sie hingeführt werden könnte, und eben so wenig kann sie mit irgend einer der vielen im Mittelmeere vorkommenden identificirt werden.

3. Chr. sycilla, Bgh., n. sp.

Tafel III. Fig. 5-13.

Hab. M. atlant. occ. (Sin. Mexicanum).

Von dieser Form hat die Blake-Expedition 16 Meile gegen Nord von den Jolbos-Inseln (an der Küste von Yucatan) ein einziges Exemplar gefischt, aus einer Tiefe von etwa 14 Faden.

Das in Alkohol gut bewahrte, nur etwas zusammengezogene Individuum hatte eine *Länge* von 2.5 bei einer Breite von 1 und einer Höhe von 1.4 Cm.; die Höhe der (zurückgezogenen) Rhinophorien fast 4, der (zurückgezogenen) Kieme 4.25 mm.; die Breite der Fusssohle 4.5 mm. — Die *Grundfarbe* des Körpers war ein sehr schönes und lebhaftes Dunkelblau. Diese Farbe war am Rücken wie an den Seiten von zahlreichen, kalkweissen, dünnen, oft zerstückelten Längslinien durchzogen; die Stückchen mitunter an dem einen oder anderen Ende kurz- schlingen- oder ösenförmig oder mit einem kurzen Seitenaste; zwischen den Linien kamen noch hier und da einzelne rundliche oder ovale Fleckchen vor. Am Rücken fanden sich etwa 9–10 solche Linien vor, an den Körperseiten 5–6. Der Mantelrand (Fig. 13) so wie der Fussrand mit einer ganz ähnlichen, ebenso unterbrochenen, kalkweissen Linie geziert. Der Rand der Rhinophorlöcher weiss; die Rhinophorien schmutzigblau. In dem theilweise weissen Rande der runden Kiemenspalte endigt die grösste Zahl der weissen Rückenlinien; die Kiemen-

---

[1] Vgl. die von mir vor einigen Jahren gelieferte Liste in Challenger Exped., Zoöl., X., 1884, pp. 65–72.

blätter sehr schön blau; ihre an der Aussenseite ziemlich breite Rhachis weiss gerändert, der schmale innere Rhachisrand mitunter auch weiss. Die Fusssohle schmutzig gelblich.

Die *Formverhältnisse* wie bei den meisten Chromodoriden; das Stirngebräme, der Mantelrand und das Schwanzsegel schmal. Ringsum an der Unterseite des Mantelgebrämes fanden sich grössere und kleinere, durchsichtig-gelbliche, kugelförmige, sessile, ungleichgrosse Blasen (Fig. 13 *aa*) von einem Durchmesser von beiläufig 0.3–2 mm; die grössten kamen am Schwanzsegel vor (Fig. 13 *aa*); jede zeigte am Scheitel eine meistens schon unter der Loupe sehr deutliche Oeffnung. Oberhalb und ausserhalb des Aussenmundes jederseits ein gleichsam eingestülpter Tentakel. Die etwas zusammengedrückte Keule der Rhinophorien mit etwa 40–45 breiten Blättern. Die Kieme weit nach hinten stehend, mit 12 schönen Blättern, von welchen das hinterste Paar kleiner, die übrigen fast gleichgross. Im Centrum des Kiemenkreises die niedrige (oben weisse) Analpapille, rechts und vorn neben derselben die Nierenpore. Der Fuss wie gewöhnlich ziemlich schmal.

Das Peritonaeum farblos oder hier und da bläulich.

Die das *Centralnervensystem* eng einhüllende starke Bindesubstanzcapsel mit der Unterseite der vorderen und mit dem Vorderende der hinteren Blutdrüse innig verwachsen. Die Ganglien an der Unterseite der ganzen Ganglienmasse deutlich geschieden. Die cerebro-pleuralen Ganglien länglich-nierenförmig, die cerebrale grösser als die pleurale Abtheilung; die rundlichen pedalen Ganglien etwas grösser als die cerebralen. Die grosse gemeinschaftliche Commissur ziemlich weit, doppelt so lang wie der Querdurchmesser des Centralnervensystems. Die proximalen und distalen Riechknoten wie gewöhnlich. Die buccalen Ganglien oval, fast unmittelbar mit einander verbunden; die gastro-oesophagalen sehr kurzstielig, etwa $\frac{1}{4}$ der vorigen betragend.

Die *Augen* mit schwarzem Pigment und schwach gelblicher Linse, durch einen kurzen N. opticus mit dem kleinen Gangl. opticum verbunden. Die *Ohrblasen* wie gewöhnlich, mit zahlreichen Otokonien gewöhnlicher Art. In den dünnen und breiten Blättern der Keule der *Rhinophorien* kamen zerstreute erhärtete Zellen, aber keine Spikel vor.

Die *Mundröhre* aussen blaugrau, innen gelblichweiss, kurz und weit; der Diam. und die Länge etwa 5 mm. betragend. — Der *Schlundkopf* sehr stark, 6 mm. lang bei einer Breite von 5 und einer Höhe von 4.75 mm., von gewöhnlichen Formverhältnissen; die 3.5 mm. lange, starke Raspelscheide längs des Hinterendes des Schlundkopfes hinaufgekrümmt; die Lippenscheibe gross, gewölbt, mit sehr starker, grünlich- olivenfarbiger *Lippenplatte*. Diese letztere einen etwa 3 mm. breiten Ring bildend oder eigentlich zwei Halbringe, die in der Mittellinie oben und unten durch ein schmaleres Zwischenstück vereinigt sind. Die Platte in gewöhnlicher Weise von ganz dicht gedrängten Stäbchen mit gebogenem hakenartigem Kopf gebildet (Fig. 5); sie erreichten eine Höhe bis zu beiläufig 0.04 mm.; die Stäbchen der erwähnten Zwischenstücke ganz klein. Die *Zunge* breit; in der grünlich- olivenfarbigen Raspel 39 Zahnplattenreihen, weiter nach hinten kamen dazu 41 entwickelte und 4 jüngere Reihen, die Gesammtzahl derselben somit 84. Die 8 vordersten Reihen mehr oder weniger incomplet. Die hintersten Reihen der Zunge enthielten (jederseits) etwa 290 Zahnplatten, und die Anzahl stieg kaum wesentlich weiter nach hinten. Die Zahnplatten schwach gelblich mit etwas grünlichem Anfluge; die Seitenzahnplatten erreichten eine Höhe bis zu 0.14 mm., die der äussersten betrug etwa 0.06–0.08 mm. Die Rhachisparthie sehr schmal, meistens mit einer seichten medianen Längsfalte. Die Zahnplatten von der allergewöhnlichsten Form (Fig. 6–9); die Haken gabelig, der obere Ast länger und mehr gebogen als die untere ; unterhalb dieses *letzteren eine Andeutung von feinen* Rundzacken, die nach aussen in den Reihen besonders etwas deutlicher wurden und selbst in feine Dentikelbildungen übergehen können (Fig. 6). Die (meistens) 5–6 äussersten Platten sind von etwas abweichender Form (Fig. 10, 11), zeigen den Haken reducirt und mit gerundetem Ende ; die 1–2 alleräussersten haben keine Auskerbung oben (Fig. 12).

Die *Speicheldrüse* sehr lang, sich über die Unterseite der vorderen Genitalmasse erstreckend, kalkweiss, dünn; vorne etwa 1.25 mm. breit, in der hinteren Hälfte kaum halb so dick ; die ganz kurzen Ausführungsgänge in die Wurzel der Speiseröhre einmündend.

Die *Speiseröhre* dünn, etwa 14 mm. lang (bei einem Durchmesser von 0.8 mm.), ganz unten am Vorderende der hinteren Eingeweidemasse eintretend und sich in die weite Leber- Magenhöhle öffnend. Der *Darm* die Leber vor der Mitte ihrer oberen Seite durchbrechend, vorwärts gehend ; sein Knie über die vordere Genitalmasse legend und dann nach hinten verlaufend ; die Länge des Darmes im Ganzen etwa 5 cm. betragend, bei einem wechselnden Durchmesser von 1.5–4 mm. Der weissliche Inhalt des Darmes (und der Leberhöhle) war unbestimmbare thierische Masse, mit langen und spitzen Spikeln vermischt.

Die hintere Eingeweidemasse (*Leber*) war 15 mm. lang bei einer Höhe von 12 und einer Breite von 9 mm. (stark zusammengezogen), hinten gerundet, vorne schief abgestutzt; ihre Substanz hell gelblichgrau. Die *Gallenblase* horizontal an der linken Seite des Pylorus liegend, 4 mm. lang bei einem Durchmesser von 1 mm., gelblichweiss.

Das Pericardium blaugrau. Das Herz wie gewöhnlich. Die grünlich-gelbgrauen *Blutdrüsen* an der oberen Seite mit hell grünlichblauem Ueberzuge, die vordere kleiner, 3 mm. breit bei einer Länge von 2.5 mm.; die hintere grösser, gerundet-dreieckig, die Spitze nach hinten kehrend, 5.5 mm. breit bei einer Länge von 3.5 und einer Dicke von 0.8 mm. — Die Niere wie gewöhnlich ; das pericardio-renale Organ birnförmig, 1.8 mm. lang.

Die gelbliche *Zwitterdrüse* als ein dünnes Lager die Leber fast vollständig überziehend ; in den Läppchen der Drüse kamen reife Zoospermien vor. — Die *vordere Genitalmasse* 8 mm. lang bei einer Höhe von 6 und einer Breite von 4 mm.; die dunkelblauen Hauptausführungsgänge noch 4 mm. lang; das Hinterende der Masse wird zum grössten Theile von der grossen Samenblase gebildet, die aber oben und an der äusseren (rechten) Seite von den Windungen des Samenleiters gedeckt wird. Die Ampulle des Zwitterdrüsenganges opak-gelblichweiss, wurstartig, etwas zusammengebogen, ausgestrèckt an Länge 6 mm. bei einem Durchmesser von beiläufig 0.75 mm. messend. Der lange, viele längere und kürzere Windungen machende, weissliche prostatische Theil des Samenleiters ausgestreckt etwa 5–6 cm. lang bei einem fast durchgehenden Diam. von 0.5 mm.; der mehr gelbliche muskulöse Theil nur beiläufig 12 mm. lang und etwas dünner. Der letztere geht in den sich nach und nach verdickenden, am Ende blauen *Penis* über, der eine Länge von 4.5 bei einem Diam. (vorne) bis zu 1.5 mm. hatte ; nur der unterste Theil desselben ist hohl (Praeputium), auch an der Innenseite blau, am Boden der Höhle die gewöhnliche, wenig vortretende papilläre glans. Die *Spermatotheke* gross, kugelförmig, von 5 mm. Diam., die Ausführungsgänge nicht lang ; die *Spermatocyste* birnförmig, 2.5 mm. lang, ziemlich kurzstielig. Die *Schleim-* und *Eiweissdrüse* kaum die Hälfte der ganzen Genitalmasse betragend, 5.2 mm. lang bei einer Höhe von 4.8 und einer Breite von 3 mm., gelblichweiss und weiss ; der weite Schleimdrüsengang aussen und innen blau.

Ringsum die Gegend der Cardia, an die Leber (Niere?) angeheftet, fanden sich vier, 1.5–2 mm. lange Individuen eines mit dem *Distoma glauci*[1] wenigstens ganz nahe verwandten Thieres.

Man kennt jetzt eine kleine Reihe von *Chromodoriden* (Chr. runcinata, pantharella, sannio (Fig. 15), picturata, camoena, elegans (Fig. 16), glauca, californiensis (Fig. 14), Marenzelleri, gonatophora, sycilla (Fig. 13)), *mit eigenthümlichen blasenartigen Drüsenbildungen am Mantelgebräme*, wozu jetzt auch die hier untersuchte Form gehört. — Sie scheint von den schon bekannten Chromodoriden specifisch verschieden.

PHLEGMODORIS, Bgh.

R. Bergh, Malacolog. Unters., Heft XIII., 1878, pp. 593–597.

Corpus molle quasi subgelatinosum, dorso tuberculoso. Tentacula pro majore parte affixa, applanata. Branchia e foliolis tripinnatis paucis formata. Podarium sat latum, sulco marginali anteriori non profundo, labio superiore capite affixo.

Armatura labialis nulla. Radula rhachide nuda, pleuris multidentatis ; dentes intimi forma simpliciori, reliqui hamati. — Penis inermis.

Die Phlegmodoriden sind von weicher Körperbeschaffenheit, der Rücken mit Knoten und Knötchen bedeckt. Die Tentakel etwas applanirt, zum grössten Theile angeheftet. Die (retractile) Kieme aus wenigen (5) tripinnaten Federn

[1] Vgl. meinen: Report on the Nudibranchiata. Challenger Exped., Zoöl., X., 1884, p. 18, Pl. X., Figs. 5–17.

gebildet. Der Fuss ziemlich breit, mit nicht tiefer vorderer Randfurche, die obere Lippe derselben an die Seiten des Kopfes angeheftet. — *Keine Lippenplatte.* Die Raspel ohne Mittelzahnplatten ; die Seitenzahnplatten ziemlich zahlreich, die innersten von einfacherer Form, die anderen hakenförmig. — Der Penis unbewaffnet.

Die Phlegmodoriden gehören den tropischen, hauptsächlich den indischen Meeresgegenden.

1. *Phl. mephitica*, Bgh.   *M. philippin.*
2. *Phl. areolata* (Ald. et Hanc.).   *M. indic.*
3. *Phl. spongiosa* (Kelaart).   *M. indic.*
4. *Phl.? anceps*, Bgh.   *M. mexican.*

Phlegmod.? anceps, Bgh., n. sp.

Tafel I. Fig. 20-26. Tafel II. Fig. 6.

Hab. M. atlant. occ.

Von dieser Form lag ein einziges, in Alkohol mittelmässig conservirtes Individuum vor, an der Long. 89° 16′ und Lat. 23° 13′ (d. h. im mexicanischen Golfe) aus einer Tiefe von 84 Faden gefischt.

Die *Länge* des Individuums betrug 10 mm. bei einer Breite bis 5 und einer Höhe bis 2 mm. ; die Länge des Fusses 7 bei einer Breite bis 2.2 mm.; die Breite des Mantelgebrämes 2 mm.; die Höhe der Rhinophorscheide 0.8, des Kiemenhügels 1 mm. — Die *Farbe* war durchgehends hell schmutzig gelblich, am Rücken mit dunklen erhabenen Punkten (Höckerchen). Die Consistenz des Körpers ziemlich weich.

Die *Form* länglich-oval, abgeplattet, mit breitem und ziemlich dünnem Mantelgebräme. Der Rücken mit Andeutung von einem medianen und jederseits einem, der Grenze des eigentlichen Rückens folgenden, lateralen Kamme; der Rücken übrigens überall mit zerstreuten spitzen Höckerchen bedeckt, die, besonders am Mantelgebräme, durch Ausläufer oft mit einander verbunden waren; am medianen Kamm so wie an den hohen Rhinophorscheiden (Fig. 20), und am hohen Kiemenhügel waren die Hökerchen höher und dichter stehend, besonders am Rande von jenen und diesem. Die Keule der Rhinophorien beiläufig so hoch wie die Rhinophorscheide, mit etwa 25 dünnen Blättern; die Kieme aus 5, bis 1.2 mm. hohen, einfach- hier und da doppelt- gefiederten Blättern gebildet, von denen die 3 vorderen höher ; die Analpapille niedrig. Die Unterseite des Mantelgebrämes eben. Die Körperseiten ganz niedrig; die Genitalpapille wie gewöhnlich. Der Fuss nicht schmal, vorn gerundet und mit Randfurche; die obere Lippe stark vorspringend, in der Mitte ausgerandet; der Schwanz nicht ganz kurz. Die Tentakel fingerförmig.

Das *Centralnervensystem* abgeplattet; die cerebro-pleuralen Ganglien ziemlich rundlich, die Grenze zwischen den zwei Abtheilungen derselben wenig ausgeprägt; die pedalen Ganglien rundlich, grösser als die pleuralen, ausserhalb derselben liegend. Die proximalen Riechknoten fast sessil, ziemlich gross; die einander fast berührenden buccalen und die gastro-oesophagalen Ganglien wie gewöhnlich; die kugelförmigen sessilen Ganglia optica kleiner als die Augen.

Die *Augen* ziemlich gross, fast sessil, mit reichlichem schwarzem Pigment. Die *Ohrblasen* etwas kleiner als die Augen, von beiläufig 0.08 mm. Diam., von Otokonien gewöhnlicher Art strotzend, die einen Durchmesser bis 0.009 mm. erreichten. In den Blättern der *Rhinophorien* lange, aber nicht stark erhärtete, auf dem Blattrand senkrecht und schiefstehende Spikel. In der *Rückenhaut* sehr zahlreiche, lange, mehr oder weniger erhärtete Spikel, die auch, und zum Theile bündelweise, in den Höckerchen vorkommen, hier aber weniger erhärtet und meistens mit den Spitzen am Scheitel der (Fig. 6, Fig. 21) Höckerchen hervorragend; eben derselben Art war das Verhältniss an den Rhinophorscheiden und am Kiemenhügel.

Die aussen weisslich, innen gelbliche (Fig. 22) *Mundröhre* stark, etwa 1.5 mm. lang ; hinten scheinen mehrere drüsenartige Körper einzumünden (Fig. 22). — Der kräftige *Schlundkopf* etwa so lang wie die Mundröhre, hinten an der Unterseite trat die Raspelscheide als eine dicke Papille hervor; die kräftige, rundliche, gelblichgraue Lippenscheibe zeigte sich von einer starken gelben Cuticula überzogen. Die *Zunge* breit und etwas abgeplattet; in der breiten gelben Raspel 7 Zahnplattenreihen, von denen die erste sehr incomplet ; weiter nach hinten 8 entwickelte und zwei jüngere Reihen; die Gesammtzahl derselben somit 17. Die Rhachis ziemlich breit, nackt; von lateralen Platten jederseits 17–18 hinten an der Zunge und weiter nach hinten 19–20. Die Zahnplatten horngelb. Die Länge der 4 innersten betrug meistens 0.06–0.08–0.1–0.11 mm.; die Höhe des Hakens der Platten übrigens bis 0.11 steigend, die der äussersten nur 0.04–0.06 mm. betragend. Die innersten (Fig. 23, 24) 4 Platten sind wenig gebogen, schlanker, mehr aufrecht; danach entwickelt sich schnell die durch die Reihe bleibende Form (Fig. 25), die allergewöhnlichste Hakenform; die äusserste oder die zwei äussersten Platten mit verkürztem Körper, mehr aufrecht stehend (Fig. 26 *aa*); die äusserste schlanker als die nächst stehenden.

Die weisslichen *Speicheldrüsen* langgestreckt.

Die *Speiseröhre* beiläufig so lang wie der Schlundkopf, ziemlich weit. Der 1.5 mm. lange, freie *Magen* und der Darm wie gewöhnlich. Die Verdauungshöhle leer. — Die hintere Eingeweidemasse (*Leber*) kurz-kegelförmig, vorne schief abgestutzt, hinten gerundet, schmutzig-weisslich.

Das Pericardium mit dem Herzen wie gewöhnlich; ebenso die abgeplatteten, gräulich-weisslichen Blutdrüsen.

Die *Zwitterdrüse* schien den grössten Theil der Leber zu überziehen, kaum etwas heller als diese ; in den Läppchen Massen von Zoospermien. — Die *vordere Genitalmasse* beiläufig 1.5 mm. lang, etwas zusammengedrückt; die Ampulle des Zwitterdrüsenganges ziemlich dick, wurstförmig, gebogen, ausgestreckt ein wenig länger als die Genitalmasse, opak gelblichweiss. Der Samenleiter nicht lang, der kurze Penis schien unbewaffnet. Die Spermatotheke kugelförmig ; die Spermatocyste sackförmig, gebogen, etwas kleiner. Die den grössten Theil der Genitalmasse bildende Schleimdrüse weisslich, die Eiweissdrüse gelblich.

Ob diese Form nun wirklich zur Gattung Phlegmodoris gehört, ist sehr zweifelhaft. Diese Thierform zeigt wie die letztere Gattung die inneren Seitenzahnplatten von einfacherer Form, hat auch eigenthümliche drüsenartige Körper hinten am Mundrohre, so wie stark vortretende Rhinophorscheiden. Die Kieme ist hier aber nicht tripinnat wie bei den Plegmodoriden, und das Vorderende des Fusses scheint von anderer Beschaffenheit.

Fam. DORIDIDÆ PHANEROBRANCHIATÆ.

NEMBROTHA, Bgh.

R. Bergh, Malacolog. Unters., Heft XI., 1877, pp. 450–461.
R. Bergh, Beitr. zu einer Monogr. d. Polyceraden, II. Verh. d. k. k. zool. bot.
  Ges. in Wien, XXX., 1880, pp. 658–663; III. Ib., XXXIII., 1883, pp. 164, 165.

Corpus limaciforme, fere laeve; tentacula breviora, lobiformia; rhinophoria retractilia clavo perfoliato; branchia paucifoliata, foliolis bi- vel tripinnatis; podarium angustius.

Armatura labialis inconspicua vel nulla. Radula sat angusta; rhachis dentibus depressis subquadratis vel arcuatis; pleurae dente laterali majori falciformi singulo et dentibus externis depressis compluribus.

Glandula hermaphrodisiaca hepate connata; prostata discreta nulla; glans penis armata.

In den Formverhältnissen stehen diese Thiere *den Trevelyanen sehr nahe*, zeigen auch den Körper Limax-artig, eben, und den Fussrand von den Körperseiten fast nicht oder nur wenig vortretend. Die Tentakel sind auch kurz, lappenförmig; die Rhinophorien retractil, mit durchblätterter Keule. Die (nicht retractile) *Kieme* auch an etwa der Mitte der Länge des Rückens stehend, aber *aus wenigen (3–5) Federn gebildet.* — *An der Lippenscheibe keine Bewaffnung* oder eine ganz schwache (N. nigerrima). Die Zungenbewaffnung gewissermassen an die der Polyceren erinnernd. An der Rhachis kommen (im Gegensatze zu der nackten Rhachis der Trevelyanen) subquadratische oder bogenförmige, niedergedrückte *Mittelzahnplatten* vor; neben der Mittelzahnplatte eine grosse unregelmässig sichelförmige Seitenzahnplatte; die äusseren Platten niedergedrückt, ohne entwickelten Haken. Die *Zwitterdrüse* ist (im Gegensatze zu dem Verhältnisse der Trevelyanen) *von der Leber nicht gesondert.* Der Penis ist in gewöhnlicher Weise mit Hakenreihen bewaffnet.

Die Nembrothen sind bisher nur aus den tropischen Meeresgegenden bekannt und zwar fast nur aus dem philippinischen und dem Stillen Meere.

Der kleinen Reihe von Arten wird die untenstehende neue aus dem mexicanischen Golfe hinzufügen sein.

1. *N. nigerrima,* Bgh.   *M. philippin., pacific.*
2. *N. Kubaryana,* Bgh.   *M. pacific.*
3. *N. gracilis,* Bgh.   *M. philippin.*
4. *N. cristata,* Bgh.   *M. philippin.*
5. *N. morosa,* Bgh.   *M. philippin.*
6. *N. diaphana,* Bgh.   *M. philippin.*
7. *N. gratiosa,* Bgh., n. sp.   *M. mexican.*
8. *N.? Edwardsi (Angas).   M. pacific.*

N. gratiosa, Bgh. n. sp.

Tafel II. Fig. 1-5.   Tafel III. Fig. 1-4.

Hab. Sinum Mexicanum.

Es fand sich nur ein einziges Individuum vor, an der Breite von 24° 26′ und Länge von 83° 16′ aus einer Tiefe von beiläufig 36 Faden gefischt.

Das *in Alcohol bewahrte* Individuum hatte eine *Länge* von 22 bei einer Höhe von 6 und einer Dicke von 4 mm., die Höhe der Kieme noch 4 mm. betragend; die Höhe der Rhinophorien 2.5, des Schwanzkammes so wie der Rhinophorkämme 1.5 mm.; die Breite des Fusses 2.5 mm. — Die *Farbe* des Thieres wird im Leben prachtvoll gewesen sein; die Grundfarbe des Körpers war jetzt hell gelblich, am Rücken wie an den Körperseiten mit zahlreich zerstreuten, runden und ovalen, grüngrauen und graugrünen Flecken von einem Diam. von meistens 0.6–0.8 mm.; die Rhinophorkämme an ihrem Grunde aussen von einer Linie von ähnlicher Farbe eingefasst, ihr oberer Rand schwarzblau, ebenso der Stirn; die Rhinophorien schwarzblau oben, gelb unten; der Rand der Becherartigen Tentakel schwarzblau, ebenso der Scheitel und der Grund der Höcker des Schwanzkammes und des Fussrandes oben, die Rhachis-Parthien der Kiemenblätter hell gelblich, das Laub schwarzblau; die Fusssohle gelb.

Das Thier war von etwas mehr zusammengedrückter *Form* und länger als andere bekannte Nembrothen. Die Tentakel wegen einer sich ihrer Länge nach erstreckenden Furche fast ohrenförmig, am äusseren Ende etwas gelöst (Fig. 2 *a*). Zwischen den Tentakeln der rundliche Aussenmund. Oberhalb des Mundes tritt der ziemlich schmale, im Vorderrande ein wenig ausgekerbte *Stirn* etwa 1.5 mm. hervor. Hinter dem Stirne erhebt sich jederseits ein starker *Rhinophorkamm* (Fig. 1 *a*) mit gebogenem, ebenem Rande; innen am Grunde des Kammes die rundliche Oeffnung der Rhinophorhöhle, der Rand derselben hinten mit einem vortretenden Zipfel (Fig. 1 *c*); die Rhinophorien kurzstielig, ihre Keule mit etwa 35–40 Blättern. Der *Rücken* schmal, gerundet in die Körperseiten übergehend; ein wenig vor seiner Mitte stand die *Kieme,* von drei doppelt-fiederigen Blättern gebildet, von denen das hin-

MUSEUM OF COMPARATIVE ZOÖLOGY.

derste an seinem Grunde noch ein kleines Blatt trug. Dicht hinter der Kieme die wenig vortretende Analpapille, an ihrem Grunde rechts die feine Nierenpore. Die Mitte des *Schwanzes* (des hinter der Kieme liegenden Körpertheils) trug (in einer Länge von 5 mm.) einen *Kamm*, der sich in mehrere, grössere und kleinere, zusammengedrückte, oben gerundete Höcker erhebt. Die *Körperseiten* ziemlich hoch ; die (zusammengezogene) Genitalöffnung in der Mitte zwischen dem Hinterrande des Rhinophorkammes und der Kieme, etwas nach oben liegend. Der *Fuss* wie gewöhnlich schmal ; der Vorderrand mit tiefer (Fig. 2 *b*) Furche ; das Fussgebräme schmal.

Die Eingeweide schimmerten am Vorderkörper undeutlich (weisslich) durch. — Das Peritonaeum farblos. Die Eingeweidehöhle sich nur bis etwa dicht hinter der Gegend der Analpapille erstreckend.

Das *Centralnervensystem* in eine dünne Bindesubstanzhülle eingeschlossen ; die Ganglien ziemlich dick. Die cerebro-pleuralen Ganglien je ein fast 8-Zahl-ähnliche Masse bildend ; die beiden Abtheilungen derselben fast gleichgross ; die rundlichen, von vorne nach hinten nur ein wenig zusammengedrückten, pedalen Ganglien etwas grösser als die pleuralen ; die gemeinschaftliche Commissur ziemlich kurz, nur noch ein halbes Mal so lang wie der Querdurchmesser des pleuralen Ganglions. Die proximalen Riechknoten fast sessil, zwiebelförmig ; die distalen ein wenig grösser, kugelförmig. Die buccalen Ganglien abgeplattet-rundlich, fast unmittelbar mit einander verbunden, etwa so gross wie die proximalen Riechknoten ; gastro-oesophagale Ganglien wurden nicht gesehen.

Die *Augen* kurzstielig, mit schwarzem Pigment, hellgelblicher Linse. Die *Ohrblasen* etwas kleiner, mit runden und ovalen Otokonien gewöhnlicher Art gefüllt. Die Blätter der Keule der Rhinophorien ohne Spikel. In der *interstitiellen Bindesubstanz* kamen erhärtete Zellen nur sparsam vor.

Die *Mundröhre* kurz und weit, an Länge und in Durchmesser 1.5 mm. messend. — Der *Schlundkopf* von gewöhnlicher Form, 2.6 mm. lang bei einer Höhe und Breite von beiläufig 2 mm. ; vom hintersten Theil der Unterseite ragt die Raspelscheide 0.75 mm. hinab ; die Lippenscheibe ziemlich gross, nur von einer, besonders im Innenmunde und oben, ziemlich starken gelblichen Cuticula überzogen. Die *Zunge* stark, etwas abgeplattet. In der hell horngelben, in der Randparthie (wegen der Aussenplatten) braungelben Raspel 10 Zahnplattenreihen ; weiter nach hinten fanden sich deren 4 entwickelte und 2 jüngere ; die Gesammtzahl der Reihen somit 16. Die vorderste Reihe war auf die mediane Platte und die letzte Aussenplatte reducirt. Die Reihen sonst an jeder Seite der medianen eine laterale und drei Aussenplatten enthaltend. Die medianen und die Aussenplatten stark horngelb, die lateralen fast farblos. Die Breite der vordersten medianen Platten 0.24, der hintersten 0.29 mm. ; die Länge der lateralen Platten hinten an der Zunge 0.56, die Länge der Aussenplatten von innen nach aussen meistens 0.2–0.18–0.14 mm. Die *medianen Platten* (Taf. II. Fig. 3) flach, mehr breit als lang ; der Vorderrand nicht umgebogen, convex, nicht oder kaum in der Mittellinie ausgerandet, mit etwas vortretendem gerundetem Ecken ; der Hinterrand mit dem Vorderrande parallel, etwas dünner als dieser (Fig. 3 *a*); die Seitenränder fast gerade, mit einander parallel. Die *lateralen Platten* (Taf. III. Fig. 1 *aa*, 2) gross, unregelmässig, sichelförmig oder eigentlich gleichsam unregelmässige, ein wenig zusammengebogene, zum Theil am Rücken ausgehöhlte, in den Rändern theilweise verdickte und oben kurz-gekluftete (Fig. 1, 2) Blätter bildend; von dem Doppelthaken der Spitze ist der untere Theil der kleinste.   Von den drei *Aussenplatten*, die alle vorne breiter waren, war die innerste fast doppelt so gross wie die folgende, subquadratisch, mit einem ziemlich starken, nach innen gerichteten Kamm (Fig. 1 *bb*).   Die folgende Platte war ziemlich convex (Fig. 1 *cc*, 3 *aa*), mit Andeutung einer Längsleiste.   Die, äusserste Platte (Fig. 1 *dd*, 3 *bb*) war auch convex, nicht halb so gross wie die vorige.

Die gelblichweissen, nicht recht dicken *Speicheldrüsen* begleiteten den über den Schlundkopf verlaufenden Theil der Speiseröhre; die Ausführungsgänge kurz.

Die *Speiseröhre* etwa 3.5 mm. lang, vorne weiter, hinten schmäler, sich oben am Vorderende der hinteren Eingeweidemasse in die Leberhöhle (den Magen) öffnend.   Der *Darm* aus der letzteren an der linken Seite der Cardia ausgehend; in seinem Verlaufe erst links, dann quer, dann rechts und nach hinten gehend, mehrere grosse Biegungen machend; ausgestreckt beiläufig 16 mm. messend bei einem Durchmesser von 1–1.5 mm., in seiner ganzen Länge (wegen seines Inhalts) kalkweiss.   Der *Inhalt des Darmes* und der weiten Leberhöhle war thierische Masse, theilweise von Bryozoen herrührend, und parenchymatöse-pflanzliche.

Die hintere Eingeweidemasse (*Leber*) 11 mm. lang bei einer Höhe und Breite von 4; sie war fast cylindrisch, hinten gerundet, vorne schief nach unten und vorne abgestutzt; ihre Farbe war aussen schwärzlichgrau, dieselbe aber zum grössten Theil von der Zwitterdrüse verdeckt; die Substanz der Leber und die Wand der weiten Höhle schwarz oder schwarzbraun.

Das Herz wie gewöhnlich.   Die *Blutdrüse* gelblichweiss, queroval, ziemlich abgeplattet, hinter dem Centralnervensystem liegend und etwa so breit wie dieses. — Die Niere wie gewöhnlich, der *Nierentrichter* birnförmig, etwa 0.55 mm. lang, mit etwa 10 Hauptfalten.

Die gelbliche *Zwitterdrüse* mit einem fast einfachen Lager von dichtstehenden meistens an einander stossenden Läppchen (Taf. II. Fig. 4), die Leber fast überall überziehend.   In den Ovarialfollikeln der Läppchen grosse Eierzellen, in der nicht sehr abgeplatteten Testicularplatte keine reife Zoospermien.   Der dünne weissliche *Zwitterdrüsengang* frei an der rechten Seite der Cardia vortretend und längs der Speiseröhre an die *vordere Genitalmasse* verlaufend.   Diese letztere, etwa 2.5 mm. lang bei einer Breite und Dicke von 2.2 mm.; die Ausführungsgänge noch 1.6 mm. vortretend; das Vorderende der Masse wird von der Schlinge der Ampulle des Zwitterdrüsenganges gebildet; hinten an der oberen Seite liegt die grosse Samenblase, und an der äusseren (rechten) Seite schlängelte sich der Samenleiter.   Die erwähnte Ampulle wurstförmig, stark zusammengebogen, ausgestreckt 3 mm. lang bei einem Durchmesser von 0.8.   Der stark geschlängelte prostatische

Theil des Samenleiters etwa 5 mm. lang; der muskulöse beiläufig 4 mm. lang, eine grosse Schlinge bildend, unten endigte derselbe als eine kleine Glans am Boden des etwas dickeren, etwa 0.7 mm. langen *Praeputiums*. In fast dem unteren Viertel des musculösen Samenleiters findet sich eine sich bis in die Glans fortsetzende *Bewaffnung*. Dieselbe besteht aus etwa 10–12 Quincunx-Reihen von kleinen gelblichen Haken, die eine Höhe bis zu beiläufig 0.02 mm. erreichen (Fig. 5). Die *Spermatotheke* (Taf. III. Fig. 4 *a*) kugelförmig, von etwa 1 mm. Durchmesser. Die (von dem Samenleiter verdeckte *Spermato-cyste* ein wenig kleiner, auch (Fig. 4 *d*) kugelförmig; ihr Ausführungsgang etwas länger als die Blase, in den uterinen Ausführungsgang der Spermatotheke (Fig. 4 *c*) übergehend. Die Schleim- und Eiweissdrüse (wie alle die übrigen der vorderen Genitalmasse gehörenden Organe) weiss und gelblich-weiss. Das Vestibulum genitale mit starken Längefalten.

Diese unzweifelhaft neue Form der Gattung Nembrotha scheint der N. dia-phana am Nächsten zu stehen, unterscheidet sich aber schon im Aeusseren deutlich genug durch die starken Rhinophorkämme und durch die ganz ver-schiedene Farbenzeichnung, noch dazu durch die etwas verschiedene Beschaffenheit der Raspel.

Fam. **PHYLLIDIADÆ**.

PHYLLIDIOPSIS, Bgh.

R. Bergh, Neue Beitr. zur Kenntn. d. Phyllidiaden. Verh. d. k. k. zool. bot. Ges. in Wien, XXV., 1875, pp. 661, 670–673, Taf. XVI. Fig. 11–15.

R. Bergh, Malacolog. Unters. (Semper, Philipp. II. ii.), Heft XVI., 2, 1889, pp. 859, 866–867, Taf. LXXXIV. Fig. 23–27.

Dorsum ut in Phyllidiis propriis. Apertura analis dorsalis.
Tubus oralis ut in Doriopsidibus; glandula ptyalina discreta (?).

Die Phyllidiopsen bilden gewissermassen ein interessantes Zwischenglied zwischen den Phyllidien und den Doriopsen. Im Ganzen sehen sie den ächten Phyllidien ähnlich aus und haben dieselbe Lage der Analöffnung. Die Tentakel sind sehr klein und wie bei den Doriopsen ihrer ganzen Länge nach angeheftet. Die Mundröhre ist wie bei den Doriopsen; es scheint, auch wie bei den Doriopsen, eine gesonderte Mundspeicheldrüse (Gland. ptyalina) vor-zukommen.

Die Gruppe ist, wie andere Phyllidiaden, nur aus den tropischen Meeres-gegenden bekannt, und umfasst bisher nur die unterstehenden Arten.

1. *Ph. cardinalis*, Bgh. *M. pacific.* (Ins. Tonga).
2. *Ph. striata*, Bgh. *M. africano-indic.* (Maurit.).
3. *Ph. papilligera*, Bgh., n. sp. *M. mexicanum.*

Phyllidiopsis papilligera, Bgh., n. sp.

Tafel II. Fig. 7-14.

Hab. M. mexicanum.

Von der Form lag nur ein einziges Individuum vor, aus einer Tiefe von 101 Faden an 25° 33′ Br. und 84° 21′ L. (d. h. im mexicanischen Golfe) hinauf gefischt.

Das in Alkohol bewahrte Individuum hatte eine *Länge* von 12 bei einer Breite bis 11 und einer Höhe bis 4.5 mm.; die Breite des Mantelgebrämes 3 mm., die Höhe der (zurückgezogenen) Rhinophorien 1.5 mm.; die Länge des Fusses 7.5 bei einer Breite bis 6 mm. — Die *Grundfarbe* des Rückens weisslich, an derselben viele runde und ovale, grosse und kleine, sammetschwarze (bis 2.5 mm. breiten) Flecken, die meistens Papillen tragen, welche theilweise auch schwarz sind; an der weisslichen Unterseite des Mantelgebrämes schimmerten die schwarzen Rückenflecke durch; die übrige Unterseite (gelblich-) weisslich. Die Rhinophorien und der Aussenmund gelblich.

Die *Form* fast rundlich, etwas gewölbt (Fig. 7), mit breitem dünnem Mantelgebräme. Die Consistenz des Thieres nicht hart, nicht recht steif. Der Rücken eben, aber mit ziemlich zahlreichen, bis etwa 1.6 mm. hohen, zusammengedrückten, mehr oder weniger, besonders an der einen (meistens vorderen) Seite, schwarzfarbigen Papillen bedeckt. Die Rhinophoröffnungen (Fig. 7 a) ziemlich weit von einander liegend, die starke Rhinophorkeule mit etwa 20–25 Blättern. Die Analpore median hinten am Rücken (Fig. 7 b). Der innerste Theil des Mantelgebrämes ist dicht mit quergehenden, meistens an der Mitte höheren, bis 1.5 langen dünnen Blättern bedeckt; hinten begegnen sich die Blätterreihen über den Schwanzgrund, vorn erstrecken sie sich bis an den Aussenmund; die Anzahl der Blätter jederseits 45–50. Keine Spur von Tentakeln wurde gesehen; der Aussenmund fand sich als eine starke durchbohrte Papille vor dem Vorderrande des Fusses. Die Genitalpore an gewöhnlicher Stelle der niedrigen (rechten) Körperseite. Der Fuss gross, breit, vorne abgestutztgerundet und mit Randfurche, das Fussgebräme nicht schmal, der Schwanz nicht kurz.

Das *Centralnervensystem* (Fig. 9) zeigte die cerebro-pleuralen Ganglien nierenförmig, schräge gegen einander liegend, nach vorne convergirend (Fig. 9 ab); die pedalen Ganglien an der Unterseite der pleuralen liegend, grösser als diese, rundlich (Fig. 9 cc); die gemeinschaftliche Commissur doppelt, dünn (Fig. 9 d). Die proximalen Riechknoten fast sessil, zwiebelförmig (Fig. 9); die distalen kugelförmig. Die buccalen Ganglien (Fig. 13 c) an gewöhnlicher Stelle, rundlich, einander berührend.

Die *Augen* fast sessil, von 0.1 mm. Diam., mit reichlichem schwarzem Pigment (Fig. 9). Die *Ohrblasen* weit von den vorigen an der Unterseite (Fig. 9) der Gehirnknoten liegend, von beiläufig 0.06 mm. Diam.; etwa 50–100 ovalen Otokonien von einem Durchmesser bis 0.013 mm. enthaltend, unter denen ein grösserer rundlicher (Fig. 12). In den Blättern der Keule der *Rhinophorien,* wie gewöhnlich, dünne, mehr oder weniger erhärtete, kürzere und längere Spikel, die letzteren zum grossen Theile auf dem freien Rande senkrecht stehend (Fig. 10). In der *Haut* des Rückens eine Unmasse von grösseren und kleineren Spikeln und Bündel von solchen, welche auch unter der Loupe schon durchschimmerten (Fig. 7); im Mantelgebräme waren dieselben zum grossen Theile senkrecht und schräg (Fig. 11) gegen den Rand geordnet; sonst lagen sie mehr ungeordnet. Die Spikel waren von den gewöhnlichen bei diesen Thieren vorkommenden Formverhältnissen (Fig. 10, 11), meistens stark erhärtet, oft glasartig; von einem Durchmesser bis 0.16 mm., von sehr wechselnder Länge, die oft bis über 0.4 mm. stieg. Bündel von ähnlichen Spikel stiegen in die Papillen bis an ihre Spitze auf (Fig. 8). In der *interstitiellen Bindesubstanz* kamen überall Massen von grösseren und kleineren meistens stark erhärteten Spikel vor, so wie verkalkte Klumpen und Kugeln.

Durch den Aussenmund war das Ende des Mundrohres etwas hervorgestülpt; unter jenem fand sich die Oeffnung der Mundröhrendrüse (Fig. 13 *g*). Die gelblichweisse *Mundröhre* (Fig. 13 *a*, 14 *a*) weit, nicht kurz, 2 mm. lang, hinten mit (Fig. 13) einer kreisartigen Einschnürung ; die Innenseite (Fig. 14 *a*) mit Längsfalten ; in das vertiefte Hinterende derselben senkt sich der gelbliche Schlundkopf, der am Boden der Mundröhrenhöhle stark vorspringt (Fig. 14 *b*). Dieser *Schlundkopf* (Fig. 13 *b*) von gewöhnlicher Form, fast cylindrisch, von starker gelblicher Cuticula an der Innenseite überzogen, etwa 2 mm. lang; am etwas engeren Hinterende des Schlundkopfes (Fig. 13 *c*) die buccalen Ganglien. Hinter den letzteren finden sich (Fig. 13 *d*) die gewöhnlichen, hier fast kugelförmigen eigentlichen (*hinteren*) *Speicheldrüsen* (Gl. salivales) (Fig. 13 *d*). Es kommt aber jederseits (?) noch eine längliche, etwas lappige, weissliche *vordere Speicheldrüse* (Gl. saliv. access.) vor, (Fig. 13 *f*), die neben dem Schlundkopf das Hinterende der Mundröhre durchbohrt; sein Hinterende geht in einen bindegewebigen Strang über. Unter dem Schlundkopf liegt die lappige, weissliche *Mundröhrendrüse* (Gl. ptyalina), welche in einen starken Ausführungsgang übergeht, die sich hier nicht in die Mundröhre, sondern unmittelbar unter dem Aussenmunde öffnete (Fig. 13 *gg*).

Das Hinterende des Schlundkopfs geht etwas enger in die gestreckt-schlauchförmige *Speiseröhre* (Fig. 13 *e*) über, welche ein wenig kürzer als der Schlundkopf ist und die obere Seite der hinteren Eingeweidemasse durchbohrt. Die in dieser letzteren eingeschlossene *Magenhöhle* nicht weit. Der *Darm* die Eingeweidemasse am Anfang des letzten Drittels durchbrechend und in gewöhnlicher Weise verlaufend. — Die Verdauungshöhle war leer.

Die hintere Eingeweidemasse (*Leber*) 5.5 mm. lang bei einer Breite von 4 und einer Höhe von 3 mm., vorne schräg abgestutzt, hinten gerundet ; die Substanz gelblichweiss.

Das querliegende Pericardium ziemlich gross; das Herz wie gewöhnlich. Die *Blutdrüse* gerundet-viereckig, gräulichweiss. Die *Niere* in gewöhnlicher Weise die obere Seite der hinteren Eingeweidemasse überziehend; die Urinkammer wie gewöhnlich.

Die *Zwitterdrüse* durch mehr gelbliche Farbe von der Leber hier und da unterscheidbar; in den Läppchen Eierzellen und reife Zoospermien. — Die *vordere Genitalmasse* gerundet-viereckig, beiläufig 3 mm. lang. Die weissliche Ampulle des Zwitterdrüssenganges wurstförmig gebogen. Die Samenblasen weisslich; die Spermatotheke kugelförmig, die Spermatocyste eiförmig. Der Samenleiter nicht lang; das Dasein einer Penis-Bewaffnung konnte nich nachgewiesen werden. Die Schleimdrüse weisslich, die Eiweissdrüse mehr gelb.

Bisher war keine am Rücken Papillen-tragende Form von Phyllidiaden bekannt worden. Diese nimmt in dieser Beziehung eine ähnliche Stellung unter den Phyllidiaden wie die Echinodoris [1] unter den Doriden ein.

[1] R. Bergh, Neue Nacktschnecken der Südsee, II. Journ. d. Mus. Godeffroy, Heft VI., 1874, pp. 19–22, Taf. III. Fig. 4–20.

TAFEL-ERKLÄRUNG.

TAFEL I.

Tethys leporina (L.).

Fig. 1. Verdauungssystem; *a* das (an der Unterseite gekluftete) Mundrohr; *b* vorderer, *c* hinterer Theil der Speiseröhre; *dd* die in den ganz rudimentären Schlundkopf einmündenden Speicheldrüsen, zwischen den Hinterenden derselben die buccalen Ganglien; *e* Hinterende des (ersten) Magens, *f* zweiter Magen, *g* Darm; *h* Zweig der rechten Nebenleber in das Rhinophor, *i* in die vorderste rechte Papille; *kl* linke Nebenleber mit ihren Zweigen, *m* Hauptausführungsgang der Hauptleber.

Fig. 2. Speicheldrüse (linke), mit Cam. gezeichnet (Vergr. 55).

Fig. 3. Otocyste, mit Cam. gezeichnet (Vergr. 350); *a* Stiel.

Chromodoris punctilucens, Bgh.

Fig. 4. Lippenscheibe mit Mundöffnung und Lippenplatte.

Fig. 5. Stück der Lippenplatte.

Fig. 6. Grösste Elemente derselben.

Fig. 7. Von der Rhachisparthie der Raspel; *a* rhachidiale Verdickung, *bb* innerste Seitenzahnplatte.

Fig. 8. Zahnplatte aus dem inneren Drittel einer Reihe.

Fig. 9. Eine der grössten Platten.

Fig. 10. Aeusserer Theil zweier Zahnplattenreihen; *a* äusserste Platte der Reihen.

Fig. 5–10 mit Cam. gezeichnet (Vergr. 350).

Chromodoris scabriuscula, Bgh.

Fig. 11. Elemente der Lippenplatte, von vorne.

Fig. 12. Aehnliche, von der Seite.

Fig. 13. Stück der Raspel; *a* mediane Platte.

Fig. 11–13 mit Cam. gezeichnet (Vergr. 350).

Fig. 14. Mediane Platte, von oben.

Fig. 15, 16. Zahnplatten vom inneren Drittel einer Reihe.

Fig. 17. Eine der grössten Platten.

Fig. 18. Aeusserste Platte einer Reihe.

Fig. 14–18 mit Cam. gezeichnet (Vergr. 750).

Fig. 19. *a* muskulöser Theil des Samengangs, *b* Praeputium mit zurückgezogener Glans; mit Cam. gezeichnet (Vergr. 55).

Phlegmodoris? anceps, Bgh.

Fig. 20. Rhinophorscheide, *a* Grund; mit Cam. gezeichnet (Vergr. 55).

Fig. 21. Höckerchen des Rückens, mit Cam. gezeichnet (Vergr. 200).

Fig. 22. Mundröhre, *aa* Retractoren, *b* Drüsen am Hinterrande des Mundrohres.

Fig. 23. Innerster Theil einer Zahnplattenreihe; *a* erste Platte.

Fig. 24. Aehnlicher von zwei Reihen, *ab* erste Platte derselben.

Fig. 25. Eine der grössten Platten.

Fig. 26. Aeusserster Theil zweier Zahnplattenreihen mit 8 und 9 Platten, *aa* äusserste.

Fig. 23–26 mit Cam. gezeichnet (Vergr. 350).

TAFEL II.

Nembrotha gratiosa, Bgh.

Fig. 1. Rhinophorkamm, in *b* den Rücken übergehend, *c* Rhinophoröffnung.

Fig. 2. *a* Tentakel, *b* Vorderrand des Fusses.

Fig. 3. Mediane Zahnplatte, mit Cam. gezeichnet (Vergr. 200), *a* Hinterrand.

Fig. 4. Läppchen der Zwitterdrüse.

Fig. 5. Haken der Penis-Bewaffnung, mit Cam. gezeichnet (Vergr. 750).

Phlegmodoris? anceps, Bgh.

Fig. 6. Stück der Rückenhaut, vom Rande gesehen; mit Cam. gezeichnet (Vergr. 200).

Phyllidiopsis papilligera, Bgh.

Fig. 7. Das Thier, von der Rückenseite; *a* Gegend der Rhinophor-Oeffnungen, *b* Gegend der Analpore.

Fig. 8. Eine der kleineren Rückenpapillen.

Fig. 9. Das Centralnervensystem, mit Cam. gezeichnet (Vergr. 55); *ab* cerebro-pleurale, *cc* pedale Ganglien; *d* gemeinschaftliche Commissur.

Fig. 10. Rand eines Rhinophor-Blattes, mit Cam. gezeichnet (Vergr. 350).

Fig. 11. Vom Mantelrande; mit Cam. gezeichnet (Vergr. 55).

Fig. 12. Otocyste, mit Cam. gezeichnet (Vergr. 350).

Fig. 13. *a* Mundröhre, *b* Schlundkopf, *c* buccale Ganglien, *d* Speicheldrüsen (Gl. salivales), *e* Speiseröhre, *f* Accessorische Speicheldrüsen (Gl. access.), *gg* Ausführungsgang der Mundröhrendrüse (Gl. ptyalina).

Fig. 14. *a* geöffnete Mundröhre, *b* Vorderende des Schlundkopfs.

TAFEL III.

Nembrotha gratiosa, Bgh.

Fig. 1. Zwei Reihen von pleuralen Zahnplatten (linker Seite) von oben; *aaa* laterale Platten; *bb* innerste, *cc* mittlere, *dd* äusserste Aussenplatte.

Fig. 2. Laterale Platte von der Rückenseite.

Fig. 3. Aeusserster Theil dreier Zahnplattenreihen mit den je zwei äussersten Aussenplatten.

Fig. 1-3 mit Cam. gezeichnet (Vergr. 200).

Fig. 4. *a* Spermatotheke, *b* vaginaler und *c* uteriner Gang; *d* Spermatocyste, *e* Diverticulum des Ausführungsganges der Spermatocyste.

Chromodoris sycilla, Bgh.

Fig. 5. Elemente der Lippenplatte.

Fig. 6. Zwei Zahnplatten aus der Mitte einer Reihe der Zunge.

Fig. 7. Zahnplatte vom inneren Zehntel einer Reihe.

Fig. 8. Zwanzigste Zahnplatte von aussen *ab*.

Fig. 9. Eine der innersten Seitenplatten.

Fig. 10. Vierte Zahnplatte, von aussen *ab*.

Fig. 11. Dritte Zahnplatte, von aussen *ab*.

Fig. 12. Die drei äussersten Zahnplatten, von innen; *a* äusserste.

Fig. 5-12 mit Cam. gezeichnet (Vergr. 350).

Fig. 13. Hinterende des Körpers, von oben (Mantelgebräme), mit den weissen Flecken; *aa* Drüsenbildungen der Unterseite des Mantelgebrämes.

Chromod. Californiensis, Bgh.

(Vgl. R. Bergh, On the Nudibr. Gaster. Moll. of the North-Pacific Ocean. II., 1880, Pl. XIV. Fig. 5. Scientific Results of the Explor. of Alaska, Vol. I. Art. vi., 2.)

Fig. 14. Hinterende des Mantelrandes, von der Unterseite, mit 6 Drüsen beutel; *a* Fuss.

Chromod. sannio, Bgh.

(Vgl. R. Bergh, Malacolog. Unters. (Semper, Philipp. II. ii.), Heft XVII., 1890, Taf. LXXXVII. Fig. 1.)

Fig. 15. Hinterende des Mantelrandes, von der Unterseite, mit 4 grossen Drüsenbeuteln; *a* Fuss.

Chromod. elegans (Cantr.).

(Vgl. R. Bergh, Untersuch. d. Chromod. elegans und villafranca. Malakozool. Blätter, XXV., 1878, Taf. I. Fig. 4.)

Fig. 16. Drüsenbeutel von der Unterseite des Mantelgebrämes.

Plate I.

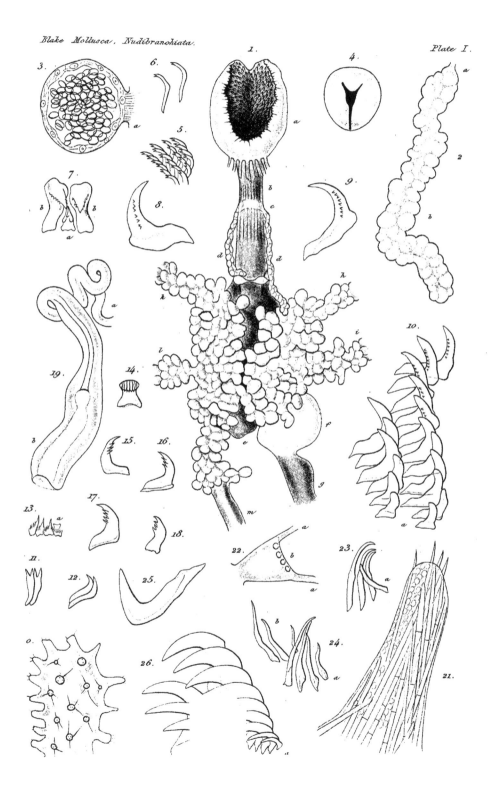

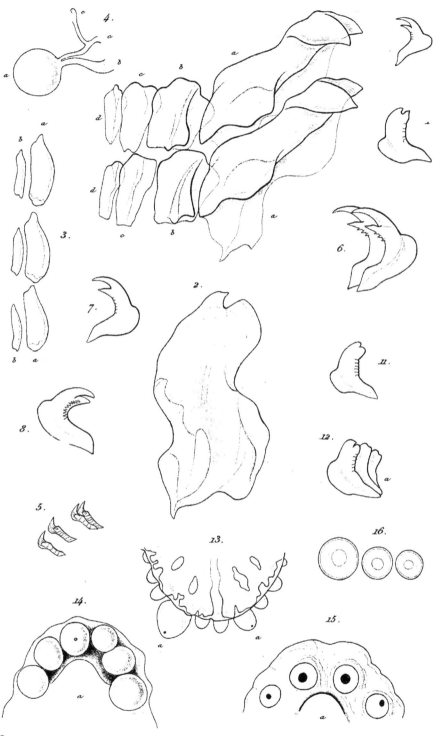

4.

3.

7.

8.

5.

2.

6.

11.

12.

13.

16.

14.

15.

R.B.

L.

No. 4. — *A Third Supplement to the Fifth Volume of the Terrestrial Air-Breathing Mollusks of the United States and adjacent Territories.* By W. G. BINNEY.[1]

As promised in the Second Supplement, the Eastern Province Species are here given, with addenda to those of the other Provinces. My purpose is to bring the subject down to this date. The "Manual of American Land Shells," published subsequently to Vol. V., must also be used in connection with the present paper. I have added figures of many species to replace those of Volume V.

BURLINGTON, NEW JERSEY, January 1, 1890.

SPECIES OF THE NORTHERN REGION.

It must be borne in mind that the Universally Distributed Species are also found here. They are: —

Patula striatella, ANTHONY.

Microphysa pygmæa, DRAP.

Placed in this genus on account of the similarity of its jaw and lingual dentition to those of other species of *Microphysa.* See 2d Suppl., p. 35.

Helicodiscus lineatus, SAY.

Vallonia pulchella, MÜLL.

Pupa muscorum, LINN.

See below, p. 186, for vars. *bigranata* and *Lundstromi.*

It may readily be doubted whether this species is not rather confined to the Northern Region.

---

[1] The Terrestrial Air-Breathing Mollusks of the United States and the adjacent Territories of North America, described and illustrated by Amos Binney. Edited by A. A. Gould. Boston, Little and Brown, Vols. I., II., 1851; III., 1857. Vol. IV., by W. G. Binney, New York, B. Westermann, 1859 (from Boston Journ. Nat. Hist.). Vol. V., forming Bull. Mus. Comp. Zoöl., Vol. IV., 1878. Supplement to same, in same, Vol. IX. No. 8, 1883. Second Supplement, in same, Vol. XIII. No. 2, 1886.

Zonites nitidus, Müll.
arboreus, Say.
indentatus, Say.

See Suppl., p. 139.

Zonites minusculus, Binn.

Dall thus describes a var. *Alachuana* (Pr. U. S. Nat. Mus., 1855, 270): —

A form of it which, at first sight, looks different from *minuscula* is rather larger than usual, and above shows no differences. On the base in the type the junction of the inner lip with the body whorl takes place, following the course of the whorl, inward from the middle line of the base of the whorl and generally about the inner third. This gives a peculiarly thimble-shaped umbilicus. In the variety under consideration, the above mentioned junction takes place outside of the middle line, or even at the outer third, while the aperture is a little dilated. The result of this is to show a much larger portion of the base of the penultimate whorl, and to alter the facies of the umbilicus. For this form, found in Alachua County, Florida, I would suggest the varietal name *Alachuana.*

Zonites viridulus, Mke.
milium, Morse.
fulvus, Drap.

These will not be repeated in the lists of the various Regions into which the Province may be divided. (See Vol. V., p. 17.)

The following are Northern Region Species: —

Vitrina limpida, Gld.
Angelicæ, Beck.
Vitrina exilis, Morelet.

The distinction between the Eastern, Central, and Pacific Provinces not being marked in these high latitudes, this species is given here. It might, perhaps, with *Patula pauper* and *Pupa borealis,* rather be considered a species of the Pacific Province.

Zonites Fabricii, Beck.
Binneyanus, Morse.
ferreus, Morse.
Zonites exiguus, Stimpson.

Plate III. Fig. 4.

The figures are copies of original drawings of Dr. Stimpson.

Zonites multidentatus, Binney.

See Suppl., p. 144.

Acanthinula harpa, Say.
Patula asteriscus, Morse.

Patula pauper, Gould.

See remarks under *Vitrina exilis,* above.

Pupa Blandi, Morse.
borealis, Morelet.

Pupa borealis,
enlarged.

See remark under *Vitrina exilis.*

The figure was drawn by me from a specimen collected at the original locality.

Pupa decora, Gould.
Höppii, Möller.
Vertigo Gouldi, Binney.
Bollesiana, Morse.

A variety *Arthuri,* from Dakota, is mentioned by Von Martens, Gesell. Nat. Freunde zu Berlin, 21 Nov., 1882, p. 140.

Very near, if not identical with, *V. milium.*

Vertigo simplex, Gould.
ventricosa, Morse.

Very near, if not identical with, *V. Gouldi.*

Ferussacia subcylindrica, Linn.

In the mountains of McDonnel Co., North Carolina, a colony of this species was found by Mr. Hemphill. He found no colony of *Vitrina,* which might be expected to exist at those high elevations.

Succinea Haydeni, W. G. B
Verrilli, Bland.
Grönlandica, Beck.
Higginsi, Bland.
Totteniana, Lea.

Dr. Westerlund, in the "Land- och Söttvatten-Mollusker" of the Vega Expedition, quoted in the Manual of American Land Shells, pp. 473, 474, also catalogues from Arctic America the following species:—

Limax hyperboreus, Westerlund. (See below, p. 205.)
Pupa arctica, Wall.
columella, Benz.
Succinea chrysis, Westerlund. (See p. 186.)
turgida, Westerlund.
annexa, Westerlund. (See p. 186.)
Vallonia Asiatica, Nevin.
Pupa edentula, Drap. ?
signata, Ms.
Vertigo Bollesiana, var. **Arthuri.**

Pupa muscorum, var. bigranata, Ross.
     muscorum, var. Lundstromi, Westerlund.
     columella, Benz., var. Gredleri, Clessin.
Krausseana, Reinh.

Of the above, descriptions and figures are given of two only, *Succinea chrysis* and *S. annexa*, which are copied here.

Succinea chrysis, Westerlund.
(Figures copied on my Plate I. Fig. 14.) .

Testa oblongo-ovata, solida, irregulariter transversim striata vel sæpe costulato-plicata, colore varia, sæpissime spira pallidiore, apice rubro, anfractu ultimo antice rotuntiore, subviolaceo-rufescente, postice pallidiore, ubique strigis transversis numerosis albidis; spira elevata, acuta, anfr. 3½, convexi, ultimus deorsum lente attenuatus, penultimus subtus tumidulus, antepenultimus transversalis, extus depressus, sutura forte excisa, anfr. ultimo minutissimo; sutura perimpressa, apertura ovata, intus aureo-micans, pariete arcuatula, obliqua; peristoma obscure marginatum, marginibus æqualiter arcuatis (exteriore superne ad insertionem forte curvato), in pariete callo tenuissimo albido conjunctis.

Long. 11½, diam. 7½, ap. 7½ mm. l., 5 mm. d.; long. 13, diam. 7½, ap. long. 9, diam. 7½ mm.; long. 10, diam. 6, ap. long. 6½, diam. 5 mm.

Asia : America, Port Clarence, Alaska.

Succinea chrysis.

I figure also a specimen from St. Michael's, Alaska (Dall), which has usually been referred to a form of *S. lineata*.

Succinea annexa.
(Figures copied on my Plate I. Fig. 15.)

Testa elongato-ovata, fragilis, intus rugas incrementales fuscas (in spec. max.) validas et extus abruptas dense striata, anfr. penultimo dense distincte spiraliter lineata, anfr. ultimo transversim irregulariter alternatim rufo- et albidostrigata ; sutura impressa ; spira exserta, apice mamillata ; afr. 4, ultimus convexus, penultimus tumidus, antepenultimus altus, exitus convexus, sutura tenui a præcedente sejunctus, summus (subtus visus) globosus; apertura ovata, pariete obliqua, columella arcuata, marginibus linea tenui alba junctis. Long. 11, diam. 8, apert. long. 8, diam. 6 mm.; long. 10, diam. 6½, apert. long. 6, diam. 4½ mm.

Fort Clarence, Alaska.

INTERIOR REGION SPECIES.

Macrocylis concava, Say.
Zonites capnodes, W. G. B.
     fuliginosus, Griffith.
     friabilis, W. G. B.

Zonites lævigatus, PFEIFFER.
Rugeli, W. G. B.

See Suppl., p. 138.

Zonites demissus, BINNEY.

The variety *acerrus* has been found near Fort Gibson, Indian Territory, by Mr. Simpson.

Zonites ligerus, SAY.

A variety *Stonei* is thus described by Mr. Pilsbry : "From Mr. Witmer Stone I have received a form of *Z. ligerus* differing from the type in having a concave, broadly excavated base, with comparatively wide umbilicus, collected by him in New Castle Co., Del. The axis in the type is barely perforated ; but in this form it is a millimeter or more wide, and the base around it broadly concave." (Nautilus, III. 4, p. 46, Aug., 1889.)

Zonites intertextus, BINNEY.
subplanus, BINNEY.

See Suppl., p. 139.

Zonites inornatus, SAY.
sculptilis, BLAND.
Elliotti, REDFIELD.
limatulus, WARD.
capsella, GOULD.
Lawæ, W. G. B.

See Suppl., p. 142, Plate II., Fig. E. The name is suggested for the shell figured by me in Vol. V. (Fig. 44) as *Z. placentula*.

Zonites placentula, SHUTTLEWORTH.

See Suppl., p. 142.

Zonites Wheatleyi, BLAND.

See Suppl., p. 141. Clingham's Peak, N. C. (Hemphill).

Zonites petrophilus, BLAND.

Habersham Co., Ga.; Clarkesville, N. C. (Hemphill). See Suppl., p. 140.

Zonites Sterkii, DALL.

Shell minute, thin, yellowish translucent, brilliant, lines of growth hardly noticeable, spire depressed, four-whorled ; whorls rounded, base flattened, somewhat excavated about the centre, which is imperforate; aperture wide, hardly oblique, not very high, semilunate, sharp-edged, the upper part of the columella slightly reflected ; upper surface of the whorls roundish, though the spire as a whole is depressed. Greater diam. 1.1, height 0.52 mm.

Zonites Sterkii, enlarged.

New Philadelphia, Ohio. Collected on a grassy slope, inclining to the northward, and covered with grass, moss, and small bushes, and so far has not been found elsewhere. Clearly not young of a *Pupilla* or *Zonites*. It is probably one of the smallest species known, and remarkable for its imperforate umbilicus.

The above forms a portion of the description by Dall of *Hyalina Sterkii*, from Proc. U. S. Nat. Mus., XI., p. 214, Figs. 1, 2, 3, 1888. The figure given by me is drawn from an authentic specimen.

Zonites gularis, Say.
suppressus, Say.
cuspidatus, Lewis.

See Suppl., p. 143.

Miss Law thus wrote from Philadelphia, Tenn., of this species : " Unlike *gularis* it seems to be a rare shell, and I find it only by scraping off the surface of the ground in the vicinity of damp mossy rocks. Its habits are more like *placentula* than *gularis*. Neither Miss Clara Bacome nor I ever mistake one for a *gularis*, even before picking it up ; the thickened yellow splotch near the lip, and the thinner spot behind, showing the dark animal through it, as well as its more globular form, particularly on the base, make it look very different when alive."

Zonites lasmodon, Phillips.

Plate III. Fig. 5.

Enlarged drawings by Miss Lawson are given of this species.

Zonites macilentus, Shuttl.

See Suppl., p. 143.

Zonites significans, Bland.

See Suppl., p. 144.

Zonites Andrewsi, W. G. B.

See Suppl., p. 144.

Zonites internus, Say.
Vitrinizonites latissimus, Lewis.

See Suppl., p. 145 ; for other localities, see Man. of Am. Land Sh., p. 231, Also in Washington Co., N. C., and in Watauga Co. at Banner's Elk (Hemphill).

Limax campestris, Binney.

Limax montanus, castaneus, occidentalis, hyperboreus, and *Hemphilli* are probably identical with this.

Tebennophorus Caroliniensis, Bosc.
Tebennophorus dorsalis, Binney.
Tebennophorus Wetherbyi, W. G. B.

See Plate VI. Fig. F.

Tebennophorus Hemphilli, W. G. B.
Plate VI. Fig. H.

See Man. of Amer. Land Sh., p. 247.

The animal is long, narrow, cylindrical, with pointed tail. Its color is black. The jaw is strongly arched, with median projection, and four or five ribs converging to the centre, all crowded on the middle third, the outer thirds being ribless. The lingual membrane has 24–14–1–14–24 teeth, all of same types as figured by Morse for that of *T. dorsalis.* Length of largest individual contracted in spirit 25 mm.

The penis sac is long, cylindrical, receiving retractor muscle and vas deferens at its summit.

Patula solitaria, SAY.
alternata, SAY.
Cumberlandiana, LEA.
perspectiva, SAY.
Bryanti, HARPER.

See Suppl., p. 147.

Helicodiscus fimbriatus, WETHERBY.

See Suppl., p. 148.

A curious form, wanting the epidermal fringe and most of the revolving ridges, was found in great numbers near Fort Gibson, Indian Territory, by Mr. C. T. Simpson. The same form has been found by Mr. Hemphill on Salmon River, Idaho. He proposes for it the name *Salmonacea.*

Strobila labyrinthica, SAY.

A form from Venezuela, without the costæ, is noticed by Dall as var. *Morsei* (U. S. Nat. Mus. Proc., 1855, p. 263).

Polygyra leporina, GOULD.
Hazardi, BLAND.
Troostiana, LEA.
fastigans, SAY.
Stenotrema spinosum, LEA.
labiosum, GOULD.
Edgarianum, LEA.
Edvardsi, BLAND.
barbigerum, REDFIELD.
stenotremum, FERUSSAC.
hirsutum, SAY.

A widely separated locality is the bank of the Yaqui River, near Guaymas (Palmer).

Stenotrema maxillatum, GOULD.
monodon, RACKETT.
Triodopsis palliata, SAY.

Triodopsis obstricta, SAY.
appressa, SAY.

It is quoted by Von Martens from the banks of the Columbia River, but from drawings and description of the single specimen found by Kraus, kindly sent me by Dr. Von Martens, it appears that the species was confounded with flattened forms of *Mullani* or *devius*.

Triodopsis inflecta, SAY.

A depauperated form of this species is about being described and figured as *T. edentula* by Mr. F. A. Sampson.

Triodopsis Rugeli, SHUTTLEWORTH.
tridentata, SAY.

The deformed specimen figured is one of *appressa*, not of this species.

Triodopsis fallax, SAY.
introferens, BLAND.
Van Nostrandi, BLAND.

Also, Jacksonville, Florida.

Mesodon major, BINNEY.

On Plate I. Fig. 2, I have figured the dentition of an individual of this species differing from that figured in Vol. V. Plate VIII. Fig. G, by wanting the

Mesodon major.

side cusps and cutting points of the central and first lateral teeth. The individual from which the lingual was extracted is labelled B in the collection given by me to the United States National Museum. Fig. 3 gives an outer lateral of the same membrane, on which the side cusp and cutting point are present. Fig. 1 gives a central tooth with side cusps and cutting points from the membrane of the specimen labelled A.

The figures show a larger range of variation in the dentition of individuals of the same species than would have been anticipated. (See also *M. Andrewsi*.)

Mesodon albolabris, SAY.
Andrewsi, W. G. B.

In the Manual of American Land Shells, p. 302, I have described and figured specimens of a larger form of this species, which would be called *major* by most collectors, but which has the genitalia and lingual dentition of *Andrewsi*. (See figure above.)

The penis sac of *Andrewsi* was described by me as constricted in the middle. Further study has convinced me that it is rather twisted than constricted. On Plate I. Fig. 4, I give a figure of the genitalia to show this; and in Fig. 5, the penis sac of still another individual.

In studying the lingual membrane of many individuals of *M. Andrewsi*, I have found some variation. I give here notes on membranes of specimens labelled as specified in the Binney collection in the United States National Museum.

AA. 60–1–60 teeth, with about 14 laterals on each side.

N. 51–1–51 teeth, with 11 laterals ; some extreme marginals have decidedly multifid cusps.

Q, from Hayesville, N. C., has also about 11 laterals.

V has 9 laterals, 60–1–60 teeth.

M. 60–1–60 teeth, with about 14 laterals. Some outer laterals have side cusps : one is figured on Plate I. Fig. 12.

G has same count as M; no side cusps to outer laterals.

N has 64–1–64 teeth, with 14 laterals. The extreme laterals have side cusps.

L has 61–1–61 teeth, with 11 laterals ; no side cusps on outer laterals.

J same. 64–1–64 teeth, with 14 laterals.

B. 60–1–60 teeth, with 16 laterals, none with side cusps.

F. All laterals, even first, have decided side cusps (see Plate I. Fig. 10) and cutting points : and marginals also (Fig. 11). 50–1–50 teeth, with 15 laterals.

K. 53–1–53 teeth, with 14 laterals.

I. 50–1–50 teeth, outer laterals with side cusps.

O. 68–1–68 teeth, with 14 laterals.

As remarked above, most collectors will refer this large form of *Andrewsi* to *major*. It differs from that species as hitherto understood very decidedly in its lingual dentition and genitalia. In its shell, also, the species differs from the generally known *major* in so marked a manner, that from it alone I could say, before examination, what were the characters of the dentition and genitalia of every specimen collected by Mr. Hemphill in the mountains of North Carolina. One of the puzzling questions to be left to future solution is the limitation of *albolabris*, *major*, and *Andrewsi*. It must be studied from the lingual dentition and genitalia, as well as from the shell. The student must also consider whether the *Helix major* of the Boston Journal and of the Terrestrial Mollusks are the same species.

Practically, the simplest way of treating specimens in collections is to refer to a variety of *albolabris* all forms more resembling that species than they do the *major* of the Terrestrial Mollusks, and to call *major* all specimens most nearly conforming to the figure and description of that species in Terrestrial Mollusks of U. S., Vols. II. and III. In the former category would be placed the *major* of the Boston Journal; in the latter, the large forms I have referred to *Andrewsi* in Manual of American Land Shells, such, for instance, as are figured in Fig. 322$\frac{1}{2}$, repeated here, ante, page 190. This variety of *albolabris* and this *major*, as above identified, would be found to differ widely in dentition and genitalia, the former in these respects resembling *albolabris*, the latter *Andrewsi*. The latter species must also be recognized as subject to variation, rendering it in some cases difficult to separate from *major*, —never from the large variety of *albolabris*.

The original specimen of *major* of the Terrestrial Mollusks was included in the collection given by Mr. J. S. Phillips to the Philadelphia Academy of Sciences. The points in which it differs from the large form of *albolabris* are pointed out in Terrestrial Mollusks, Vol. II. p. 98.

Mesodon multilineatus, SAY.
Pennsylvanicus, GREEN.
Mitchellianus, LEA.
elevatus, SAY.
Clarki, LEA.
Christyi, BLAND.
exoletus, BINNEY.
Wheatleyi, BLAND.
dentiferus, BINNEY.

In a specimen collected by Mr. Hemphill, at Banner's Elk, N. C., I found the retractor muscle of the penis sac near its junction with the vas deferens, not at half the length of the latter. There was no constriction to the penis sac.

Mesodon Wetherbyi, BLAND.
thyroides, SAY.
clausus, SAY.
Downieanus, BLAND.
Lawæ, LEWIS.
profundus, SAY.
Sayi, BINNEY.

Pupa pentodon, SAY.

The enlarged view of the aperture gives on the left *P. Tappaniana*, on the right *P. curvidens*.

Under the name of *Pupilla Floridana*, Mr. Dall has described what I consider as a form of this species in Proc. U. S. Nat. Mus., 1885, p. 251, Plate XVII. Fig. 11.

Shell greenish spermaceti-white; when living, the tissues of the animal show with pale salmon-color through the shell in the apical whorls; surface smooth or lightly striated, with a tendency to retain dirt upon itself; form subcylindrical, with a rather obtuse apex, the last whorl forming nearly half the shell; suture evident; whorls five, neatly rounded; aperture longer than wide; lip white, thin, reflected; teeth about nine, of which there are generally three larger than the rest, their tips nearly meeting, and their bases mutually nearly equidistant; one is on the pillar, one on the body whorl, and one on the anterior margin; on either side of the latter are two generally subequal much smaller denticles. Lon. 1.60, lat. 0.75 mm.

Habitat. — Under loose oak bark, oak hamak, Archer, Alachua County, Florida, April, 1885, W. H. Dall, sixteen specimens.

This is one of our smallest species, and is related to *P. pentodon* and *P. pellucida.* It is about half the size of the former and much more slender. Its teeth recall those of *P. curvidens,* Gould, in their arrangement, but the shell is more cylindrical and smaller than it is in *P. pellucida* (*servilis*) as figured by Gould. The teeth are more numerous than in the latter shell, and set, as in *P. pentodon,* in one series ; not, as in *pellucida,* partly deeper in the throat.

I describe this with some hesitation, for the condition in which the Pupidæ and Vertigos of North America are is most unsatisfactory, and offers an excellent field to some careful student who shall be able to examine and figure large series of authentic specimens. Still, as there is absolutely no other form with which I feel able to unite this one, it is better to give it a name than to leave it erroneously with some other species.

The above description is copied from that of Dall, while the figure, Plate XVII. Fig. 11, is copied in my Plate III. Fig. 2. I have seen no specimen of it.

Pupa fallax, SAY.
armifera, SAY.
contracta, SAY.

Pupa Holzingeri, STERKI.

Shell narrowly perforated, turrited-cylindrical, vitreous (or whitish), very minutely striate, shining; apex rather pointed; whorls 5, regularly increasing, well rounded, especially the upper ones, the last somewhat narrowed and a little ascending towards the aperture, compressed at the base but not carinated, at some distance from the outer margin provided with an oblique, rather prominent, acute crest corresponding in direction to the lines of growth, extending from the base to the suture, formed by a whitish callosity; behind the crest the whorl is flattened, and corresponding to the lower palatal lamella, impressed; aperture lateral, scarcely oblique, relatively small, inverted subovate, with a slight sinus at the upper part of the outer wall, margins approximated ; peristome moderately reflected; lamellæ 6; one parietal, rather long, very high, in its middle part curved outward, towards the aperture bifurcated, the outer branch reaching the parietal wall ; one columellar, longitudinal, rather high, its upper end turning in nearly a right angle towards the aperture, but not reaching the margin; basal exactly at the base, short, high, dentiform; 3 in the outer wall, viz.: the lower palatal long, ending in the callus, highest at about its middle ; the upper short, rather high on the callus ; above the upper, one supra-palatal, quite small, dentiform, nearer the margin.

Length 1.7 mm., diam. 0.8 mm. = .068 × .032 inch.

As already stated, our species ranges beside *P. armifera* and *P. contracta,* Say, standing nearer the latter. Yet it is different from this species by the shape of the aperture, the wanting callus[1] connecting the margins on the

[1] In many specimens of *P. contracta* so strongly developed that the peristome is rendered continuous.

body whorl, by the longer crest behind the aperture, which in *contracta* disappears in about the middle of the (height of the) whorl, and by the wanting constriction, especially in the columellar wall, not to speak of the size and shape of the whole shell. The lamellæ also show some marked differences, such as the presence of a high basal, the shorter columella not reaching the base, but with relatively larger horizontal part, the bifurcation of the parietal and the presence of a supra-palatal, the last just as it is in *P. armifera.*

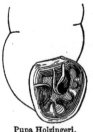

It must be added here that the specimen first obtained from Minnesota in several respects differs from those found in Illinois and Iowa, which I consider as typical; by its size which is one third smaller, by the basal lamella developed in a peculiar way, being rather longer at the truncated top than at its foot, and by the stronger, thicker palatal lamellæ. Yet, as there was only one specimen, it was liable to be an individual peculiarity, — even then of interest. Should, however, more specimens be found with the same configuration, they would represent a distinct and well characterized variety; possibly it is a peculiar northern form.

Pupa Holzingeri, enlarged

New Philadelphia, Ohio, June, 1889.

The above is a description by Dr. V. Sterki[1] of a Pupa received by him from Winona, Minn., and Northern Illinois. He kindly furnished me the above figure.

Pupa rupicola, S<small>AY</small>.
corticaria, S<small>AY</small>.
Vertigo milium, G<small>OULD</small>.
ovata, S<small>AY</small>.
Succinea retusa, L<small>EA</small>.
ovalis, S<small>AY</small>.
avara, S<small>AY</small>.
aurea, L<small>EA</small>.
obliqua, S<small>AY</small>.

SOUTHERN REGION SPECIES.

Glandina Vanuxemensis, L<small>EA</small>.
truncata, S<small>AY</small>.
bullata, G<small>OULD</small>.
decussata, P<small>FEIFFER</small>.
Texasiana, P<small>FEIFFER</small>.

Lingual membrane as usual in the genus. Teeth 35–1–35. Central small, narrow, with a single blunt rounded cutting point. See Plate IX. Fig. G.

[1] The Nautilus, Vol. III., No. 4, p. 37, August, 1889.

Zonites caducus, PFEIFFER.
cerinoideus, ANTHONY.
Gundlachi, PFEIFFER.

Found also in Texas, at Hidalgo, by Dr. Singley.

Zonites Singleyanus, PILSBRY.

Shell minute, broadly umbilicate, planorboid, the spire scarcely perceptibly exserted; subtranslucent, waxen white, shining, smooth, under a strong lens seen to be slightly wrinkled by growth-lines; whorls three, rather rapidly increasing, separated by well impressed sutures, convex, the apex rather large; body whorl depressed, slightly descending, indented below around the umbilicus; aperture small, semilunar, oblique; peristome simple, acute. Umbilicus nearly one third the diameter of the shell, wide, showing all the whorls.

Alt. 1, diam. 2 mm.

New Braunfels, Comal Co., Texas.

Allied to *Z. minusculus*, but much more depressed, more shining, smoother, smaller, with broader umbilicus and a complete whorl less than *minusculus*.

This species, one of the most distinct of the smaller forms of *Hyalina*, was communicated to me by Mr. J. A. Singley, in whose honor it is named. I have also found a few specimens among the shells collected by myself in Central Texas, during the winter of 1885–86. With *Z. Singleyanus* at New Braunfels are found quantities of *Z. minusculus*. The latter species exhibits some variation, being often more depressed than more northern specimens. This depressed form has been noticed in Mexico by Strebel, who proposes for *Z. minusculus* the new generic title of *Chanomphalus*, which is, of course, completely synonymous with *Pseudohyalina*, Morse, 1864, and this, again, is not different enough from *Hyalina* to warrant the erection of a new genus or subgenus. There is some variation in the width of the umbilicus in Texan specimens of *Z. minusculus*, but I have not seen specimens with it so wide as Dr. Dall indicates for his var. *Alachuana* from Florida. *H. elegantulus*, Pfr., is about the size and form of my *Zonites Singleyanus*, but it is a strongly sculptured species.

The above description was published by Pilsbry, Proc. Phil. Acad., N. S., 1889, p. 84, Plate XVII. Figs. 6, 7, 8. A specimen kindly furnished me by Dr. Singley for the purpose is drawn in my figure.

Zonites Singleyanus, enlarged.

Zonites Dallianus, SIMPSON.

Shell minute, depressed, narrowly umbilicated, fragile, pale straw-colored, somewhat shining; under a lens seen to be marked with delicate growth-lines above, smoother beneath. Spire a little convex; apex subacute; sutures scarcely impressed. Whorls three and one half, scarcely convex, the last wide. Aperture oblong-lunate, oblique, upper and lower margins sub-parallel, slightly converging; peristome acute. Alt. 1½, diam. maj. 3, min. 2½ mm.

Zonites Dallianus, enlarged.

West Florida, at Shaw's Point, Manatee Co., and Little Sarasota Bay.

Differs from *Z. arboreus*, Say, in the smaller spire and wider last whorl; fewer whorls; differently shaped aperture. It is about half the size of *Z. arboreus*, and the sculpture is the same as in that species. The *Helix Ottonis* of Pfeiffer, of which specimens from Cuba and Hayti are before me, has no special relationship to this species, but is undoubtedly a synonym of *Z. arboreus*, as Pfeiffer himself concluded. *H. Ottonis* differs from *arboreus* in nothing but the lighter color; the form and dimensions are precisely as in *arboreus*. (See Pfr. in Wiegm. Archiv für Naturgeschichte, 1840, p. 251; the species was never described in the Monographia Heliceorum.)

The aperture in *Z. Dallianus* is less lunate than in *Z. arboreus*, embracing less of the penultimate whorl; seen from beneath, the greater portion of the aperture lies outside of the periphery of the penultimate whorl; whilst in *Z. arboreus* the reverse is the case. The much smaller size of *Dallianus* also separates it from *Z. arboreus*.

This species was sent me under the above name by Mr. Charles T. Simpson, the well known student of Floridian shells. The same form I find in the museum of the Academy, collected by Mr. Henry Hemphill.

The above description was published by Mr. Pilsbry in Proc. Phil. Acad., N. S., 1889, p. 83, Plate III. Figs. 9, 10, 11. A specimen kindly furnished me for the purpose by Mr. Pilsbry is also figured above.

Microphysa incrustata, POEY.

vortex, PFEIFFER.

All the specimens received from West Florida collected by Mr. Hemphill, and from East Florida by Mr. G. W. Webster, are heavily incrusted with dirt.

Microphysa (?) dioscoricola, C. B. ADAMS.

Microphysa
dioscoricola,
enlarged.

Shell minute, subperforate, conic globose, thin, very delicately striate, horn-colored; spire elevated, obtuse; whorls 3–3½, convex, the last medially subimpressed; aperture lunately rounded; peristome simple, acute, the columellar margin subvertically descending, very slightly reflected, diam. greater 1¾, lesser 1⅜, height 1½ mm. (Pfr.).

This species is placed by Von Martens (Die Heliceen, p. 73) in *Conulus*, a subgenus of *Hyalina*, with *fulvus, Gundlachi*, and others. Mr. Dall tells us (Nautilus, III. 25) that it belongs to *Microconus*. This last is synonymous with *Microphysa*, a subgenus of *Zonites*, according to Tryon, Syst. Conch., III. 24.

Mr. Dall says also that the species was originally described from Jamaica by Adams, and subsequently from Trinidad by Guppy as *cœca*. In its jaw and lingual dentition it seems to agree with most of the other species of *Microphysa* which I have examined. I retain it, therefore, in that genus.

The species seems widely distributed in Florida. St. Augustine; Blue Spring, St. John's River; Lake Worth to Hawk's Park along the east coast; Hilo River, emptying into Mosquito Inlet, east coast, not Hillsborough River, emptying into Tampa Bay, as stated by Dall. The specimens examined by me were collected by G. W. Webster at Hawk's Park, "widely distributed in dry places, where other species are not found." Also at Hidalgo, Texas (Singley).

The shell is figured on preceding page.

The jaw (Plate III. Fig. 6) is high, strongly arched, with acuminated ends ; it is very thin, membranous, light horn-colored and transparent ; there are numerous — some fifteen on each side the median line — narrow, delicate ribs, running obliquely to this line, denticulating either margin ; on the upper median portion the ribs meet before reaching the lower margin, leaving upper, median, triangular plates as in *Orthalicus*. The jaw is quite such as I have described and figured for *Macroceramus* in Terr. Moll., V. 384. It also resembles that of *Microphysa turbiniformis* (Ann. N. Y. Acad. Sci., III., Plate XV. Fig. C), excepting that the latter wants the upper median triangular plates. A greatly magnified view of the central portion of the jaw is given.

The lingual membrane is long and narrow. Owing to its small size, it was very difficult to determine the shape of any but the lateral teeth. Three of these last are figured on Plate II., Fig. 5, drawn by camera lucida. They have wide, square bases of attachment, bearing, as usual, two cusps, both stout and blunt, and bearing short, stout cutting points ; the centrals appear of the same shape and tricuspid, but I failed to distinguish them clearly enough to draw by camera ; the laterals are separated, low, wide, quadrate, with long irregularly serrated cusp. I failed also to distinguish these clearly enough to draw by camera. I have represented them in the figure as they appeared to me. The laterals seem like the teeth of *Pupa*, the marginals much like those of *Cionella subcylindrica*. The dentition is somewhat similar to what I have figured of *vortex* on page 356 of the Manual of American Land Shells. There are about 15–1–15 teeth, with six perfect laterals on each side the median line.

Mr. Dall says of this species that the shell is much smaller than that of *granum*, olive-greenish, with a silky lustre and few inflated whorls, the first of which is usually finely punctate. The suture is very deep, and the umbilicus is proportionally larger than in *granum*.

The figure of the dentition of an undetermined species found by Dr. W. M. Gabb, in Costa Rica, published by me in the Annals of the New York Academy of Science, Vol. III. p. 261, Plate XI. Fig. G, is said by Mr. Pilsbry to represent that of this species, — he having identified the shell from which the lingual was extracted to be *H. cœca*, Guppy.

Hemitrochus varians, Menke.
Strobila Hubbardi, Brown.

Polygyra auriculata, Say.

Dall (U. S. Nat. Mus. Proc., 1855, p. 263) thus characterizes a variety *microforis* : —

This form is quite well marked, and when fully adult shows as a rule little variation from the form figured by the Binneys, and generally regarded as typical. A quite uniformly characterized variety was found, however, by me at Johnson's

Sink, Alachua County, Florida, where it was abundant. Some twenty specimens were picked up in a few moments during a hurried visit made with other ends in view, and a quart could easily have been gathered in half an hour. This form is distinguished by its generally smaller size (max. diam. 12.0, min. diam. 10.0, alt. 6.0 mm.) as compared with the type (15.0, 12.0, and 7.9 mm.), and by being more closely rolled, thus having not only an actually smaller umbilicus, but one in which one third less of the preceding whorl is visible. The specimens were uniform in this, and in all other respects were like the typical *auriculata*.

Polygira uvulifera, Shuttleworth.
auriformis, Bland.
Postelliana, Bland.
espiloca, Bland.
avara, Say.
ventrosula, Pfeiffer.
Hindsi, Pfeiffer.
Texasiana, Moricand.
triodontoides, Bland.
Mooreana, W. G. B.
hippocrepis, Pfeiffer.

Through the kindness of Mr. Singerly, I have the opportunity of examining the jaw and lingual membrane.

Jaw long, low, ends blunt; anterior surface with over 14 ribs denticulating either margin.

Lingual membrane long and narrow (Plate III. Fig. 8, *a*, *b*). Centrals tricuspid, laterals bicuspid, marginals low, wide, irregularly denticulate. Teeth 30–1–30, the ninth lateral having its inner cutting point bifid.

Polygyra Jacksoni, Bland.

A form was found abundantly near Fort Gibson, Indian Territory, by Mr. C. T. Simpson, who thus describes it in Proc. U. S. Nat. Mus., 1888, p. 449.

Instead of the bicrural tooth on the body whorl, at the aperture there is a heavy elevated deltoid callus, which is joined to the upper and lower margins of the peristome, and which occupies about the same area as the tooth in the type. The number of whorls is 5; greater diam. 7, lesser 6, height 3 mm. In examining several hundred specimens, I have found none which approach the type, and I would therefore propose for it the varietal name of *deltoidea*.

Polygyra oppilata, Moricand.
Dorfeuilleana, Lea.
Ariadnæ, Pfeiffer.
septemvolva, Say.
cereolus, Muhlfeldt.
Carpenteriana, Bland.

Polygyra Febigeri, BLAND.

pustula, FÉRUSSAC.

pustuloides, BLAND.

Triodopsis Hopetonensis, SHUTTLEWORTH.

Levettei, BLAND.

See 2d Suppl. This species may perhaps be considered one of the Central Province. A variety, however, approaches very nearly the Indian Territory shell lately described as *Mesodon Kiowaensis*. This variety is toothless. It is smooth, like *Levettei*, and has six whorls.

Triodopsis vultuosa, GOULD.

Copei, WETHERBY.

See 2d Suppl. To the synonymy add *Triodopsis Cragini*, Call, Bull. Washburne Coll. Library, I., No. 7, p. 202, Fig. 5, Dec., 1888, Topeka, Kansas. I have seen an authentic specimen, given by Mr. Call to the National Museum. It is figured here.

Mesodon Romeri, PFEIFFER.

divestus, GOULD.

Triodopsis Cragini, enlarged.

The typical form has few separated, very stout ribs ; a variety from Eufala, Indian Territory, sent me by Mr. C. T. Simpson, has numerous fine ribs and revolving microscopic lines. One individual is 24 mm. in greater diameter.

Mesodon jejunus, SAY.

See Manual of American Land Shells, p. 390.

Mesodon Kiowaensis, SIMPSON.

Shell umbilicated, orbicularly depressed, solid, dark brown in color ; whorls 5, with rather coarse striæ, and fine revolving impressed lines, which are much more conspicuous on the last whorl. Suture deeply impressed, leaving the whorls well rounded ; aperture oblique, somewhat transversely rounded, forming fully three fourths of a circle ; peristome thick and solid, whitish or purplish, evenly reflected, with a slight constriction behind it ; umbilicus moderate, deep, exhibiting but little more than one of the whorls. Greater diam. 15, lesser 13, height 7 mm.

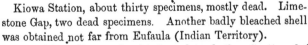

Mesodon Kiowaensis.

Kiowa Station, about thirty specimens, mostly dead. Limestone Gap, two dead specimens. Another badly bleached shell was obtained not far from Eufaula (Indian Territory).

Jaw with 9 ribs; teeth with fewer laterals than *Sayii*, and the inner cusp is bifid on the marginals, while in *Sayii* it is entire (Simpson).

The foregoing description is copied from the Proceedings of the U. S. National Museum, 1888, p. 449, while the figure is drawn from a specimen kindly furnished by Mr. Simpson.

The shell appears to me to be a toothless form of some *Triodopsis*, rather than a *Mesodon* (see above, under *Triodopsis Levettei*). It also resembles nearly some of the toothless forms of *Triodopsis Mullani*.

Acanthinula granum, Strebel and Pfeiffer.

Acanthinula granum,
enlarged.

Shell small, umbilicated, thin, scarcely shining, light horn-colored, with rib-like striæ of growth, crossed obliquely with rib-like folds, in fresh specimens hirsute or with punctate epidermis. Whorls 4½, four of them broad, rounded, regularly increasing in size, rapidly in elevation, the last descending, impressed at the umbilicus. Peristome simple, broadly reflected at its columellar margin, partially covering the deep umbilicus, within with whitish, light thickening. Greater diam. 2.8, lesser 2.6, height 2.8 mm.; of aperture, height 1.2, breadth 1 mm. (Strebel and Pfeiffer.)

Acanthinula granum, Strebel and Pfeiffer, Beitrag zur Kennt. der F. Mex. L. und S. W. Conch., IV., 1880, p. 31, Plate IV. Fig. 13, not Plate IX., as quoted in text.

A Mexican species, found also in Florida ; Archer, Alachua Co.; Evans Plantation, Rogers River ; Lake Worth (Dall).

Mr. Dall says the shell, when perfect, is nearly the size of *labyrinthica*, very thin, reddish brown, with very deep sutures and a rather deep, small tubular umbilicus. It is covered with beautiful deep oblique epidermal ridges, which are easily lost, and do not agree with the lines of growth.

The figure is drawn from a specimen kindly furnished by G. W. Webster.

Dorcasia Berlandieriana, Moricand.
griseola, Pfeiffer.

Bulimulus patriarcha, W. G. B.
alternatus, Say.

I am assured by Dr. Singerly and Mr. Simpson that the form known as *alternatus* does not always have a dark aperture, and the intermingling of the forms leads an observer on the spot to believe *alternatus*, *Schiedeanus*, *Mooreanus*, and *dealbatus* varieties of one and the same species. They were so treated by my father in Vol. II.

Bulimulus Schiedeanus, Pfeiffer,
var. Mooreanus, W. G. B.
dealbatus, Say.

Bulimulus serperastrus, SAY.
multilineatus, SAY.
Dormani, W. G. B.

Bulimulus Floridanus, PFEIFFER.

I have already in Terr. Moll., IV., Plate LXXIX. Fig. 3, figured the front view of the typical specimen in Mr. Cumings's collection, drawn by Mr. G. B. Sowerby. The back view is now offered (Plate III. Fig. 7), received from the same source.

A comparison of the front view of Mr. Sowerby's drawing referred to above, with the figure of *Bulimulus Hemphilli* (Plate III. Fig. 9), recently received from Mr. George W. Webster, will lead one to believe the two to be identical. I so suggested in Manual of American Land Shells (p. 408), when treating the variegated shell figured in Fig. 449 of that work, here repeated. There appear to be two varieties of coloring, one corresponding to Pfeiffer's description, and one to Sowerby's figure.

Bulimulus
Floridanus.

I give the description of *B. Hemphilli* in full, though I believe it to be identical with *Floridanus*.

Shell imperforate, very thin, transparent, amber-colored and marked by coarse lines of growth; body whorl with six revolving and slightly interrupted brownish red bands, the lower two being close together and upon the rounded base, spire obtuse, whorls five, slightly convex, the body whorl constituting two thirds of the entire length of the shell. Suture slight, base uniformly and gracefully rounded. Aperture direct and oval, peristome thin. Length, 19 mm.; diameter, 8 mm. Hab. both coasts of South Florida.

Remarks. Mr. Henry Hemphill, of San Diego, Cal., first found a few dead and badly preserved specimens of this shell in 1884, at Marco, west coast of Florida. These Mr. Binney thought identical with *B. Floridanus*, Pf. (See Manual of American Land Shells, 1885.) Numerous specimens collected during the past summer by the author and Mr. G. W. Webster and son, prove beyond a doubt that this is not identical with the shell figured and described on page 407 of Mr. Binney's Manual. The *B. Hemphilli* is more ventricose, not angular at base, imperforate, differs in color, and in fact there is a general difference.

Mr. Berlin H. Wright describes the above species in the West American Scientist, San Diego, April, 1889, p. 8. He found also a variety of uniform light brown or russet color, bandless, which I have figured on Plate III. Fig. 9. This form had a jaw and lingual membrane the same as in *B. Marielinus* and *Dormani*

Bulimulus Marielinus, POEY.
Cylindrella Poeyana, D'ORBIGNY.
jejuna, SAY.

Macroceramus pontificus, GOULD.

I give here, for comparison, a figure of the true *M. Kieneri*, from a type in Dr. Pfeiffer's collection, from Honduras.

Macroceramus Kieneri.   Macroceramus Gossei.

Macroceramus Gossei, PFEIFFER.

The figure given represents the species.

Pupa variolosa, GOULD.
modica, GOULD.
pellucida, PFEIFFER.
Strophia incana, BINNEY.
Holospira Römeri, PFEIFFER.

Pfeiffer says "allied to *Goldfussi*, but from all species easily recognized by the basal carina of the last whorl, and its singular twist, which at first sight gives a sinistral appearance to the shell."

Holospira Goldfussi, MENKE.
Stenogyra octonoides, D'ORB.
subula, PFEIFFER.
gracillima, PFEIFFER.
Cæcilianella acicula, MÜLLER.
Liguus fasciatus, MÜLLER.

See p. 435 of Manual of American Land Shells for still another variety of coloring of this species.

Orthalicus undatus, BRUGUIÈRE.
Succinea Concordialis, GOULD.
luteola, GOULD.
effusa, SHUTTLEWORTH.
Salleana, PFEIFFER.
campestris, SAY.
Veronicella Floridana, BINNEY.

Onchidium Floridanum, DALL.

See Plate VI. Figs. B, C, for a drawing of an original specimen, enlarged three times.

To Mr. Hemphill is due the credit of adding this genus to the fauna of Eastern North America. The specimens arrived as this paper is going through the press, and a detailed description must be deferred. The following notes, however, will indicate its external characters : —

When living, the creature is of a uniform slaty blue, the under parts bluish white, with a greenish tinge to the veil. The surface appears beautifully smooth and velvety without dorsal tubercles ; just within the slaty margin of the mantle is a single row of about (in all) one hundred whitish elongated tubercles. When crawling, it is of an oval shape, about an inch long, and two tentacles extend forward beyond the mantle margin, resembling the oculiferous ones of *Vaginulus Floridanus*. In spirits the surface is still smooth, but numerous circular hardly elevated domelets cover the back, each appearing to contain one of the dorsal eyes described by Semper. The tentacles are entirely retracted ; a narrow veil, with lightly escalloped edge, precedes the head ; the muzzle is not prominent, is indented in the middle, and puckered at the edges. The foot is about one third wider than the mantle at each side of it. There is no jaw. The penis resembles that of *Siphonaria* in form and position. The animal exudes very little mucus. It was found on rocks between tides associated with *Chiton piceus*. Fifteen specimens were found at Knight's Key by Hemphill.

Onchidium indolens of Couthouy (Rio) and *O. armadillo* of Mörch differ from the above in coloring. The latter, described from St. Thomas, has a very different dorsal surface. No others are known from East America. It would seem as if the small Northern species, possessing a jaw like *O. boreale*, Dall, and *O. Celticum*, Cuvier, might appropriately be separated from the agnathous tropical forms as a subgenus, for which the name of *Onchidella* might be revived in a restricted sense.

The above description is by Dall (Proc. U. S. Nat. Mus., 1885, p. 288). Specimens received by him have the lingual dentition of the genus. (See my Plate III. Fig. 10, where a central tooth and adjacent lateral are given.) There are numerous rows of over 97–1–97 teeth.

The following are to be added to the species treated in the Second Supplement.

PACIFIC PROVINCE SPECIES.

Microphysa Stearnsi, BLAND.
Lansingi, BLAND.

It must be borne in mind that the other species of *Microphysa* examined by me have quadrate marginal teeth, while *Stearnsi* and *Lansingi* have the aculeate marginal teeth of the *Vitrininæ*. Thus they can hardly be classed in *Microphysa*. The name *Pristina* has been suggested by Ancey (Conchologists'

Exchange, I. 5, p. 20, Nov., 1886).   As a substitute for this preoccupied-name, Mr. Pilsbry suggests *Anceyia*.   (See same, I. 6, p. 26, Dec., 1886.)   Mr. Ancey's description is: —

Pristina, Anc. (nov. subg. Hyalinæ).   Testa parvula, imperforata, cornea, nitens, multispirata; spira depresse conica.   Apertura interdum lamellis radiantibus sub-serratis in palato sitis insignis.

Geographical Distribution: Western and Arctic North America.

Types: *Hyalına Stearnsi*, Bland, and *Lansıngı*, Bland.

Mr. W. G. Binney put these species, but with doubt, in Microphysa, while other authors consider them as Hyalinæ; they differ from the latter by anatomic features, and from the former by the form of the shell.   Altogether I am inclined to place the group in Hyalina, as a series nearly allied to *Conulopolita*, Boettger (type, *C. Raddei*, Boettg.); I am confident the presence or absence of internal laminæ or tooth-like processes within the aperture of *Helices* are not generic characters; in some instances they are either present or absent in closely allied species. I established this fact when at work (Le Naturaliste, 1882) on the New Caledonian forms, and I now repeat this as my opinion in regard to *Pristina* and *Gastrodonta*. In the latter the teeth are frequently absorbed by the animal when growing larger.

Macrocyclis Duranti, NEWC.

To the synonymy add : —

Selenites cœlatura, MAZYCK, Proc. U. S. Nat. Mus., 1886, p. 460, with figures of that form and of typical *Duranti*.   Also, Proc. Elliott Soc., Feb., 1886, p. 114, same figures.

Mr. Mazyck's description and figures are repeated here : —

Shell small, depressed, brownish horn-color, with very coarse, rough, crowded, sub-equidistant, irregular ribs, which are obsolete at the apex; whorls four, rounded, somewhat inflated below, gradually increasing, the last not descending at the aperture; suture impressed; umbilicus wide, clearly exhibiting all of the volutions; aperture almost circular, slightly oblique; peristome simple, its ends approaching and joined by a very thin, transparent, whitish callus, through which the ribs are distinctly seen.   Greater diameter, 4 mm.; height, 1¾ mm.

Santa Barbara, California, Dr. L. G. Yates.   Hayward's, Alameda County, California, W. H. Dall, U. S. National Museum.

Macrocyclis Duranti,
var. cœlata,
enlarged.

Newcomb's description of this little shell (*M. Duranti*) is as follows : —

"Shell depressed, discoidal, pale corneous, under the lens minutely striated, opaque, broadly and perspectively umbilicated; whorls 4, the last shelving but not descending (at the aperture); suture linear; aperture rounded, lunate, lip simple, the external and internal approaching.

"*Habitat.* — Santa Barbara Island."

Mr. Binney's description, which is repeated in each of his works above named, differs in this important particular.   For Newcomb's "Under the lens minutely striated," he substitutes the contradictory words "with very coarse, rough striæ."

In a note written in answer to an inquiry addressed to him regarding this singular discrepancy, he says, "My description and figure are from an individual, not from the species. I am absolutely sure my specimen was one of the original find." His figure, drawn by Morse, rather represents a comparatively smooth, semi-transparent shell.

Limax hyperboreus.

See Manual of Amer. Land Shells, p. 473. I have figured on Plate VIII. Fig. F, an individual from British Columbia. Here I give the dentition.

Jaw arched, smooth, with blunt median projection. Lingual membrane with 42–1–42 teeth ; centrals tricuspid ; laterals bicuspid, 12 in number on each side; marginals about 30 on each side, aculeate, simple, without bifurcation or side spur.

The figure shows a central tooth with its adjacent lateral, and three extreme marginals.

Limax montanus, L. castaneus, L. occidentalis, and *L. campestris* all have side spurs to their marginal teeth. Otherwise, their dentition shows no specific distinction from that of *hyperboreus.* Until the genitalia of the last is shown to vary, I am inclined to believe all four to be one and the same species.

Limax hyperboreus.

Limax Hemphilli.

Mr. Henry Hemphill has sent me in spirits from Julian City, California, a small, slender, smooth, dark species of *Limax,* 20 mm. long in its contracted state. It does not outwardly resemble *Limax agrestis,* nor does it seem probable that that species would have been accidentally introduced from the Eastern cities.[1] The dentition, however, agrees with that of *agrestis* by its having the peculiar side spur to the larger cutting point of all the lateral teeth. I venture to propose a specific name for it, in hopes of having an opportunity later to fix its specific position by an examination of the genitalia. It is figured on Plate VIII. Fig. E.

The jaw is as usual in the genus.

There are 50–1–50 teeth to the lingual membrane, of which ten on each side are laterals. Centrals tricuspid; laterals bicuspid, the larger cutting point having a well developed side cutting point on its inner side ; the laterals have also an inner, slightly developed, horizontal side cusp, bearing a small, stout cutting point (see Plate I. Fig. 13); marginals simple, without side spur.

The figure on Plate II. Fig. 3, shows one central with its adjacent laterals, an outer lateral, and several extreme marginals.

A specimen, apparently of the same species, from British Columbia, has 53–1–53 teeth, of which 13 on each side are laterals.

I have the same species, with similar dentition, from San Tomas, Lower California (Hemphill).

---

[1] It is, however, found in San Francisco.

Limax Hewstoni, J. G. Cooper.

On Plate II. Fig. 4, will be found a better figure of the dentition of this species than is given in Terr. Moll., V. It will be seen that the inner side cusp of the lateral teeth is quite distinct from the side spur found in *Limax Hemphilli* and *agrestis*. (See line third of p. 223.)

I have figured (Plate VIII. Figs. D and I) individuals received from Dr. Cooper, drawn by Mr. Theo. D. A. Cockerell.

Limax campestris, var. occidentalis.

The specimen figured on Plate VIII. Fig. H, was kindly furnished by Dr. Cooper. I have already expressed my belief in the identity of this with the Eastern form.

Arion foliolatus, Gould.

It is with the greatest pleasure that I announce the rediscovery by Mr. Henry Hemphill of this species, which has hitherto escaped all search by recent collectors. It has till now been known to us only by the description and figure of the specimen collected by the Wilkes Exploring Expedition, almost fifty years ago, and given in Vols. II. and III. of Terrestrial Mollusks. A single individual was found in December, 1889, at Olympia, Washington, and sent to me living by Mr. Hemphill. It can thus be described. (See Fig. A of Plate VIII.)

Animal in motion fully extended over 100 millimeters. Color a reddish fawn, darkest on the upper surface of the body, mantle, top of head, and eye-peduncles, gradually shaded off to a dirty white on the edge of the animal, side of foot, back of neck, and lower edge of mantle, and with a similar light line down the centre of back ; foot dirty white, without any distinct locomotive disk ; edge of foot with numerous perpendicular fuscous lines, alternating broad and narrow ; mantle minutely tuberculated, showing the form of the internal aggregated particles of lime, the substitute of a shell plate, reddish fawn color with a central longitudinal interrupted darker band and a circular marginal similar band, broken in front, where it is replaced by small, irregularly disposed dots of same color ; these dots occur also in the submarginal band of light color. Body reticulated with darker colored lines, running almost longitudinally, scarcely obliquely, toward the end of the tail, and connected by obliquely transverse lines of similar color, the areas included in the meshes of this network covered with crowded tubercles, as in *Prophysaon Andersoni*, shown in Plate IX. Figs. I, J. Tail cut off by the animal. (See page 207.)

What appears to be the same species, or a very nearly allied one, was found by Mr. Hemphill at Gray's Harbor, Washington, on the banks of the Chehalis River, near its mouth. This form is figured on Plate VIII. Fig. C. When extended fully, it is 70 millimeters long. It is more slender and more pointed at the tail than the large form. The body is a bright yellow, with bluish black reticulations. The edge of the foot and the foot itself are almost black; shield irregularly mottled with fuscous; the body also is irregularly mottled with fuscous, and has one broad fuscous band down the centre of the back, spreading as it joins the mantle, with a narrower band on each side of the body. The other characters, external and internal, are given below. This smaller form loses its colors on being placed in spirits, becoming a uniform dull slate color.

The large Olympia form is surely *Arion foliolatus*, Gould, agreeing perfectly with his description in Vol. II., and with his figure in Vol. III., excepting that the latter is colored with a deeper red.

Mr. Hemphill writes of it : " I have to record a peculiar habit that is quite remarkable for this class of animals. When I found the specimen, I noticed a constriction about one third of the distance between the end of the tail and the mantle. I placed the specimen in a box with wet moss and leaves, where it remained for twenty-four hours. When I opened the box to examine the specimen, I found I had two specimens instead of one. Upon examination of both I found my large slug had cut off his own tail at the place where I noticed the constriction, and I was further surprised to find the severed tail piece possessed as much vitality as the other part of the animal. The ends of both parts at the point of separation were drawn in as if they were undergoing a healing process. On account of the vitality of the tail piece, I felt greatly interested to know if a head would be produced from it, and that thus it would become a separate and distinct individual." The animal on reaching me still plainly showed the point of separation from its tail. (See Plate VIII. Fig. A.) The tail piece was in an advanced stage of decomposition. I noticed the constriction towards the tail in one of five individuals of *Prophysaon cœruleum* from Olympia. (See page 209.) Another individual of the same lot had a truncated tail, having undergone the operation. The edges of the cut were drawn in like the fingers of a glove.

The tail of the *Arion foliolatus* having been cut off, I was unable to verify the presence of a caudal pore from this individual. On the only living one of the lot from Gray's Harbor, the pore was distinctly visible, and is figured on Plate VIII. Fig. C. Usually, it seemed more " a conspicuous pit " than a longitudinal slit, as in *Zonites*. At one time I distinctly saw a bubble of mucus exuding from it. It opened and shut, and is still plainly visible on the same individual, which I have preserved in alcohol and added to the Binney Collection of American Land Shells in the National Museum at Washington. Another individual from Seattle plainly shows the pore.

Five specimens of the Gray's Harbor lot had, concealed in the mantle, a group of particles of white limy matter which it was impossible to remove as one shell plate. In the large Olympia individual these irregularly disposed particles of lime, of unequal size, seemed attached to a transparent membranous plate. With care, I removed this entire, and figure it. It is suboctagonal in shape (Plate VIII. Fig. B). Under the microscope it appears that the particles of lime do not cover the whole plate ; at many points they are widely separated. This aggregation of separate particles is the distinctive character of the subgenus *Prolepis*, to which *A. foliolatus* belongs.[1]

The genitalia of the large individual from Olympia is figured on Plate IX. Fig. D. The ovary is tongue-shaped, white, very long and narrow; the oviduct is greatly convoluted ; the testicle is black in several groups of cœca ; the vagina is very broad, square at the top with the terminus of the oviduct, and the duct of the genital bladder entering it side by side ; the genital bladder is small, oval, on a short narrow duct ; the penis sac is of a shining white color, apparently without retractor muscle; it is short, very stout, blunt at the upper end where the extremely long vas deferens enters, and gradually narrowing to the lower end. There are no accessory organs. The external orifice of the generative organs is behind the right tentacle.

The form from Gray's Harbor (Plate IX. Fig. H) has its generative system very much the same as described above. The ovary is much shorter and tipped with brown, and is less tongue-shaped. The penis sac tapers to its upper end. The vagina is not squarely truncated above. The system much more nearly resembles that of *Prophysaon Andersoni* (see Terr. Moll., V.) than that of the Olympia *foliolatus*.

. The jaw of both forms is very low, wide, slightly arcuate, with ends attenuated and both surfaces closely covered with stout, broad separated ribs, whose ends squarely denticulate either margin. There are about 16 of these ribs in one specimen from Gray's Harbor, and over 20 in that of the true *foliolatus* from Olympia (see Plate IX. Fig. B). The lingual membrane in each form is long and narrow, composed of numerous longitudinal rows of about 50-1-50 teeth, of which about 16 on each side in the true *foliolatus* (Plate IX. Fig. C), and 19 in the other form, may be called laterals. Centrals tricuspid, laterals bicuspid, marginals with one long inner stout cutting point, and one outer short side cutting point. The figure shows a central tooth with its adjacent first lateral, and four extreme marginals.

I have figured both the true *foliolatus* from Olympia (Plate VIII. Fig. A) and the smaller form from Gray's Harbor (Plate VIII. Fig. C) of natural size. Should the latter prove a distinct species or variety, I would suggest for it the name of *Hemphilli*, in honor of the discoverer of it and the long lost *foliolatus*.

Prophysaon Hemphilli.

See Plate VII. Fig. D, drawn by Cockerell from the living animal.

Prophysaon Andersoni, J. G. Cooper.

Figure 1 of Plate III. was drawn from a specimen received from Dr. Cooper. It represents the true *Andersoni*, distinguished by a light dorsal band, and by genitalia such as I have described for *P. Hemphilli*. The same form, also re-

---

[1] Mr. Theo. D. A. Cockerell, finding the slug not to be a true *Arion*, is about to suggest for it the generic name of *Phenacarion*.

ceived from Dr. Cooper, is drawn by Mr. Cockerell on Plate VII. Fig. C. Mr. Cockerell has shown me that I have confounded with it another species, which he proposes to call *P. fasciatum.* See next species.

Prophysaon fasciatum, COCKERELL.

This species is described by Mr. Cockerell as distinct from *Andersoni*, with which I have formerly confounded it. (2d Suppl. to Vol. V., p. 42.) It has a dark band on each side of the body, running from the mouth to the foot. To this must be referred the descriptions of animal, dentition, jaw, and genitalia formerly published by me as of *Andersoni.*

I am indebted to Mr. Theo. D. A. Cockerell for a figure and description of this species. The former is given on Plate VII. Fig. A, while the latter is given here in the words of Mr. Cockerell, whose name must consequently be associated with it as authority : —

Length (in alcohol), 19 mm. Mantle black, with indistinct pale subdorsal bands, — an effect due to the excessive development of the three dark bands of the mantle. Body with a blackish dorsal band, commencing broadly behind the mantle and tapering to tail, and blackish subdorsal bands. No pale dorsal line. Reticulations on body squarer, smaller, more regular, and more subdivided than in *P. Andersoni,* Cooper. Penis sac tapering, slender. Testicle large. Jaw ribbed.

Prophysaon cœruleum, COCKERELL.

Plate VIII. Fig. I, J.

In the Nautilus, 1890, p. 112, it is thus described : —

Length (in alcohol), 22½ mm.; in motion, 43 mm. Body and mantle clear blue-gray, paler at sides, sole white. Mantle finely granulated, broad, without markings. Length of mantle, 7 mm.; breadth, 5 mm. Respiratory orifice, 2½ mm. from anterior border. Body subcylindrical, tapering, pointed. (In one specimen eaten off at the end.) Distance from posterior end of mantle to end of body, 10¾ mm.

The reticulations take the form of longitudinal equidistant lines, occasionally joined by transverse lines, or coalescing. Sole not differentiated into tracts. Jaw pale, strongly ribbed. Liver white.

Mr. Binney sends me colored drawings of the living animal; the neck is long and white, or very pale. Mr. Binney has examined the jaw and lingual, and finds them as usual in the genus.

Several specimens were sent from Olympia, Washington, by Mr. Hemphill to Mr. Binney.

P. cœruleum is an exceedingly distinct species, distinguished at once by its color and the character of its reticulations.

Prophysaon cœruleum, var. dubium, n. var., COCKERELL.

Length (in alcohol), 8 mm. Length of mantle, 4 mm. Distance from posterior end of mantle to end of body, 3½ mm. Mantle broad, with four bands composed of coalesced black marbling, very irregular in shape, and running together anteriorly. Body dark, tapering. Sole pale, its edges gray. Liver white.

With the *P. cœruleum* is a small dark slug, probably a variety of it, but differing as described above. It will easily be distinguished by its blackish color and the peculiar markings on the mantle.

Prophysaon Pacificum, COCKERELL.

Plate VII. Figs. B, E, F, H.

Mr. Theo. D. A. Cockerell gives the following in the Nautilus of February, 1890, pp. 111–113 : —

Length (in alcohol), 17¼ mm. Body and mantle ochrey brown, head and neck gray. Mantle granulated, rather broad, with a black band on each side not reaching the anterior border; these bands are farthest (2¼ mm.) apart near the respiratory orifice, from which point they converge posteriorly, and anteriorly by the bending of the band on the right side. Length of mantle, 7¾ mm.; breadth, 4 mm. Respiratory orifice 3¼ mm. from anterior border. Body cylindrical, rounded and very blunt at end, not conspicuously tapering. Distance from posterior end of mantle to end of body, 8 mm. Body dark grayish-ochre above, with an indistinct pale dorsal line; sides paler. Reticulation distinct, with indistinct "fóliations." Sole somewhat transversely wrinkled, but not differentiated into tracts.

Jaw dark, strongly curved, blunt at ends, with about ten well-marked ribs (Plate VII. Fig. F). Lingual membrane with about 35–1–35 teeth; centrals tricuspid, the side cusps very small, laterals bicuspid, marginals with a large sharp straight inner point and a small outer one. Compared with *P. humile* the centrals are slightly shorter and broader. Liver dark gray-brown.

Found by Mr. H. F. Wickham under logs in ditches by the roadside and damp places at Victoria, Vancouver Island, 1889.

This is a very distinct species, easily recognized by its color, the absence of dark bands on the body, the pale dorsal line, and the blunt posterior extremity.

Prophysaon flavum, COCKERELL.

Plate VII. Fig. K.

From the Nautilus, 1890, p. 111: —

Length (in alcohol), 25 mm. Body and mantle dull ochreous, head and neck ochreous. Mantle tuberculate-granulose, grayish ochre, pale at edges, and with black marbling or spots in place of the bands of *P. Pacificum*. Length of mantle, 11 mm.; breadth, 5½ mm. Respiratory orifice 5 mm. from anterior border. Body cylindrical, hardly tapering, and blunt at end. Distance from posterior end of mantle to end of body, 14 mm. Body dark grayish-ochre above, with a pale ochreous dorsal line not reaching much more than half its length; sides paler. Reticulations distinct, "foliated." Sole with well marked transverse lines or grooves, those of either side meeting in a longitudinal median groove, which divides the foot into two portions. Liver pale grayish.

Uniform tawny, as is *Limax flavus*. It stretches itself out in a worm-like shape unlike other species. Internal shell plate, jaw, and tongue as in *Andersoni*.

Gray's Harbor, Washington. (Hemphill, 1889.)

This is probably a variety of *P. Pacificum*.

Prophysaon humile, COCKERELL.

Plate VII. Figs. F, G, L, M.

From Nautilus, 1890, p. 112.

Length (in alcohol), 16½ mm. Body above and mantle smoke-color, obscured by bands. Mantle wrinkled, and having a broad dorsal and two lateral blackish bands, reducing the ground-color to two obscure pale subdorsal bands. Length of mantle, 7 mm.; breadth, 5¼ mm. Respiratory orifice 2¾ mm. from anterior border. Body subcylindrical, somewhat tapering, rather blunt at end. Distance from posterior end of mantle to end of body, 8 mm. Back with a blackish band reaching a little more than half its, length, and lateral darker blackish bands reaching its whole length. Reticulations distinct, "foliated." Sole strongly transversely striate-grooved, but not differentiated into tracts.

Jaw pale, strongly striate, moderately curved, not ribbed. (See Fig. F.) Lingual membrane long and narrow. Teeth about 35-1-35. Centrals tricuspid, laterals bicuspid, marginals with a large inner point, and one (sometimes two) small outer points. Liver pale chocolate.

Found by Mr. H. F. Wickham under the bark of rotten logs in the woods around Lake Cœur d'Alene, Idaho, 1889.

In its reticulations, and general external characters, this species resembles *P. Andersoni*, of which it is possibly a variety.

Hemphillia glandulosa.

(See also p. 216.)

From Olympia and Gray's Harbor, Washington, Mr. Hemphill sent me living specimens of this species, both young and mature. Several of the young had the horn-shaped process to the tail noticed in the original description of the genus. The shell in these young individuals is very slightly attached, apparently simply by having its posterior margin lightly covered by the mantle. It often becomes detached. In these young, the mantle is proportionally smaller, and the neck much longer. I have figured an enlarged view of a young individual, Plate IV. Fig. D.

Ariolimax[1] Columbianus, GOULD.

Found also by Mr. Hemphill on Santa Cruz Island.

Plate VI. Fig. A, represents the mottled variety, found recently by Mr. Hemphill in the State of Washington. Mr. Cockerell suggests for it the varietal name *maculatus*. This form shares with the type the peculiar penis sac (Fig. G) distinguishing it from the next species.

Ariolimax Californicus, COOPER.

See Plate V. Fig. E, for the animal in motion, and a portion of the genital system (Fig. H), showing variation from that of *A. Columbianus*.

[1] The name should be *Arionilimax*.

Ariolimax Andersoni.

See Plate V. Fig. F, showing the typical specimen in spirits restored.

Ariolimax Hemphilli.

Plate V. Fig. B, G.

A variety *maculatus*, Cockerell, is figured in B. The Figure G is drawn from a typical specimen, with the tail, the pore, and the locomotive disk.

Ariolimax niger, J. G. Cooper.

Plate V. Fig. A, gives a lighter-colored form ; Fig. I, the typical form; Figs. C and D, the caudal pore.

Triodopsis inflecta, Say.

This has erroneously been quoted from the Pacific Province, at the mouth of Columbia River. It is difficult to decide what species Middendorff had in view. His words are thus translated : —

Let it not be objected that *Helix clausa* up to this time has not been discovered west of the Rocky Mountains. The Northwest Coast of America is almost wholly unexplored conchologically, and I do not doubt that *H. clausa* will be there found, just as I can now assert with reference to *H. planorboides*. Even the American authors know this hitherto only from the Ohio and Missouri. Its distribution nevertheless appears to extend over the whole of North America, since I have received a great number of specimens of the same through Mr. ——, from Sitka, whereby it becomes incorporated with our Russian Fauna. Southwards it extends to the west coast of America, at least to Upper California, where they were likewise collected by Mr. ——. It appears to have undergone no alteration whatsoever, and presents in Sitka a considerable size, as the ordinary representations show (up to 22, etc.). Moreover, Binney in the Boston Journal, III., Plate XIV, has them copied equally large.

Polygyra Roperi, Pilsbry.

Shell umbilicated, plane above, slightly inflated below, shining, pellucid, light horn-color, with delicate wrinkles of growth ; spire flattened ; whorls $5\frac{1}{2}$,

Polygyra Roperi, enlarged.

scarcely rounded, very regularly increasing, the last flattened above, abruptly deflected at the aperture, deeply constricted behind the peristome ; aperture transversely lunar, gaping, much contracted, tridentate ; peristome thickened, broad, white, gradually thinning and scarcely reflected at its edge, and not extending beyond the surface of the whorl, its ends approached, joined by a light callus, on which is a heavy white callus bearing a stout, white, broad, blunt, transverse tooth, slightly curving inward, its basal margin with an erect conical, short tooth, separated by a small circular sinus from another rather more deeply seated similar tooth on its upper margin. Umbilicus broad, showing the volutions clearly.   Greater diameter, 9 mm. ; lesser, 7 mm. ; height, 2½ mm.

Helix (Triodopsis) Roperi, PILSBRY.   The Nautilus, Vol. III. No. 2, June, 1889, p. 14.

Redding, Shasta Co., California, in drift of the Sacramento River, three dead shells were collected by Mr. Edward W. Roper, of Chelsea, Mass.

The above description is drawn from one of the original specimens, kindly lent me by Mr. Roper, while another in the collection of the Academy of Natural Sciences of Philadelphia, from which Mr. Pilsbry drew his description, is figured above.   The third specimen was given by Mr. Roper to Mr. Henry E. Dore of Portland, Oregon.

Never having seen a specimen of *P. Harfordiana*, I cannot say if this species is identical with it.   At least, it must be nearly allied.

Aglaja fidelis, GRAY.

New figures of several forms of this species are given.   Plate X. Fig. A represents the black elevated form approaching *infumata*.   Its sculpturing is given in Fig. B.   The small, black, elevated form is given in Fig. C, with its sculpturing in D ; the small, depressed form, in E.

Aglaja infumata, GOULD.

Plate X. Fig. F, gives an enlarged view of the hirsute surface.

Arionta arrosa, GOULD.

Plate XI. Fig. A gives this species.   A form of *arrosa* nearly approaching *A. exarata* is given in Fig. B, its sculpturing in Fig. C.

Arionta exarata, PFEIFFER.

The typical form and its sculpturing are given on Plate XI. Figs. D and E.

Arionta Mormonum, PFEIFFER.

The typical form is given on Plate XI. Fig. F.   The variety (Vol. V. p. 141) approaching *Aglaja Hillebrandi*, is given in Figs. G and H ; the sculpturing of the same form, on Plate X. Fig. G.   The genitalia of this form are the same as of the type.

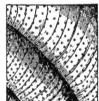

Enlarged sculpturing of Arionta sequoicola.

Arionta sequoicola, J. G. COOPER.

A figure of the sculpturing of this species is here given, greatly enlarged.

Arionta Californiensis, Lea.

I give here new figures of two forms of this species, *Arionta Diabloensis* and the depressed variety of *A. Bridgesi,* the former drawn from a shell received from Dr. Cooper.

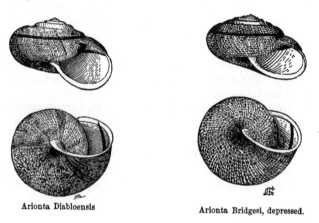

Arionta Diabloensis   Arionta Bridgesi, depressed.

Onchidella Carpenteri, Dall.

An alcoholic specimen received from Mr. Dall is figured on Plate VI. Figs. D, E, enlarged twice.

Veronicella olivacea, Stearns.

I have failed to receive Californian specimens. That figured on Plate IX. Figs. E, F, is one of the original lot from Folvon, Central America.

CENTRAL PROVINCE SPECIES.

Limax montanus, Ingersoll.

A specimen is figured on Plate VIII. Fig. G.
The species is surely identical with *L. campestris.*

Patula solitaria, Say.

Mr. Hemphill found this species very abundant at Old Mission, Cœur d'Alene, Idaho. There was an albino variety, a depressed form, and one very much more elevated than that which I figured in the Second Supplement, Plate I. Fig. 10.

Patula strigosa.

Among the shells recently collected by Mr. Hemphill at Old Mission, Cœur d'Alene, Idaho, was a marked variety of this species, for which Mr. Hemphill suggests the name *subcarinata*. The specimens vary greatly in elevation of the spire, and in the number and disposition of the revolving bands, often quite wanting. All have a very heavy shell, the body whorl of which has an obsolete carina which is well

P. strigosa, var. subcarinata, Hemphill.

marked at the aperture, modifying the peristome very decidedly. See the figure.

In examining the genitalia I find the base of the duct of the genital bladder greatly swollen along a fifth of the total length of the duct.

On the banks of the Salmon River, Idaho, Mr. Hemphill found a form like var. *Gouldi*, but distinctly carinated. None of the Utah individuals of this form are so characterized.

Patula strigosa, var jugalis, Hemphill.

Another form of *strigosa* from the same locality is very large, flat, with a transversely oval aperture, the ends of the peristome so nearly approached as almost to touch, and often joined by a heavy callus, which forms a continuous rim around the aperture. Mr. Hemphill has called this var. *jugalis*.

Microphysa pygmæa.

Found by Mr. Hemphill at Old Mission, Cœur d'Alene, Idaho.

Microphysa Ingersolli, Bland.

A better figure of this species is here given.

Microphysa Ingersolli, enlarged.

Triodopsis Hemphilli.

Mr. Tryon has suggested the name *binominata* for this species, though *Hemphilli* is not preoccupied in *Triodopsis*.

Triodopsis Sanburni.

Triodopsis Sanburni, enlarged.

The cut is drawn from one of the original specimens.

Mesodon ptychophorus.

At Old Mission, Cœur d'Alene, Idaho, Mr. Hemphill found a form of this species characterized by a heavy, dead white shell with scarcely any trace of ribs or wrinkles of growth which are usually so characteristic of the species. On the banks of the Salmon River he found a small form, the lesser diameter of which is only 12 mm. See figure.

Mesodon ptychophorus, var.

Triodopsis Harfordiana.

Ancey suggests *commutanda*, and Tryon *Salmonensis*, as a substitute for the name *Harfordiana*. I retain the last name, it not being preoccupied in the genus *Triodopsis*.

Prophysaon Andersoni?

Specimens collected by Mr. Hemphill at Old Mission, Cœur d'Alene, Idaho, appear to agree with specimens of this species received from Dr. Cooper. The jaw is low, wide, slightly arcuate, with over 12 broad, stout ribs, denticulating either margin. The lingual membrane is given in Plate II. Fig. 2. The central and lateral teeth are slender and graceful. The latter have, apparently, a second inner cutting point, as is found in *Limax agrestis*. I have so figured it, hoping to draw attention to it, and thus settle the question of its being there.

Hemphillia.
Plate IV.

From Old Mission, Cœur d'Alene, Idaho, Mr. Henry Hemphill has sent me fine large specimens of *Hemphillia* alive. From these I am able to give the outward characteristics of the animal in drawings by Mr. Arthur F. Gray.

The animals are larger and much lighter in color than those originally found at Astoria. They do not while in motion differ from other slugs, though my former figure of the animal in spirits shows a very great difference, owing to the contraction being resisted by the internal shell. The rear end of the mantle seems swollen and blunt, separated from the back, however, and thus alone does there seem to me any difference in its appearance from *Limax*, whose mantle lies flat upon the back. The slit in the mantle is sometimes open, sometimes closed, and the slit seems to extend quite to the rear of the mantle. There is a profuse flow of mucus from over the slit. There seem on the mantle to be little protuberances, rather than the elongated reticulation of the rest of the animal. The caudal pore opens and shuts, and exudes mucus in bubbles sometimes, which occasionally form a solid lump of mucus on the tail. The horn-like process of the tail so prominent in the first specimens from Astoria — contracted in alcohol — does not exist in these living specimens, though occasionally there is a kind of hump above the pore. (See Plate IV. Fig. D.)

Mr. Hemphill writes: "*Hemphillia* has a peculiar habit when removed from its resting place of switching its tail, so to speak, quite rapidly, — a habit I never noticed in any of our other slugs. I find them hibernating in old rotten logs."

The viscera are enclosed under the mantle.

Mr. Gray in drawing the animal called my attention thus to the characters of the outward markings of the slug : —

"You are right in saying that the slit in the mantle extends to the back margin. The central pit seems flooded with mucus at all times, but does not change its form; the slit, however, seems to widen and show a little ridge on either margin when the animal is fully expanded. The little tubercles, or small pimples as it were, seem to cover the posterior portion of the mantle, while the elongated tubercles seem to cover the anterior half, though these at times disappear and the anterior portion runs into folds, which break up the surface, and starting from the margin of the mantle run to its centre in parallel lines like miniature waves. They move steadily inward from both margins, disappearing before reaching the little mucous pit in the centre of the mantle, little wavelets rising at the margins and keeping up a constant rhythmic motion toward the centre."

The jaw of this specimen has about 25 ribs, denticulating either margin. It is low, wide, slightly arcuate, with slightly attenuated ends. (See Plate IX. Fig. A.)

The lingual membrane is as described and figured by me in Vol. V.; there are, however, in this form, 57–1–57 teeth, with some eleven true laterals.

The genitalia I have figured in Plate III. Fig. 3. It agrees with my figures in Vol. V. of the genitalia of the original specimens, excepting that the penis sac, as represented there in Plate XII. Fig. K, is here doubled on itself.

Pupa hordeacea, GABB.

An authentic specimen of this species is figured in the Second Supplement, Plate III. Fig. 10, referred by mistake to *P. Arizonensis* in the explanation of Plate III.

Pupa Arizonensis, GABB.

The reference to *hebes* in Second Supplement should be Fig. 12, not Fig. 10.

LOCALLY INTRODUCED SPECIES.

Tachea nemoralis, LINN.

Fine large specimens of this species have been sent me by Prof. James H. Morrison, found by him living during the last three years at Lexington, Virginia. They form part, no doubt, of a colony descended from living individuals introduced from Europe around plants.

Zonites cellarius, MÜLLER.

Also at San Francisco (Cooper).

Limax maximus, LINN.

Also at New Braunfels, Texas (Singerly).

A drawing of the lingual dentition on Plate II. Fig. 1, shows the cutting points of central and lateral teeth to be trifid. This is not shown in my figure in Vol. V.

SINCE the foregoing was written, the following species have been de. scribed : —

Zonites selenitoides, PILSBRY.

This species is similar in form and general appearance to *Z. minusculus*, Binn., though decidedly larger. The umbilicus is broad, as in the latter species. The

shell is thin, light yellowish horn-color, almost white. Surface shining, covered with close strong oblique rib-striæ, like *Patula striatella;* these striæ, while generally regular, sometimes bifurcate, or separate to give room for another to be intercalated. The spire is flatter than *minusculus,* nearly plane. The earlier 1¾ to 2 whorls are smooth, polished, not striate ; the sutures are well impressed. There are 3¼ whorls in all, convex, gradually widen-

Zonites selenitoides, enlarged.

ing, the last proportionately wider than in *Z. minusculus.* Aperture slightly oblique, lunate, narrower

Sculpturing, enlarged.

than in *Z. minusculus,* its margins thin, acute, scarcely converging, the columellar shortly subreflexed.

Alt. 1.2 mm., diam. 3 mm.

The specimens were presented to me by Mr. W. G. Binney, who, regarding them as new, kindly permitted me to describe them. They were gathered by Hemphill, prince of collectors! at Mariposa Big Trees, California. The name *selenitoides* is given because of a certain resemblance to the little *Selenites Duranti* of Southern California.

The above description was published by Pilsbry in Proceedings of Academy of Natural Sciences of Philadelphia, 1889, p. 413, Plate XII. Figs. 13–15.

I give a figure of the original specimen, and of its sculpturing.

Zonites Simpsoni, PILSBRY.

This species belongs to that group of *Hyalina* comprising *capsella,* Gld., *Lawæ,* W. G. Binn., and *placentula,* Shutt., — species with narrow umbilicus, numerous closely coiled narrow whorls, and without a callus or thickening within the base of the last whorl. *Z. Simpsoni* differs from *placentula* in its much smaller size, nearly straight, instead of arcuate, basal lip, seen from beneath, proportionately wider last whorl, and the more trigonal, wider aperture. With *Z. Lawæ* I need not compare it, as that species is much larger and more elevated. *Z. capsella* is about the same size, color, and texture as *Simpsoni,* but has a narrow umbilicus and very much narrower aperture, narrowly semilunar instead of trigonal in outline. *Z. Simpsoni* has 5 whorls. Alt. 2, diam. maj. 4½, min. 4 mm.

The specimens before me were collected by Mr. C. T. Simpson, at Limestone Gap, Indian Territory. The trigonal form of the aperture is so peculiar that the species may be separated from *Z. capsella* at a glance. My comparisons were made with specimens of *capsella* received from Gould, and *placentula* from W. G. Binney. The figures are camera lucida drawings.

From Proc. Acad. Nat. Sci. Phila., 1889, p. 412, Plate XII. Figs. 8–10.

Pupa calamitosa, PILSBRY.

Shell minute, cylindrical, very blunt at apex, chestnut-colored; whorls $4\frac{1}{2}$, the first one and a half smooth, the following regularly costulate striate, the costulæ separated by spaces wider than themselves; last whorl abruptly turning forward, rounded beneath, encircled by a slight central constriction or furrow; aperture about one third the total length of shell, rounded, truncated above, contracted within; peristome thin, expanded, without crest or callous thickening behind; columellar margin rather dilated; parietal wall bearing two entering lamellæ, one arising near the termination of the outer lip, the other more deep seated, elevated, entering less obliquely; columella with a strong white deep-seated obliquely entering fold; outer lip with two short white lamellæ.

Alt. 1.70, diam. 0.80 mm.

Two trays of this tiny species are before me. One received from Henry Hemphill, collected near the mouth of San Tomas River, Lower California, the other collected by Orcutt near San Diego, California. Most specimens show the widening inward of the outer lip shown in the figure. Several specimens have only one lamella on the outer lip, and are rather larger than the typical form described, measuring 1.90 mm. alt. The second parietal lamella is usually much larger than the first, but in one or two specimens before me this is not the case. The umbilical rimation terminates in a tiny depression, perhaps minutely perforated at the axis. The formula of denticles or folds (according to Dr. Sterki's scheme [1]) AA B D E or AA B E. The species is of a decidedly different type from any known American *Pupa*. *P. hordacea, Californica,* and *Rowelli,* abundant Western forms, belong in quite diverse groups; the first being allied to *P. corticaria* and *pellucida*, the last two grouping with *P. decora, Rowelli,* and *corpulenta.*

From the *Pupæ* of the Mexican fauna, *leucodon, pellucida,* and *chordata,* the present species is quite distinct in every respect.

The inward continuation of the parietal and columellar folds is shown in Figure 17. They are white, regularly veined with darker, like polished plates of agate.

From Proc. Acad. Nat. Sci. Phila., 1889, p. 411, Plate XII. Figs. 16, 17.

Mr. Hemphill sends me the following descriptions, which must be fully credited to him : —

Helix tudiculata, var. Binneyi.

This beautiful variety belongs to the globosely depressed forms of *H. tudiculata,* Binn. It is of a uniform greenish yellow color, without blotches or markings, except a very faint trace of a band at the periphery. *H. tudiculata* is very variable in form, size, and sculpture, and with the umbilicus either open or closed, but it is very constant in its dark chestnut-color in Southern California. North of Merced County, however, it becomes a shade lighter, and passes towards the light, thin form of *H. arrosa,* which I regard as the

---

[1] See Proc. U. S. Nat. Mus., 1888, p. 369. I have repeated the letter presenting the parietal fold, as the two seem to be of equal importance.

progenitor of *tudiculata*, *arrosa* in turn having evolved ·from its northern
neighbor, *H. Townsendiana*, Lea, and *Townsendiana* from the· form we now
call *H. ptychophorus*, Brown, found in Eastern Oregon and Idaho.

Habitat. Mountains of San Diego County, California. Only one specimen
found.

Helicodiscus fimbriatus, var. Salmonensis.

This variety varies from the Eastern or typical forms in the absence of the
revolving lines; otherwise the shells are alike.

Habitat. Banks of Salmon River, Idaho, Old Mission, Idaho, and Oakland,
California.

Helix Kelletti, var. albida.

This is a beautiful clear white translucent variety, with no markings or
stains of any kind. It is quite thin and frail, and a trifle smaller than the
average size of *Kelletti*.

Habitat. Santa Catalina Island, California. Two specimens only found
by me.

Helix Kelletti, var. castanea.

Among the numerous patterns of coloring assumed by *H. Kelletti*, none are
more conspicuous than this well marked variety. The body whorl is of a
deep shiny chestnut-color above the periphery, and becomes lighter as it follows the whorls of the spire to the apex. The band at the periphery is quite
variable in the different speeimens; it is generally light, and well defined
above, but below it is irregular and spreads over the base of the shell more
or less.

Habitat. Santa Catalina Island, California. This variety is not rare.

Patula strigosa, var. Buttonii.

Shell umbilicated, elevated, or moderately depressed, nearly white, sometimes stained with light chocolate; whorls five, convex, with numerous oblique
striæ ; suture impressed, aperture circular ; peristome thickened, not reflected,
darker than the body of the shell; extremities nearly approached and joined
by a callus; with or without a basal tooth ; tooth when present very variable,
generally consisting of a single tubercle ; in some specimens it is nearly or
quite square, as high as long; in other specimens it is long and bifid·

Diameter of the largest specimen, $\frac{7}{8}$ inch ; height, $\frac{1}{2}$ inch. Diameter of the
smallest specimen, $\frac{1}{2}$ inch ; height, $\frac{2}{8}$ inch.

Habitat. Box Elder Co., Utah.

I dedicate this interesting form of *strigosa* to my friend, Mr. O. Button, of
Oakland, California.

Selenites Duranti, var. Catalinensis.

Shell widely umbilicate, depressed, white, transparent when fresh ; whorls 4, flattened above and below, with fine oblique striæ; spire planulate ; aperture transversely rounded ; peristome simple, acute ; extremities approached and joined by a very thin callus in fully matured specimens.

Greatest diameter, $\frac{1}{4}$ inch ; height, $\frac{1}{16}$ inch.

Habitat. Santa Catalina Island, California.

My little shell differs from the typical *Duranti* in its greater size, smoother surface, broader umbilicus in specimens of the same size, but principally in its transparent shining surface. It is larger than the largest *Duranti* that I have seen, but not so large as the costate variety of that species described by Mr. Mazyck as distinct under the name of *S. cœlata*, which I have in my possession. My specimen of that species is larger than his measurements.

I can add the following to his locality : Los Angeles and San Diego, California, Point Abunda, and banks of San Tomas River, Lower California; thus giving it a range of about two hundred miles up and down the coast. I have collected the typical *S. Duranti* at the following places : Etna Springs, Napa Co., Healdsburg, Sonoma Co., Bolinas and San Rafael, Marin Co., Oakland, Alameda Co., Santa Cruz, Monterey, Santa Barbara Island, Santa Catalina Island, and San Clemente Island, a range of over one hundred miles north and south. It is confined to the Coast Range as far as we know at present.

EXPLANATION OF THE PLATES.

PLATE I

Fig. 1. Central tooth of lingual membrane of *Mesodon major*, the specimen la.
belled A (see p. 190).

Fig. 2. Central tooth, two adjoining lateral teeth, and two marginal teeth of lingual membrane of *Mesodon major*, the specimen labelled B (see p. 190).

Fig. 3. Same: an outer lateral tooth bearing a side cusp and cutting point (see
p. 190).

Fig. 4. *Mesodon Andrewsi*: the genitalia.

    ov.   oviduct.

    g. b.   genital bladder.

    d. g. b. duct of same.

    v. d.   vas deferens.

    r.   retractor muscle of penis sac.

    p. s.   penis sac.

    or.   common orifice.

    p.   prostate gland.

Fig. 5. Penis sac of another specimen of same.

Fig. 7. Lingual dentition of same, from specimen labelled E. Two central teeth,
with an adjoining lateral tooth.

Fig. 8. Same: marginal teeth.

Fig. 9. Same: extreme marginal teeth.

Fig. 10. Same: first lateral tooth of specimen labelled F (see p. 191).

Fig. 11. Same: marginal tooth (see p. 191).

Fig. 12. Same: specimen labelled M (see p. 191), an outer lateral tooth.

Fig. 13. The fourth lateral tooth of *Limax Hemphilli* (see p 205).

Fig. 14. *Succinea chrysis*, Westerlund, copied from the " Vega Expedition," Plate
III. Fig. 10.

Fig. 15. *Succinea annexa*, Westerlund, copied from the same, Fig. 11.

PLATE II.

Lingual dentition of —:

Fig. 1. *Limax maximus*. A central tooth with two adjacent laterals; an outer
lateral; two marginals, the left hand one the last.

Fig. 2. *Prophysaon* (see p. 216). A central tooth with its adjacent lateral tooth;
an outer lateral tooth; an extreme marginal tooth.

Fig. 3. *Limax Hemphilli.* A central tooth with two adjacent laterals; an outer lateral tooth; two outer marginal teeth.

Fig. 4. *Limax Hewstoni.* A central tooth with adjacent lateral on either side; incorrectly numbered on the plate; two extreme marginals.

Fig. 5. *Microphysa dioscoricola* (see p. 196).

PLATE III.

Fig. 1. *Prophysaon Andersoni,* J. G. C., received from Dr. Cooper.

Fig. 2. *Pupilla Floridana,* Dall, from original figure.

Fig. 3. Genitalia of *Hemphillia,* from Old Mission, Cœur d'Alene, Idaho (see p. 217) : —

    t.   testicle.

    ep.   epididymis.

    ov.   ovary.

    ovid. oviduct.

    pr.   prostate.

    g. b. genital bladder. .

    d. g. b. duct of same.

    v. d. vas deferens.

    r.   retractor muscle of penis.

    p. s. penis sac.

    or.   common orifice.

Fig. 4. *Helix exigua,* from an original drawing by Dr. Stimpson.

Fig. 5. *Zonites lasmodon,* Phillips, enlarged. Drawn by Miss Helen E. Lawson.

Fig. 6. Central portion of jaw of *Microphysa dioscoricola,* greatly enlarged.

Fig. 7. *Bulimus Floridanus* (see p. 201). Drawn from original specimen in Mr. Cumings's collection, by G. B. Sowerby.

Fig. 8. Lingual dentition of *Polygyra hippocrepis.*

    a. central and two lateral teeth.

    b. marginal teeth.

Fig. 9. *Bulimus Hemphilli.*

Fig. 10. Dentition of *Onchidium Floridanum.*

PLATE IV.

Fig. D was drawn by W. G. Binney, the other figures by Arthur F. Gray: all from life.

Fig. A. *Hemphillia glandulosa,* twice the natural size.

Fig. B. The same; animal in motion, natural size; the slit on the mantle partially open.

Fig. C. The same; partially contracted and at rest.

Fig. D. The same; the very young animal.

Fig. E. The same; dorsal view of posterior portion of the animal, twice the natural size; pore closed

Fig. F. The same; lateral view, pore closed.

Fig. G.   The same; dorsal view, pore open.
       a. mucus beads exuding.
       b. slit widely opened, the walls or lips rolled out.
       c. mucus accumulations.
Fig. H.   The same; lateral view, pore open.
Fig. I.   The same as last.
Fig. J.   The same; the internal shell plate.

PLATE V.

Figs. F, H, drawn by W. G. Binney; A, C, D, by Arthur F. Gray; B, E, G, I, by T. D. A. Cockerell, of West Cliff, Custer Co., Colorado: all from life.

Fig. A.   *Ariolimax niger*, fully extended.
Fig. B.   *Ariolimax Hemphilli*, var. *maculatus*, Cockerell; animal contracted in alcohol.
Fig. C.   *Ariolimax niger;* the caudal mucus pore, twice the natural size, dorsal
      view, the pore open.
       a.   mucus exuding.
       b. b. ridges each side of slit or channel.
       c.   mucus channel or pore.
       d.   little channels conducting mucus from back of animal into channel *c.*
Fig. D.   The same; posterior view.
Fig. E.   *Ariolimax Californicus*, in motion, natural size.
Fig. F.   *Ariolimax Andersoni*, restored from an alcoholic specimen.
Fig. G.   *Ariolimax Hemphilli*, in motion, with end of tail and pore.
Fig. H.   Portion of genitalia of E.
       p. s. the penis sac.
       f.   the flagellum.
       r.   the retractor muscle.
       v. d. the vas deferens.
Fig. I.   *Ariolimax niger*, partially extended.

PLATE VI.

Figures B, C, D, E, H, were drawn by A. H. Baldwin, the last from life, the others from specimens preserved in spirits; Figures F, G, by W. G. Binney, from life; A, from life, by Arthur F. Gray.

Fig. A.   *Ariolimax Columbianus*, var. *maculatus*, Cockerell, natural size; from a
      specimen collected by Mr. Hemphill.
Fig. B, C.   *Onchidium Floridanum*, three times natural size; **from type.**
Fig. D, E.   *Onchidella Carpenteri*, twice natural size.
Fig. F.   *Tebennophorus Wetherbyi;* from type.
Fig. G.   Portion of genitalia of A.
       p. s. the penis sac.
       r.   the retractor of same.
       v. d. the vas deferens.
Fig. H.   *Tebennophorus Hemphilli;* from the type.

PLATE VII.

All the figures drawn by T. D. A. Cockerell, excepting I, which was drawn by Miss Annie Roberts.

Fig. A. *Prophysaon fasciatum.*
Fig. B. " *Pacificum.*
Fig. C. " *Andersoni.*
Fig D. " *Hemphilli.*
Fig. E. " *pacificum*, jaw.
Fig. F. " *humile*, jaw.
Fig. G. " " the animal contracted in spirits, and the surface.
Fig. H. " *Pacificum;* the same views as last.
Fig. I. " *cœruleum.*
Fig. J. " "
Fig K. " *flavum.*
Fig. L. " *humile.*
Fig. M. " "

PLATE VIII.

Figure C was drawn by F. W. Earl, from life; A, from life, by W. G. Binney; B, D, G, I, from life, by T. D A. Cockerell; E, F, H, were restored by Mr. Cockerell from specimen in spirits

Fig. A. *Phenacarion foliolatus*, natural size ; the tail eaten off.
Fig. B. Internal shell of A.
Fig. C. The same, var. *Hemphilli*, natural size.
Fig. D. *Limax Hewstoni;* in motion and at rest.
Fig. E. " *Hemphilli;* same views as last, and surface.
Fig. F. " *hyperboreus;* same views as last.
Fig. G. " *montanus;* same views.
Fig. H. " *occidentalis ;* same views.
Fig. I. " *Hewstoni;* a larger individual.

PLATE IX.

Figures A, B, C, D, G, H, were drawn by W. G. Binney; E, F, by T. D. A. Cockerell; I, J, by Arthur F. Gray.

Fig. A. Jaw of *Hemphillia glandulosa.*
Fig. B. Jaw of *Phenacarion foliolatus.*
Fig. C. Lingual membrane of same; one central tooth, with its adjacent lateral and three extreme marginals.
Fig. D. Genitalia of same; one half of natural size.
  ov. ovary.
  ovid. oviduct.
  t. testicle.
  g. b. genital bladder.
  p. s. penis sac.
  v. d. vas deferens.

Fig. E, F. *Veronicella olivacea*, from one of original lot from Folvon.

Fig. G. Lingual membrane of *Glandina decussata*.

Fig. H. Genitalia of *Phenacarion foliolatus*, var. *Hemphilli*; same references as in D; one half of natural size.

Fig. I. *Prophysaon Andersoni*; surface magnified sixteen times.

    a. a. a. reticulations of the body.

    b. b. foliolated spaces between reticulations.

    c.   lower edge of the body.

    d.   locomotive disk.

Fig. J. The same, magnified eight diameters; upper surface; same references as the last.

PLATE X.

Drawn by A. H. Baldwin, Smithsonian Institution.

Fig. A. *Aglaja fidelis*; the large, elevated black variety.

Fig. B. Sculpturing of same.

Fig. C. The same; small, black, elevated form.

Fig. D. Sculpturing of last.

Fig. E. The same; small, depressed form.

Fig. F. *Aglaja infumata*; sculpturing.

Fig. G. *Arionta Mormonum*; sculpturing of the form figured on Plate XI. Figs. G, H.

PLATE XI.

Drawn by A. H. Baldwin.

Fig. A. *Arionta arrosa*.

Fig. B. Variety of last, approaching *A. exarata*.

Fig. C. Sculpturing of last.

Fig. D. *Arionta exarata*; type.

Fig. E. Sculpturing of last.

Fig. F. *Arionta Mormonum*.

Fig. G, H. Variety of last, connecting with *Hillebrandi*.

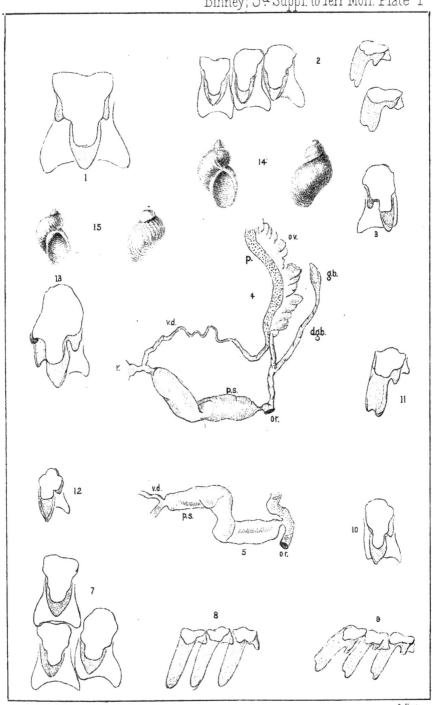

W.G.B.del.

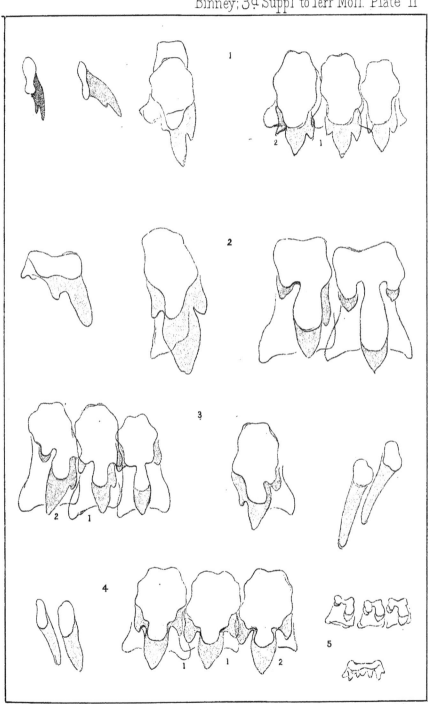

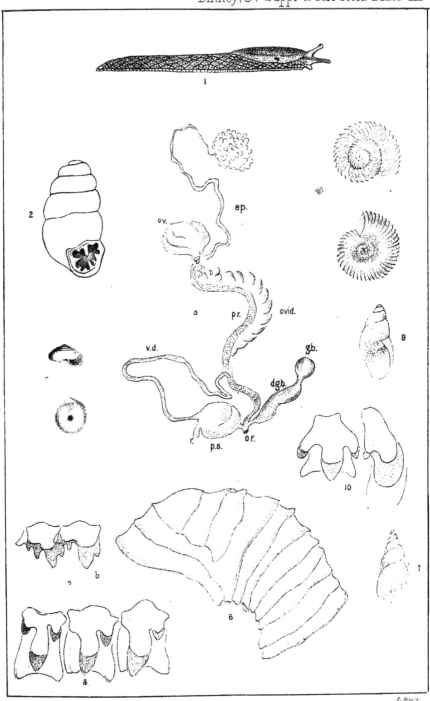

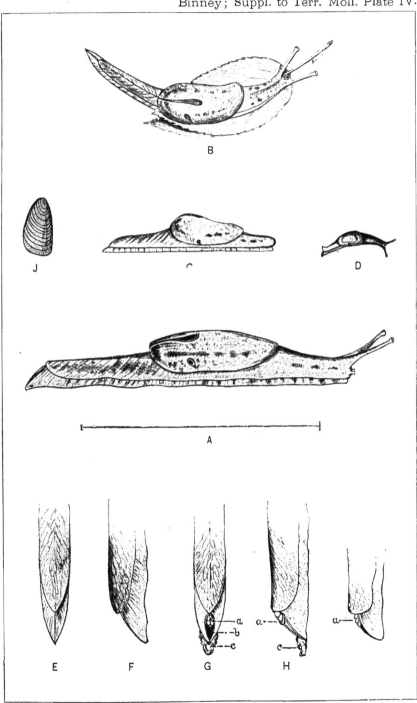

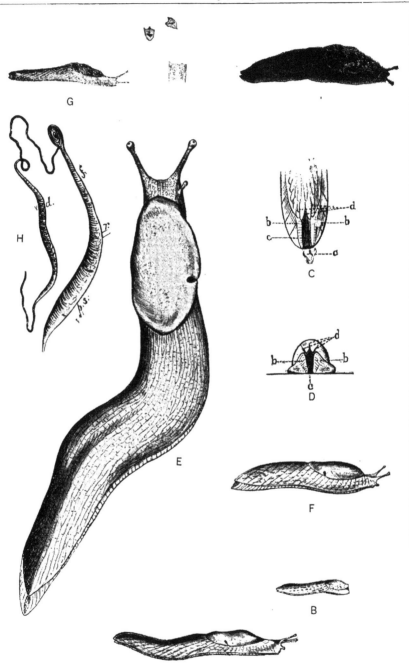

G

H

C

D

E

F

B

A

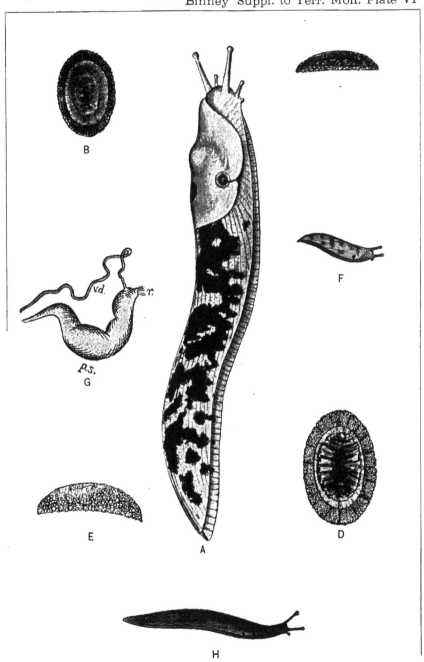

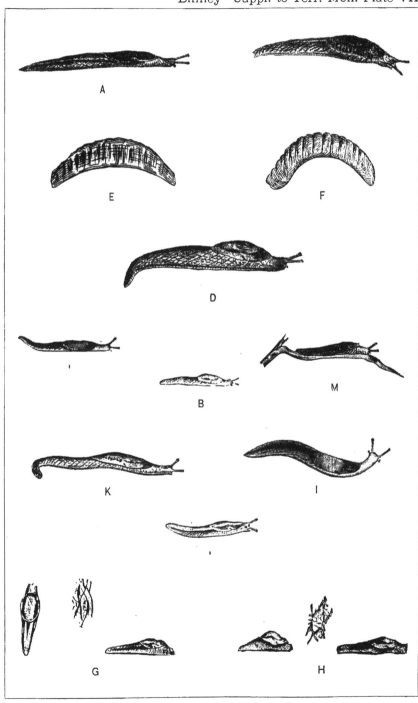

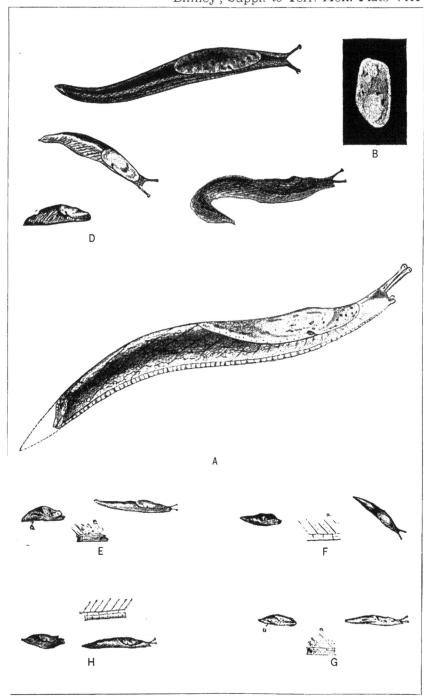

B

D

A

E

F

H

G

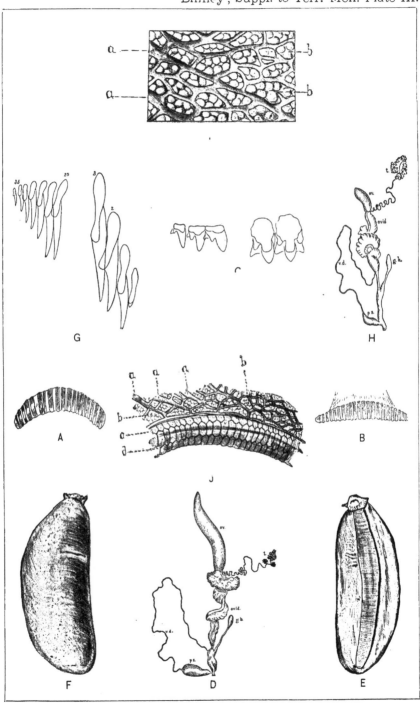

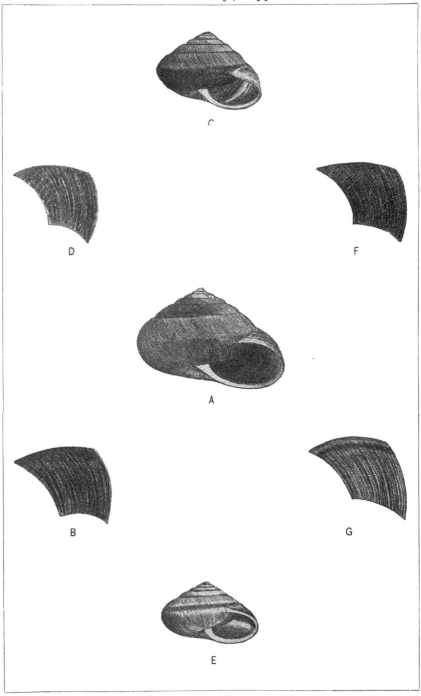

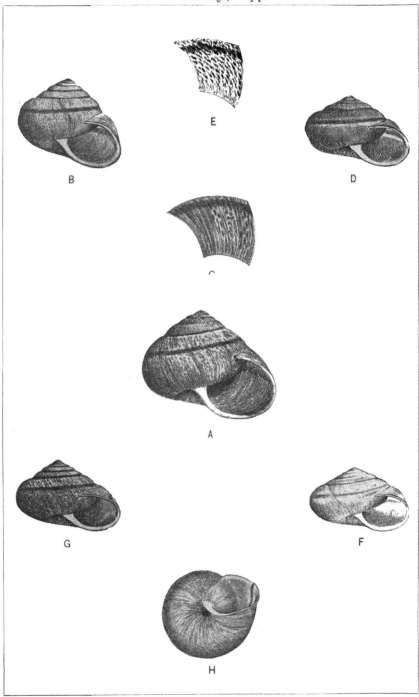

BULLETIN

OF THE

MUSEUM OF COMPARATIVE ZOÖLOGY

AT

HARVARD COLLEGE, IN CAMBRIDGE.

VOL. XX.

———

CAMBRIDGE, MASS., U. S. A.
1890–1891.

Uɴɪᴠᴇʀsɪᴛʏ Pʀᴇss:
Joʜɴ Wɪʟsoɴ ᴀɴᴅ Soɴ, Cᴀᴍʙʀɪᴅɢᴇ, U. S. A.

CONTENTS.

No. 1. — *The Histology and Development of the Eye in the Lobster.* BY G. H. PARKER.[1]

INTRODUCTION.

THROUGH the kindness of Mr. Alexander Agassiz it was my privilege to spend the greater part of the summer of 1887 at the Newport Marine Laboratory. During the preceding winter I had been interested in the structure of the eyes in Arthropods, especially in the inversion of the retina in Arachnoids and my instructor, Dr. E. L. Mark, had called my attention to the importance of ascertaining whether the retina in the compound eyes of Crustaceans was inverted or not. At about this time Kingsley ('86ª) published his preliminary account of the development of the compound eye of Crangon, and claimed that in this crustacean, as in spiders, the retina was inverted. For reasons which I shall mention in the course of this paper, Kingsley's account did not seem fully satisfactory to me, and consequently I decided to study for myself the development of the eye in a crustacean. My visit to the Newport Laboratory offered an excellent opportunity to collect embryological material for such a study. During August and September spawning lobsters were easily obtained, and I therefore determined to study the eye in the lobster, *Homarus americanus*, Edwards. A series of lobsters' eggs were collected, and before leaving Newport my observa-

[1] Contributions from the Zoölogical Laboratory of the Museum of Comparative Zoölogy, under the direction of E. L. Mark, No. XVII.

tions had been carried far enough to satisfy me that the retina in the lobster was a simple ectodermic thickening. On returning to Cambridge from Newport, the study of the lobster's eye was continued in the Embryological Laboratory at Harvard College, under the direction of Dr. Mark. Here I completed the observations on the development of the eye, and studied its histology. In the fall of 1888 a brief preliminary account of the results which are now presented in full was published in "The Proceedings of the American Academy of Arts and Sciences," Vol. XXIV. pp. 24, 25.

In procuring at Newport the necessary stages in the development of the lobster I proceeded as follows.

Female lobsters with eggs were obtained from the fishermen, and kept in floating latticed boxes which were anchored in the small cove beside the Laboratory. A few eggs were taken daily from each lobster. The reagents which I employed in killing the eggs were Kleinenberg's picro-sulphuric acid, Perenyi's fluid, a saturated aqueous solution of corrosive sublimate, and hot water. The eggs which were prepared with corrosive sublimate were rendered almost useless by the subsequent formation of a fine precipitate. Those which were killed in Kleinenberg's picro-sulphuric acid and in Perenyi's fluid gave fair results; the latter reagent left the yolk in good condition for cutting. The best results, however, were obtained by the use of hot water. Eggs which had been prepared in this way could be easily shelled, and the embryos could be readily dissected from the yolk. The separation of the embryo from the yolk proved to be a great advantage, and obviated the necessity of cutting the yolk, a tedious process in an egg as large as the lobster's.

In the following account of the development of the lobster's eye, the stages which it is necessary to describe are taken from different sets of eggs. These sets were from different lobsters, consequently I cannot state with exactness their relative ages. I shall therefore characterize them by their most evident structural peculiarities. Beginning with the earliest stage and proceeding to the later ones, I have lettered them A, B, C, D, E, and F. Set A is in the stage of the "egg-nauplius"; in this set the characteristic three pairs of appendages are easily distinguishable. In set B the thoracic appendages have begun to form. This stage corresponds very closely to what Reichenbach ('86, Plate III. Fig. 11) has designated in the crayfish as stage H. In stage C the first trace of pigment in the retina is visible. Stage D is from the same series of eggs as stage C, but is seven days older than C. In both stages C and D, the abdomen of the embryo is recurved, and reaches forward covering the space between the optic lobes. Stage E corresponds to the time of hatching. Stage F is represented by a young lobster one inch in length.

The younger stages which follow the hatching of the lobster are obtained with considerable difficulty, and I am under obligations to several of my friends for material which covers this period. For some lobsters in the "Schizopod" stage I am indebted to Mr. Sho Watase. Mr. H. H. Field and Mr. Carl H. Eigenmann kindly collected for me some young lobsters one inch in length. From Mr. F. L. Washburn I received the eyes of several half-grown lobsters, six to eight inches in length. The material which I used in studying the histology of the eye in the adult was very kindly supplied to me by A. T. Nickerson and Company, of Charlestown, Mass.

Methods.

The methods of staining, embedding, etc., which I have employed, are those known to all students of modern histology. In one case, the staining of nerve-fibres, I have used a method which I accidentally discovered while experimenting with Weigert's hæmatoxylin.

In employing this method it is necessary to stain the sections on the slide. The way in which I have stained sections on the slide has already been described ('87, p. 175). Further experience has shown, however, that the successful employment of this method necessitates a careful observance of certain precautions. These I have not sufficiently emphasized in my former account, and I therefore redescribe the method, calling especial attention to the precautions. The method consists in a cautious use of Schällibaum's fixative. The fixative which I have employed is composed of clove oil three parts and Squibb's flexible collodion one part. The mixture before being used should be allowed to stand for about a week. After several months it may become ineffective. When working, I usually employ the fixative frequently enough to follow its changes, and at the first signs of failure I make a new mixture. If for any reason I have not used the fixative for some time, I test it with a few waste sections before employing it with valuable material. In using it a moderate amount is applied to the slide, and the sections in paraffine are placed on it. The slide and its sections are now subjected to a temperature of 58° C. for fifteen minutes. It is important to observe carefully both the length of time during which the slide is heated and the temperature to which it is raised. At the end of fifteen minutes, the slide, while warm, is thoroughly washed with flowing turpentine. This can be applied conveniently from a small wash-bottle. All of the paraffine should be removed from the slide before it becomes cool, otherwise on cooling some paraffine may solidify. This is liable to loosen the film of collodion. The wash of turpentine should be continued not only till the paraffine is thoroughly removed, but till the slide is cool. Then, and not till then, can the turpentine be safely replaced by alcohol, first 95%, then 70%, 50%, and 35%, and finally it can be immersed in water. After once having got the slide with its sections into water, the subsequent treatment with alcohol and water seems to have no effect in loosening the sections, although the film of collodion will dissolve easily in ether. I have very generally employed this method of staining for two years, and as it obviates the difficulties which arise from maceration or partial penetration of dyes, I use it in preference to staining *in toto*. I have lost very few sections by it, and such accidents as I have had were due, I believe, to a neglect of some of the precautions which have been mentioned.

The method of staining nerve-fibres which I have employed consists of a modified use of Weigert's hæmatoxylin. The tissue which was stained by this method was for the most part killed in hot water, although I have also successfully stained nerve-fibres which were killed in chromic acid and Kleinenberg's picro-sulphuric acid. Sections of the optic nerve which had been mounted on the slide and carried into water were treated for about half a minute with an aqueous solution of potassic hydrate $\frac{1}{10}$%. They were then *thoroughly* rinsed in distilled water and transferred to Weigert's hæmatoxylin. Here they remained for about three hours at a temperature of 50° C. They were then rinsed again in distilled water, carried through the grades of alcohol, and after being dehydrated with alcohol of about 99%, they were cleared in turpentine and mounted in benzole balsam. Each nerve-fibre when so treated had a distinct blue-gray outline. The sections do not overstain even when they are kept in the dye for a prolonged period, and there is of course no subsequent decoloring. This method yields fair results when applied to nerves from any part of the lobster's body, but it is especially successful in treating that portion of the optic nerve which intervenes between the retina and the optic ganglion.

THE HISTOLOGY.

The two movable eye-stalks of the lobster are situated one on either side of the rostrum, at the angle which that structure makes with the anterior edge of the carapace. The form of the eye-stalk approaches that of a short cylinder terminated by a hemisphere. The cylindrical part of the stalk resembles the general surface of the body in that it is covered with a firm, calcified cuticula. Excepting a portion of the surface next the rostrum, the whole of the hemispherical part during life is black, and covered with a flexible cuticula. The black area defines the position of the retina. That portion of the hemispherical surface which is not black, and which faces the rostrum, is covered with a peninsula-shaped piece of inflexible cuticula. A broad isthmus of the same kind of cuticula connects this with the shell of the cylindrical part. The absence of the retina from the peninsula-shaped portion of the hemisphere is due in all probability to the fact that the field of vision for this part of the hemisphere is cut off by the rostrum. The remainder of the hemisphere, that part on which the retina is developed, faces away from the lobster's body, and its field of vision is not permanently obstructed by any part of the animal.

A section perpendicular to the surface, and cutting the eye-stalk in a region where the cylindrical and hemispherical parts unite, is shown in Figure 26. The thick, calcified cuticula of the cylindrical part is indicated at *cta*. On the inner surface of this cuticula is a thin hypodermis (*h d.*). The hypodermis is bounded on its inner face by a basement membrane (*mb.*). The cuticula of the hemispherical part (*crn.*) is thin and flexible. It can be designated by the name corneal cuticula. (Compare Patten, '86, p. 544.) Resting on the deep face of the corneal cuticula is the thick cellular layer, named by Lankester and Bourne the ommateum (*omm¹.*). The proximal face of the ommateum is limited by a basement membrane, which is continuous with that bounding the corresponding face of the undifferentiated hypodermis. The ommateum is continuous with the hypodermis, and in fact can be regarded as a thickening of that layer. Carrière ('85, p. 169) has already pointed out in the eye of Astacus a similar relation between the hypodermis and ommateum, and he believes that this relation holds good for all Decapods.

On inspecting the external face of the corneal cuticula, one finds it divided into an immense number of square facets, one of which is shown in Figure 2. Although as a rule the outline of the facet is square, it is not invariably so; for on the margin of the retinal area close to where the ommateum passes over into the undifferentiated hypodermis, the outline often becomes somewhat irregular, and more frequently presents the form of a hexagon than of a square (Fig. 59). The number of facets in each eye of an adult lobster is about 13,500.

66 BULLETIN OF THE

In the ommateum the cells are arranged in specialized groups or ommatidia. There is a single ommatidium under each corneal facet, consequently in any given eye the number of ommatidia equals the number of facets. The cellular composition of each ommatidium is best understood from a comparison of longitudinal and transverse sections. Figure 1 represents a longitudinal section through an ommatidium. The thick lamellated layer (*crn.*) at the distal end is the corneal cuticula. Directly below this is a thin layer of cells, the corneal hypodermis (*crn. h d.*). Following on the corneal hypodermis are the cone-cells (*cl. con.*). They are very long, and extend from the corneal hypodermis inward till their proximal ends disappear in the deep part of the retina. In reality they terminate upon the basement membrane. Their distal ends in the region of the crystalline cones are surrounded by pigment-cells, to which I give the name distal retinulæ (*rtn'. dst.*). These, like the cone-cells, extend to the deeper part of the retina. Here the proximal retinulæ and accessory pigment-cells occur. The proximal retinulæ are elongated cells (*rtn'. px.*), and contain black pigment. They surround the rhabdomes (*rhb.*). The accessory pigment-cells are irregular cells, which fill the space between the deep ends of the proximal retinulæ. They contain a pigment which is whitish by reflected and yellowish by transmitted light. Their nuclei are shown at *nl. pig.*, Figure 1. The last two kinds of pigment-cells described rest upon the basement membrane (*mb.*); below this membrane the fibres of the optic nerve can be seen (*n. fbr.*).

From this description it will be seen that the ommateum lies between the corneal cuticula and the basement membrane, and is composed of the following kinds of elements: cells of the corneal hypodermis, cone-cells, distal retinulæ, proximal retinulæ, and accessory pigment-cells. The numbers and positions of these cells are best made out from transverse sections. The several kinds of cells will be discussed in the order named.

The Corneal Hypodermis.

That the corneal cuticula in Decapods is separated from the cone-cells by an intervening layer of cells is a view which has been held only by recent investigators. Grenacher ('79, p. 123), in his account of the eyes in Decapods makes no mention of such a layer, and leaves one to conclude that the cone-cells abut against the cuticula. Claus ('86, p. 57) suspected the presence of a corneal hypodermis in Decapods, Schizopods, and Stomatopods, but his search for it was in vain.

The view that the cuticula and cone-cells are in contact, is strongly contrasted with that maintained by Patten ('86, pp. 626, 642). According to this writer, the corneal cuticula is due to the activity of a layer of cells, the corneal hypodermis, which intervenes between the cuticula and the cone-cells. Patten has identified the corneal hypodermis in the following genera of Decapods : Penæus, Palæmon, Pagurus, and Galathea. It has also been described by Kingsley ('86, p. 863) in the eyes of Crangon, and by Herrick ('86, p. 43) in the eyes of Alpheus. Carrière ('89, p. 225) has recorded it in the eye of Astacus, and there is now good reason for believing that a corneal hypodermis exists in the eyes of all Decapods.

Patten's statement ('86, pp. 665, 666) that the corneal hypodermis " has been invariably overlooked by Grenacher," and Kingsley's assertion ('86, p. 863) that the existence of the corneal hypodermis "was utterly ignored by Grenacher," are perhaps a trifle too strong. It seems much more probable that Grenacher confused the nuclei of the cone-cells and corneal hypodermis. He evidently never saw both kinds of nuclei in the eye of the same Decapod. In some cases he may have described the nuclei of the cone-cells, in other cases those of the corneal hypodermis. In both instances what he described he took to be the nuclei of the cone-cells. In the eye of Mysis, I believe that he ('79, p. 118) described the nuclei of both the cone-cells and corneal hypodermis, although in this case he was of the opinion that both sets of nuclei belonged to the cone-cells. Where only one set is figured, it is difficult to decide whether he has given the nuclei of the cone-cells or of the corneal hypodermis. So far as I am aware, there are always in each ommatidium of a Decapod *two* hypodermal nuclei, and *four* nuclei in the cone-cells. This numerical relation is sufficient to distinguish the groups of nuclei, but it can only be employed satisfactorily where transverse sections at the proper niveau are given. Unfortunately, in the Decapods, Grenacher did not figure any such sections, and it is therefore difficult to decide in particular cases which kind of nuclei he has described.

In the lobster a well differentiated corneal hypodermis has already been pointed out (Fig. 1, *crn. h d.*). In transverse sections this presents the appearance of squares of granular protoplasm (Fig. 3). Each square contains two nuclei, and is bounded by a membrane. A narrow space filled with granular substance separates the membranes of adjacent squares. From the longitudinal section (Fig. 1) it will be seen that these squares are relatively thin, so that their proportions are somewhat like those of square tiles. The outer face of the tile is flat; its inner face is hollowed, however, so that its centre is the thinnest part. In a few cases the corneal hypodermis has appeared as cubical blocks, rather than as tiles. This thickened condition probably indicates an increased functional activity, and the more frequently occurring tile-like condition may correspond to a quiescent stage.

The two nuclei contained in each square are placed some distance apart, and on one of the diagonals of the square (Fig. 3). Their long axes are approximately parallel to the other diagonal. In a given eye all the squares agree in having the nuclei on parallel diagonals. The presence of two nuclei in a square indicates that the square consists of two cells. Any membrane separating the two cells must necessarily pass between the two nuclei, but all attempts to discover such a membrane have failed. However, for a reason which will be given shortly, I believe that the protoplasm of the hypodermal square is divided by the diagonal which lies between the nuclei. In the centre of each square several oval or round outlines are usually visible (Fig. 3). These are vesicular bodies which occur in the distal ends of the cone-cells, and which can be seen through the very thin corneal hypodermis.

The corneal cuticula is the result of the activity of the corneal hypodermis. Viewed from the surface, the cuticula is divided by narrow bands into square facets (Fig. 2). Each facet is external to a hypodermal square. The proximal and distal faces of each facet, as can be seen in the transverse section (Fig. 1), are very nearly flat, the proximal face only being a trifle convex. This convexity, however, is so slight that one cannot attribute to the facet the character of a lens.

When a piece of corneal cuticula is cleaned by treating it with potassic hydrate, and is then examined in water, the markings which are visible with difficulty in preparations mounted in balsam are easily seen. Each facet in addition to its narrow limiting bands has a faintly marked diagonal band which divides the square into two equal triangles (Fig. 2). In the different facets of a given eye the diagonal bands are parallel. Newton ('73, p. 327, Plate XVI. Fig. 3), in describing the structure of the eye in the lobster, states that each facet is crossed by *two* diagonals at right angles to each other. This statement I cannot confirm, for, although I have searched with care, I have never succeeded in finding more than a single diagonal in each facet. In the middle of the diagonal there is an irregular hazy patch. This at times has a distinctly marked cross in it. When the cross is present, one of its axes lies in the diagonal band, the other extends at right angles to the band (Fig. 2).

Whether all of these markings extend through the substance of the cuticula, or whether they are confined to its surface, is difficult to say. The production of the cuticula is such a uniform process that one would naturally expect to find that the marking extended through it, for the successive layers would be similarly marked, and thus bands would be established extending from its deep to its superficial face. Concerning the vertical extension of the bands between the facets there is no question, for in transverse sections of the cuticula (Fig. 1, x) they reappear in their proper positions, and extend from one surface to the other. Owing to the roughness of the cut face, they are much less readily detected in sections than when viewed from the outer surface of the cuticula (compare Figs. 1 and 2). The diagonal band and its central spot have not been observed in transverse sections, even when the plane of section is in the most advantageous position for demonstrating these structures. Notwithstanding their apparent absence, both may be present, although indiscernible. For even in the superficial view, when the outline of the facet was so readily visible, the diagonal band was only faintly seen. In transverse sections, where the distinct boundary of the facet is visible with difficulty, one should not expect to see the much fainter diagonal. On comparing the diagonal band and the boundary of the facet by focusing through the corneal cuticula, I was unable to distinguish a greater vertical extension in the one than in the other. Since it has been shown that the boundary of the facet extends through the cuticula, this observation supports the conclusion that the diagonal band also extends through it.

Patten ('86, pp. 626, 627) has described in the facet of Penæus a band which has many resemblances to the diagonal band in the lobster. It is not diagonal, however, but transverse, and divides the square facet into two equal rectangles, in which the sides are in the proportion of one to two. I have already given my reason for believing that the diagonal band in the cornea of the lobster extends through the substance of the cuticula. Patten states that the transverse band in Penæus is only a superficial structure, and says ('86, p. 627) that in cleaning the cuticula " when the treatment with caustic potash has been carried to excess, all markings disappear except the contours of the facets." I have subjected the corneal cuticula of a lobster to a boiling solution of potassic hydrate (75%) for a quarter of an hour, and, although the potash completely cleaned the cuticula, the outlines of the facets, the diagonal band, and its spot were as readily visible after this treatment as before. A second and third trial with the same piece of cuticula did not noticeably effect the markings. In this respect, then, the diagonal band in the lobster is materially different from the transverse band in Penæus, and I conclude that in the cornea of the lobster the limiting and diagonal bands are essentially similar in that they both extend through the cuticula.

In all probability the bands between the facets were produced during the secretion of the cuticula by the interference of the partitions which separate the hypodermal squares. If this be true, it is probable that the diagonal bands represent a like interference. It is important to notice that the diagonal band in the cuticula corresponds to the imaginary diagonal which lies *between* the nuclei of each hypodermal square, never to the diagonal which crosses the nuclei (compare Figs. 2 and 3). This diagonal then corresponds to the position in which one would look for a membrane between the pair of hypodermal cells; and although such a structure has not been observed, the diagonal band in the cornea is a strong indication of its presence.

Admitting this to be the significance of the diagonal band, it is but natural to expect that, if deeper cells touch the cuticula, they would pass outward *between* the hypodermal cells. The fact that the hazy patch which lies in the middle of the facet is always on the diagonal band, and directly external to the distal tips of the cone-cells, leads to the belief that this patch marks the place where the cone-cells pass between the cells of the hypodermis and touch the cuticula. I am not of opinion that the patch is produced by the secretion of the cone-cells, although I have no evidence that the cone-cells cannot produce cuticula at their distal tips. It seems to me more probable that they have given rise to the patch by a series of interrupted interferences with the activity of the corneal hypodermis. If such be the case, a distinct cross might be produced when the area of interference was definitely circumscribed. When the area was not so sharply bounded, a hazy patch with indistinct outlines might be the result.

From the facts which have been presented, I conclude that each hypodermal square consists of two flattened cells, triangular in outline, and very intimately applied on their longest sides.

The Cone-cells.

One of the most important questions in the anatomy of the cells of the crystalline cones (retinophoræ) concerns the relation which these cells bear to the rhabdome. Max Schultze was the first to maintain ('67, p. 407) that the cone-cells and rhabdomes were separate structures. Grenacher's researches lead to the same conclusion. As an opponent of this view, Patten ('86, p. 670) has claimed that the cone-cells and rhabdome were continuous, and in fact that the rhabdome of the compound eye was only an enlargement of the proximal end of the cone-cell. Kingsley ('86, p. 863) in his description of the eye in Crangon supported Patten's view.

Of those authors who maintain the separateness of the cone-cells and rhabdome, no one, I believe, has given a fully satisfactory account of the way in which the proximal ends of the cone-cells terminate. Grenacher, in describing the eye in Palæmon said ('79, p. 123) : " Die fein ausgezogene Spitze dieser Pyramide [the cone-cells] durchsetzt, bevor sie in contact mit der Retinula tritt, zuerst eine in Form eines Hohlcylinders sie umhüllende Pigmentmasse um sich dann in das Vorderende der Retinula eine Strecke weit einzusenken." A more detailed account was given by Schultze, who, after stating ('68, p. 10) that in some crustaceans the cone-cells appeared to terminate a little in front of the distal end of the rhabdome, said that in the crayfish "geht der Krystallkegel nach unten in vier Spitzen aus, welche sich aus den vier Kanten der Oberfläche entwickeln und das obere Ende des nervösen ebenfalls vierkantigen Sehstabes umschliessen. Die vier Spitzen legen sich dabei an die Kanten des letzteren an und laufen als lange feine Fäden auf der Oberfläche des Sehstabes herab, diesen umklammernd und mit ihm oberflächlich verbunden aber durch Maceration isolirbar. Gegen das Ende spitzen sie sich fein zu und verlieren sich auf der Oberfläche des Körpers, den sie umfassen." This account is the most complete of any that I have seen, and yet that Schultze was not fully satisfied that he had seen the proximal termination of the cone-cells is probable from the fact that he says the fibres *are lost* on the surface of the rhabdome.

The relation of the rhabdome to the cone-cells, and the way in which these cells terminate in the lobster, is as follows. As in other Decapods, each ommatidium in the eye of the lobster contains four crystalline cone-cells. Together these cells form an elongated pyramid, with its base next the corneal hypodermis and its apex on the basement membrane (Fig. 1, *cl. con.*). At the distal end of the ommatidium, in the region which corresponds to the base of the pyramid, the four cells are closely applied to each other. This condition is maintained till the deeper part of the ommatidium is reached. Here the four cells, reduced to fibres, separate and end independently on the basement membrane.

A transverse section of the distal ends of the cone-cells is shown in Figure 4. On the external faces of each group of four cells there is a distinct bounding membrane (*mb. pi ph.*). This can be called the peripheral membrane. The four cells in each group are separated one from another by delicate membranes (*mb. i cl.*), which often show undoubted continuity with the peripheral membrane. These membranes, since they lie between the cone-cells, can be called the intercellular membranes. The distal end of each cell contains coarsely granular protoplasm and a nucleus (Fig. 4, *nl. con.*). The nuclei usually lie in the external angles of the cells, and do not readily take up coloring matter. The terminal granular protoplasm of the four cells forms a distal cap (Fig. 1, *cap.*). This cap fills the concavity on the proximal face of the corneal hypodermis, and its central distal tip probably passes between the pair of hypodermal cells and touches the cuticula. The spot or cross which is thus probably produced in the centre of each facet has already been described.

Below the cap of granular protoplasm is the crystalline cone. The firm peripheral membrane of the cap is continued over the cone and proximal part of the cone-cell. The distal end of the cone in cross-section is a square with rounded angles. At this end there is no indication of a division of the cone into four segments corresponding to the four cells. A transverse section midway the length of the cone (Fig. 5) shows no features essentially different from those of the section across the distal end. The proximal end of the cone in cross-section is nearly circular (Fig. 6). On the sides of this end of the cone one often notices small re-entrant angles (Fig. 6, *x*). The peripheral membrane dips into these. The angles are usually four in number, never more, and occupy positions which indicate the planes of separation between the four cone-cells. In some cases delicate membranes originating from the angles divide the substance of the cone into its four constituents (Fig. 6, *mb. i cl.*). These membranes correspond in position to the intercellular membranes at the distal end of the cone-cells. The substance of the cone is very finely granular. The four constituents of each cone terminate very nearly at one level. In passing in a proximal direction through a series of sections, the substance of the cone is last seen as a thickening which flanks the cell membranes, especially the intercellular membranes. (Compare Figs. 1 and 6.)

Below the cones the outlines of the four cone-cells are well marked by both peripheral and intercellular membranes (Fig. 7). The intercellular membranes are continuous with those seen in the proximal ends of some cones. In this region the cells contain coarsely granular protoplasm. In passing from the deep ends of the cones to the proximal

retinulæ, the most striking difference noticed in the cone-cells is a diminution in their diameter. (Compare Figs. 1, 6, and 8.) On a level with
the distal ends of the proximal retinulæ, the groups of cone-cells still
retain their four-parted character (Fig. 9, *cl. con.*). Each group is easily
traced between the retinulæ till it approaches the distal end of the
rhabdome. The change which here takes place is represented in Figure 12. Of the four ommatidia which are shown in this section, the one
indicated at *a* is cut slightly above the rhabdome. In the case of ommatidia *b*, *c*, and *d*, the plane of section passes through the end of the
rhabdome. In each of these three, it will be noticed that the rhabdome
is surrounded by four bodies, which correspond in position to the four
cone-cells. The four ommatidia which are drawn in Figure 12 are in
no way exceptional, but represent a very usual condition. Many such
cases have been examined, and whenever the tip of the rhabdome was
in the section, it was invariably surrounded by the four bodies previously
mentioned. When, on the other hand, the plane of section did not pass
through the rhabdome, only the four cone-cells were present. The
round bodies at the sides of the rhabdome can be traced from section
to section, and I therefore believe them to be fibres. Moreover, there
is no break observable in their continuity with the cone-cells, and I
therefore further believe that they are the fibrous prolongations of the
cone-cells. They have one peculiarity which is worthy of comment.
As the cone-cell passes over into the fibre, a considerable diminution in
its diameter takes place. This is accomplished at the distal end of the
rhabdome, and within a space equal to the thickness of one or at most
two sections (7.5–15 μ). Occasionally there is to be seen a group, in
which one or two cells have been reduced to fibres, and the remaining
ones are as large in transverse section as an individual in ommatidium *a* (Fig. 12). The conclusions which are arrived at from the study of
sections are confirmed by isolation-preparations. Figure 28 represents
a portion of an isolated group of four cone-cells from a single ommatidium. In the distal part of the specimen the four cells are intimately
bound together, but at the proximal end they appear as four separate
fibres. As in the transverse section (Fig. 12), the continuity of the
fibrous and thicker portion of the cone-cell, and the rapid reduction of
the cone-cell to form the fibre, are plainly seen. The cone-cells are usually
somewhat separated before they reach the rhabdome. It can scarcely
be said that they touch the rhabdome, although this is the region in
which they are nearest to it. As the fibres pass into the deeper part of
the retina they are found to lie nearer the periphery of the ommatid-

ium. They lie between the proximal retinulæ, but at some distance from the rhabdome (Fig. 14, *cl. con.*). They still retain, however, the same relative positions in the ommatidium. Their peripheral location is maintained (Fig. 17) till they are very close to the basement membrane.

The changes which the fibres undergo as they approach the basement membrane is shown in Figure 19. In this section the plane of cutting was slightly oblique to the basement membrane. Of the three ommatidia which are here represented, those indicated at *c* and *d* are cut at about the same level, and very close to the basement membrane. Ommatidium *b* is cut farther from the membrane than either *c* or *d*. The fibres of the cone-cells are closer to each other in *c* and *d* than in *b*. As this condition is generally true in other sections, it follows that, as we approach the basement membrane, the fibres of the cone-cells converge. The convergence is also shown in Figure 21. Of the ommatidia here figured, *b* is cut farthest from the basement membrane. In it the four cone-cells (*cl. con.*) can be seen, and between them a dot which represents the proximal end of the rhabdome (*rhb.*). Nearer the basement membrane is ommatidium *a*, in which the four cone-cells can be recognized, and to one side a fibrous area. The rhabdome does not extend as deep as this. The fibrous area represents a region in which the plane of section passes through an elevation on the distal face of the basement membrane. Ommatidium *d* is still nearer the membrane. The cone-cells are here brought more closely together, and are surrounded by the fibrous substance of the elevation. The form of the elevation is now seen to be that of a cross. At *x* is shown a basal section of the cross-shaped elevation surrounded by four large openings through the basement membrane. The cone-cells are no longer visible at this level, and I therefore believe that without penetrating the basement membrane they terminate in these elevations. This belief is further supported by the fact that in transverse sections of the basement membrane the cone-cells distinctly end in the substance of these elevations (Fig. 29, *cl. con.*).

The facts obtained from a study of the lobster's eye support the claim made by Schultze and Grenacher, that the cone-cells and rhabdomes are separate structures. In the case of the crayfish, Schultze, moreover, saw the prolongations of the cone-cells, and traced them into the deeper part of the retina. It is probable that in the crayfish, as in the lobster, the fibrous ends of the cone-cells terminate in the basement membrane, but this Schultze did not see. Such an omission is by no means surprising; for when we reflect upon the methods at his command, it is remarkable what success he had in tracing the course of the fibres, and in demonstrating the relation of the cone-cells and rhabdome.

Patten has advanced the view that the cone-cells are provided with an axial nerve-fibre, and that the cone itself is the true perceptive element. I shall defer the consideration of this topic till I describe the innervation of the retina.

The Distal Retinulæ.

Surrounding each crystalline cone are two pigment-cells, the distal retinulæ. These cells not only surround the cone, but extend as fibres into the proximal part of the retina (Fig. 1, *rtn'. dst.*).

The relation which the distal retinulæ sustain to the cone can be studied most readily in transverse sections. In a section passing through the distal end of the cone (Fig. 4, ommatidium *a*), it will be observed that of the four lateral faces which the cone presents, two, the lower and left-hand ones, are covered by a single retinula (*rtn'. dst.*). The retinula is thickest at the lower left-hand angle of the cone, and becomes thinner the farther it extends on the two adjacent faces. At the more distant edges of these two faces the retinula terminates. Thus the retinula is composed of a central portion and two blade-like extensions. Each blade covers one face of the cone. The second retinula is essentially like the one just described, but lies at the upper right-hand angle of the cone and covers its upper and right-hand faces. In this way the four faces of the cone are sheathed by a pair of retinulæ.

On inspecting the arrangement of the retinulæ in adjoining ommatidia (Fig. 4), it is evident that they are so placed that the thick end of each blade-like portion is opposite the thin end of the blade of a neighboring retinula, and that, in passing along the space between the cones, as one retinula becomes thicker the other becomes thinner. The delicate membranes which separate the blades consequently extend obliquely across the spaces between the cones (Fig. 4). In the space which is left between the angles of four adjoining cones the membranes of the retinulæ are very much thickened. That this is a thickening in the membrane of the retinula, and not due to substance produced by the cone-cells, seems probable for two reasons. First, the membrane is often somewhat thickened in regions between two retinulæ, and where the cone-cells could not well touch it. Such thickenings are directly continuous with the larger ones already mentioned (Fig. 4).

Secondly, in isolation-preparations the thickenings always remain attached to the retinulæ; the cones, on the other hand, are covered with membranes of uniform thickness. The thickening in the membrane is not characteristic of the whole length of the retinula, but is peculiar to the region corresponding in level to the distal end of the cone (Fig. 1).

The foregoing description applies to the structure of the distal retinulæ as seen in the plane of Figure 4. This plane passes through the outer ends of the cones. In other regions the retinulæ present somewhat different conditions. The relation of the retinulæ to the hypodermal squares is shown in a section (Fig. 3) which is slightly more superficial than that just described. The two retinulæ which were located at the two angles of the cones here occupy the corresponding angles of the hypodermal squares. They do not, however, entirely cover the four lateral faces of the square, as they did those of the cone, but from the angles at which they are located they extend over half of each of the adjoining faces. It follows from this that together they flank only one half of the lateral exposure of the square. The blades are now no longer wedge-shaped in transverse section, nor do they overlap neighboring blades, but each one stretches completely across the space in which it lies. Of the lateral surface of the square. that half which is not sheathed by its own pair of retinulæ is covered by the arms of four adjacent retinulæ. Consequently, six retinulæ in all touch each hypodermal square. Two of these belong to the ommatidium which is represented by the square; four belong to adjoining ommatidia. The relation of these will be readily seen by referring to Figure 3.

In passing from the plane in which each cone is surrounded by its own pair of retinulæ to the one in which the corresponding hypodermal square is surrounded by six retinulæ, the blades of the two retinulæ proper to the cone undergo a gradual narrowing; so that, instead of each blade covering the whole of one face of the cone, it covers less and less, and eventually sheathes only one half of the corresponding face of the hypodermal square. As the blades of the retinulæ become narrower, they expose the surface of the cone, but this is still kept covered by the retinulæ of adjoining ommatidia. In any ommatidium there are four blades which become narrow, consequently there are four regions in which adjacent retinulæ touch the cones; and as there is a separate retinula for each region, it follows that four additional retinulæ here come in contact with the cone.

The distal retinulæ touch the cuticula along the band which marks the boundaries of the facets. That the retinulæ contribute to the formation of the cuticula is very improbable, although I believe that it is largely through their interference that the outlines of the facets are produced.

The lateral surfaces of each cone are completely enclosed by retinulæ; the pair of retinulæ belonging to the cone play the principal part. That portion of each retinula which encloses the proximal two thirds of the cone is densely pigmented (Fig. 1). In transverse sections through the pigmented region one can see that each cone is *completely* surrounded by a pigment band. On a level with the middle of the cone each retinula contains a nucleus (Figs. 1 and 5, *nl. dst.*). This is imbedded in pigment. The membranes of the retinulæ are less distinct in the pigmented region than near the distal end of the cone. The only membrane which was observed (Fig. 5) was one which corresponds to the thickened membrane shown in Figure 4. At the proximal end of the cone, the retinulæ rapidly contract till they are reduced to fibres (Figs. 6 and 7, *rtn'. dst.*). The pigment is present for only a slight distance below this level. The fibres of the retinulæ are grouped in pairs, and in this relation extend to the proximal part of the retina. It is noticeable that the two fibres which constitute a pair are derived, not from a single ommatidium, but from two adjacent ommatidia. These fibres when seen in longitudinal sections were probably mistaken by Newton ('73, pp. 328, 329) for an investing membrane. At least, in all attempts to demonstrate the existence of such a membrane I have failed, and there is so strong a resemblance between the fibres of the distal retinulæ and the structure which Newton figured ('73, Plate XVII. Fig. 15) as the cut edge of an investing membrane, that I am inclined to think them identical. Transverse sections from the proper region would have settled the question whether these bodies were fibres or membranes, but unfortunately Newton has not figured any such sections.

The pair of fibres in passing from the basal ends of the cones to the proximal retinulæ retain the same relative position, and are only slightly reduced in diameter. (Compare Figs. 7 and 8, *rtn'. dst.*) Deeper than this, they are still identifiable, and can be distinguished from the fibres of the cone-cells by their slightly greater diameter, and by the fact that they are always in pairs. They lie in the space between four ommatidia. (Compare Figs. 12, 15, and 18.) Till within a very short distance of the basement membrane they maintain the condition shown in Figure 18, *rtn'. dst.* Beyond this I have not been able to trace them with certainty. The groups are no longer observable, and it is probable that the fibres have separated. I know that in this region the other cells suffer a very considerable rearrangement; and such being the case, it would be a very difficult matter to identify single fibres, especially fibres as small as these are. I have not found any satisfactory method of staining the fibres so as to distinguish them, as in the case of the fibrous ends of the cone-cells. I can therefore claim to have traced the fibres only to within about 20 μ from the basement membrane.

As I have already mentioned, the distal face of the basement membrane has cross-shaped thickenings on it. In the angles which the arms of the cross make with each other, the basement membrane is perforated. There are consequently around each cross four openings through the membrane. Each opening, however, lies between two crosses, so that in reality only one half of each opening belongs to a given cross, or, if one counts whole openings only, half of the four openings, i. e. two openings, belong to each cross. The crosses correspond in number and position to ommatidia, hence there are also two openings for each ommatidium. In each opening, beside three or four large fibres which will be described later, one finds a single small fibre (Fig. 21, *rtn'. dst.*). That this fibre represents the continuation of the fibrous end of a distal retinula seems probable, for two reasons. First, the diameters of this fibre and of the fibrous part of the distal retinulæ are so nearly the same as to be undistinguishable. Secondly, the number of fibres which pass through the basement membrane, two for each ommatidium, agrees with the number of distal retinulæ in each ommatidium. I therefore believe that the small fibres which are seen in the openings through the basement membrane are the proximal continuations of the fibres of the distal retinulæ. If this explanation be true, then it is only natural to expect that, as a pair of fibres approaches the basement, the individual fibres should separate, one passing through each opening. As I have already explained, the fibres, while separated, could be identified only with great difficulty.

If the fibres pass through the basement membrane, as I believe they do, they terminate only a short distance below it. For at about 15 μ below the membrane all of the fibres are of nearly the same size, i. e. somewhat larger than the large fibres which pass through the membrane (Fig. 22). In this region, then, the smaller fibres have either increased to the size of the larger ones, or diminished till no trace of them is left. The fibres here are in groups, however, and these are directly continuous with the groups which pass through the membrane. Each group of large fibres as it passes through the membrane consists of either three or four individual fibres. If the smaller fibres disappear, the groups below the membrane should consist of three or four fibres also ; if, on the other hand, the smaller fibres increase in size, the deeper groups should consist of four or five fibres. By either method of change there would be groups of four fibres, so that it is the groups of three or five fibres which will be decisive. As a matter of fact, the fibres are very commonly in groups of three, and not in groups of five ; consequently I conclude that the smaller fibres dwindle out a short distance below the basement membrane.

The distal retinulæ have not been identified in many Decapods. Carrière ('85, p. 169) has described them in the eye of Astacus. In Penæus, Patten ('86, p. 634) has observed four cells which belong to the pigmented collar of the retinophora. Two of these, the inner ones, evidently correspond to the distal retinulæ of the lobster. They surround the cones. The other two, the outer ones, appear to have no homologue in the lobster's eye.

The Intercellular Spaces of the Retina.

In the region of the retina which lies between the proximal ends of the cones and the distal border of the deeper band of pigment, the groups of cone-cells and the pairs of distal retinulæ are separated by considerable intervening space (Fig. 1, *spa. i cl.*). This space is filled with a fluid which contains a very small amount of albuminoid substance. Patten ('86, Plate 31, Fig. 73, *x*) has figured a similar fluid-filled space in Penæus. On the application of heat this albuminoid substance in the lobster coagulates and forms larger or smaller vesicular bodies, which vary much in size. They are usually loosely attached to the cone-cells and the fibres of the distal retinulæ. They readily take up coloring matter. They have never been observed in fresh retinas when teased in normal salt solution, nor in maceration-preparations. It was probably these bodies which Newton ('73, p. 329, Fig. 15, *c'*) described as the nuclei on the investing membrane.

In addition to the albuminoid substance which I have described, one occasionally meets with a thin layer of homogeneous material which lies slightly in front of the rounded ends of the proximal retinulæ. This forms a dividing membrane which separates the retina into a proximal and distal portion. The membrane is of course pierced in many places. There is an opening in it for each pair of distal retinulæ, and each group of cone-cells. In many cases the membrane has not been observed. It was noticed by Newton ('73, p. 328), and as it is non-cellular it is probably a feeble representative of what Herrick ('86, p. 44) has described as a "chitinous" framework in the deeper part of the retina in Alpheus.

The Proximal Retinulæ.

The proximal retinulæ are pigment-cells which closely invest the rhabdome. With the brownish accessory pigment-cells they constitute the proximal band of brownish black pigment on the distal side of the basement-membrane (Fig. 26, *pig. px.*). In some cases they appear to terminate distally in rounded knobs, each of which contains a nucleus (Fig. 1, *rtn'. px.*). In other instances, and these are of frequent occurrence, their distal ends, in addition to having a swollen nucleated part, are prolonged into delicate fibres (Fig. 30). These fibres when present extend toward the outer surface of the retina, and are applied, not to the cone-cells, but to the fibrous portion of the distal retinulæ. The fibres have been traced only a short distance beyond the rounded ends of the cells from which they originate. As the region into which they extend is one readily studied in both sections and maceration preparations, and as these methods of study have given no evidence of fibres other than that of the very short ones already mentioned, it seems fair to conclude that the distal retinulæ terminate as fine fibres a short distance in front of their nucleated portions.

In transverse sections the distal retinulæ first clearly appear in the plane represented in Figure 9. Here each group of four cone-cells is surrounded by a circle of seven retinulæ. The section from which this figure was drawn is in a slightly oblique plane. In moving from right to left, one passes into deeper and deeper regions. In the more superficial part of the section, the right-hand half, each retinula contains a nucleus, which is surrounded by a small amount of pigmented cell-substance. In the deeper part of the section, the left-hand half, the plane is below the region of most of the nuclei, and one sees the seven retinulæ densely filled with pigment. In the next section (Fig. 10), the retinulæ are broader in transverse section. In their expansion they have so far encroached on the space which they surround that it is only large enough to allow the passage of the four cone-cells. The contraction of the space within the circle of retinulæ takes place almost in one plane, as can be seen in the longitudinal section (Fig. 1). In the plane of Figure 10, the retinulæ show a tendency to group themselves.

One can recognize an odd, usually larger retinula, which occupies the lower right-hand corner of each group. The remaining six retinulæ are disposed in pairs. In Figure 11, which represents a plane of section deeper than that shown in Figure 10, the retinulæ, although somewhat reduced in thickness, present nevertheless the same method of grouping as was pointed out in Figure 10. In this plane one also notices next the odd retinula, a nucleus. This is remarkably constant in its occurrence, both as to position and as to the fact that there is always a *single* nucleus. When compared with the nuclei of the surrounding retinulæ, it is found to resemble them very closely. The nuclei of the seven retinulæ are characterized by their sharply marked oval outlines, and by the possession of one or two very distinct nucleoli (Fig. 30, *nl. px.*). In both of these respects the single nucleus agrees so closely with the nuclei of the retinulæ, that, were it not for its somewhat smaller size and deeper position, it could not be distinguished from them. The regularity with which it occurs, and its structural peculiarities, incline me to believe that it represents a reduced retinula in which pigment has never been developed. This belief is further supported by the fact, that the additional nucleus is always found next the larger retinula, which from its great size seems to have replaced a second cell. It is therefore probable that each ommatidium of the lobster's eye possesses eight proximal retinulæ rather than seven, and that one of these is rudimentary.

Below this additional nucleus, the proximal retinulæ pass around the rhabdome. In this region they are deeply pigmented, and so completely envelop the rhabdome that I am of opinion that no appreciable amount of light gains access to it except through the cone-cells. The cone-cells, it will be remembered, extend through the central region of each group of proximal retinulæ, until they almost impinge on the distal end of the rhabdome. Thus, by excluding the retinulæ, they form a transparent shaft, which leads to the distal tip of the rhabdome, and by which that structure can receive light. The rhabdome in transverse section has a four-sided outline. Three sides of the rhabdome are occupied each by a pair of retinulæ. These pairs are composed of the same couples which were previously noticed. (Compare Figs. 10 and 14.) The fourth side of the rhabdome is occupied by the seventh or odd retinula. Thus, again, it is noticeable that this retinula occupies a position where, if perfect symmetry were shown, we should expect two retinulæ. The rhabdome in transverse section is broadest midway of its length. In this position the retinulæ are small, as if closely pressed

against its sides (Fig. 14).   Below its middle transverse plane the
rhabdome becomes gradually smaller and smaller, till finally it terminates about 15 μ from the basement membrane.   As the rhabdome contracts in size, the retinulæ enlarge.   (Compare Figs. 14 and 17.)

As I have already mentioned, the retinulæ are definitely arranged
around the rhabdome, and this arrangement persists nearly to its
proximal termination; but between the end of the rhabdome and the
basement membrane the retinulæ rearrange themselves.   This rearrangement of the retinulæ is a step preparatory to their passing
through the apertures in the basement membrane, the general structure of which has already been described.   It will be remembered that
under each ommatidium the distal face of the membrane presents a
cross-shaped thickening, and that in each of the four angles which the
arms of the cross make with each other there is an opening (Fig. 21).
The openings are oval in outline, especially on the distal face of the
membrane.   One end of a given oval lies in the angle of the cross,
and the crosses are so close to each other that the other end of the
same oval lies in the angle of a neighboring cross.   Each opening then
lies in the angles of two adjoining crosses, and through it pass two
groups of retinulæ, one from each of the two ommatidia to which the
crosses correspond.

The four groups into which the retinulæ of a single ommatidium
are divided pass one through each of the four surrounding apertures.
Three of the groups consist of pairs of retinulæ; the fourth group is
represented by only a single retinula.   Although these groups agree
numerically with the groups of retinulæ, which, as I have already
shown, surround the rhabdome, they are not composed of the same
individual retinulæ.

For convenience of comparison, numbers can be assigned to the different retinulæ.   This has been done in ommatidium c (Fig. 15), where
the large odd retinula is numbered 1, and the remaining retinulæ, proceeding in a circle to the left, are numbered 2 to 7.   On this plan of
numbering, the four groups of retinulæ which have been already indicated as surrounding the rhabdome are composed as follows.   What
may be called the first group is formed of retinulæ 2 and 3, the second
group contains retinulæ 4 and 5, the third retinulæ 6 and 7, and the
fourth retinula 1.   It will be observed (Fig. 15) that the seven retinulæ
are also divided into four other groups by the fibres of the cone-cells.
Three of these groups are composed of pairs of retinulæ, the individuals
of which lie nearest the angles of the rhabdome.   In Figure 15, omma-

tidium *c*, these three groups are represented by retinulæ 1 and 2, 3 and 4, and 5 and 6. The fourth group is composed of the single retinula number 7.

The four groups thus defined are identical with the ones which pass through the four openings in the basement membrane. In Figure 21 the four openings which surround the cross (*x*) can be designated from their positions as the upper, lower, right-hand, and left-hand openings. The upper and lower openings each present the transverse section of four large fibres. The right- and left-hand openings are each occupied by three large fibres. The source of these fibres can be ascertained by comparing Figures 20 and 21. In ommatidium *c* (Fig. 20) retinulæ 5 and 6 unite with retinulæ 1 and 2 of ommatidium *d*, and thus constitute the four fibres which pass through the upper aperture (Fig. 21). In a similar way, retinulæ 1 and 2 of ommatidium *c* unite with 5 and 6 of an ommatidium which lies below *c*, and pass as four fibres through the lower opening (Fig. 21). Retinulæ 3 and 4 of ommatidium *c* (Fig. 20) unite with retinula 7 of an ommatidium to the left of *c*, and emerge as three fibres through the left-hand opening (Fig. 21). Retinulæ 7 of *c*, and 3 and 4 of *b* (Fig. 20), unite and form the three large fibres of the right-hand opening (Fig. 21). This plan of distribution is repeated in each ommatidium, and thus brings about the groups of three or four fibres which occur in each opening through the basement membrane. The groups of fibres are distinguishable for only a very short distance below the basement membrane. The individual fibres of each group soon separate, and in the deeper part of the optic nerve they never again present this grouping. The description of the termination of the fibres will be deferred to a later part of this paper.

The relation of the rhabdome to the cone-cells in Homarus has already been described. That they are separate structures, as Schultze and Grenacher have asserted, I believe there can be no doubt. I have seen nothing which favors the view held by Patten, namely, that the rhabdome is an enlargement in the proximal part of the cone-cells. In Homarus the rhabdome has the general form of a spindle. In transverse section, however, it is not circular, but square. Its four sides are thrown into ridges, the crests of which extend across the sides at right angles to its longest axis. The inner face of each proximal retinula is thrown into corresponding undulations. The rhabdome and retinula are so adjusted to each other that a crest on one fits into a furrow on the other. (For a similar condition in Penæus, compare Patten, '86, Fig. 72.) The retinulæ and rhabdome are thus intimately bound

together. On comparing opposite faces of the rhabdome, it will be seen
that the crests of one side are in the same horizontal plane as those of
the other ; but on comparing adjoining faces, it will be observed that the
crests of one correspond to the furrows of the other.

In the rhabdome of the lobster I have not found a complicated system
of plates, such as Patten describes in Penæus. The substance of the
rhabdome in the fresh condition is apparently homogeneous, but in hardened preparations it is finely granular and stratified. The strata are
at right angles to the long axis of the rhabdome, and the rhabdomes
often break transversely. The stratified condition of the rhabdome, and
the close relation which the proximal retinulæ bear to it, support the
conclusion that the rhabdome is the product of the proximal retinulæ.

In the structure of the rhabdome there is one peculiarity which,
although I cannot explain it, requires some comment. In transverse sections the square area of the rhabdome is often divided into four smaller
squares by two intersecting lines (Figs. 13 and 43). I made this observation before I had studied the relation of the distal tip of the rhabdome to the cone-cells, and I concluded then, that, if Patten was correct
in believing that the cone-cells and rhabdome were continuous, these
four divisions of the rhabdome must correspond to the four cone-cells.
I recognized the fact, however, that, if such was the case, the group of
cone-cells in the region of the rhabdome was turned through an angle of
45° as compared with its position at the surface of the retina (contrast
Figs. 4 and 13). After having satisfied myself that the cone-cells and
rhabdome were separate structures, I was forced to the conclusion, that
the four segments of the rhabdome were independent of the four cone-
cells. That there can be no question of the independence of these two
structures is shown by the condition of the cone-cells and rhabdome
in Mysis. In all Decapods, so far as I am aware, the cone is formed of
four cone-cells, and the rhabdome has four segments; in Mysis, however, the cone is formed of only *two* cells, although the rhabdome has
four segments.[1]

I can offer no explanation of the cross lines which occur in the rhabdome. As Grenacher ('79, p. 124) has observed, one might at first take
them for the outlines of such parts of the rhabdome as were produced by
individual retinulæ. There are, however, seven retinulæ, and only four
segments. Not only do the numbers disagree, but the position of the
lines in the rhabdome is difficult to explain. If the lines are related
to the retinulæ, it would be natural to expect that they would coincide

---

[1] My attention was called to this fact by my friend, Mr. H. H. Field.

with the junction of two retinulæ. As a matter of fact, three lines do come between retinulæ, but the fourth one abuts against the middle of the large odd retinula (Fig. 34). This relation of the cross-lines and retinulæ persists from the earliest stage in the production of the rhabdome, and although the lines are doubtless formed during this process they show a strange independence of the retinulæ.

An exactly similar relation between the cross-lines and retinulæ has been described by Grenacher ('79, p. 124) in Palæmon. I cannot agree with Grenacher in believing that the four lines are due to the fact that only four retinulæ are concerned in the production of the rhabdome. The relations of the retinulæ and rhabdome are the same in the young lobster (Fig. 58) in which the rhabdome is being produced as in the adult. This fact was not known to Grenacher. It shows, I believe, that the rhabdome is the product of the surrounding seven retinulæ, and that the problematic lines have some other significance than that of indicating regions of production.

The Accessory Pigment-cells.

These cells occupy the open space at the base of the ommatidia. They are characterized by possessing a pigment which, as I have before stated, is brownish by transmitted and whitish by reflected light. The cells are bounded proximally by the basement membrane, and their distal ends rarely reach beyond the middle of the rhabdomes (Fig. 1). They are extremely irregular in form, and seem to fit themselves to a cavity of almost any shape. Their function seems to be that of filling what would be otherwise an unoccupied space, as though to lend solidity to the tissue in the base of the retina. (Compare Figs. 13 and 16.) Their nuclei are irregular in form and size (Fig. 18, *nl. pig.*). Judging from the number of nuclei, two or three cells are associated with each ommatidium. The number, however, is variable. The physical properties of the pigment which these cells contain are very characteristic. Streaks of this pigment, and even whole cells, are to be met with in the open space on the proximal side of the basement membrane.

Cells similar to those which I have called accessory pigment-cells have been described by Carrière ('85, p. 169) in Astacus, and by Patten ('86, p. 636) in Penæus. It is highly probable that the yellow pigment which Grenacher ('79, Fig. 114) figured in the base of the retina of Mysis represents accessory pigment-cells as well as the dark pigment which he described ('79, p. 124, Fig. 117) in Palæmon.

The Innervation of the Retina.

The study of the termination of the nerves in the retina is of particular importance, since it affords a means of identifying the perceptive
elements. Wherever these elements may be, the ultimate branches of
the nerve-fibres must unquestionably lead to them ; hence the importance of discovering the termination of the nerve-fibres.

Students who have investigated the compound eyes of Arthropods
have held two opinions as to the position in which the nerve-fibres terminate. One school has maintained that the fibres terminate in the
crystalline cones, and that therefore these bodies are the perceptive
elements. The other school has endeavored to show that the fibres
end in the region of the rhabdome, and that for this reason the rhabdome is the perceptive element. I shall not attempt to give an historical account of this subject, but only call attention to the fact, that
of late years the majority of writers have expressed the opinion that the
rhabdome is the perceptive element, and that the cone is merely dioptric
in function. This conclusion has been recently criticised by Patten,
who believes the cone to be the perceptive body.

Many of Patten's statements are based upon observations which were
possible only by his methods of investigation. On this account, as well
as for the reason that his paper is the most important recent contribution to the study of nerve-termination in compound eyes, I shall not
refer to the older publications, but limit myself to what he has presented in his article on " Eyes of Molluscs and Arthropods," and to such
papers as have appeared since the publication of that work.

When comparing Patten's statements on nerve-termination with
those of other recent investigators, I was inclined to believe, since
Patten used new and probably better methods of study, that his results
were more trustworthy. The contrast between his views and those
which are more generally accepted is so striking, however, that in beginning a study of the nerve-terminations the first step to be taken was
necessarily one of confirmation. I was unable to obtain the same
species of Crustaceans as Patten had used, but I believe that I was safe
in assuming that the difference which may exist between the innervation of the retina in Penæus and Homarus could not be a fundamental
one, and that the more important feature which had been demonstrated
in one genus could be shown in the other. I consequently prepared
sections of the retina of Homarus according to the methods which Patten
had recommended, and although I was careful in both preparation and

examination of these sections, I was entirely unsuccessful in discovering any trace of nerve-fibres such as Patten had described. I therefore resolved to try his methods of maceration. This I did, following closely his directions as to the strength of solutions, length of time during which reagents should be employed, etc., but my results were again negative. Still adhering to the reagents which he had recommended, especially chromic and sulphuric acids, I varied the time during which the eyes were treated, hoping thereby to obtain a combination more favorable for Homarus. The separation of the elements of the retina was often very successful, but I never saw in any of my preparations systems of nerve-fibres which resembled those figured by Patten. In almost all cases the isolated parts of the retina presented many delicate fibrous projections. These projections might be interpreted as shreds of broken nerve-fibres, although in no case did they show a systematic arrangement. Moreover, they were found on all kinds of tissue. For these two reasons, I believe that they were not broken nerve-fibres, but simply shreds of tissue. The substance of the cones was finely granular, and was never penetrated, so far as I could discover, by any fibres. My results from both sections and isolation preparations were invariably negative; and as my observations had been made upon somewhat over sixty lobsters' eyes, I concluded that in Homarus there was no evidence in favor of the method of nerve-termination which Patten had described. As I have previously mentioned, it is highly improbable that the methods of innervation in the retinas of Penæus and Homarus are fundamentally different, and since I have found in the retina of Homarus no confirmation of Patten's views, I am of opinion that he must be mistaken as to the method of nerve-termination in Penæus. Many of Patten's figures of the individual nerve-fibres in Penæus ('86, Plate 31, Figs. 69, 70, 71) resemble so closely the fibres which I have seen in all of my isolation preparations, and which, for reasons already given, I am persuaded are not nervous, that I am forced to believe that Patten has mistaken for nerve-fibres shreds of non-nervous tissue.

My criticism of Patten's results refers only to those which he obtained from a study of the Crustacea. No one, so far as I am aware, has fully confirmed his views concerning the nerve-terminations in this group. Kingsley ('86, p. 864) claims to have seen the axial nerve-fibre in Crangon, but he was unable to trace its finer ramifications in the cone. He states ('87, p. 57), however, that the method of preparation which he employed was not intended especially for the fibres, and that therefore it is not surprising that they were not identifiable. Herrick

('89, p. 168) could not distinguish fibres in the cones of Alpheus, although, like Kingsley, he admits that his failure may be due to the method which he used. Relative to nerve-terminations, the evidence which the work of these two investigators presents can scarcely be called critical, and I therefore hold to my former conclusion, namely, that the fibres which according to Patten represent the nerve-terminations are in reality not nerve-fibres at all.

After having reached this conclusion, I was naturally led to look elsewhere for the true nerve-fibres. It occurred to me that, in order to be certain that I was dealing with nerve-fibres, it was safer to begin studying them in regions where their identity was beyond question. I therefore examined maceration-preparations of parts of the larger nerves from the lobster's body. These nerves were readily resolved into a number of fibres, which in transverse section were enormous when compared with such fibres as the axial nerve-fibre figured by Patten ('86, Plate 31, Figs. 72, 74, 108). I then studied in a similar way the optic nerve. The fibres in this nerve were smaller than those from the other nerves which I had macerated, but they were much larger than those figured by Patten. Each fibre (Fig. 36) possessed a distinct sheath, and its contents were marked by lines which extended parallel to its long axis. These lines I interpreted as indications of fibrillæ which composed the fibre. In addition to the large fibres, the optic nerve, as well as the other nerves, showed, when macerated, the fibrous shreds which I have previously mentioned. They were very insignificant in amount, forming, I should judge, not more than a fraction of one per cent of the whole optic nerve, and I was never able to trace them as continuous fibres for any considerable distance. I believe that here, as in the retina, they arose from an artificial tearing of the tissue.

At about this time I happened to find the modification of Weigert's method of straining nerve-fibres, which I have described in the Introduction. As soon as I was aware of the results which could be obtained by this method, I applied it to a series of transverse sections of the optic nerve. The series extended from the retina to the optic ganglion, and demonstrated conclusively, I think, that the proximal ends of the seven proximal retinulæ, after passing through the basement membrane, as I have described, continued inward as fibres, and finally passed into the substance of the optic ganglion. Figures 21, 22, 23, 24, and 25 illustrate steps in the passage from the retina to the ganglion. In Figure 21 the groups of three or four proximal retinulæ are seen as they pass through the basement membrane. Figure 22 is taken at a level immediately below the basement membrane. Here two groups of four fibres, and one of three, can be distinguished. Below this level the groups of three and four fibres are no longer to be recognized. The fibres diminish rapidly in diameter as they leave the retina. Figure 23 represents a transverse section of the fibres at one fourth the distance from the basement membrane to the distal face of the ganglion. Figure 24 represents a similar section midway between retina and ganglion. Figure 25 represents the fibres as they enter the ganglion. The same kind of fibres have been identified in the substance of the ganglion. These fibres, I believe, are the true fibres of the optic nerve, and, as I have shown, they connect the seven proximal retinulæ of each ommatidium with the optic ganglion.

The termination of the nerve-fibres in the proximal retinulæ is so directly opposed to the method of termination which Patten has described, that, before making a final statement based on a study of the lobster only, it seemed prudent to seek confirmation in other species. This I have done, and I can now state that the termination of the nerve-fibres in the retinulæ has been demonstrated in Eupagurus, Cambarus, and Gammarus. From this I conclude that it is highly probably that in the compound eyes of all Crustacea the nerve-fibres terminate in the retinulæ.

This method of nerve-termination, namely, the direct continuity of the nerve-fibre and the perceptive cell, has also been demonstrated by Watase ('89, pp. 34–36) in the eyes of Limulus.

There are some interesting phases in the transition from the nerve-fibre to the retinula. On examining the transverse sections of nerve-fibres at a level slightly below the basement membrane, one observes that it is in the form of a transparent cylinder the periphery of which has scattered over it a few pigment granules. As the fibre passes through the basement membrane the pigment increases in quantity, and when the retinula is reached the nerve-fibre is represented by a transparent axis in its centre (Fig. 19, $ax.\ n.$). The nervous axis is transparent, because it contains no pigment granules, while the peripheral portion of the retinula is densely pigmented. In this way the nerve-fibre proper is continuous as a transparent axial shaft from below the basement membrane to a level in the retina, which corresponds to the middle of the rhabdome. From this level to its distal end the retinula is completely filled with pigment, no trace of a transparent central axis being visible. (Compare Figs. 19 and 14 with Figs. 11, 10, and 9.) The transparent nervous axis of each retinula terminates, as I have said before, on a level with the middle of the rhabdome. When it is cut very close to its distal end, the nervous axis is seen to lie almost next the rhabdome. I have never seen a case, however, in which the axis and rhabdome were not separated by a line of pigment granules. In transverse sections which have been thoroughly depigmented, the distal half of each rhabdome is surrounded by a great number of bodies which resemble coarse granules (Fig. 32, *fbr'.*). These bodies are limited almost entirely to the distal half of the rhabdome. At first I supposed that they were the colorless skeletons of pigment granules, but they were easily distinguished from the latter by their larger size and sharper outline. It then occurred to me that, since these bodies were found only about that portion of the rhabdome which was distal to the nervous axis, they might therefore be the cut ends of the finer fibrillæ into which that axis had been resolved. If such was the case, these bodies were in reality fibres, not granules. In order to determine whether they were fibres or granules, I examined oblique sections of the rhabdome. Figure 33 is taken from one of these sections. The granular body in the figure is the rhabdome, and the transverse bands are the strata in its substance. The projection of the long axis of the rhabdome in this figure would be a vertical line. The lower end of the figure is proximal, the upper end distal. Owing to the fact that the rhabdome is cut obliquely, what is seen at its proximal end belongs on one side of it, and what is seen at its distal end belongs on the opposite side. The first proximal dark band in the substance of the rhabdome is very nearly midway between its ends. At the proximal end of the rhabdome three distinct fibre-like bodies are seen. These are in reality longitudinal sections of the greatly compressed retinulæ, which have been noticed in transverse sections. (Compare Figs. 15 and 18.) At the distal end of the rhabdome, in place of the three retinulæ, there are a great number of fine fibres. The fibres occur only around the distal half of the rhabdome, and I believe that in transverse sections these fibres are represented by the small round bodies previously described. From the distribution of these small fibres and their relation to the distal end of the nervous axis, I am of opinion that they represent the fibrillæ of the nerve-fibre.

The entrance of the fibres of the optic nerve into the proximal retinulæ is of itself strong evidence that the rhabdome, not the cone, is the perceptive body. This conclusion is further supported by the ultimate distribution of the optic fibrillæ. If it be admitted that the rhabdome is immediately concerned in the perception of light, it is only natural to expect that this structure would be accessible to the light. As I have already shown, the cone-cells form a transparent axis, which leads directly to the rhabdome, and through which light could readily reach that structure. Once having penetrated to the rhabdome, it is probable that, either in the substance of the rhabdome itself, or in the superficial layer of pigment which immediately surrounds the rhabdome and in which the fibrillæ are, the light is transformed into that kind of energy which is transmitted by nerve-fibres.

The Development.

In discussing the development of the eyes in Arthropods, the more recent investigators have given much attention to the general plan of these organs. One of the objects of their researches has been the reduction of the eyes of these animals to a single structural type. If such a type were found to exist, it would very probably reproduce the essential structural feature which the eye of the ancestral Arthropod possessed. The desirability of ascertaing if there is a common type for the eyes of all Arthropods is evident, for upon the nature of the answer to this question must depend to some extent the conclusions concerning the phylogenetic relationship of the group, and its different classes. Possibly in the phylogeny of two classes of Arthropods the eyes may have originated independently. Of course organs independently developed could not be homologous, and they might be so differently constructed that it would be impossible to reduce them to a common type. Notwithstanding the difficulty in homologizing the eyes of one class of Arthropods with those of another, the homologies among the members of a single class are much more readily determined, and many important comparisons can be safely made. It would, therefore, seem more prudent to limit investigations to the eyes of a single class until they were well understood, rather than institute comparisons between the eyes of different classes, where, from the limitations of our knowledge, such comparisons must be more or less hazardous.

The Plan of the Eye.

The work of different investigators has already suggested several structural types for the compound eyes of Arthropods, but the differences which some of these types present are of such a fundamental character, that to accept one is to reject another. It was with the hope of gaining confirmation for some one type that I was led to study the development of the eye in a Crustacean, and, as I have previously explained, the species most available for such a study was the lobster.

The results which have been presented in the more recent papers on the embryology of the compound eyes require a brief notice before the development of the eye in the lobster is described. What is said in this connection is purely introductory; no criticism of the views of different authors will be made until after the development of the lobster's eye has been described.

The writers who have thus far published accounts of the development of compound eyes can be grouped under four heads, depending upon the type of eye which their researches indicate. The first type is that represented by the eye of Peripatus. Patten ('86, p. 688, '87, p. 211) is of the opinion that the compound eyes of Hexapods, as well as Crustaceans, are constructed upon this plan. Each eye should then consist of a closed vesicle which was produced by an involution of the hypodermis. The eye would be composed of three layers, which in the order of their positions are as follows: first, the superficial hypodermis; second, the outer wall of the vesicle; and third, the inner wall of the vesicle. Patten is of opinion that in the eye of an adult individual these three layers are modified in the following way: the superficial hypodermis becomes the corneal hypodermis; the outer wall of the vesicle is so far reduced as to be inconspicuous, and the inner wall gives rise to the retina. The retina includes the crystalline cones and the pedicels (rhabdomes). Patten supports these conclusions mainly from theoretical grounds, but he believes that he has found evidence of the existence of this type in the development of the compound eyes of Vespa, Blatta, and the Phryganids.

The second structural type which I shall mention has been advocated by Kingsley ('87, p. 51) in his description of the development of the compound eye of Crangon.[1] In this type the eye results from a vesicular infolding, as in that which was proposed by Patten, but it differs from the latter in the fate which is ascribed to the two walls of the vesicle. According to Kingsley, the outer wall of the vesicle is not reduced, but gives rise to the retina. In it are developed the crystalline cones and the pedicels (rhabdomes). The inner wall of the vesicle, instead of forming the retina, as Patten believed, is converted into a part of the optic ganglion.

The third structural type is that which is presented by Reichenbach ('86, pp. 85–96) in his account of the development of the crayfish.

---

[1] See the note on page 41.

Here a hypodermal involution takes place, and a vesicle is produced, but the cavity of the vesicle is soon obliterated. The mass of cells which results from the fusion of the walls of the vesicle now divides into an outer and an inner layer. These two layers do not necessarily correspond to the outer and inner walls of the original vesicle. The superficial hypodermis and outer layer of the infolded mass fuse, and give rise to the retina. The crystalline cones are developed in that part of the retina which is derived from the superficial hypodermis; the rhabdomes probably originate in that part which is derived from the outer layer of infolded cells; the inner layer of cells becomes ganglionic.

Bobretsky's ('73) account of the development of the compound eyes in Astacus and Palæmon agrees in its essential features with the description which Reichenbach has given for the eyes in the crayfish. Bobretsky, however, does not describe an involution in the optic disks. As Reichenbach suggests, the fact that Bobretsky did not have the opportunity of studying very early stages may explain his failure to observe the involution. In other respects, the accounts are essentially alike, and there is little doubt that the plan of eye which Bobretsky's researches indicate is the same as that suggested by Reichenbach's studies.

Each of the three structural types which have thus far been described are dependent upon the formation of a hypodermal vesicle. The fourth type is simpler than any of the three preceding ones, in that a vesicle is not necessarily produced, the retina being supposed to originate as a simple thickening in the superficial hypodermis. This type has been advocated by Herrick ('86, p. 43) in his account of the development of Alpheus. The researches of Nusbaum ('87, pp. 171–186) on the development of Mysis indicate the same type. Neither in Grobben's ('79) account of the development of the eyes in Moina, nor in Claus's ('86, pp. 307–324) description of those in Branchipus and Artemia is any mention made of an involution. These authors might, therefore, be cited as favoring the fourth type of eye, although it is to be observed that the special question of the vesicular origin of the eye is not discussed by them.

The advocates of the fourth type find support, not only in the embryology of the Crustacea, but also in that of the Hexapods. According to Carriere ('85, pp. 181–186), the compound eyes of some Hymenoptera and Lepidoptera develop as simple thickenings of the hypodermis.

Of the four types which have been mentioned, the one with which the development of the lobster's eye accords will be seen from the following description.

The first traces of the eyes in the development of the lobster are the optic disks. These disks lie one on either side of the median plane, and are for a considerable time the most conspicuous structures in the anterior region of the embryo. In the early stages of development the disks face ventrally, but as the head of the animal becomes differentiated they come to face almost in the opposite direction, i. e. dorsally. At stage A (Fig. 37) the optic disk, so called, is oval in outline rather than disk-shaped; its longer axis is transverse to the principal axis of the embryo. From that portion of each disk which is near the median plane a band of tissue extends posteriorly, and connects the disk with the ventral plate of the embryo. The disk and the band with which it is connected to the ventral plate are distinguished from the surrounding tissue by their greater number of nuclei. The disks comprise the tissue from which both retina and optic ganglia develop.

A section passing through an optic disk in a plane perpendicular to the longitudinal axis of the embryo is shown in Figure 38. In this case the section is from the right optic disk. The left side of the section is farthest from the median plane; the right side is near that plane. Since *at this stage* the disk faces ventrally, and since the ventral edge of the section is uppermost in the figure, it is the posterior face of the section which is presented. To the right and left of the disk one can see the undifferentiated ectoderm with its occasional nuclei. This ectoderm is directly continuous with the tissue of the disk; in fact, the disk is only a local thickening in the ventral ectodermic layer of the embryo. It is due to its greater thickness that the disk contains more nuclei than the surrounding ectoderm. The deep face of the ectoderm, both of that which is undifferentiated and that which forms the disk, is limited by a delicate but distinct basement membrane (Fig. 38, *mb.*).

Some idea of the method of growth in the optic disk can be gained from a study of its nuclei. It will be noticed that the nuclei in the deeper part of the disk are rather small and irregularly grouped when compared with those which are found next its outer surface. These superficial nuclei form an almost regular series, which extends from one margin of the disk to the other. They are usually also characterized by having their long axes perpendicular to the surface of the disk. When they increase by division, their planes of separation are in most cases either parallel or perpendicular to the outer surface of the disk. When the plane of separation is perpendicular to the surface of the disk, the tendency for such divisions is to increase the diameter of the disk. When, on the other hand, the plane of separation is parallel to the surface of the disk, the tendency is to thicken it. In order to determine the distribution of these two methods of division, the planes of separation in the superficial nuclei of four disks were carefully observed. Each disk can be divided into halves by a plane parallel to the sagittal plane of the embryo. That half of the disk which lies near the median plane can be called the proximal half; that which is farther from the median plane, the distal half. Among the superficial nuclei of the distal halves of the four disks which were examined there were seen thirteen cases of division. In all of these cases the tendency was to broaden the disk. There was no case where the division of the nucleus would have thickened it. In the proximal halves of the disk there were six cases of division observed; five of these tended to thicken the disk, and one would have broadened it. It is therefore apparent that in the superficial layer of each disk the proximal half is becoming thicker, while the distal half is becoming broader.

The deep nuclei of each disk, those lying below the row of superficial nuclei, divide in different planes. Among these nuclei in the four disks which were examined, seven instances of division were noticed. Five of these were in planes which would have thickened the disk; the remaining two would have broadened it. This part of the disk consequently shows a tendency both to become broader and thicker. Of these two tendencies, that which would thicken the disk is the stronger.

The method of growth which has been described for different parts of the optic disk can be easily distinguished only in its earlier stages. The subsequent changes which affect the structure of the disk render it rather difficult to follow in detail the growth of the disk as a whole; but in general the broadening of the distal superficial region, the thickening of the proximal superficial part, and the broadening and thickening of the deeper parts are continued.

The most important changes in the differentiation of the optic disks are the following: first, the separation of the retina and optic ganglion by the formation of the basement membrane; second, the production of the optic nerve; and third, the differentiation of the ommatidia. These three changes will be considered in the order named.

The first step in the differentiation of the retina and optic ganglion occurs at stage B. Figure 39 represents a section from the optic disk of an embryo of this stage. The plane of section in this case corresponds to that of Figure 38. The chief difference observable between the disk at stages A and B is its greater thickness in the older embryo. Not only has the disk thickened, but it has spread laterally, so that the small angle which is seen on the outer margin of the disk in stage A (Fig. 38, x), and which indicates a tendency on the part of the disk to grow over the adjacent ectoderm, is represented in stage B by a very acute angle (Fig. 39, x), while the disk itself forms an actual fold, which covers a portion of the undifferentiated ectoderm.

The separation of the retina and its ganglion is accomplished by the production of a basement membrane. This is gradually developed in the substance of the disk between the regions of the retina and the optic ganglion. In order to distinguish the newly formed membrane from the original basement membrane which bounds the under surface of the disk, I shall speak of the former as the *intercepting membrane*. The intercepting membrane (Fig. 39, *mb. i cpt.*) takes its origin from the basement membrane on a line a little within the lateral edge of the disk. From this position it extends at this stage as a delicate lamella for a short distance through the tissue of the disk. The direction of its course is approximately parallel to the outer surface of the disk, and it divides the distal portion of the disk into two masses, one of which is superficial, the other deep. The superficial mass is the first portion of the retina to be differentiated, and, as I have previously stated, the growth in this region is chiefly lateral. The deep mass is the beginning of the optic ganglion. The intercepting membrane does not extend so far as to separate the superficial portion of the disk from the deeper part in the proximal half. This condition is what one might expect, since in the proximal region of the disk the superficial nuclei divide in such planes as to thicken the disk, and a membrane which would separate the deep and superficial parts would be a source of interference in the process of thickening.

The intercepting membrane is not an involution of the original basement membrane, but is produced by the activity of the ectodermic cells between which it lies. There is no reason for doubting, I believe, that at this early stage (B) it is strictly an ectodermic secretion. At this stage it is fan-shaped, the handle of the fan being formed by that part of the membrane which is in contact with the original basement membrane. The plane of the fan is parallel with the outer surface of the disk. In sections which are either anterior or posterior, but parallel to that shown in Figure 39, portions of the intercepting membrane are often visible, and may be apparently unconnected with the basement membrane. This is due to the fact that the plane of section cuts the anterior or posterior edge of the fan without including the handle.

The chief difference between the intercepting membranes at stages

B and C is the greater extension of the membrane at stage C. (Compare Figs. 39 and 41.) The retina is now much more completely separated from the ganglion than formerly. The superficial and deep portions of the proximal part of the disk are, however, still continuous.

In stage D (Fig. 45) an important step is taken in the development of the membrane. It splits at this stage into two layers, one of which adheres to the retina, and the other to the optic ganglion. The retinal and ganglionic layers of the membrane have probably been distinct from the time of their formation; but until stage D is reached, the double nature of the membrane is not apparent. Occasionally in stage C one notices at the point where the intercepting and basement membranes unite a re-entrant angle (Fig. 41, x). This in some cases extends a short distance between retina and ganglion, and doubtless represents the first step in the separation of the components which form the intercepting membrane.

In stage E (Fig. 46) the membrane has completely severed the superficial from the deep ectoderm, and the separation of its retinal and ganglionic constituents extends over a broader area than in the previous stage.

The subsequent changes in the intercepting membrane consist in a complete separation of its retinal and ganglionic portions. This separation is effected by the withdrawal of the optic ganglion from the superficial ectoderm. In the adult, the ganglionic membrane remains relatively thin, but the retinal membrane becomes much thickened. This membrane, which has already been described as the basement membrane of the adult retina, is not uniformly thickened, but presents local elevations, each of which is in the form of a cross. The four apertures which pierce the membrane in the angles of the cross-shaped elevations, and the relation which adjoining crosses and apertures bear to one another, have already been described. The proximal face of the basement membrane is nearly flat (Fig. 29). The cross-shaped elevations occur on its distal face. The substance of the membrane is apparently homogeneous, and contains no traces of cells or nuclei. The fact that its substance is alike throughout favors the idea that it has been derived from a single source. From stage E to the adult condition a few mesodermic cells have been noticed next its proximal face (Fig. 1, cl. ms d.). These cells are not intimately attached to it, and I am of opinion that they contribute little or nothing to its composition.

From the foregoing account I draw the following conclusions concerning the growth of the basement membrane of the eye. In its earliest stages the basement membrane is strictly ectodermic in origin. In the adult condition it is much thicker than in its early stages, and the greater part of its substance has probably come from the ectoderm, although mesodermic cells rest against its proximal face, and possibly may contribute to its formation.

From the description which I have given, it is evident that the optic disks are thickened regions in the superficial ectoderm, and that these disks are cut by an intercepting membrane into two parts, one deep, the other superficial. The deep part is converted into the optic ganglion; the superficial part becomes the retina. So far, then, as the development of the retina in the lobster is concerned, it supports the view that the compound eye in Crustaceans is developed from a simple thickening of the ectoderm.

In describing the formation of the retina and optic ganglion in the lobster, I have made no mention of an involution. Both Reichenbach and Kingsley have described an infolding in the formation of the eyes, — the former in Astacus, the latter in Crangon, — and it is therefore only natural to look for a similar condition in the eyes of Homarus. Any evidence of an involution in the production of the lobster's eyes is to be sought in the early stages of development. I regret that in the very early stages my material is deficient, and I have not grounds enough to warrant the statement that no involution occurs. All that I can state is, that in all stages which I have examined I have not been able to find any evidence of an involution. The youngest individual which I have studied was one in which the optic disk was about two thirds as thick as that represented in Figure 38. Excepting its thinness and the smaller number of its nuclei, it presented essentially the same appearance as the one which is figured. It will be noticed that near the centre of the disk in Figure 38 there is a space devoid of nuclei. It occurred to me that such a space might represent the last traces of an involution, and I therefore plotted carefully the nuclei in five pairs of disks, some of which were less mature than the disk shown in Figure 38. The result of the plotting was that the light space which is seen in Figure 38 proved to be an individual peculiarity, and I did not find in the arrangement of the nuclei in the other disks any evidence of an involution.

The plotting of the nuclei, however, brought to light the method by which the disks increased in size. This has already been described, and offers, I believe, an explanation of the fact that in some cases, as for instance in the crayfish, the formation of the eye is attended with an involution, while in other instances, as in Alpheus, no involution is present.

It will be remembered that the superficial layer of nuclei in the optic disk of the lobster was divided into two regions. The one farther from the median plane has been called the distal region; the one nearer, the proximal region. The broadening of the distal region produced the retina. By a proliferation of its cells the proximal region resulted in the formation of the optic ganglion. It is my opinion that this proliferation of cells represents what is produced in the case of some Crustaceans by an involution, and that either an involution or proliferation, or possibly a combination of both processes, occurs in the eyes of all Crustaceans. Whichever process characterizes the development of a given eye, it must be borne in mind that the involution or proliferation is connected with the formation of the ganglion only, and takes no part in the production of the retina. The latter is a simple thickening in the ectoderm; the optic ganglion is developed either as a proliferation or involution of the ectoderm which lies close beside the retina. The results at which various investigators have arrived, different as they may at first appear, can be harmonized, I believe, by this interpretation of the origin of the retina and optic ganglion.

In his account of the development of the eye in the crayfish, Reichenbach ('86, p. 85) has described an ectodermic involution, which occurs nearly in the centre of the optic disk. That portion of the disk which is farther from the median plane than the region of involution gives rise to the retina, so that the region of involution in the crayfish occupies a position which corresponds to the region of proliferation in the lobster. Not only do the two regions correspond, but the masses of tissue which are developed from each have certain peculiarities common to both animals. In the case of the crayfish the mass of tissue which results from the involution becomes divided into an outer and an inner wall. These two walls are separated from each other by a band of nuclei, which are larger and lighter in color than the surrounding nuclei. In the lobster the ganglionic tissue which arises by proliferation is divided into an outer and an inner part. The separation is effected by a band of nuclei, which in position and structure resemble the band figured by Reichenbach. (Compare Reichenbach, '86, Plate XII. Fig. 174, and Plate III. Fig. 41 of this paper.) The similarity presented by the bands of nuclei in the lobster and crayfish supports the conclusion that the involution in the crayfish and the proliferation in the lobster are homologous structures.

An objection to this comparison might be raised on the ground that, according to Reichenbach's statement ('86, p. 93), the involution in the case of the crayfish gave rise to a mass of tissue which formed on the one hand the deeper part of the retina, and on the other a portion of the optic ganglion, while the cells which arise by proliferation from this region in the lobster produce only a part of the optic ganglion. But, as Patten ('87, pp. 208, 209) has shown, Reichenbach himself was not certain that any part of the infolded ectoderm contributed to the formation of the retina. Reichenbach found it difficult to locate exactly the position which the developing rhabdomes occupied. On page 92 in his account he describes a part of the outer wall (Ausserwand) of the infolded ectoderm, and states his belief that in it the rhabdomes develop; in fact, he describes certain red bodies which he says are without doubt the rhabdomes themselves. On page 96 he admits that the layer in which he supposed the rhabdomes originated may be a layer of nerve-fibres. Granting this interpretation, it is no longer possible to consider the previously described red bodies as rhabdomes. Reichenbach does not make this last statement, but his description implies it when he states that, although the region of the red bodies may not be the region of the rhabdomes, yet the rhabdomes doubtless originate in a somewhat more superficial part of the outer wall. Apparently he has not identified the rhabdomes in their new position : at least, he makes no such statement in his text or description of plates. Since he has also admitted that the red bodies may not be rhabdomes, I cannot see that he has positively identified any structure as a rhabdome. Such being the case, it is difficult to understand on what grounds he can maintain the assertion that the rhabdomes develop in the outer wall. If this assertion cannot be defended, then it is possible that they may develop in the superficial hypodermis. This would be analogous to the condition presented in the lobster.

If the rhabdome in the eye of the crayfish is developed, as I believe it is, in the superficial hypodermis, and not in the outer wall of the infolded hypodermis, the objection which was suggested in homologizing the involution in the eye of the crayfish with the proliferation of cells in the eye of the lobster has no weight.

In the development of the eye of Crangon, according to Kingsley's ('86ª) account, there is also an optic invagination. If what I have attempted to show in regard to the eyes of the crayfish be true, then this invagination in Crangon should be connected with the formation of the ganglion only. This, as I have already stated, is not the view held by Kingsley, for he maintains that the outer wall of the invaginated pocket gives rise to the retina, and only the inner wall is concerned in the production of the ganglion. This is fundamentally different from the condition found in the lobster. The difference, however, is due, I believe, to the fact, that in describing the later stages of development Kingsley has pointed out a cavity which he believed to be the cavity of the invagination, but which in reality is not. The cavity which he has marked *oc* in his Figures 3, 4, and 5, is unquestionably the cavity of involution, but the space marked *oc* in his other figures is, in my opinion, a part of the body cavity.

My reason for this belief is as follows. The cavity of an involution such as is found in the anterior median eyes of spiders or the median eyes of scorpions is, when it has lost its connection with the exterior, a closed ectodermic sac. That such a cavity should be occupied by migrating mesodermic cells seems to me extremely questionable, for in order to enter the cavity it would be necessary for the cells to penetrate one wall of the vesicle. This of course is not impossible, but it is not borne out by analogy with the Arachnoids. The cavity which Kingsley has marked *oc* in Figure 7 is occupied by several mesodermic nuclei, and what is more important, perhaps, is that it is apparently connected with other cavities in the embryo. These cavities also contain mesodermic tissue. Excepting Figures 3, 4, and 5, the cavities marked *oc* I believe to be homologous, and I further believe that they represent, not the cavity of involution, but the space which intervenes between the infolded pocket and the superficial ectoderm. This is a part of the embryonic body cavity, and is of course readily accessible to mesodermic cells. The fate of the real cavity of involution is not so easily discovered. Probably, as in the case of the crayfish, it is obliterated in the mass of tissue from which the retinal ganglia arise.[1]

If the interpretation which I have suggested in the preceding paragraph be admitted, the development of the eye in Crangon is essentially the same as in the lobster and crayfish. The proliferation in the optic disks of the lobster is represented by an involution in the disks of Crangon. The cavity of the involution disappears in Crangon, and its walls give rise to ganglionic tissue. The part of the superficial ecto-

[1] Since this paper was written, I have received a copy of the third part of Kingsley's studies on the development of Crangon. As the following quotation will show, Kingsley ('89, p. 20) has materially changed his views as to the formation of the retina : "I may say here that I am inclined to believe that I fell into error in my account of the development of the Compound Eye of Crangon, and that the invagination or inpushing which I there described as giving rise to the ommatidial layer of the eye, in reality gives rise to the ganglion of the eye which in the adult is contained within the ophthalmic stalk."

derm which is immediately external to the ganglionic involution thickens
and produces the retina.

The mode of development which Patten has suggested for the compound eye of Arthropods has not received any support, so far as I am
aware, from the embryology of the Crustacean eye.   The only observations which go to confirm Patten's opinion are those of his own on
Vespa, Blatta, and the Phryganids.   Possibly the compound eyes of
Crustaceans may be developed upon a different plan from those of Hexapods.   Certainly the evidence which Patten has given for the Peripatus-
like type of the compound eye of Hexapods has not been found in any
of the Crustacea.   According to Patten's view of the origin of the compound eyes, the corneal hypodermis should arise on the sides of the
optic area, and spread over the retina until the latter is entirely covered.   This constitutes the closing of the shallow vesicle.   When I
come to describe the differentiation of the ommatidia, I believe it can be
shown beyond a doubt that in the lobster the corneal hypodermis arises,
not by any lateral growth from the edge of the optic area, but by a
simple process of delamination.   The cells of the corneal hypodermis are
the differentiated superficial cells of that thickening in the hypodermis
which produces the retina.   Thus the plan which Patten has suggested
for the compound eyes in Arthropods is not supported by the evidence
derived from the development of the lobster.   Patten himself ('87,
p. 202) admits that in Vespa he did not see the closure of the hypodermis over the retina, and that in Blatta and the Phryganids ('87,
pp. 208 and 211) the process is very obscure.

The researches of Carrière point to a type of compound eyes for the
Hexapods which is similar to that exhibited by the lobster.   It is possible that the vesicular origin of the compound eyes of Hexapods, owing
to their obscure method of formation, may have been overlooked by
Carrière, but I am inclined to believe, after a consideration of both sides
of the question, that the evidence favors the simpler method of origin,
and therefore that the compound eyes of Hexapods as well as Crustaceans arise as simple hypodermal thickenings.

In Mysis, according to Nusbaum ('87, pp. 171–185), the development of the eye follows essentially the same course as that which I have
described in the lobster.   The optic disks are formed and the ganglionic
and retinal portions are differentiated in the same manner in both cases.
The development of the retina is slightly complicated in Mysis by the
fact that, instead of lying in one plane, or very nearly so, the retina is
strongly folded on itself, so as to give the appearance of two lamellæ, an

internal and an external one. The retinal elements are differentiated earliest in the internal lamella. As the eye becomes more distinctly separated from the body, the internal and external lamellæ are unfolded so that they are no longer distinguishable as separate parts. The retina is developed from the thickened layer of hypodermis. So far as Nusbaum's observations extended, the ganglion is produced without an involution.

The development of the compound eye in Alpheus has been studied by Herrick ('86, '88, and '89). The course of development is almost identical with that of the lobster.[1] The optic disks after they thicken are cut by a basement membrane into a ganglionic and a retinal portion. There is no involution connected with the formation of the eye.

In the introduction to the development of the lobster's eye, mention was made of four structural types which the work of different investigators indicated as possible plans for the compound eyes of Crustacea. I have given reasons for excluding three of these. The type of eye which Patten has advocated is unsupported by the embryology of the Crustacea. Reichenbach and Kingsley misinterpreted structures in the eyes which they studied, and were consequently led to erroneous conclusions. If the interpretations which I put on the work of these two investigators be admitted, all studies on the development of the compound eyes of Crustacea point to one conclusion, namely, that in these eyes the retina originates as a thickened layer of hypodermis, and is not modified by any form of involution. The involution when present is connected with the formation of the optic ganglion only. In the production of the ganglion, the involution can be replaced by a proliferation of the cells.

The Optic Nerve.

The development of the optic nerve[2] is intimately connected with the formation of the intercepting membrane. Before the formation of this

[1] I have had an opportunity of examining Dr. Herrick's unpublished plates on the development of Alpheus, and Dr. Herrick has kindly looked over my figures of the lobster. The correspondence between the method of growth in the eye of Alpheus and Homarus is certainly very close. The few differences that were noticed were such as might be expected between different species. I take this opportunity of thanking Dr. Herrick for his kindness in extending to me the use of his plates.

[2] The optic tracts of a lobster consist of four principal parts. The first of these is the retina, from which nerve-fibres lead to the optic ganglion. These three

membrane the retina and ganglion is one continuous mass of cells (Fig. 38). When the intercepting membrane is formed, the retina and ganglion are apparently separated (Fig. 39). I am of opinion, however, that this separation is only apparent, and that in reality the two structures are still in connection. At least in this stage and in stage C (Fig. 41) nuclei are frequently found lying directly across the membrane. These nuclei present so normal an appearance, and their occurrence is so frequent, that I cannot believe that their position is due to accidental displacement. My only way of accounting for the place which they occupy is by supposing that the basement membrane is perforated where they are found. The membrane was probably produced in the form of a net, through the meshes of which the retina and ganglion retained their original connection. Either this is true, or it must be admitted that the retina and ganglion were first connected, then separated, and finally reconnected ; a supposition which seems to me unnecessary as well as improbable. From stage C to the adult condition the retina and ganglion are unquestionably connected by nerve-fibres. If the conclusion arrived at concerning the origin of the optic nerve is correct, it follows that the optic nerve cannot be properly described as an outgrowth either from the retina or from the ganglion, but it must be considered as a remnant of the original connection which existed between retina and ganglion. Patten ('87, p. 196) has described essentially the same method of origin for the optic nerve of Vespa. The formation of the optic nerve from tissue which represents the original connection of a portion of the optic ganglion with the superficial ectoderm is doubtless a reproduction of the method by which that nerve arose phylogenetically.

The fact that the fibres of the optic nerve are from the outset attached to the proximal ends of the retinulæ is of significance in determining the plan of the eye. Of the four structural types of compound eyes which have been suggested, that which Kingsley has presented involves the inversion of the middle or retinal layer. Mark ('87, pp. 91, 92) has shown that when the retina is inverted, as in the anterior median eyes of spiders, the nerve-fibres are at first attached to the morphologically deep ends of the retinal cells, and that the attachment afterwards migrates toward the other ends of these cells. From the fact that in

parts, retina, nerve, and ganglion, are contained in the eye-stalk. The fourth part is a second bundle of nerve-fibres which connect the optic and cephalic ganglia. In using the term optic nerve I refer to that collection of fibres which unites the retina and optic ganglion.

Crustaceans the nerve-fibres are always attached to the proximal ends of the retinulæ, it can be argued that the retina in this group has never been inverted, but retains its original position, and that any structural plan which involves the inversion of the retina is therefore probably wrong.

The Differentiation of the Ommatidia.

In the development of the compound eye in the lobster, the deposition of pigment and the differentiation of the ommatidia take place at about the same time. These changes occur at stage C. (Compare Figs. 39, 41, and 42.) At this stage (Fig. 41) it will be observed that the retinal layer is thickest at the lateral margin of the disk (the extreme left in Fig. 41). The retina becomes thinner as one proceeds from the margin toward the median plane (from left to right in the figure). The thickest part of the retina, it will be recalled, was the part first to be separated from the ganglion by the intercepting membrane. As it was the first part of the retina tó be separated, so it is the first part in which the ommatidia are differentiated and pigment is deposited.

The first steps in the differentiation of the ommatidia are seen in Figure 42. Here the nuclei in the thicker part of the retina have separated into two bands, one distal (y), and one proximal (x). The distal band, as I shall presently show, can be further separated into a superficial and deep layer. These two layers are close to the outer surface of the retina, and approximately parallel with it.

The arrangement of the nuclei which make up the distal band is most easily observed when the retina is viewed from the surface. The grouping of these nuclei, from the time when the ommatidia are differentiated to the adult condition, is so characteristic that the fate of the individual nuclei can be easily traced. For the sake of simplicity I shall therefore designate the different nuclei, from their first appearance in groups, by the names of the cells in which they are ultimately found, although it is to be borne in mind that in the early stages the cell walls are not as yet differentiated.

Viewed from the surface, the superficial layer of the distal band of nuclei presents the appearance which is shown in Figure 43. In this layer there are two kinds of nuclei, one elongated, the other roundish. The elongated ones are always in pairs. They ultimately become the nuclei of the corneal hypodermis. These hypodermal nuclei arise from among the superficial nuclei of the distal band, and do not originate, as Patten believed ('86, p. 645), from a fold which grows over the retinal area. Of course the number of pairs of hypodermal nuclei equals the number of ommatidia. In the earliest stages in which the ommatidia have been seen there were always three or four pairs of hypodermal nuclei, so that it is probable that the first step in differentiation is the simultaneous production of three or four ommatidia.

The second kind of nuclei in the superficial layer are roundish and arranged in circles of six around each pair of hypodermal nuclei. The circles are so combined that each nucleus plays a part in three circles; and as there is one circle for each ommatidium, it follows that only one third of each nucleus belongs to a given ommatidium, or, if one estimates the nuclei as units, only one third of each circle of six nuclei, or two nuclei, belong to an ommatidium. These nuclei represent the cells which in the adult have been called the distal retinulæ, and of which there was a single pair to each ommatidium.

The deep layer in the distal band of nuclei lies directly below the superficial layer (Fig. 42). The nuclei which constitute this layer are all of one kind, and are arranged in groups of four (Fig. 44). They represent the cells of the crystalline cones. The centre of a given group of cone-nuclei is directly below the centre of a pair of hypodermal nuclei. At this stage the cone-cells can be observed as elongated pyramids, which lie with their bases in the region of their four nuclei, and their apices extending into the deeper part of the retina (Fig. 42).

The nuclei of the proximal band show at this stage no special arrangement. They migrate to the deeper part of the retina, and there undergo further change. Between them and the basement membrane the pigment is deposited.

The most noticeable changes which the retina as a whole now undergoes are two. First, it thickens until it is throughout nearly as thick as in the region where the ommatidia were first differentiated. (Compare Figs. 45 and 46.) Second, the number of ommatidia greatly increases. These two changes, the thickening of the retina and the production of new ommatidia, go hand in hand and spread over the general surface of the eye, from the region in which the first ommatidia appear. It is worthy of notice that the new ommatidia are constructed from the undifferentiated cells which immediately surround the area of ommatidia already formed. Cells once incorporated in a given ommatidium never in any way contribute to the formation of other ommatidia. Moreover, in the differentiation of an ommatidium no cells are left between it and the neighboring ommatidia, so that ommatidia once adjoining each other remain so. In other words, new ommatidia are not produced between old ones, but only on the edge of the ommatidial area.

The changes in the ommatidia themselves can be studied most readily in longitudinal and transverse sections of the retina. Figure 47 is an enlarged drawing of that portion of the retina which is marked by a bracket in Figure 46. It will be noticed that in this stage, E, the nuclei are limited for the most part to the distal half of the retina. The middle of the retina is occupied by a band of pigment, which gradually fades as it approaches the basement membrane. The space between the basement membrane and the ganglion is relatively narrow, and is occupied chiefly by nerve-fibres. Returning now to the nuclei in the distal half of the retina, it is to be observed that the general arrangement which was pointed out in the last stage also persists here. The nuclei of the distal retinulæ are still in circles of six (compare Figs. 47 and 48, *nl. dst.*), and are very close to the external surface of the eye. In the centre of each circle is seen a round pink body, the tip of the cone-cells (Fig. 48 *con.*). Directly below the level of the nuclei in the distal retinulæ are the pairs of nuclei belonging to the corneal hypodermis (Figs. 47 and 49, *nl. crn.*). Each pair of corneal nuclei surrounds the distal end of the cone.

Below this level one meets the four nuclei of the cone-cells (Figs. 47 and 50, *nl. con.*). From the side view (Fig. 47) it will be seen that the groups of cone-cells are spindle-shaped in outline, and have their nuclei arranged in a transverse plane at the thickest part of the spindle. The nuclei which in stage C formed the proximal band are scattered in this stage between the deep ends of the cones (Fig. 47, *nl. px.*). They are not definitely arranged. In order to estimate the number of nuclei in the proximal band for each ommatidium, I counted these nuclei as seen in a series of tangential sections. In the outermost section in which the proximal nuclei occur there were six nuclei around each cone. These nuclei, however, were arranged in circles similar to the circles of the corneal nuclei; consequently for each ommatidium there were only two nuclei in each circle of six. The remaining nuclei were all embraced in the two succeeding deeper sections. In each of these two sections there were about nine nuclei around each cone. The arrangement of these nuclei was extremely irregular, and it was consequently difficult to estimate the number of nuclei which belonged to one ommatidium. Most of the nuclei were situated between three cones, therefore, about one third of the nuclei around a given cone can be considered as belonging to the ommatidium represented by that cone. As there were in

these two sections about eighteen nuclei around each cone, it follows
that one third of this number, or six, represents approximately the number of nuclei for each ommatidium.   If then the deeper sections contain six nuclei for each ommatidium and the outermost section two, the
total number of proximal nuclei for each ommatidium must be eight.
I do not mean to imply that this estimate can be insisted upon as absolutely invariable; but I wish to show that, as these nuclei represent the
proximal retinulæ in an adult lobster's eye, and as there are eight
such retinulæ and about eight of these nuclei to an ommatidium, the
change which the eight embryonic cells undergo in becoming adult
retinulæ is chiefly that of arrangement, and certainly does not involve
any considerable increase in numbers.

At stage E, which was the one last described, the young lobster was
about to escape from the egg-shell.   The next stage, F, is that of a
lobster about one inch in length.   At this stage the optic lobes are
represented by optic stalks, and the distal rounded end of each stalk is
occupied by the retina.   Figure 51 represents a longitudinal section of
a single ommatidium from this stage.   The distal end of the ommatidium is covered with a well developed corneal cuticula.   The cuticula
is marked out into hexagonal corneal facets (Fig. 52).   The facets of
course indicate the arrangement of the ommatidia.   This arrangement
at the first differentiation of ommatidia was such that hexagonal and
not square facets would have resulted if a cuticula had been then
produced.

Directly below each corneal facet is a pair of crescentic nuclei, those
of the corneal hypodermis (Fig. 53, *nl. crn.*).   These nuclei have in all
preceding stages shown a tendency to become elongated and crescentic
in outline, but it is in this stage that this peculiarity reaches its highest
development.   Between each pair of hypodermal nuclei the distal end
of the cone-cells is usually seen (Fig. 53, *con.*).   The four cone-cells
with their nuclei occur immediately below the corneal hypodermis
(Figs. 51 and 54, *nl. con.*).   The cone itself is already formed in part,
and lies below the nuclei of the cone-cells.   Proximally the four cone-
cells can be traced to near the middle of the retina; here they are no
longer distinguishable.   The proximal end of the cone itself terminates
as four processes, one on the outer lateral wall of each cone-cell (Fig. 56).
Surrounding the cone about midway on its length are the nuclei of the
distal retinulæ (Figs. 51 and 55, *nl. dst.*).   In transverse sections of
stage F, these nuclei show the same grouping in circles of six as they
showed in previous stages.   The outlines of the retinulæ are not visible.

The substance in which the nuclei lie contains a few granules of pigment. Below the proximal end of the cone the nuclei of the proximal part of the retina can be seen. These are not so definitely arranged that they can be counted. They occur on several planes. Fortunately, their cells in this stage have very distinct outlines, and a short distance below the nuclei one can see with perfect clearness the seven retinulæ which surround each rhabdome (Fig. 57, *rtn'. px.*). Occasionally, as is seen in Figure 57, the nucleus of the retinula occupies a position in its cell as deep as the plane of this section. The proximal retinulæ contain a few pigment granules. On account of the great similarity of the groups of proximal retinulæ, I have not been able to plot and superimpose sections with certainty; and as the nuclei of the proximal retinulæ are placed at different levels I have not succeeded in identifying an eighth nucleus, which, it will be remembered, was pointed out in the histology of the adult eye as probably representing a degenerate retinula.

At this stage the first trace of the rhabdome appears (Fig. 57, *rhb.*). It is a cylindrical thickening in the centre of each group of proximal retinulæ, and extends from a short distance below the crystalline cone very nearly to the basement membrane. From its earliest appearance it is divided into four segments, which bear the same relation to the surrounding retinulæ as they do in the adult. (Compare Figs. 58 and 34.) Nothing has been observed in the development of the rhabdome which indicates the significance of the four lines by which the segments of the rhabdome are separated.

Owing to a lack of distinctness in the tissue near the basement membrane, I have been unable to identify the individual retinulæ in that region. What I have observed is, that fibres pass from the retina through the basement membrane and into the optic ganglion. Presumably these fibres are from the proximal ends of the retinulæ, and are grouped as I have described in the retina of mature lobsters.

At this stage the only observation which I have made bearing on the question of nerve termination is as follows. In each proximal retinula near the rhabdome there are one or two fibres which extend nearly the whole length of the retinula. In transverse sections they of course appear as dots (Fig. 58, *fbr'.*), and might be mistaken for the remains of pigment granules were it not for their sharper outlines and the regularity of their arrangement. I am of opinion that they are the first indications of nerve-fibrillæ, which, as I have pointed out in the section on Histology, lie in the adult eye next the rhabdome.

The cells which I have thus far described in the differentiation of the ommatidia are unquestionably ectodermal in origin. In stage F certain nuclei appear which may have another origin. It will be recalled that in stage C (Fig. 41), although the intercepting membrane is well developed, the retina and optic ganglion are still closely applied to each other. In stage D (Fig. 45) the retina and ganglion have separated enough to form an intervening space of considerable extent. In stage E (Fig. 46) this space not unfrequently contains several nuclei. These are smaller than the ectodermic nuclei in the retina, and of about the size of those in the ganglion. Their chromatine differs from that of both the retinal and ganglionic nuclei, in that it has the form of very distinct particles which give the nucleus a decidedly granular appearance. Moreover, these nuclei, which I believe to be mesodermic in origin, are variable in shape, whereas the different kinds of ectodermic nuclei possess characteristic forms (Fig. 47). In stage F the space between the retina and ganglion also contains a few mesodermic nuclei, and similar ones are noticeable in the base of the retina (Fig. 51, *nl. pig.*). The latter are different from the nuclei of the proximal retinulæ (Fig. 51, *nl. px.*), and resemble so closely the nuclei which in the adult have been described as belonging to the accessory pigment cells, that I believe them to be the nuclei of those cells.[1]

From stage F to that of the fully grown lobster the changes in the retina are, with one exception, rather insignificant. The parts of the retina increase considerably in size, especially at the distal end of the ommatidium, and additional pigment is deposited in the distal and proximal retinulæ. The only change of importance which the eye undergoes before full maturity is reached is a rearrangement of the ommatidia. It will be remembered that up to stage F the ommatidia were so arranged in relation to each other that the resulting corneal facets were hexagonal in outline. In the adult lobster the facets are square. The change from the six- to the four-sided facet is effected, I believe, by a partial slipping of rows of ommatidia on each other. Imagine, in Figure 55, a row of ommatidia extending in the direction of the arrow and on either side, and parallel to this another row. Only one ommatidium in each of these two lateral rows is given in the figure.

[1] When the preliminary notice of this paper was written, I was somewhat in doubt as to the origin of the accessory pigment-cells. I had not then had the opportunity of studying stage F, and I was of opinion that these cells were probably of ectodermic origin. I believe that the evidence which is now at hand indicates that they are cells derived from the mesoderm.

Suppose the individual ommatidia of the three rows to be arranged in reference to one another as the four in Figure 55; i. e. the ommatidium of one row covering the open spaces between two ommatidia in the adjoining row. Under these conditions hexagonal facets would be produced. But now imagine the middle row to move in the direction of the arrow through the distance of half the thickness of an ommatidium. After the completion of this movement, the ommatidia of the middle row are directly opposite the ommatidia of the adjacent rows. With this arrangement the ommatidia can be grouped in *square* blocks of four, nine, etc. This is the grouping which obtains in the adult retina, and the facets resulting from it are square in outline. In the more primitive arrangement, that with hexagonal facets, the ommatidia can also be grouped in blocks of four, nine, etc. These blocks, however, are never square in outline, but *lozenge-shaped* (Fig. 55).

The process of rearranging the ommatidia is accomplished at a period in the growth of the young lobster much later than that at which the nerve-fibres have arranged themselves in relation to the openings in the basement membrane. Such being the case, one would expect to find in the deeper part of the adult retina traces of the more primitive arrangement. This is seen, I believe, in the lozenge-shaped outline which groups of four ommatidia present in the deeper part of the retina. A good instance of this is to be seen in Figure 14, although it is apparent in almost all of the transverse sections which include the proximal retinulæ. One might say that the ommatidia were rooted to the basement membrane when the hexagonal system prevailed, and that the rearrangement fully affected only their free distal ends.

The movement suggested as a means of rearranging the ommatidia not only explains the new position of the ommatidia, but also accounts for the situation in which the nuclei of the distal retinulæ occur. In Figure 55 each ommatidium is surrounded by a circle of six nuclei. These belong to the distal retinulæ. Instead of describing them as being in circles of six, it might be said that there is one nucleus at each corner of every cone. Thus, cone x (Fig. 55) has nuclei 1, 2, 3, and 4 at its four corners, and cone y has in a corresponding way nuclei 5, 6, 7, and 8. If the cones were to move as I have already described, and were to carry their surrounding nuclei with them, the result would be that nucleus 5 would come to lie next to nucleus 2, and 7 next to 4, and so on. In other words, a pair of nuclei would occupy each space between the adjoining angles of four adjacent cones. This is the position which the nuclei of the distal retinulæ occupy in the retina of an adult lobster (Fig. 5).

Several investigators have already described the development of the
ommatidia in the compound eyes of the higher Crustaceans. The different accounts disagree especially in two particulars; first, as to the source
of the different structures in the retina, and, secondly, as to the number
and arrangement of the cells in an ommatidium. In describing the
development of the retina, I have already discussed the first difference,
and need here only recall my conclusion; namely, that the retina, including the corneal hypodermis, crystalline cones, retinulæ, and rhabdome,
originates as a simple hypodermal thickening, and that no part of it is
derived from the deeper ectoderm, which becomes the central nervous
system. As to the number of cells which constitute an ommatidium, it
will be recalled that in the lobster there are at least sixteen in each
ommatidium; two in the corneal hypodermis, four cone-cells, two distal
retinulæ, eight proximal retinulæ one of which was rudimentary, and a
small, but variable number of accessory pigment-cells. The last are
probably mesodermic in origin; all of the others are derived from the
ectoderm.

The several accounts of the corneal hypodermis given by various
authors differ principally in the number of cells which are said to be
found in each ommatidium. Reichenbach ('86, p. 91) and Nusbaum
('87, p. 179) state respectively that in Astacus and Mysis there are
four hypodermal cells in each ommatidium. Nusbaum's statement
is further supported by Grenacher ('79, p. 118), who describes four cells
under each facet in Mysis. Herrick ('89, p. 167) has found two hypodermal cells in the ommatidium of Alpheus. Patten states that the
number of corneal cells in the ommatidia of all Decapods which he has
examined is two.

All the direct evidence that I have seen points to the conclusion that
the ommatidia of the Decapods possess two cells in the corneal hypodermis. Reichenbach's observation directly opposes this view. I have
not had the opportunity of examining the same species as Reichenbach
did, but I have studied a representative of the fresh-water crayfishes,
Cambarus Bartoni, and there is no question that in the ommatidium
of this species only two corneal cells are present. The nuclei of these
two cells lie in the angles of the hypodermal squares, each one directly
above a nucleus of a cone-cell. When viewed from the surface, it is
difficult to say whether there are two or four hypodermal nuclei, because the four nuclei of the cone lie so near to the hypodermis and
resemble its nuclei so closely. It seems to me possible that in
his surface view Reichenbach ('86, Fig. 226) may have drawn with

the hypodermal nuclei those of the cone-cells, thus giving four instead of two.

The four nuclei in the corneal hypodermis of each ommatidium, as described by Nusbaum in the case of Mysis, had already been seen by Grenacher, who further stated that the nuclei were arranged in pairs at two levels. Grenacher does not describe nuclei in the two segments of the crystalline cone. In the retina of *Mysis stenolepis*, the four nuclei which were described by Grenacher are easily identified, but I cannot agree with Grenacher's statement ('79, p. 118) that all four belong to the same category. The more superficial pair unquestionably belong to the corneal hypodermis, but the deeper pair, I feel confident, are the nuclei of the crystalline cone. The conclusion to be drawn from this interpretation of the work of Reichenbach, Nusbaum, and Grenacher is, that in Decapods and Schizopods each ommatidium possesses two cells in the corneal hypodermis and only two.

Concerning the number of cone-cells in the ommatidium of Decapods, different writers very generally agree. Reichenbach, Kingsley, Herrick, and Patten all state that there are four cone-cells in the Crustaceans which they have studied.

In the cone-cells the number four, according to Grenacher, is characteristic not only of Decapods, but, excepting the Schizopods, it is the distinguishing feature of the Podophthalmata. In Mysis, Grenacher states that the cone has two, instead of four segments. I have studied the eyes of *Mysis stenolepis*, Smith, and my observations confirm this statement. Notwithstanding Grenacher's assertion, Nusbaum ('87, p. 179) claims that the nuclei of the cone-cells in Mysis are grouped in fours. I am confident that there are only two nuclei in the adult cone, and having seen no evidence of a suppression of two nuclei, I must consequently side with Grenacher in his belief that the cone of Mysis is composed of only two cells.

The differentiation of the retinulæ into distal and proximal groups is more complete in the lobster than in the majority of other Decapods studied. As a result of this incomplete differentiation, it is often impossible to get exact statements from the descriptions of different authors concerning the number and position of the retinulæ.

Reichenbach ('86, p. 93) maintains that in Astacus, after the four cone-cells, and as he believes the four hypodermal cells, were differentiated, the remaining cells became pigment-cells (retinulæ). Essentially the same account is given by Nusbaum ('87, p. 180) for Mysis. Kingsley ('87, p. 53) states that, in addition to the corneal hypodermis,

each ommatidium in Crangon at first consists of four vertical series of
nuclei, each series containing five nuclei. This would give a total of
twenty cells for each ommatidium. After deducting four, the number
of cone-cells, from twenty, the original number of cells, there remain
sixteen cells to be accounted for, presumably as pigment-cells. I have
examined the eye of an adult *Crangon vulgaris*, and I find in it, as in
the lobster's eye, two distal retinulæ and seven proximal retinulæ. This
can scarcely be reconciled with Kingsley's account, unless one admits
an extensive suppression of cells. Such a suppression seems to me
scarcely probable, and I am therefore inclined to believe that there
has been some error in Kingsley's method of counting. Possibly
the series of nuclei were so placed that they were shared by neighboring ommatidia, and did not all belong to one ommatidium. This,
however, could only be settled by re-examining the early stages of
Crangon.

Herrick ('86, p. 44) describes seven retinulæ in the ommatidium of
Alpheus, and makes the statement that they do not possess nuclei. He
then describes some undifferentiated ectodermic cells, the nuclei of
which can be seen in the space between the cones. As this is the
position which in the young lobster is occupied by the nuclei of the
proximal retinulæ, and as Herrick has not identified any nuclei for
the proximal retinulæ in Alpheus, I am inclined to regard the deeper
nuclei of this group as belonging to these cells. The more superficial of
the nuclei described by Herrick are apparently arranged in circles of
six around each cone. (Compare Herrick, '86, Figs. 1 and 2.) As in the
early stages of the lobster this arrangement was characteristic of the
nuclei in the distal retinulæ, it is possible that these superficial nuclei
in Alpheus may represent the distal retinulæ.

Reichenbach ('86, p. 92), Kingsley ('87, p. 52), Nusbaum ('87, p. 179),
and Herrick ('89, p. 167) describe an ingrowth of mesodermic tissue
between the retina and ganglion, and in Mysis, according to Nusbaum
('87, p. 180), these cells, as in the lobster, give rise to what I have
called the accessory pigment-cells.

From the investigations which have been summarized in the preceding pages, it is difficult to draw any general conclusion concerning the
number of retinulæ in the ommatidia of the higher Crustacea. Reichenbach and Nusbaum make no statement as to the number of these cells
in Astacus and Mysis. Kingsley's enumeration of them in Crangon
seems to be erroneous. Admitting the undifferentiated ectodermic
nuclei in Alpheus to be the nuclei of the retinulæ, Herrick's statement

that there are seven retinulæ in the ommatidia of this Crustacean coincides fairly with the results obtained from the lobster.

From the histological evidence of the adult, on the other hand, writers are generally agreed that the ommatidia of Decapods possess seven proximal retinulæ. It is probable, however, that this statement requires some qualification. It will be recalled that, in describing the proximal retinulæ of the lobster, I referred to an additional nucleus which apparently represented an eighth rudimentary retinula. I have identified this eighth nucleus in the ommatidia of Cambarus, where, as in the lobster, it lies in a plane different from that of the other seven nuclei. If the eighth nucleus should be present in approximately the same plane as the other nuclei, it could be identified only with great difficulty. It is my belief that it often occurs in this position, and probably for this reason it has generally escaped the attention of investigators, for I am of opinion that it is present in the ommatidia of all Decapods. When, therefore, the statement is made that the ommatidia of a certain Decapod contains seven proximal retinulæ, the probabilities are that the ommatidium in reality contains eight proximal retinulæ, one of which is rudimentary.

Concerning the Schizopods, Grenacher states ('79, p. 119) that in Mysis there are more than four proximal retinulæ, but how many more he is not certain. In *Mysis stenolepis* my own observations have shown me that there are certainly seven pigmented proximal retinulæ. This number agrees with the number of functional retinulæ in Decapods, and my opinion is that in Mysis, as in Decapods, an eighth rudimentary proximal retinula may be expected.

The presence of distal retinulæ in the ommatidia of Decapods seems to have generally escaped attention. Patten and Carrière, however, describe these cells in Penæus and Astacus, and the fact that they are easily recognized in the eyes of Homarus, Cambarus, and Eupagurus inclines me to the belief that they form a constant element in the ommatidia of all Decapods. They have also been seen in the retina of Mysis. In the eyes of this genus their nuclei occupy the position indicated by *d* in Grenacher's Figure 12. It is of interest to observe that their permanent position in Mysis is an early and transitory one in the lobster. (Compare Grenacher, '79, Fig. 112 *d*, with Figs. 48 and 55 of this paper.) In both the Decapods and Mysis the number of distal retinulæ is two.

The Types of Ommatidia.

With the conclusions arrived at in the foregoing account as a basis, an ommatidium can be constructed which will serve as a type for the ommatidia of all Decapods. Omitting the accessory pigment-cells, this typical ommatidium would be composed of sixteen cells as follows : two cells in the corneal hypodermis, four cone-cells, two distal retinulæ, and eight proximal retinulæ. In a similar way a typical ommatidium can be constructed for the Schizopods. This type would differ from that proposed for the Decapods, in that its crystalline cone would be formed of two, instead of four cells.

The ommatidia of the Schizopods and Decapods, as the development of the lobster shows, are closely related. The arrangement of the ommatidia by which hexagonal facets are produced is permanent in Mysis, and temporary in the lobster. The distal retinulæ are grouped in the adult Mysis in a way which is reproduced only in the early stages of the lobster. In Mysis the outline of the nuclei in the corneal hypodermis, as seen in sections tangential to the retina, is strikingly crescentic. This form is temporarily assumed by the corresponding nuclei in the young lobster (Fig. 53). Thus it is evident that the ommatidia of the lobster pass through a stage in which they closely resemble the permanent condition of the ommatidia in Mysis. For this reason, I believe that the ommatidium of Mysis represents a type ancestral to that of the lobster.

The only important difference which exists between the ommatidium of Mysis and that of the lobster in its early stages is that in the latter the cone consists of four cells, while in Mysis it is composed of only two. If one admits that the ommatidium of Mysis is the forerunner of that of the lobster, this difference in the number of cone-cells can easily be explained on the supposition that the cells which form the cone in Decapods divided once more than those in Schizopods. If these two types of ommatidia are thus related, it is only natural to expect that this process of cell-division may connect the ommatidium of Mysis with that of some lower Crustacean.

The ommatidia in the eyes of Amphipods are constructed upon a type which seems thus connected with that of Mysis. In Gammarus, for instance, I have found that the ommatidia possess an undifferentiated corneal hypodermis, a crystalline cone of two cells, and five retinulæ. If one desires to convert this type into that of Mysis, or through Mysis into that of the Decapods, the necessary change merely involves　.

the division and differentiation of cells. The two cone-cells in the simpler type would remain unaffected in Mysis; in the Decapod they would divide once, thus producing a cone of four cells. By division, the five retinulæ of the Amphipod would become the ten retinulæ of the higher type. These ten retinulæ are differentiated into two sets, eight proximal retinulæ which surround the rhabdome, seven of them possessing nerve-fibres, and two distal retinulæ which envelop the cone, but have no nervous connections. If the ten retinulæ of the higher type were developed from the five retinulæ of the lower type, it follows, since all of the five retinulæ possessed nerve-fibres, that those of the ten retinulæ which do not possess nerve-fibres must be considered as degenerate. The degenerate cells of the higher type of ommatidium are consequently the eighth proximal retinula and the two distal retinulæ. The structural condition of these three cells favors this view. The eighth proximal retinula is identifiable only through its nucleus. Each distal retinula, as I have previously described, possesses a proximal fibre. This fibre passes through the basement membrane with the nerve-fibres of the proximal retinulæ, but it is very much smaller than these fibres, and terminates without reaching the optic ganglion. This condition is easily interpreted as one of degeneracy.

The process by which distal and proximal retinulæ were probably differentiated from unmodified retinulæ is easily suggested. A characteristic distinction between the ommatidia of the higher and lower Crustacea, as, for instance, between Gammarus and Homarus, is that in the former the cone and rhabdome are very close to each other, whereas in the latter the cone proper and rhabdome are separated by considerable space. Without attempting to assign physiological reasons, the statement may be made that it seems necessary that the sides of the cone should be sheathed with pigment. In the ommatidia of Gammarus this is accomplished by the distal ends of the five retinulæ; these reach beyond the rhabdome and envelop the cone. They thus form a funnel, in the large central cavity of which the cone rests, while the neck of the funnel is occupied by the rhabdome. Imagine now the division of these five retinulæ into ten, and the separation of the cone from the rhabdome. It is easy to understand how two of the ten retinulæ may retain their connection with the cone, while the remaining eight adhere to the rhabdome. The two retinulæ connected with the cone would lose their nervous function and become simple pigment-cells. The retinulæ which remain attached to the rhabdome would continue as perceptive cells. The separation of the rhabdome and cone not only offers an explanation of the way in which the distal and proximal retinulæ have arisen, but it also accords well with the fact that the proximal retinulæ end distally in fine fibres which stretch toward the cone, and are applied to the fibres of the distal retinulæ.

If what has been said of the growth of ommatidia be true, the course which the development of these structures takes in the case of the lobster must be somewhat different from that by which they arose phylogenetically. In the lobster the division of the nuclei is entirely completed before the ommatidia are differentiated. Consequently, after differentiation has occurred, no further cell-division ensues among the elements of an ommatidium. This fact, however, is by no means a serious objection to the view which I have expressed, for of the two processes, cell-division and the differentiation of ommatidia, it is only necessary to imagine that in successive generations the differentiation was retarded until a stage was reached in which all cell-division was accomplished before the differentiation of ommatidia began. It is of interest to observe, however, that in the lobster the planes of division among the nuclei, which eventually enter into the formation of ommatidia, correspond in direction to the plane in which the nuclei of the ommatidium in Gammarus would divide, should this ommatidium be converted into that of the lobster. Thus the plane of nuclear division which was so characteristic of the distal superficial part of each optic disk in the lobster may have a phylogenetic significance.

In how far the ommatidia of all the Crustacea can be brought into relation by a process of cell-division such as I have outlined, and what constitutes the simplest form of an ommatidium, are questions which require careful and extensive comparative study, and which I am not able to discuss here. Suffice it to say, that between the ommatidia of the higher and lower Crustacea there is reason to conclude that such a relation as I have pointed out probably exists.

Cambridge, October 1, 1889.

BIBLIOGRAPHY.

Bobretsky, N.
'73. Development of Astacus and Palæmon. Kiew, 1873. (The substance of this paper, so far as it refers to the eye, is known to me through the abstract in Balfour's Comparative Embryology, Vol. II. pp. 397, 398. London, 1881.)

Carrière, J.
'85. Die Sehorgane der Thiere vergleichend-anatomisch dargestellt. München u. Leipzig: R. Oldenbourg. 1885. 6 + 205 pp., 147 Abbildg. u. 1 Taf.
'89. Bau und Entwicklung des Auges der zehnfüssigen Crustaceen und der Arachnoiden. Biologisches Centralblatt, Bd. IX. No. 8, pp. 225–234. 15 June, 1889.

Claus, C.
'86. Untersuchungen über die Organisation und Entwickelung von Branchipus und Artemia nebst vergleichenden Bemerkungen über andere Phyllopoden. Arbeit. Zool. Inst. Wien, Tom. VI. Heft 3, pp. 267–371, 12 pls. 1886.

Grenacher, H.
'79. Untersuchungen über das Sehorgan der Arthropoden, insbesondere der Spinnen, Insecten und Crustaceen. Göttingen: Vandenhöck u. Ruprecht. 1879. 8 + 188 pp., 11 Taf.

Grobben C.
'79. Die Entwickelungsgeschichte der Moina rectirostris. Arbeit. Zool. Inst. Wien, Tom. II. Heft 2, pp. 203–269, 7 pls. 1879.

Herrick, F. H.
'86. Notes on the Embryology of Alpheus and other Crustacea and on the Development of the Compound Eye. Johns Hopkins Univ. Circulars, Vol. VI. No. 54, pp. 42–44. Dec., 1886.
'88. Development of Alpheus. Johns Hopkins Univ. Circulars, Vol. VII. No. 63, pp. 36, 37. Feb., 1888.
'89. The Development of the Compound Eye of Alpheus. Zool. Anzeiger, Jahrg. 11, No. 303, pp. 164–169. 25 March, 1889.

Kingsley, J. S.
'86. The Arthropod Eye. Amer. Naturalist, Vol. XX. No. 10, pp. 862–867. Oct., 1886.
'86a. The Development of the Compound Eye of Crangon. Zool. Anzeiger, Jahrg. 9, No. 234, pp. 597–600. 11 October, 1886.

Kingsley, J. S.
    '87· The Development of the Compound Eye of Crangon. Journ. of Mor-
      phology, Vol. I. pp. 49–67, 1 pl. 1887.
    '89· The Development of Crangon vulgaris. Third Paper, with Plates I.,
      II., III. Bull. Essex Inst., Vol. XXI., pp. 1–42. 1889.
Mark, E. L.
    '87· Simple Eyes in Arthropods. Bull. Mus. Comp. Zoöl. at Harvard Col-
      lege, Vol. XIII. No. 3, pp. 49–105, 5 pls. Feb., 1887.
Newton, E. T.
    '73· The Structure of the Eye of the Lobster. Quart. Jour. of Micr. Sci.,
      Vol. XIII., New Series, pp. 325–343. 1873.
Nusbaum, J.
    '87· L'Embryologie de Mysis chameleo (Thompson). Arch. Zool. exper.,
      sér. 2, Tom. V. pp. 123–203. 1887.
Parker, G. H.
    '87· The Eyes in Scorpions. Bull. Mus. Comp. Zoöl. at Harvard College,
      Vol. XIII. No. 6, pp. 173–208, 4 pls. Dec., 1887.
    '88· A Preliminary Account of the Development and Histology of the Eyes
      in the Lobster. Proc. Amer. Acad. Arts and Sci., Vol. XXIV. pp. 24, 25.
      NOTE. — Copies of this paper were distributed in November, 1888, although the volume in which it
    appeared was not issued till December, 1889.
Patten, W.
    '86· Eyes of Molluscs and Arthropods. Mittheilungen a. d. zool. Station
      zu Neapel, Bd. VI. Heft 4, pp. 542–756, Pls. 28–32. 1886.
    '87· Studies on the Eyes of Arthropods. 1. Development of the Eyes of
      Vespa, with Observations on the Ocelli of some Insects. Journ. of Mor-
      phology, Vol. I. pp. 193–227, 1 pl. 1887.
Reichenbach, H.
    '86· Studien zur Entwicklungsgeschichte des Flusskrebses. Abhandl.
      Senckenb. Naturf. Ges., Bd. 14, Heft 1, pp. 1–137. 1886.
Schultze, M.
    '67· Ueber die Endorgane des Sehnerven im Auge der Gliederthiere. Arch.
    f. mikr. Anat., Bd. III. pp. 404–408. 1867.
    '68· Untersuchungen über die zusammengesetzten Augen der Krebse und
      Insecten. Bonn: Max Cohen & Sohn. 32 pp., 2 Taf. 1868.
Watase, Sho.
    '89· On the Structure and Development of the Eyes of the Limulus. Johns
      Hopkins Univ. Circulars, Vol. VIII. No. 70, pp. 34–37. March, 1889.

EXPLANATION OF FIGURES.

All the figures were drawn with the aid of an Abbé camera. Unless otherwise stated, the figures are from specimens stained with Grenacher's alcoholic borax-carmine and mounted in benzol-balsam. Where sections have been depigmented the reagent employed was an aqueous solution of potassic hydrate $\frac{1}{4}\%$ (see Parker, '87, p. 175). Figs. 1 to 36 refer to the *histology* of the adult lobster's eye. Figs. 37 to 59 inclusive deal with the *development* of the eye.

PLATE I.

All figures on this plate illustrate the *histology* of the lobster's eye.

Fig. 1. A longitudinal section of an ommatidium. This is a composite figure, its parts having been drawn to one scale from various sections, and afterwards combined. The distal retinula on the right side of the cone contains its natural pigment; that on the left side has been depigmented. The letter x indicates the position of the band which limits the corneal facet. The numbers to the right of the figure correspond to the numbers of the following figures of transverse sections, and mark the level at which the latter were taken. × 105.

The remaining figures on this plate (Nos. 2–25) are transverse sections either of ommatidia or bundles of nerve-fibres. In each case the sections were studied and drawn from their distal faces. Figs. 2 to 20 are magnified 375 diameters; Figs. 21 to 25, 575 diameters.

" 2. A corneal facet. This specimen was cleaned in a boiling solution of potassic hydrate and studied in water.

" 3. Four pairs of cells from the corneal hypodermis. Around the lower right-hand pair of cells the outlines of the six surrounding distal retinulæ are indicated in part by dotted lines.

" 4. Four groups of cone-cells. The ommatidia to which the cone-cells belong are indicated by the letters a, b, c, and d. These letters are used to indicate corresponding ommatidia in deeper sections.

" 5 to 20 represent transverse sections through ommatidia in the various regions indicated as follows : —

" 5. The middle region of four cones. Completely depigmented.

" 6. The proximal ends of four cones. The re-entrant angle on the surface is indicated at x. Completely depigmented.

" 7. Slightly below the proximal ends of the cones.

" 8. Beyond the distal ends of the proximal retinulæ.

" 9. The distal ends of the proximal retinulæ.

" 10. The thick distal portion of the proximal retinulæ.

" 11. The proximal retinulæ immediately above the distal termination of the rhabdome.

" 12. At the distal ends of the rhabdomes. Completely depigmented.

Fig. 13.   The rhabdome between its distal end and middle. In this section the accessory pigment-cells of each ommatidium are apparently separated by an intervening space from those of neighboring ommatidia. This space is probably the result of shrinkage and subsequent rupture.

" 14.   In the same plane as that shown in Fig. 13. Partially depigmented.

" 15.   The middle of the rhabdome. The distal retinulæ of ommatidium c have been numbered. Completely depigmented.

" 16.   The proximal end of the rhabdome. The accessory pigment-cells in this section have probably been ruptured, as in Fig. 13.

" 17.   In the same plane as that shown in Fig. 16. Partially depigmented.

" 18.   The proximal tip of the rhabdome. Completely depigmented.

" 19.   The proximal retinulæ close to the basement membrane. Partially depigmented.

" 20.   A diagram of Fig. 19. The proximal retinulæ of the different ommatidia have been numbered to correspond with those of ommatidium c in Fig. 15. The retinulæ of ommatidium c are tinted pink.

" 21.   Transverse section of nerve-fibres as they pass through the basement membrane. The line $y z$ shows the plane of section for Fig. 29; x indicates the cross-shaped thickening in the basement membrane. Completely depigmented and stained in Weigert's hæmatoxylin. (See page 4.) × 575.

" 22, 23, 24, and 25 represent transverse sections of the fibres in the optic nerve. Weigert's hæmatoxylin. × 575. The planes at which these sections were taken are as follows : —

" 22.   Directly below the basement membrane. The fibres are in groups of threes and fours.

" 23.   At one fourth the distance from the basement membrane to the optic ganglion.

" 24.   At half the distance between the membrane and ganglion.

" 25.   At the surface of the ganglion.

ABBREVIATIONS.

ax. n.	Nervous axis of retinula.	mb. pi ph.	Peripheral membrane.
cap.	Protoplasmic cap of cone-	n. flr.	Nerve-fibre.
cl. con.	Cone-cell. [cell.	nl. con.	Nucleus of cone-cell.
cl. ms d.	Mesodermic cell.	nl. crn.	" corneal hypodermis.
con.	Cone.	nl. dst.	" distal retinula.
crn.	Corneal cuticula.	nl. pig.	" accessory pigment-cell.
crn. h d.	Corneal hypodermis.	nl. px.	" proximal retinula.
cta.	Cuticula.	omm'.	Ommateum.
enc.	Brain.	pig. dst.	Distal band of pigment.
flr'.	Fibrillæ.	pig. px.	Proximal band of pigment.
gn. opt.	Optic ganglion.	r.	Retina.
h d.	Hypodermis.	rhb.	Rhabdome.
mb.	Basement membrane.	rtn'. dst.	Distal retinula.
mb. i cpt.	Intercepting membrane.	rtn'. px.	Proximal retinula.
mb. i cl.	Intercellular membrane.	spa. i cl.	Intercellular space of retina.

The other abbreviations which occur on the plates are explained in the description of the figures with which they are found.

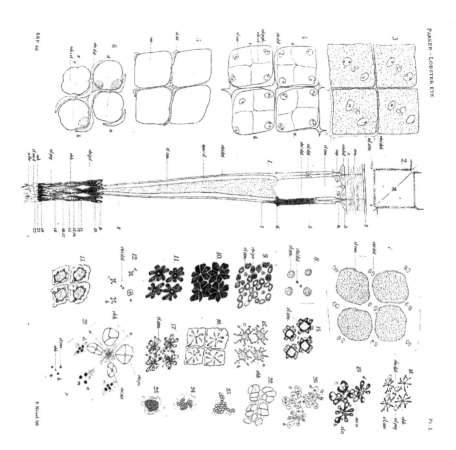

Pl. I.

PLATE II.

Figs. 26 to 36 deal with the *histology* of the lobster's eye ; Figs. 37 to 39, with its *development*.

Fig. 26. Vertical section through that portion of the eye-stalk where the transition from the undifferentiated hypodermis to the ommateum is accomplished. The open space x is due to shrinkage. × 31.

" 27. A group of the four cone-cells of an ommatidium and the attached corneal hypodermis. Isolated and studied in Muller's fluid. × 365.

" 28. Proximal end of a group of four cone-cells where they separate as fibres to pass around the rhabdome. Isolated and studied in Müller's fluid. × 365.

" 29. Transverse section of the basement membrane. The distal face is uppermost; the proximal face below. The section is taken in a plane which would be represented in Fig. 21 by the line $z\,y$. × 575.

" 30. A rhabdome and its seven surrounding proximal retinulæ. At the distal end the free tips of the seven retinulæ can be seen. At the proximal end the rhabdome and the four groups of retinulæ which pass through the basement membrane are visible. Isolated and studied in chromic acid $\frac{1}{50}$%. × 200.

" 31. An individual proximal retinula. Isolated and studied in chromic acid $\frac{1}{50}$%. × 200.

" 32. Transverse section of a rhabdome and its surrounding cells. The plane of section is about half-way between the middle and distal end of the rhabdome. Completely depigmented with potassic hydrate. × 460.

" 33. Oblique section of a rhabdome from the same series of sections as Fig. 32. The upper end of the figure is distal; the lower, proximal. × 460.

" 34. Transverse section of a rhabdome and its surrounding retinulæ. The nervous axis of each retinula is distinctly stained. Completely depigmented; stained with Kleinenberg's alum-hæmatoxylin. × 460.

" 35. Longitudinal section of a bundle of nerve-fibres extending through the basement membrane toward the optic ganglion. Depigmented; Weigert's hæmatoxylin. × 575.

" 36. Optic nerve-fibre. Isolated and studied in Müller's fluid. × 575.

" 37. Superficial view of a left optic lobe. Enough of the right lobe is drawn to indicate the position of the median plane ($x\,y$). x is anterior; y is posterior. Stage A (see page 2). × 280.

" 38. Posterior face of a section from a right optic disk cut transversely to the longitudinal axis of the embryo (see page 34). At x is an angle formed by the growth of the retinal cells over the undifferentiated ectoderm. Stage A. × 280.

" 39. A section from the optic disk of an embryo in stage B. The plane of cutting corresponds to that in Fig. 38. In this figure x indicates, as in the preceding one, the angle between the undifferentiated ectoderm and the growing retina. × 280.

PLATE III.

All figures on this plate illustrate the *development* of the lobster's eye.

Fig. 40. A section through the left optic lobe and left half of the supra-œsophageal ganglion. The plane of section is tangential to that part of the surface of the egg on which the embryo rests. The position of the median plane is indicated at xy. The surfaces tinted with deeper pink in the figure represent areas containing nuclei in the specimen; those in lighter pink, areas in which no nuclei were present. The optic lobe is divided into two parts by a band of large, faintly colored nuclei, which, with the smaller surrounding nuclei, are shown in the figure. To the right of the nuclei the broad tinted marginal area represents the retina, r. The remainder of the optic lobe gives rise to the optic ganglion. Stage C (see page 2). × 280.

" 41. Posterior aspect of a transverse section of a right optic lobe. The plane of section corresponds to that in Fig. 38; x is the angle which indicates the separation of the retinal and ganglionic constituents of the intercepting membrane. Stage C. × 280.

" 42. This figure is taken from a region which corresponds to the left-hand portion of Fig. 41. Although from the same set of eggs the embryo from which Fig. 42 was drawn was somewhat more advanced than that from which Fig. 41 was taken. At x the proximal band of retinal nuclei can be seen; at y the distal band is shown. Stage C. × 460.

" 43. The superficial layer from the distal band of retinal nuclei; seen from the external surface of the retina. Stage C. × 460.

" 44. The deep layer of the distal band of nuclei. These are seen in optical section somewhat within the outer face of the retina. Stage C. × 460.

" 45. A transverse section of an optic lobe from a lobster at stage D. The plane of section corresponds to that of Fig. 38. As in Fig. 40, the deeply tinted areas were nucleated; the lighter areas were without nuclei. × 145.

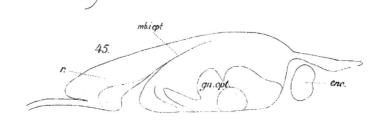

40.

gn opt.

x.

enc.

x.

y.

43.

nl.crn.
nl.dst.

44.

nl.con

enc.

41.

mb.i.opt.

gn. opt.

n.

n.fbr.

x.

nl.dst.
nl.crn.
nl.con.

y.

42.

x.

n.

gn.opt.

45.

mb.i.opt

n.

gn.opt.

enc.

PLATE IV.

All figures on this plate, except Fig. 59, illustrate the *development* of the
lobster's eye.

Fig. 46. A transverse section of an optic lobe at stage E (see page 2). The plane
of section and the method of coloring the figure are the same as in
Fig. 45. × 145.

" 47. An enlarged drawing of that portion of the retina which is in brackets in
Fig 46. Stage E. × 460.

" 48. A view of the external surface of the retina. The distal ends of four
ommatidia are seen. Stage E. × 460.

" 49. A transverse section of four ommatidia in the region of the hypodermal
nuclei. (Compare Fig. 47.) Stage E. × 460.

" 50. A transverse section of four ommatidia in the plane which the nuclei of
the cone-cells occupy. Stage E. × 460.

" 51. Longitudinal section of a single ommatidium. Stage F. × 460.

" 52. Four corneal facets seen from the external surface. Stage F. × 460.

" 53 to 58 represent transverse sections of four ommatidia at Stage F. The
numbers on the left side of Fig. 51 indicate the heights at which these
sections were taken, and correspond to the numbers of the following
figures. In Figs. 53 to 58 the magnification is 460.

" 53. A transverse section in the region of the corneal hypodermis.

" 54. A transverse section through the region in which the nuclei of the cone-
cells occur.

" 55. A transverse section in the same plane as the nuclei of the distal
retinulæ.

" 56. A transverse section of the proximal ends of two cones.

" 57. A transverse section through the rhabdomes and proximal retinulæ.

" 58. A transverse section of a rhabdome from Fig. 57. Fig. 58 was drawn
with a higher magnification than Fig. 57 in order to show the relation
of the proximal retinulæ to the segments of the rhabdome. × 640.

" 59. A corneal facet from near the periphery of the retina in an adult lobster.
The hexagonal outline is noteworthy. This specimen was cleaned in
boiling potassic hydrate and examined in water. × 280.

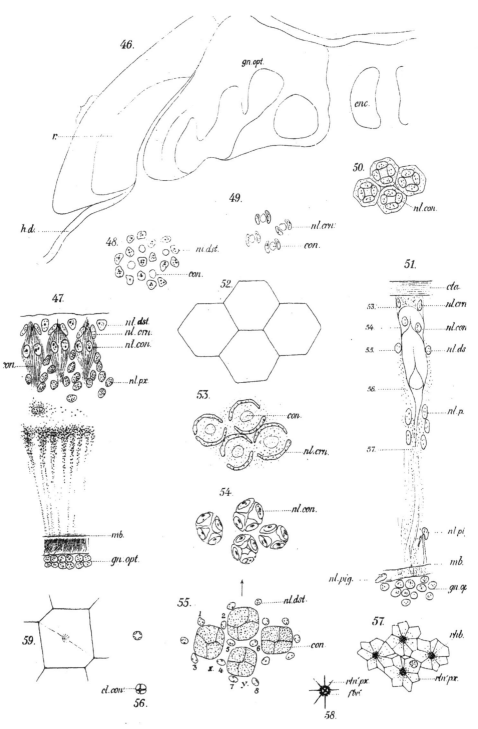

46.

gn.opt.

enc.

r.

h.d.

50.
nl.con.

49.
nl.crn.
con.

48.
nl.dst.
con.

51.

47.
nl.dst.
nl.crn.
nl.con.
con.
nl.px.

52.

53.
cta.
nl.crn.
54
nl.con.
55.
nl.ds
56.
nl.p.

57.

53.
con.
nl.crn.

54.
nl.con.

mb.
gn.opt.

nl.pi.

mb.

nl.pig.
gn.o.

59.

55.
nl.dst.
1
2
5 6
con.
3 4
7 y. 8

57.
rhb.
rdn'px.
rdn'px.
fbr.
58.

cl.con.
56.

G H F del

B Meisel. lith

No. 2. — *On the Rate of Growth of Corals.* By ALEXANDER AGASSIZ.

WE know as yet comparatively little regarding the rate of growth of corals under different conditions. Dana has given, in his "Corals and Coral Islands,"* a *résumé* of our knowledge on the subject, so that it is only necessary for me here to refer the reader to his account of the statements of Darwin, Stuchbury, Duchassaing, Verrill, and others, relating to this subject.

The specimens figured in this communication have been kindly sent me by Lieut. J. F. Moser, commanding the U. S. Coast and Geodetic Survey steamer "Bache." They were all taken (as stated by Mr. Hellings, the cable manager) off the cable laid between Havana and Key West, in June, 1888, from a portion of the cable repaired in the summer of 1881; so that the growth is about seven years. Lieutenant Moser writes: "Taken from the shore end of the International Cable; the specimens were taken between the triangular buoys and the outer reef, the shore end being that portion between Key West and the outer reef." The Coast Survey maps indicate a depth of from six to seven fathoms, and this portion of the cable is most favorably situated as regards food supply, being directly in the track of the main flow of the tide as it sweeps in and out from the outer reef into Key West Harbor, and over the flats to the northward.

Some of the specimens belong to different species from those of which the rate of growth was already known.

Orbicella annularis (Plates I. and II.) shows a much greater increase in the thickness of coral formed than the case mentioned by Verrill, where the thickness formed in sixty-four years was not more than about eight inches. The specimens sent by Lieutenant Moser grew to a thickness of two and a half inches in about seven years.

* Coral and Coral Islands, by James D. Dana. Third edition. New York, 1890. (Pp. 123, 253, 418.)

The *Manicina areolata* (Plate III.) shows also a very rapid rate of increase.   This corresponds to the rate of growth of allied genera, (*Mæandrina labyrinthica*) observed by Pourtalès at Fort Jefferson, Tortugas.

The *Isophyllia dipsacea* (Plate IV.) shows a still more rapid increase.

Of course, we are unable to state that these corals began to grow the first season the cable was laid ; but, judging from the favorable locality in which the corals were found, it is not probable that more than a few months passed before some of the swarms of pelagic coral embryos which must have floated past the cable found a place of attachment.

The specimens have all been figured of the natural size.

The figures all show, with the exception of those of Manicina, the size of the cable to which the corals were attached.

CAMBRIDGE, August, 1890.

EXPLANATION OF THE PLATES.

PLATE I.

Orbicella annularis Dana (natural size).

The thickness of the coral at the edge of the mass varies from $\frac{3}{8}$ to $\frac{3}{4}$ of an inch. The greatest height of the mass above the cable is $2\frac{1}{2}$ inches.

PLATE II.

Orbicella annularis Dana (natural size).

The thickness of this specimen is very much less than that of Plate I. It varies at the edge of the mass from $\frac{1}{8}$ to $\frac{1}{4}$ of an inch. The greatest height above the cable is $2\frac{1}{4}$ inches,

PLATE III. •

Manicina areolata, Ehrenb. •

1. Seen in profile. The thickness above the cable is one inch.
2. Same, seen from above.

Both figures natural size.

PLATE IV.

Isophyllia dipsacea Ag. (natural size).

The greatest thickness is $2\frac{1}{2}$ inches

PLATE I.

ORBICELLA ANNULARIS *Dana.*

PLATE I.

PLATE II.

ORBICELLA ANNULARIS *Dana.*

PLATE III.

2.

MANICINA AREOLATA *Ehrenb.*

PLATE IV.

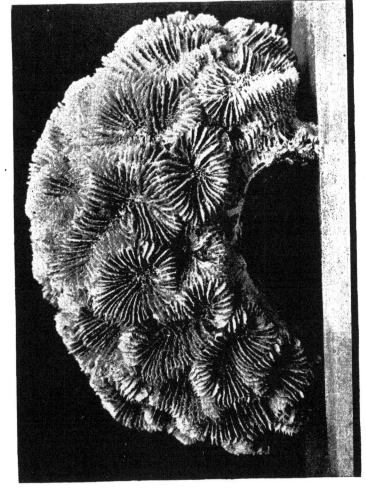

ISOPHYLLIA DIPSACEA *Ag.*

No. 3. — *Preliminary Account of the Fossil Mammals from the White River and Loup Fork Formations, contained in the Museum of Comparative Zoölogy.* Part II. *The Carnivora and Artiodactyla by* W. B. Scott. *The Perissodactyla by* Henry Fairfield Osborn.

This paper, the second upon the Fossil Mammals of the Museum of Comparative Zoölogy, is a continuation of the one published by the writers[1] in August, 1887, upon the White River Mammalia, and includes a number of additions to and corrections of the results there described. It is, however, especially devoted to a consideration of the upper Miocene or Loup Fork mammals collected in Nebraska by Messrs. Garman and Clifford, and in Kansas by Mr. Sternberg. The specimens from these different localities exhibit a considerable range of specific variation.

The Loup Fork species here described have for the most part been long established, but these collections add much to our knowledge, and enable us to determine very fully the structure of forms which have been known hitherto only from fragments. Of such new observations we may mention : (1) the determination of the foot structure of *Merycochœrus;* (2) of *Blastomeryx;* (3) the restoration of *Cosoryx;* (4) discovery of the mandible of *Ælurodon hyænoides;* (5) the discovery of an exceedingly large feline animal; (6) observations upon the molars of the equine series; (7) the manus and pes of *Aceratherium;* (8) the skeletal characters and restoration of *Aphelops fossiger;* (9) the homologies of the elements of the molar teeth in the rhinoceroses; (10) the brain characters of *Aphelops* and *Mesohippus;* (11) the discovery of a Loup Fork species of *Chalicotherium.*

We have again to express our thanks to Dr. F. C. Hill, Curator of the Geological Museum at Princeton, for his skilful excavation and mending of the specimens, and to Mr. R. Weber for the very accurate series of drawings which accompany this paper.

Geological Museum, Princeton, N. J., July 8, 1890.

[1] The authors, as initiated in their Memoir upon the Uinta Mammalia, have divided the subjects for their present and future joint papers.

CARNIVORA.

CANIDÆ.

ÆLURODON, Leidy.

(Syn. *Epicyon*, Leidy. *Canis*, Leidy, in part. *Palhyæna*, Schlosser.)

The dogs of this genus are the most abundant of the Loup Fork *Canidæ*, and, as their relations and systematic position have been very generally misunderstood, it will be well to describe them in some detail. The special peculiarity of the genus is to be found in the development of a large anterior basal lobe on the superior sectorial, as in the cats. The postero-internal cone (metaconid) of the lower sectorial is much reduced, and in some species almost disappears. The talon of this tooth is rather short, and consists of an internal and external cone or tubercle, being of the basin-like character. The premolars are remarkably heavy, and possess well developed basal conules. There are four well marked species of this genus, of which the best known is

Ælurodon sævus, Leidy (Cope).

(Syn. *Canis sævus*, Leidy. *Ælurodon ferox*, Leidy. *Ælurodon sævus*, Cope)

This species is characterized by the very small size of the internal cusp of

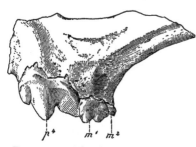

the upper sectorial, and by the nearly straight and slender mandible; the incisors are rather small, and the first upper molar is very large and subtriangular in shape. The skull as figured by Cope (American Naturalist, XVII.) presents a rather short, narrow muzzle, and is in general quite bear-like in appearance. Notwithstanding its peculiarities of dentition, this animal is an unmistakable dog, and the structure of the skull, vertebræ, limbs, and feet is characteristically cynoid. The metapodials are, however, somewhat less elongated proportionally than in existing dogs.

FIGURE 1.—*Ælurodon sævus*, fragment of right superior maxillary × ⅔.

Ælurodon Haydeni, Leidy.

(Syn. *Canis Haydeni*, Leidy. *Epicyon Haydeni*, Leidy.)

This species is very large, and is remarkable for the short, massive mandible and the strong upward curvature of the posterior portion of the alveolus, so that the inferior tubercular molars may almost be said to be inserted in the ascending ramus. In Dr. Leidy's type of the species (Ext. Mam. Fauna, Dak. and Neb., Plate I. fig. 10) the third lower molar is inserted by two fangs, and in the Cambridge specimen by only one; but this does not appear to be a constant character. The postero-internal cusp (metaconid) of the lower sectorial is reduced to a rudiment, and the talon is much shortened antero-posteriorly. $\overline{\text{Pm}}$. 1 and 2 are relatively quite small, while $\overline{\text{pm}}$· 3 and 4 are quite high and massive.

Ælurodon Wheelerianus, Cope.

(Syn. *Canis Wheelerianus*, Cope. *Ælurodon Wheelerianus*, Cope.)

This species is nearly as large as the preceding one, but differs from it (1) in the much less strongly curved alveolar region, and (2) in the very large size of the external upper incisor, which at the base is nearly as large as the canine. The species is represented in the collection by the facial region of a very old individual, the teeth of which are worn down to mere stumps. The face appears to be proportionately longer than in *Æ. sævus*, the orbit lying somewhat farther back; it is also very deep, and encloses an unusually large nasal chamber.

Ælurodon hyænoides, Cope.

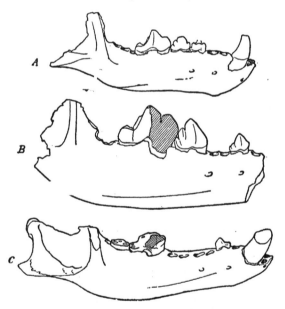

FIGURE 2.—Mandibles of *Ælurodon* × ½. A. *Æ. hyænoides*; B. *Æ. Haydeni*; C. *Æ. sævus*.

This, the smallest species of the genus, has been described by Cope from the superior dentition: "The second and third premolars are robust and somewhat swollen at the inner base. Each has a short heel, but no median posterior lobe. The principal lobe is robust, in the third [pre]molar as wide as long at the base.

The internal anterior lobe of the superior incisor [sectorial] is very large, and its apex is distinct from the inner side of the rest of the tooth. It is relatively larger than in *Crocuta brunnea*. . . . The first true molar is somewhat wider near the inner extremity of the crown than at the external extremity." (Bulletin U. S. Geological and Geographical Survey of the Territories, Vol. VI. p. 388.)

The Cambridge collection contains a mandible which should almost certainly be referred to this species. It is proportionately short, stout, and of nearly uniform depth, not tapering anteriorly as in *Æ. sævus;* the symphysis is very obliquely placed and the chin abruptly rounded, giving the jaw a somewhat cat-like appearance. The first and second premolars are small, and the latter is implanted by a fang which is but imperfectly divided into two; the third and fourth premolars are low but strong, and differ from the corresponding teeth of the other species in the presence of a small *anterior* basal cusp. The sectorial is large, and has a well developed metaconid; the talon is obliquely worn upon its outer side, showing a different mode of opposition of the teeth from that which obtains in *Æ. sævus*. The incisors are very closely crowded together, and the median one is pushed very far back out of the line of the other two.

MEASUREMENTS.

	m.			m.
Length, inferior molar series	.032	Blade of sectorial (ant. post.)		.013
" " premolar series	.031	Talon " "		.006
" sectorial (m. 1)	.019			

? Ælurodon ursinus, COPE.

(Syn. *Canis ursinus*, Cope.)

Some large specimens agree best with the figures and descriptions of the *Canis ursinus*, but they are so damaged as to render any final reference of them impossible. Indeed, it is by no means clear that the species here named can be regarded as distinct.

The systematic position of *Ælurodon* has been somewhat disputed. Leidy placed it provisionally among the *Felidæ* (*op. cit.*, pp. 68 and 367). Cope, though referring it to the *Canidæ*, has regarded it as the forerunner of the hyænas. " I nevertheless suspect that this genus is the ancestor of the *Hyænidæ*, through the intermediate forms *Ictitherium* and *Hyænictis*." (American Naturalist, Vol. XVII. p. 244) Professor Cope has, however, informed us that he does not attach much importance to this view. Schlosser has adopted the same opinion, but believes that *Æ. sævus* should be generically separated from the other species. " Der *Canis sævus*, Leidy, wird von Cope zur Gattung *Ælurodon* gestellt, indess offenbar ohne hinreichenden Grund, denn sowohl der Schädelbau, als auch die Beschaffenheit der einzelnen Zähne, namentlich des oberen Pr. 1 sprechen sehr für die Zugehörigkeit zu den echten Caniden, wäh-

rend die beiden übrigen *Ælurodon*-Arten sich höchst wahrscheinlich als Vor-
läufer der Hyänen erweisen werden." (Beitr. z. Paläont. Oesterr. Ungarns,
Bd. VIII. p. 252.)

In these statements Schlosser has been misled by the fact that the specimen
of *Ælurodon sævus* which was figured by Cope is very old, and the teeth so
much worn down that the anterior lobe of the upper sectorial is hardly distinguishable. The specimens before us demonstrate clearly that Cope's reference
of the species is correct, and that *Æ. Wheelerianus* and *hyænoides* cannot be
generically distinguished from it. The only characters of *Ælurodon* which in
any way resemble those of the hyænas are (1) the massive premolars, (2) the
presence of an anterior basal cusp on the upper sectorial, and (3) the reduction
(in some species) of the postero-internal cusp of the lower sectorial. These
resemblances are obviously merely analogical, and are of far less importance
than the characters of the skull and limbs, which are distinctively cynoid.
These animals are genuine dogs, if somewhat peculiarly modified, and to regard
them as ancestors of the hyænas is to ignore the close connection between the
latter and the viverrines, besides being improbable on geographical grounds.

CANIS.

? Canis vafer, LEIDY.

This small alopecoid is represented in the collection by a mandible with
broken teeth and some other fragments. It agrees almost exactly with Leidy's
type (*op. cit.*, Plate I. fig. 11), except that the diastema between the canine and
p̄m̄. 1 is shorter. The small size of the sectorial places the species in the mi-
crodont division of the alopecoid series. M. 2 is very elongate antero-posteriorly,
and m. 3 is implanted by two fangs. The mandible is very slender, much

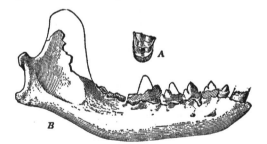

FIGURE 3. — *? Canis vafer* × ¾. A. First superior molar; B. Mandible.

curved, and non-lobate. The first upper molar is nearly quadrate in shape, the
metaconule being almost as large as the protocone, and placed upon nearly the
same antero-posterior line. The cingulum is internally very greatly enlarged
and thickened, and is disposed symmetrically around the inner side of the
crown, instead of being confined to the postero-internal angle, as is usual
among the recent *Canidæ*.

The head of the radius is less transversely extended and more discoidal than in recent dogs, apparently indicating the retention in some degree of the power of supination.

FELIDÆ.

FELIS.

? Felis maxima, sp. nov.

This species is founded upon a well preserved humerus from the Loup Fork of Kansas. The chief peculiarity of the specimen is its great size, which very

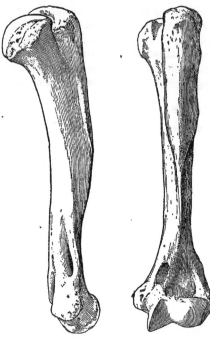

FIGURE 4. — Humerus of ? *Felis maxima* × ⅓; internal and anterior views.

much exceeds that of any living feline. In construction it closely resembles the humerus of the lion, with some minor differences. The external tuberosity rises high above the head, and is somewhat less rugose; the deltoid ridge is exceedingly broad and massive, and descends far down upon the shaft; the outer condyle for the capitellum of the radius is less decidedly convex; the internal epicondyle is very prominent and massive, and is surmounted by a large epicondylar foramen. The presence of this epicondylar foramen shows that the specimen before us cannot be referred to *Smilodon*, for the humerus of *S. necator* figured by Cope (American Naturalist, Vol. XIV. p. 857) has no such foramen. The supinator ridge is somewhat broken, but it appears to have been proportionately less robust than in the lion. The following table will exhibit the great size of this specimen. The measurements of the humerus of *Smilodon* are taken from Cope's figure.

MEASUREMENTS.

	Smilodon necator. m.	Felis leo. m.	F.? maxima. m.
Humerus length	.384	.313	.429
" width of distal end	.087	.054	.072
" antero-posterior diameter proximal end	—	.088	.118

Another cat, perhaps a smaller individual of the same species, is represented by a phalanx of the median row, which is of the characteristically asymmetrical shape, so as to allow the retraction of the claws. It agrees best in shape with the median phalanx of the fourth posterior digit of the lion, but is much larger, measuring 37 mm. in length, and the proximal end is 20 mm. wide; in the lion these dimensions are 27 and 14 mm.

A third very large feline is indicated by the proximal end of a radius from the Loup Fork of Nebraska: it agrees closely in shape and size with that of the lion.

Still another cat is represented by the third and fourth metatarsals from the same horizon and locality. The mode of interlocking, the shape and character of the proximal articular surfaces, are very cat-like, but the bones are short and massive, showing strikingly different proportions from those to be observed in the recent forms.

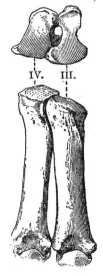

MEASUREMENTS.

		Felis leo.	?
		m.	m.
Metatarsal III., length	. . . . .	.119	.089
"	" width proximal end .	.021	.022
"	IV., length	.115	.093
"	" width proximal end .	.015	.018

These specimens show that the number of cats occurring in the Loup Fork formation is much more considerable than has hitherto been supposed. Unfortunately, however, these remains are not associated with teeth, so that they cannot be referred to their proper genera and species.

FIGURE 5. — Third and fourth metatarsals of unknown feline × ½.

? Pseudalurus intrepidus, LEIDY.

This species is doubtfully indicated by a humerus, lacking the proximal end, which is distinctly feline in character, but remarkable for the very weak development of the supinator ridge.

MUSTELIDÆ.

Carnivora of this family are not certainly known to occur in any American formation older than the Loup Fork, and they are very rare even in that formation. The mustelines are represented in the collection by only a fragment of a lower jaw supporting pm. 4. In the absence of the molars, it is impossible to determine to what genus this specimen should be referred; but it would appear to agree best with the *Mustela parviloba* of Cope.

ARTIODACTYLA.

OREODONTIDÆ.'

MERYCHYUS, LEIDY.

Merychyus elegans, LEIDY.

The genus *Merychyus* is abundant in the Loup Fork, but has been known hitherto chiefly from the dentition. The Garman collection contains some portions of the skeleton, which are therefore of great interest. These specimens show that the genus has departed but little from the type of the family, *Oreodon*, but present, nevertheless, some important approximations to the ruminants.

The ulna and radius show no tendency to coalesce, and the former has the shaft considerably more reduced than in *Oreodon*. The radius differs in many ways from the ordinary oreodont type; the groove for the intertrochlear ridge of the humerus is narrower, the inner flange of the head smaller and less oblique, the outer larger and more concave, and the upward projection from the anterior edge much better developed, almost as in a true ruminant. The shaft is broader and more flattened, the walls much thinner, and the medullary cavity larger. The distal end is less expanded and thickened, and the tendinal sulcus barely indicated. The facets for the scaphoid and lunar are very distinctly separated; the former is shaped much as in *Oreodon*, but more deeply incised and more obliquely placed; between the two is a very deep notch, which penetrates from the posterior side through nearly half the thickness of the radius. This notch is indicated in *Oreodon*, but is not nearly so deep. The only bone of the manus which is preserved is the magnum, the shape of which, however, shows that it has moved entirely beneath the scaphoid, and has a deeply concave facet upon its ulnar side which embraces the side of the lunar almost in a semicircle. No facet for the second metacarpal is to be seen upon the radial side of the magnum, whence it follows that the third metacarpal was in contact with the trapezoid, and that an adaptive reduction of the manus had commenced, which, except in *Merycochœrus*, is unknown in other oreodonts.

Of the tibia only the distal end is preserved, and this portion differs but little from that of *Oreodon;* the astragalar facets are somewhat more deeply grooved and of more unequal size, and the fibular surface is deeper, as if the distal end of the fibula had commenced to wedge itself between the tibia and the calcaneum. The pes is higher and more slender than in *Oreodon*, but shows few important changes. The middle and external cuneiforms are united, but the limits of the two elements are plainly shown by the step cut in the distal surface. Metatarsal II. occupies the whole of the distal surface of the mesocuneiform, and abuts against the side of the ectocuneiform, while metatarsal III. is confined to the latter alone. *Merychyus* thus presents the curious condition of an adaptively reduced manus and an inadaptively reduced pes. The metatarsals are relatively very long and slender, more so than in any other member of the entire family, though far from reaching the elongation seen in the true ruminants; the lateral digits are especially slender, though proportionately as long as in *Oreodon*. The phalanges are likewise long and slender, but the unguals are still plainly of the true oreodont pattern.

Another specimen of this species is the skull of a very young animal with the milk dentition, which shows some interesting differences from that of *Oreodon*. In the latter genus, as in the Tragulina and the older selenodonts generally, the third upper milk molar, d. 3, is of a triangular shape, having only the posterior crescents developed, with the anterior portion elongated and trenchant, while *Merychyus* agrees with the true ruminants in the fact that this tooth is like a permanent molar, consisting of four crescents.

Of all known oreodonts *Merychyus* is perhaps the one which most closely approximates the true ruminant type. This is apparent in the elongated and more or less prismatic crowns of the true molars, in the increased size and complexity of pm. 2, in the character of the milk dentition, in the structure of the long bones of the skeleton with their large medullary cavities and thin walls, as well as in the adaptive method of reduction assumed by the manus.

MERYCOCHŒRUS.

Merycochœrus cenopus, Scott.

The type of this species is the specimen consisting of a beautifully preserved manus and pes contained in the Garman collection from the Loup Fork of Nebraska, which unfortunately are not associated with teeth. It is therefore possible that they may belong to some already described species, though they do not agree well in size with any of them.

The foot structure of this genus has been briefly noticed by Cope, who states that the feet are tetradactyle, and that "the os magnum is entirely beneath the scaphoid, and there is a distinct trapezium. The posterior foot is constituted as in *Eucrotaphus*." (Proc. Am. Ass. Adv. Sci., 1884, p. 484.) The manus in *Merycochœrus* agrees much more closely with that of *Merychyus* than with that of any other genus of the family. The carpus is higher in proportion to its breadth than in *Oreodon*, and very much higher as compared with the height of the metacarpus, thus giving the manus very different proportions in the two genera. When the individual carpal bones are compared, we find many differences of detail. The scaphoid has lost its cuboidal shape and become higher, narrower, and deeper (antero-posteriorly); the proximal surface has a convex anterior ridge which is very oblique, rising to a high point on the ulnar, and dying away on the radial side. The distal surface is anteriorly much narrower than in *Oreodon*, broadening however behind; the magnum facet is much larger, and the trapezoidal smaller and more lateral. The trapezium appears not to have been in contact with the scaphoid. The lunar is very peculiar, especially in the great downward prolongation of the beak-like process, which is wedged in between the unciform and magnum, and almost reaches the metacarpals. The proximal surface has not the simply convex shape seen in *Oreodon*, but rises high towards the ulnar side, and is much depressed on the radial. The distal surface is occu_pied by the long concave and obliquely placed facet for the unciform, that for the magnum being altogether lateral. This is the culmination of a tendency already noticeable in *Protoreodon* of the Uinta formation, the earliest known member of the family, and more marked in *Oreodon*, namely, the movement of the magnum away from the lunar and under the scaphoid. In *Merycochœrus* (and *Merychyus*) the lunar does not rest upon the magnum at all, touching it only laterally. The cuneiform is much like that of *Oreodon*. The pisiform is very different from that seen in the earlier genera of the family, and shows a tendency to assume the form characteristic of the pigs, though relatively much larger than in those animals. Compared with that of *Oreodon*, it is shorter, heavier, and especially much more expanded at the free end. No trapezium is preserved in connection with this specimen, and as no facets for it are clearly distinguishable on the other carpals, it may not have been developed. The trapezoid is very different from that of *Oreodon*, in being very much higher, nar_rower, and deeper; the facet for the scaphoid is oblique and almost as much posterior as superior; behind, the bone is drawn out into a projecting process, not abruptly truncated by the facet for the trapezium, as it is in *Sus*. The sig_nificant characters of the trapezoid are shown by the distal surface, which is constituted as in the pigs, having a large facet for mc. II. and a small one for mc. III.; in the pig the two facets are of nearly the same size. The magnum is very peculiar; as Cope has shown, it lies entirely beneath the scaphoid and internal to the lunar; its proximal surface is occupied by a large, slightly con_vex facet for the scaphoid, very different in shape from the same facet in *Oreo_don*, as it lacks the abruptly rounded posterior rising; the ulnar side is even more deeply concave than in *Merychyus*, encircling the convex lunar. The unciform differs but little from that of *Oreodon*, except that the proportions of the proximal facets have changed, that for the lunar being considerably the larger.

The metacarpals are relatively much shorter and broader than in *Oreodon*, the lateral digits are somewhat reduced, though not very much, while the me_dian ones have greatly increased in thickness. In proportions the metacarpus is quite like that of *Sus*, though as in all the oreodonts the keels of the distal trochleæ are confined to the palmar surface. Mc. II. is short, stout, and com_pressed; it articulates by a narrow surface with the trapezoid, but is excluded from the magnum. Mc. III. is very suilline in appearance, but its proximal end is not much extended transversely; on each side of the magnum surface is a facet for the trapezoid and unciform, the latter considerably the larger, while in the pig they are of nearly equal size. Mc. IV. is of about the same breadth and thickness as mc. III.; its proximal end is transverse, as in the pig, not oblique, as in *Oreodon*. Mc. V. is not preserved in the specimen, but the facet for it on the unciform shows that its head was flatter than in *Oreodon*, and that it did not rise so much upon the external side of the unciform.

The phalanges resemble those of *Oreodon*, except for their greater stoutness, and, as compared with the metacarpus, their greater length, for the three pha-langes of the fourth digit are together as long as the metacarpal. The unguals are of the same general shape in the two genera, but broader, more depressed, and with the ends less pointed in *Merycochœrus*.

The pes in the species of *Merycochœrus* from the John Day and Deep River beds differs less from that of *Oreodon* than does the manus, but the species be-fore us appears to show an important departure from that type. The astragalus is shorter, broader, and more massive than in *Oreodon*, and the distal trochlea has a broader surface for the cuboid. The calcaneum is not preserved. The cuboid is low and broad; the surface for the astragalus is broader than that for the calcaneum, reversing the proportions seen in *Oreodon;* the calcaneal surface is also of a different shape, as it does not project outwards, and its external mar-gin is straight, not rounded; unlike the pigs, this facet is not notched on its outer margin. The astragalar surface is not so deeply concave as in the White River genera, and another difference lies in the presence of a broad shallow groove which separates the articular surface into anterior and posterior portions. The peroneal sulcus is shallow. The distal end is almost entirely taken up by the large facet for mt. IV., that for mt. V. being very small and more lateral than distal; in *Oreodon* it is entirely distal. The navicular does not differ sufficiently from that of the earlier genera to require description. The ento-cuneiform is relatively large, and in general resembles that of *Oreodon*, but has a larger bearing upon mt. II. The ento- and meso-cuneiforms are missing, but they were doubtless ankylosed together as in all the other members of the family.

As a whole, the tarsus has changed in an opposite sense to the changes in the carpus, having become lower and broader, while the carpus has become narrower and remarkably high.

The metatarsus is suilline in general appearance ; the median digits are short and massive, while the laterals are reduced, especially in length, being not only proportionally but absolutely shorter than in *Oreodon Culbertsoni*. Mt. II. has an exceedingly small surface for the mesocuneiform, but the head is not oblique as in *Sus*. Mt. III. has a minute facet upon the tibial side of the head, which appears to encroach upon the mesocuneiform; and if this is the case, we have here the beginnings of an adaptive reduction of the pes, which is not known to occur in any other member of the family. Except for its heavier propor-tions, mt. IV. is like that of *Oreodon;* mt. V. has a smaller, more concave and obliquely placed facet for the cuboid than in the latter genus.

Merycochœrus and *Merychyus* thus agree with each other, and differ from other oreodonts in which the foot structure is known in the adaptive reduction of the manus, and it is interesting to note that this adaptive method has been independently assumed in several distinct lines of artiodactyles, e. g. the true ruminants, the pigs, and the camels. A study of the oreodonts shows that they are not closely connected with any existing artiodactyles, and it is difficult to see how the same result could be so often reached independently, unless it be the effect of the similar mechanical conditions to which the extremities are subjected.

MEASUREMENTS.

	Merycochœrus cenopus.	*Merychyus elegans.*
	m.	
Carpus, height	.030	
" breadth	.044	
Lunar, height	.025	
" breadth proximal end	.014	
Metacarpal II., length	.049	
" " breadth proximal end . .	.008	
" III., length	.064	
" " breadth proximal end . .	.019	
" IV., length	.057	
Phalanges of IV. digit, length	.057	
		m.
Astragalus, height	.039	.027
" breadth	.023	.014
Metatarsal II., length	.050	.056
" III. "	.062	.067
" " breadth proximal end . .	.015	.009
" IV., length	.066	
" V., "	.051	

SUIDÆ.

DICOTYLES.

Several species of peccaries have been described from the Loup Fork beds.

p^4 m^1 m^2

FIGURE 6. — Fragment of mandible of
Peccary × ⅔.

The Clifford collection contains two jaw fragments, apparently of different species. One of these differs from existing species in the fact that the last lower premolar is of much simpler construction than the molars, and more perfect specimens would probably show that this represents a distinct genus; but it would be premature to propose a name for it in the absence of more complete material.

GELOCIDÆ.

BLASTOMERYX, COPE.

This genus of true ruminants is abundantly represented in the collection. The type species, *B. gemmifer*, Cope, is from the Loup Fork, and differs from the closely allied *Cosoryx* chiefly in the brachyodont dentition. The later described species from the John Day formation not improbably belong to *Palæomeryx*, from which *Blastomeryx* is distinguished by the absence of the characteristic fold on the lower molars, and the greater narrowness and compression of the molar crowns.

THE SKULL.

A little of the superior wall of the cranium is preserved in one of the specimens, which probably belonged to a young animal, as the horn is a mere rudiment. The frontals are extended back of the orbits and form a considerable part of the cranium, but they are shorter than in *Cariacus*. Distinct though not prominent ridges converge from the back of the orbits, and probably unite behind in a sagittal crest, though as this part of the cranium is broken away the existence of a sagittal crest cannot be certainly affirmed. If present at all, it must have been a mere indication. The orbits are large, and have sharp superior borders. In *Cosoryx* and *Antilocapra* the horn arises directly over the orbit, and the same is probably true of the John Day species of *Blastomeryx;* but in the Loup Fork species of the latter genus the base of the antler has shifted its position somewhat, so as to spring from the posterior portion of the orbit, and it is also directed obliquely backwards, which apparently is the beginning of a process which results in the position of the pedicels observed in *Cariacus*. So much of the frontals as is preserved shows no trace of any sinuses, only the ordinary diploetic structure of the cranial bones. The bases of the antlers are much farther apart than in the deer, and are not connected by any intervening ridge. The coronal suture is nearly straight. As usual in ruminants, the parietals have coalesced into a single large bone, which clearly makes up most of the roof of the cranium. In the anterior portion the supra-orbital ridges are carried over from the frontals and converge to a point. Only the anterior part of the parietal is preserved. The inferior surface of this, and of the frontals as well, is deeply channelled by the winding and complex cerebral convolutions. Of the *dentition* we possess only a superior molar, and several inferior molars of a smaller species. The upper molar is very cervine in structure. The crown is brachyodont, and nearly as broad as long, while in *Cosoryx* it is strikingly narrow as well as hypsodont. The valleys are deeper and the external crescents more flattened than in *Palæomeryx*, while the internal crescents are somewhat simpler. The cingulum has almost disappeared, but a small basal pillar occurs between the inner lobes, as in many deer. The lower molars have low and rather narrow crowns; the valleys are shallow, and disappear after a comparatively short time of attrition. The cingula are but faintly indicated. As compared with the molars of *Cariacus*, those of *Blastomeryx* are simpler in the uncomplicated inner crescents of the upper teeth and the shallower valleys.

THE SKELETON.

The *scapula* is much like that of *Cosoryx*, though with some difference, the neck is more contracted, the coracoid more prominent, and the acromion more overhanging. The glenoid cavity is small, nearly circular in shape, and quite deep; the anterior or coracoid border is thin and curved, the glenoid border much thickened and nearly straight. The spine is not very high and divides the blade into unequal fossæ, the postscapular being much the larger, and the acromion overhangs the neck, but does not nearly reach the margin of the glenoid cavity.

The *humerus*. The proximal end of this bone is not preserved in any of the specimens; the shaft is rather short and slender, and shows a distinct sigmoid curve; an indistinctly marked deltoid ridge runs for some distance down the shaft. The distal end is moderately expanded and thoroughly cervine in appearance; the inner condyle is much the wider, and the intercondylar ridge is sharp and prominent. The anconeal fossa is deep and narrow, but does not perforate the bone. A moderate internal tuberosity forms a downward projection at the postero-internal angle. The ridges for muscular attachment are but feebly developed.

The *radius* is entirely distinct from the ulna, no co-ossification between the two occurring at any portion of their length. The proximal end is much expanded, as this bone carries nearly the entire weight of the fore limb, and covers the whole of the distal end of the humerus, the ulna being confined to the posterior aspect. The groove for the intercondylar ridge is deep, and emarginates the anterior and posterior edges. Two small facets for the proximal end of the ulna occur on the posterior side, the inner one very small, the outer larger and quite deeply concave. The shaft is long, slender, and considerably flattened, forming in section a transversely directed oval. The distal end is expanded and thickened, and is deeply grooved in front by the tendinal sulcus. On the external side there is a strong and roughened extension, which fits into a corresponding depression in the side of the ulna. The facets for the carpus are separated by a strongly defined ridge, and are placed very obliquely to the axis of the bone. · That for the scaphoid is deeply concave in front and as markedly convex behind, and is continued well up on the posterior side of the bone. This portion has the greatest antero-posterior diameter. The lunar facet is smaller and less deeply incised. External to it is a small oblique surface which articulates with the cuneiform.

The ulna is much reduced, though still retaining its independence. The olecranon is rather short and much compressed, though of considerable fore and aft depth. As the radius has usurped the entire distal end of the humeral trochlea, the sigmoid notch of the ulna is shallow, and the internal radial facet has become minute, though the external one forms quite a protuberance. The shaft is exceedingly slender and compressed; for most of its length hardly more than a thread of bone. The distal end is somewhat expanded, though very small, and deeply excavated on the inner side for the protuberance of the radius. The cuneiform facet is saddle-shaped, and sends down a well marked process on the outer side.

The *pelvis*, so far as it is preserved, is much like that of *Cosoryx*, though the ilium has a longer neck, and was apprently even less everted than in that genus. The ischium is rather short.

Little of the *femur* is preserved. The head is small and compressed, and rises little above the ridge connecting it with the great trochanter. The rotular trochlea is very broad, with rounded and somewhat more prominent internal edge; the outer edge is lower and sharper. The condyles are rather small and widely separated; above the inner one there is a small but distinct plantaris rugosity.

The *tibia* has a large trihedral head, with large external and small internal surfaces for the femoral condyles, and prominent bifid spine. The cnemial crest is very well developed, and just posterior to it on the external side is a very deep tendinal sulcus. The shaft is quite long and stout, with oval section and broad distal end. The surfaces for the astragalus are deeply incised, and the external one is somewhat the larger. The tongue is broad and thick, corresponding to the breadth of the groove in the astragalus. The internal malleolus is very long, and forms a tongue-like projection from the antero-internal corner.

The *fibula* is as completely reduced as in any ruminant. The proximal end is ankylosed with the tibia, where it forms a short sharp process. The distal end is not represented in any of the specimens, but from the structure of the tibia it is plain that it was a small nodule wedged in between the distal end of the tibia and the fibular process on the calcaneum. Between the two distal fibular facets of the tibia is a groove for the reception of the rudimentary shaft.

The *carpus* is like that of recent deer; the bones of the proximal row are high and narrow, those of the distal row low and broad. The scaphoid is deep antero-posteriorly, and broader in front than behind, where it is much narrowed by the great lateral extension of the lunar; the proximal surface is directed very obliquely backwards and inwards, and is deeply incised so as to form a very firm interlocking joint with the radius; this facet is divided into a strongly convex anterior portion, and an as strongly concave posterior portion. The lunar is curiously shaped; it is broadest in front and behind, and contracted in the middle; the anterior surface is transverse, the posterior very oblique. The radial surface is directed obliquely inwards parallel to that of the scaphoid. The distal surface is divided nearly equally between the magnum and the unciform, which meet at a very open angle, and the " beak " is barely indicated.

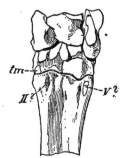

FIGURE 7. — Carpus of *Blastomeryx*, nat. size, posterior view.

The inner face of the lunar is nearly vertical, the outer very oblique, as the upper part of the bone is considerably wider than the lower, corresponding to the reduction in width of the proximal portion of the cuneiform. The latter bone has a broad distal surface and small saddle-shaped facet for the ulna, which is prolonged well down upon the postero-external surface. The cuneiform rises above the level of the lunar, and presents a small oblique surface, which articulates with the radius. The pisiform facet is high and very narrow. The trapezium is not preserved in any of the specimens, but its presence is demonstrated by a small facet on the postero-internal angle of the trapezoid. It was obviously very small and did not reach the scaphoid. The trapezoid and magnum have coalesced; the proximal surface of the compound bone is mostly occupied by the scaphoid, the facet for which is low and concave in front, rising behind to a low broad convexity; the distal surface is nearly flat. The unciform is narrower and higher than the trapezo-magnum ; its proximal surface is divided obliquely into facets for the lunar and cuneiform. The unciform projects below the level of the trapezo-magnum, and so presents upon the radial side a small facet for the corresponding projection of mc. III.

The metacarpus consists of the III. and IV. metacarpals, which have coalesced to form a cannon-bone, and the II. and IV., which are very slender, styliform bones, perhaps interrupted in the middle of the shaft. The cannon-bone is more slender than the anterior one of *Cosoryx* figured by Cope (Am. Nat., 1881, p. 547). The proximal surface is unequally divided, considerably the larger part belonging to mc. III., which projects above the level of mc. IV., and so comes into contact with the unciform. This arrangement occurs also in *Dremotherium* (see Gaudry, Enchainements, Fig. 142), and to a much less degree in *Antilocapra*. The shaft of the cannon-bone is broader and flatter proximally, becoming narrower and more rounded distally, and the distal trochleæ are completely encircled by sharp keels, as in existing ruminants. On the posterior side of the proximal end are two small facets, probably for the heads of the lateral digits. At all events, much of the shafts of the metacarpals II. and V. were preserved in the shape of very slender and compressed splint bones.

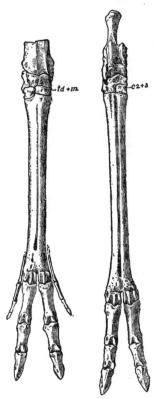

FIGURE 8. — Manus and pes of *Blastomeryx* × ⅔.

The phalanges are long and slender, and so asymmetrical as to produce a decided convergence of the toes. The proximal ends of the first row are deeply grooved for the keels of the metapodials, but are not emarginated in front as are those of the recent Pecora. The ungual phalanges differ from those of the deer and antelopes only in their greater slenderness. The phalanges of the lateral digits are of about the same proportionate size as in existing *Cervidæ*.

The tarsus is also cervine in character, and differs little from that of *Palæomeryx (Cervus) Flourensianus* as figured by Fraas (Fauna von Steinheim, Taf. VIII. fig. 24). The astragalus is high, narrow, and deeply grooved, and the distal end shows hardly more than an indication of the ridge which passes between the navicular and cuboid. The calcaneum is long and much compressed, though with considerable depth, antero-posteriorly in the lower third (in *Palæomeryx* the calcaneum is thicker and more rounded); the cuboidal facet is narrow, pointed in front and quite concave from before backwards; the sustentaculum is short, but broad and thick, and the fibular facet is high, but short and narrow. As compared with the calcaneum of recent deer, that of *Blastomeryx* has less antero-posterior diameter below, and a somewhat less thickened tuberosity at the free end, quite unimportant differences. The cuboid and navicular are firmly co-ossified, as in *Palæomeryx* and all existing Pecora ; these bones are quite low, and the navicular rises but little in front to fit the distal groove of the astragalus, but on the postero-internal side it sends up a strong and high process, which makes the astragalar facet very deeply concave ; distally the navicular shows two facets, a large one for the compound cuneiform, and a much smaller one for the entocuneiform. On the distal surface of the cuboid, besides the large facet for mt. IV., is seen a minute, oblique infero-lateral one, obviously for a rudimentary fifth digit. The only other tarsal bone preserved in the specimens is the compound cuneiform, which is rather low, narrow, and deep ; in front it is nearly on a level with the cuboid, but behind descends somewhat below it, and thus affords a lateral attachment to mt. IV. The presence of a distinct entocuneiform is demonstrated by the facets for it upon the navicular and cannon-bone.

The metatarsus presents some features of much interest. Rosenberg (Zeitschr. f. wiss. Zool., Bd. XXIII.) has shown that in the sheep embryo there are at one stage four complete metatarsals ; he states, however, that the lateral ones are ultimately absorbed. In *Blastomeryx*, as in *Amphitragulus*, and probably all existing Pecora and Tylopoda, there are at least three elements which enter into the formation of the poste-rior cannon-bone ; viz. mt. III. and IV., and the proximal por-tion of mt. II. The latter, though ankylosed with mt. III., shows its limits distinctly ; it has a small facet for the entocunei-form, and ends below in a point. Mt. V. was obviously present, as upon the postero-external side of the cannon-bone there is a shal-low groove, and upon the cuboid, as already stated, there is a small

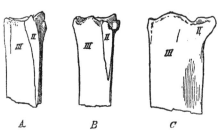

A   B   C

FIGURE 9. — Proximal end of posterior cannon-bones × ⅔, internal view ; A. *Amphitragulus* ; B. *Blastomeryx* ; C. *Antilocapra*.

facet for the head of it. An examination of the metatarsus of a modern rumi-nant seems to show that the portion articulating with the entocuneiform is the head of mt. II. ; whether mt. V. be also present is more difficult to decide, but in existing forms there is no portion which can be identified with it, while in *Blastomeryx*, though undoubtedly present, it does not coalesce with the cannon-bone. There is nothing in any of the specimens to indicate that any portion of the distal ends of the lateral metatarsals were retained, though doubtless phalanges were preserved, as in the deer.

In this brief, but fairly comprehensive review of the osteology of *Blastomeryx*, we have seen nothing which can be opposed to the view expressed by Professor

Cope, that this genus should be placed in the ancestral line of the distinctively American deer. *Alces, Tarandus,* and *Cervus* are really immigrants from the Old World, and do not belong in this category; but the truly American types, of which *Cariacus* is the chief example, have a peculiar skull structure, first pointed out by Garrod, which seems to show that the American deer were separated from those of the Old World at a comparatively early date, though it is very questionable whether both series could have independently acquired the extraordinary peculiarity of the deciduous antler.

COSORYX, LEIDY.

Cosoryx furcatus, LEIDY.

This very interesting animal is represented in the collections of Garman and Sternberg by several specimens, which enable us to add materially to the descriptions hitherto published. These descriptions are so brief that the relationships of this genus have been very generally misunderstood. Schlosser says: " In Nordamerika finden sich im oberen Tertiär zwar Geweihe von Hirschen, Kiefer derselben sind indessen noch nicht mit Sicherheit ermittelt, wenigstens nähern sich die Gebisse des *Dicrocerus* Cope, *Merycodus* Leidy, *Cosoryx* Marsh, zweifellos eher den Antilopen, besonders dem lebenden nordamerikanischen Genus Antilocapra, als den Hirschen. Sie sind zugleich viel einfacher gebaut und schliessen sich namentlich die Marken sehr bald, was bei den Hirschen erst in einem ziemlich späten Stadium der Abkauung auftritt. Das Gleiche dürfte wohl auch der Fall sein bei *Blastomeryx* Cope = *Cosoryx gemmifer,* trotzdem Cope denselben als Stammvater von *Cervus* und *Cariacus* betrachtet. Die von Marsh behauptete Existenz von Seitenzehen bei *Cosoryx* dürfte wohl mit Recht bezweifelt werden; die Hand hat Cope abgebildet und zeigt dieselbe keine Spur von etwaigen Griffeln. Wenn auch die systematische Stellung dieser Formen noch nicht völlig klar gelegt erscheint, so können wir doch mit grosser Wahrscheinlichkeit annehmen, dass wir hier einen eigenen Seitenzweig der Ruminantier vor uns haben, als dessen letzter Rest die merkwürdige nordamerikanische Gabelantilope zu betrachten ist. Die Verästelung des Geweihes ist bisweilen fast so stark wie bei den echten Hirschen. Wahrscheinlich war es von Hornmasse überzogen — den verwachsenen Haaren des Bastgeweihes. Für diese Annahme spricht die auffallende Glatte der von Cope und Leidy abgebildete Geweihfragmente." (Morph. Jahrb., Bd. XII. p. 70.)

As we shall see, some of these inferences are probably quite correct, others are equally probably misleading. This group of closely allied species is not confined to the "upper Tertiary" or Loup Fork beds, but appears first in the lower middle Miocene of Oregon in the John Day beds, where the genus *Blastomeryx* is abundantly represented by some large species. Now *Blastomeryx* is, so far as we can at present determine, almost identical with the type variously named in Europe *Palæomeryx* and *Dremotherium,* about the only difference of importance being the absence of the characteristic "Palæomeryx fold" on the lower molars. *Cosoryx* is very closely allied to *Blastomeryx*, and is distinguished from it chiefly by the much more hypodont molars. The bones of the various Loup Fork species of this genera cannot be distinguished apart in the absence of associated teeth, and it is quite probable that the John Day species of *Blastomeryx* will prove to belong to a different genus from the Loup Fork species.

THE DENTITION.

One undoubted specimen of *Cosoryx* contained in the Cambridge collection consists of a fragment of the superior maxillary containing one molar, the lower jaw with first and third molars, an antler, the sacrum, all the lumbar and the five posterior dorsal vertebræ in unbroken succession, the scapula, humerus, pelvis, and posterior cannon-bone. The resemblance of these bones to *Antilocapra* is very striking, and fully justifies what Schlosser has said with regard to the relationships of the two genera. The second upper molar is not much extended in the antero-posterior direction, and has a fairly high crown, though not hypsodont to the same degree as in the prong-buck; the median fold of enamel on the external wall, or, more properly speaking, the projecting anterior horn of the postero-external crescent is less strongly developed than in the recent form, and the corresponding horn of the anterior crescent hardly projects at all. The valleys are shorter and wider than in *Antilocapra*, and though the tooth is in an advanced state of wear, they are still quite deep, in contrast to what occurs in the lower molars. The lower incisors and canines are all broken away, but from the alveoli and remaining fangs it may be seen that they were of the ordinary ruminant pattern, probably not very long; they decrease in size from the median incisor outwards, and the canine is the smallest of the series. The premolars, three in number, are represented only by their alveoli, which shows them to have been very small. The most anterior is implanted by a single root, the others by two. Leidy's figure (*Merycodus necatus*) shows them to possess considerable complication, but they are less molariform and more trenchant than in *Antilocapra*. The true molars are more truly hypsodont than in the upper jaw; the first is very small, but the third resembles that of the modern genus exceedingly closely.

The same may be said as to the form of the mandible itself; the horizontal ramus is very long, compressed, and rather shallow, and with an extremely long diastema between the canine and premolar 3; the ramus is less rounded on the external side than in *Antilocapra*, and in that genus there is no such descent of the upper margin in front of the premolars as occurs in *Cosoryx*. The symphysis is short (much shorter than in *C. trilateralis*, Cope) and much contracted, and on a level with its posterior edge is a large single mental foramen. The antler is branched like the one figured by Leidy with the name of *Cervus Warreni*, but with a much longer beam, and the tines meeting at a more open angle. The beam is longer and the tines shorter than in any of the antlers figured by Cope, except, perhaps, the imperfect specimen named *Cosoryx* (*Dicrocerus*) *teres* (Wheeler, Pl. LXXXII. fig. 6). The antler is composed of dense bone, with a smooth and here and there furrowed surface, a texture which, as Schlosser has remarked, is very different from that of a deer antler. The burr is very large and prominent, but a vertical section shows that the beam passes into the pedicel without any perceptible break or change in the tissue. In *Cosoryx* the burr is very variable, as may be seen from Cope's figures. In this collection are some specimens without any burr, others with a single burr, and some with two or three. They can hardly be regarded as an evidence that the antler was deciduous.

THE SKELETON.

The *vertebræ*, so far as they are preserved, resemble very much those of *Antilocapra*. Owing to the fact that only the posterior part of the column is preserved in the specimen, it will be most convenient to describe them from behind forwards. The only caudal represented is like the second of the prong-buck, but a little more complete, and clearly shows that the tail was short, as may also be inferred from the sacrum. This caudal is short and narrow, especially in front, with short wide transverse processes near the posterior end. There are a pair of rudimentary prezygapophyses, and an exceedingly minute neural canal, which will just allow the passage of a needle, and a corresponding neural spine. In the prong-buck the second caudal has neither canal nor spine, and the transverse processes are wider.

The *sacrum* consists of four completely ankylosed vertebræ. The first has a broad depressed centrum, well developed prezygapophyses, and much enlarged pleurapophyses, which occupy most of the sacral surfaces of the ilia. The spine is coalesced with the others into a high and arched ridge. In the prong-buck the spines are more distinct. The other sacrals have expanded pleurapophyses, but only the second has any contact with the ilium. The centra decrease rapidly in size from the first posteriorly, and that of the last is exceedingly depressed and thin. The whole sacrum is quite strongly arched from before backwards. The lumbar region is quite long, and consists of six vertebræ, which are slenderly constructed; the centra are anteriorly comparatively narrow and trihedral in section, posteriorly they are broader and more depressed. The spines are low and comparatively broad, and are inclined well forward, with concave anterior borders. The transverse processes on the first lumbar are short, depressed, but comparatively broad; these processes lengthen as we pass backwards, but are very slender as compared with those of *Antilocapra*, and the neural spines are lower than in that genus. The zygapophyses are of the interlocking cylindrical type usual among artiodactyles, and there are no metapophyses. We may infer with considerable confidence that the number of dorsal vertebræ was thirteen; on this assumption, the most anterior dorsal of this specimen is the ninth. In this the centrum is short and trihedral in section, with the inferior border sharp and arched from before backwards; the spine is rather short, and directed very obliquely backwards; the transverse processes are short and slender, and have well marked facets for the tubercles of the ribs; the prezygapophyses are flat and placed on the pedicels of the neural arch, and, separated from them by a short interval, arises a pair of small metapophyses. The tenth is the anticlinal vertebra; the spine is at first very oblique, but curves, and in its upper portion is vertical. In other respects this vertebra is like its predecessor. On the eleventh the spine is directed slightly forwards, but the end is rounded like that of the anterior dorsals; the metapophyses have approached the median line so as to touch the post-zygapophyses of the tenth, while the post-zygapophyses of the eleventh have assumed the cylindrical shape found in the lumbar region. The twelfth and thirteenth vertebræ are much like lumbars in their construction, and are distinctly longer than the three antecedent vertebræ; the spines have the nearly straight thickened free ends seen in the lumbars, and the metapophyses have disappeared. The transverse processes, however, are very short, though they still retain the rib-facets, even on the thirteenth.

The *ribs,* so far as can be judged from the fragments, are narrow and very slender. Of course this may be true only of the posterior part of the series.

The *scapula* is characteristically ruminant. The glenoid cavity is nearly round and quite shallow, the coracoid process is prominent, recurved and thickened at the end; the neck is very long and much contracted, the borders sloping away from it very gradually; the coracoid border is thin and rounded at the edge, it curves gently forwards and upwards from the neck; the glenoid border is very much thickened and somewhat overhanging, from the neck it is nearly straight, and forms a right angle with the very thin suprascapular border. The spine rises abruptly from the neck into the high acromion; the latter overhangs very slightly, in sharp contrast to the condition found in *Antilocapra.* The spine divides the blade into unequal fossæ, the prescapular being much the smaller, as is ordinarily the case among the ruminants. Except for the nearly straight inferior edge of the spine, and the consequent lack of an overhanging acromion, this scapula very closely resembles that of the prong-buck.

The *humerus* has a broad and flattened head, which projects but little beyond the shaft. The external tuberosity is large, and curves over the deep bicipital groove; the internal tuberosity very small; both are much less developed than the corresponding processes in *Antilocapra.* Proximally the shaft is broad and compressed, below it is rounded and slender. No ridges for muscular attachment are more than very faintly indicated. The distal end is broken away, but in all probability it was like that of *Blastomeryx* described above.

The *pelvis* is also entirely ruminant in character. The ilium has a short, deep, and much compressed neck, expanding into a curved and strongly everted plate, which projects a considerable distance in front of the sacral attachment. The ilium is somewhat trihedral in section, the median rounded ridge of the plate being more prominent, and the expansion itself smaller than in the prong-buck. The ischium is very long; above the acetabulum its superior border shows the convexity so usual in the recent ruminants, though in a less marked degree. The tuberosity of the ischium is very long and prominent, and directed straight outwards; behind the tuberosity the ischium is prolonged further than in the prong-buck. The cannon-bone belonging to this specimen is broken, and its proximal end obviously diseased, so that it does not merit description; the only fact of importance which it shows is the comparative slenderness of the bone.

So far as the material will enable us to judge, the feet of *Cosoryx* differ in no important respect from those of *Blastomeryx*, and the same statement applies to the long bones of the limbs.

RESTORATION OF COSORYX FURCATUS.

(See Plate I.)

This drawing is made from the specimen already described, completed by fragments of others, while the feet are drawn from *Blastomeryx;* the cervical vertebræ are represented only by the axis, the others being conjectural, as are also the anterior dorsals. The skull is taken chiefly from that of the closely allied European genus, *Palæomeryx*, and from specimens of the large *Cosoryx teres*, Cope, belonging to the Smithsonian Institution. The fortunate association of the mandible in the same specimen with the vertebræ, pelvis, scapula, etc., gives a very useful standard as to the length and character of the skull, position of the molars, etc. It may be assumed with some confidence that the drawing gives a fairly accurate representation of the animal. Marsh's account of the feet of *Cosoryx* shows that they were constructed much like those of *Blastomeryx*. In general appearance *Cosoryx* seems to have had the same light, graceful build as *Antilocapra*, but with a very different skull and deer-like antlers. The proportions of the limbs also differ somewhat, the hinder cannon-bone being considerably longer than the fore, while in the prong-buck they are of nearly the same length. *Cosoryx* was a much smaller animal, the bones are all more slender than in *Antilocapra*, and the carpal and tarsal bones are much higher and narrower proportionately.

The view held by Cope that *Cosoryx* is the ancestor of *Antilocapra* is very probably the true one. So far as the dentition, the vertebræ, and the limbs are concerned, the differences between the two genera are only such as might be expected to occur between a Miocene and a recent ruminant. A distinction of some importance, however, consists in the character of the horns. In *Cosoryx* they are branched, but probably not deciduous antlers; in *Antilocapra*, a core with a horny sheath, which, however, differs strikingly from the horn of the typical Cavicornia. But the unique branched horn of *Antilocapra* not improbably indicates, as has been suggested by Cope, a remnant of a former branching of the bony core itself, and so this difference does not preclude a genetic connection between the two forms. In *Cosoryx* the antler was almost certainly covered with skin; its smooth surface, as Schlosser points out, shows that it could not have been naked, as in the true deer.

Both *Blastomeryx* and *Cosoryx* are probably to be derived from the species referred to the former genus which occur in the John Day beds, but there is no form yet known in the White River which could have given rise to these John Day ruminants. The latter are most probably descended from some *Palæomeryx* of the Old World, which migrated to this continent. The very close connection between these American genera and the *Amphitragulus, Dremotherium,* etc. of St. Gérand le Puy is obvious from the most superficial comparison.

The collection contains specimens probably indicative of other species of *Cosoryx,* some of them much larger than *C. furcatus;* but in the absence of associated teeth, it is not possible to refer them to their proper categories.

PERISSODACTYLA.

ANCHITHERIIDÆ.

MESOHIPPUS, MARSH.

THE BRAIN.

Mesohippus had a large and well convoluted brain. The length and breadth indicate that it weighed about one third as much as the brain of the recent horse, while if we estimate the body weights of the fossil and recent animals by the relative size of the humeri, the brain of the Miocene species was proportion-

ally heavier. The cerebrum of the horse is, however, much more highly convoluted, and the frontal lobes are relatively broader. The *Mesohippus* brain is distinguished in a marked manner by the longitudinal direction of the parietal and occipital sulci, and by the deep transverse frontal sulci, as contrasted with the oblique sulci of all recent ungulates. In fact, in this respect it bears a marked general resemblance to the brain type of recent Carnivora, and conforms with the higher Ungulata of the Eocene.

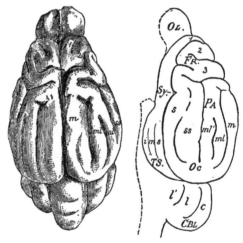

FIGURE 10. — Brain of *Mesohippus Bairdii* × ⅔. From above, and from side.

On either side of the longitudinal fissure is a long deep fissure forking anteriorly and marking off the median gyrus, *m*, of the parieto-occipital region. Parallel with this is a short fissure, which separates the two medilateral gyri, *ml, ml'*. The third fissure extends to the posterior transverse, and thus entirely separates the supersylvian gyrus, *ss*, from the medilateral. The fourth fissure is shallower. There are three transverse frontal fissures (FR. 1, 2, 3) which divide this lobe into three gyri; the median fissure extends almost to the longitudinal fissure, and suggests the crucial sulcus of the Carnivora. The sylvian fissure is very shallow. The temporo-sphenoidal lobe is very prominent, and is divided into three gyri (*s*, *m*, *i*) by two sulci. Beneath the third frontal gyrus is a vertical sulcus, parallel with the sylvian.

The cerebellum has a large central lobe with transverse simple furrows.

THE DENTITION.

There are a few new points to be noted in regard to the teeth of *Mesohippus*, which bear upon the dentition of the horses in general, and are clearly shown in

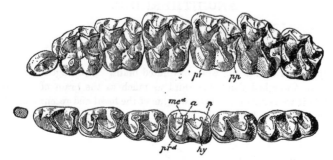

FIGURE 11. — Superior and inferior molars of *Mesohippus Bairdii* × ⅓.

a series of unworn crowns of the upper and lower jaws. Scott has already pointed out that the incisors in this genus are simple, there being no indication of the infolding of the enamel, such as is seen in *Anchitherium aurelianense*. In some of the John Day species of *Anchitherium* the enamel is not infolded, as observed in the lower jaw of a specimen referred to *A. equiceps*, Cope.

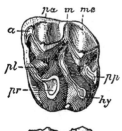

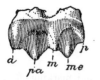

FIGURE 12. — Superior molar of *Anchitherium longicriste* × ⅓. Superior and external view. Cope collection.

The upper molars of *Mesohippus* clearly show the first step in the formation of the *posterior pillar*, *pp*, which is so conspicuous a feature in *Anchitherium*, in the posterior valley. This can also be observed in a still simpler stage in a specimen of *Anchilophus* from the French Phosphorites. Step by step with the development of this cusp appears the *posterior pillar*, *p*, in the lower molars, behind the entoconid; this accessory cusp can be traced back to the teeth of *Epihippus*. When it finally unites with the entoconid, in *Hipparion*, it forms the posterior twin cusp (*b*, *b*, Rütimeyer), which is analogous to the anterior pair formed by the union of the metaconid and *anterior pillar*, *a* (*a*, *a*, Rütimeyer).

Thus the transition from the *Mesohippus* to the *Anchitherium* molars is very gradual, as shown in the accompanying figures. By tracing back the rise of the eleven elements which compose the upper *Equus* molar, we find that six belong to the primitive sextubercular bunodont crown. Two elements of the ectoloph, the *anterior pillar* and *median pillar*, rise from the simple primitive basal cingulum of the *Hyracotherium* molar; the same mode of development, we have just seen, is true of the *posterior pillar*. The eleventh element, the fold of the postero-external angle of the crown, *p*, is not prominent until we reach *Equus*. The term "posterior pillar" is taken from Lydekker; the other terms, "median" and "anterior," are applied to parts which have an analogous origin from the basal cingulum. The remaining coronal cusps are readily identified with their homologues in the primitive tritubercular molar.

? Anchitherium parvulus, Marsh.

(Syn. *Equus parvulus*, Marsh.)

Among the Loup Fork specimens collected by Clifford are found two lower molars, m_1 and m_3, which are almost identical in size with those of *Mesohippus Bairdii*. The crown of m_1 measures: antero-posterior, .011 m.; transverse, .009 m. Unlike the *Mesohippus* molars, there is no external cingulum. The "posterior pillar" has the same degree of development as in *Anchitherium*. The fangs are separate. There is no trace of cement. Marsh has described a diminutive horse (*Equus parvulus*), estimated at two feet in height, from the same beds, and it is highly probable that these teeth belong to this species. The generic reference is of course very uncertain. The brachydont crowns point either to *Merychippus* or *Anchitherium*, but the stage of development of the coronal pattern approximates most closely that in the latter genus, being a little more advanced than in *Mesohippus*.

RHINOCERIDÆ.

ACERATHERIUM.

The Manus and Pes.

The characteristics of the pes of *Hyracodon* from the lower White River beds have been fully enumerated by us.[1] They are principally as follows: cuboid not supporting astragalus anteriorly; lateral digits reduced and not spreading; ectocuneiform not articulating laterally with mts. II. We may subsequently find that the feet of the later species of *Hyracodon* varied in some of these respects, although this is not probable, owing to the fixity of foot-types once established. We have, however, no present means of distinguishing between the *Metamynodon* and *Aceratherium* foot-bones.

On page 169 of the first Bulletin a high, rather slender tarsus was described,

[1] See Scott, E M. Museum Bulletin, No. 3, May, 1883, p. 19. Also, Osborn, Mammalia of the Uinta Formation, May, 1889, Part IV. "Evolution of the Ungulate Foot," p. 549.

which probably belongs to the *Aceratherium* of the lower beds. It differs widely in its proportions from other specimens found in this collection, which belong either to the *Aceratherium* of the higher beds, or to *Metamynodon*. The

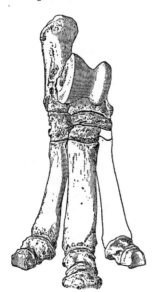

best preserved specimen of this second type (marked a^3) is comparatively short and broad, with spreading digits and rugose surfaces for muscular attachment (Figure 13). The proportions of the metapodials to the tarsals are similar to those in *Ceratorhinus*. The calcaneum has a powerful tuber; the ectal astragalar facet is very convex; the sustentaculum is narrow, and its oval facet is continuous with the inferior; the cuboidal facet is nearly horizontal. About one fifth of the astragalus rests upon the cuboid. The relations of the cuboid, navicular, and ectocuneiform repeat those observed in *Rhinocerus*. The mesocuneiform is very short, giving mts. II. a wide articulation with the ectocuneiform. The metatarsals are powerful, the lateral pair having approximately the same length as in *R. indicus*. This type of foot is related directly to that of *Aphelops*.

The manus and pes of a third specimen (marked a^6) show several interesting differences. In the pes, the metatarsals are of the same proportions,

FIGURE 13 — Right pes of *Aceratherium* × ¼.

but the calcaneo-cuboidal facet is oblique and narrow, resembling that in *Hyracodon*, and the sustentaculum is very small. The remains of the carpus show that the species to which this specimen belonged had a greatly reduced fifth digit, constituting a functionally tridactyl manus. The evidence for this is in the greatly reduced lunar-magnum facet, which is invariably characteristic of tridactylism.[1]

It may be noted here that among the carpals of *Titanotherium* there is a well preserved lunar, which has its magnum facet much reduced anteriorly, so there is little question that we shall yet discover a tridactyle species of the genus.

THE RHINOCEROS MOLARS.

The peculiarities of the molars of *Aphelops* will be made more clear by a few observations upon the molars of the rhinoceroses in general. The three main crests of the lophodont crown may now be distinguished in part by terms which express their homologies with the elements of the sextubercular superior and quadritubercular inferior molars of the primitive ungulate, *Phenacodus*. In the upper molars, the outer crest is formed by the union of the primitive paracone

---

[1] See Osborn, Mammalia of the Uinta Formation, p. 567. It is possible that these feet belong to *Metamynodon*.

and metacone, to which is joined the anterior pillar (see *Mesohippus*, p. 88); it may be called the *ectoloph*. As the anterior crest is formed by the union of the protocone, protoconule, and paracone, it may be termed the *protoloph*. The posterior crest, which unites the primitive metacone, the metaconule, and the hypocone, may be termed the *metaloph*. The

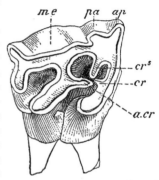

outer surface of the ectoloph in the primitive molar of the Rhinoceros is marked by three vertical ridges corresponding to its three primitive component elements, *me, pa, ap;* one or all of these disappear in the flattening of the surface. It will be observed that nothing corresponding to the 'median pillar' of the superior molar of the horse is developed. In the lower molars (the paraconid disappearing), the union of the metaconid and protoconid forms the anterior crest, or *metalophid*, while the hypoconid and entoconid unite to form the *hypolophid*.

FIGURE 14. — Superior molar of Rhinoceros (sp. indet.) × ½. After De Blainville.

The secondary enamel folds, which are developed from the three crests, bear a most interesting analogy to those observed in the horse series, beginning with *Protohippus;* they are outgrowths of the same regions of the crown and subserve the same purpose. They are moreover of like value in phylogeny.. The useful descriptive terms introduced by Busk, Flower, and Lydekker, should be adopted in part.[1] These secondary elements consist, first, of three folds projecting into the median valley, one from the ectoloph, the *crista;* one from the protoloph, the *crochet;* one from the metaloph, the *anticrochet.* Secondly, the ectoloph unites with the posterior cingulum and metaloph. Thus the anterior and posterior valleys may be cut off by the union of these folds into from one to three 'fossettes,' precisely analogous to the 'lakes' in the horse molar, except that they are not filled with cement.

The accompanying diagram is taken from a fossil molar figured by De Blainville. (Osteogr. Gen. Rhin, Plate XIII.) It is remarkable in exhibiting all the primary and secondary elements, for they are very rarely combined in a single tooth. Similar accessory folds are frequently developed in the lower molars.

[1] The terms 'protoloph' and 'metaloph' are, however, substituted for 'anterior collis' and 'posterior collis' of Lydekker. The term 'anterior pillar' = 'first costa,' and 'paracone' = 'second costa.' The mode of evolution of the 'pillar' must have been similar to that in the horses, where Lydekker has proposed this term for the 'posterior pillar.' It is very appropriate, because the pillars in their earliest development can be shown to rise independently from the cingulum (see *Mesohippus*, p. 88), and not as folds of the main elements of the crown, as we should infer from their fully developed stage.

APHELOPS, Cope.

The generic characters of *Aphelops* have been given by Cope as follows. Den-
tition, I. $\frac{2-1}{1}$, C. $\frac{0}{1}$, P. $\frac{4-3}{3}$, M. $\frac{3}{3}$; post-glenoid and post-tympanic processes
in contact but not co-ossified; digits, 3–3; nasals hornless. To these charac-
ters may be added: magnum not supporting lunar anteriorly; absence of the
'crista' and invariable presence of the more or less strongly developed 'crochet'
and 'anticrochet' in the superior molars.

The specific nomenclature of *Aphelops* is in confusion. The type of *A.* (*Rhinoceros*) *crassus*, Leidy,[1] is a last upper molar, which is closely similar to that of
A. megalodus; the characters of the milk molar associated with this type cannot
be used in definition.[2]   The penultimate upper molar, the type of *A. meridianus*, Leidy,[1] corresponds in the development of the two 'crochets' to the same
tooth in *A. fossiger*, Cope, but the posterior 'fossette' is not enclosed by the strong
cingulum as in the latter species.   *A.* (*Aceratherium*) *acutum*, Marsh, is identical
with *A. fossiger*.   *A. malacorhinus*, Cope, resembles *A. meridianus* in the open
posterior fossette and the development of the 'crochets.'   It is impossible, however, to clear up this synonymy without bringing the original types together
for comparison.   General characteristics of all these types are the invariable
development of the 'crochet,' absence of the 'crista,' usual development of the
'anticrochet.'   The specific names proposed by Cope are here adopted because
they are established upon a very complete knowledge of the skull as well as
of the teeth.

Aphelops fossiger, Cope.

Dentition: I. $\frac{1}{1}$, C. $\frac{0}{1}$, P. $\frac{4}{4}$, M. $\frac{3}{3}$.   First premolar simple, conical, sometimes
absent; nasals not overhanging premaxillaries; foramen lacerum medium
confluent with foramen ovale; occiput broad and low; limbs short and bulky;
molars with well developed 'crochet' and 'anticrochet.'

In the figure given by Marsh (Am. Journ. Sci., Oct., 1887, p. 3) and by Cope
(Am. Nat., Dec., 1879, p. 771 *e*), the third and fourth premolars have both the
'crochet' and 'anticrochet.'   There is some ground for the supposition that the
skull here described belongs to a different species, since the 'anticrochet' is not
developed in the premolars.   This reference is therefore provisional.

This is apparently the only species which is represented in this collection.
All the specimens are from Kansas, and include several skulls and well preserved bones from all parts of the skeleton, enabling us to give a complete
description and restoration of the animal.

---

[1] See Ext. Mamm. Fauna, Dak, p. 228.

[2] Cope has nevertheless employed the 'cristæ' developed in this milk molar in
his definition of *A. crassus*.   "On the Extinct Species of Rhinoceriidæ of North
America," etc., Bull. U. S. Geol. Survey, Vol. V. No. 2, p. 237.

THE BRAIN.

One of the most interesting features of *Aphelops* is the very large size of the brain. The walls of the cranium are solid. There are no vacuities or air-cells in the diploë of the mid-region of the brain-case, such as attain from 1 to 1½ inches in thickness in *Ceratorhinus*. Thus the brain is relatively much larger

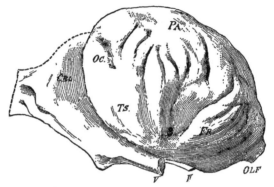

FIGURE 15 — Brain of *Aphelops fossiger* × ⅓. Lateral view of intracranial cast.

than that of the recent rhinoceros, and presents a marked advance upon that of *Aceratherium occidentale*. The bulk of the fore- and mid-brain, or the divisions in front of the cerebellum, is approximately as follows :—

   Aceratherium, 420 c.c.   *Aphelops,* 1240 c.c.   *Ceratorhinus,* 720 c.c.

The bulk of the entire brain is : *Aphelops,* 1470 c c.   *Ceratorhinus,* 850 c.c. The relative body weight of the two animals can be roughly estimated from a comparison of the femora as *Aphelops* 4, *Ceratorhinus* 3. It thus appears that the steady brain growth of the ungulates during the Eocene and early Miocene periods reached its highest point in some families of the later Miocene, and was followed by a degeneration.

The cerebellum in *Aphelops* is small and partly overhung by the hemispheres. The lateral view of the hemispheres shows a very marked predominance of transverse sulci, which radiate from the vertical sylvian fissure, *S*, so that in the basal view of the frontal lobes the fissures are antero-posterior. The dorsal surface of the cast is somewhat imperfect, giving an incomplete reproduction of the parietal and occipital regions. The superior

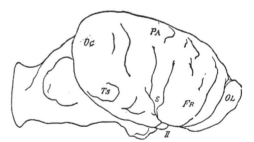

FIGURE 16 — Brain of *Ceratorhinus Sumatrensis* × ⅓. Lateral view of cast.

sulci of the frontal lobe are directed obliquely backwards to the longitudinal fissure, thus reversing the direction observed in the recent ungulates.

The Skull and Dentition.

The *skull* (Plate III.) is broad in relation to its length, owing to the shortening of the ant-orbital region and the recession of the nasals. The maxillaries spread very widely for the powerful series of molars, while the *premaxillaries* are slender. The orbit is placed above the first molar. The *nasals* are compressed anteriorly, and extend only so far as to overhang the premaxillary suture. A marked feature of the skull is that the upper surface is in a nearly straight line from the supra-occipital ridge to the tip of the nasals, while in *A. megalodus* it is concave. The orbit is very slightly overhung by the supra-orbital process. The zygomatic arch is deep vertically, but compressed laterally. The post-glenoid process is deep and narrow; it has contact with the post-tympanic of variable length. The remarkable feature of the post-tympanic is its extension into a broad flat plate behind the auditory meatus. The occiput is broad and low, and does not overhang the condyles; it is deeply cleft in the median line. On the base of the skull, the foramina rotundum and spheno-orbitale are confluent, as observed by Cope. The foramen ovale is either confluent with or separated by a slender ridge of bone from the foramen lacerum medium.

The *molars* and *premolars* are remarkable for the extreme flattening of the outer surface of the ectoloph, all trace of the three vertical ridges having disappeared. The first premolar is a simple conical tooth implanted by a single fang; it is apparently inconstantly developed, for Marsh makes no mention of it in his description of *A. (acutum) fossiger*. The inner angles of the protoloph and metaloph unite by the 'crochet' in pm^2 and pm^3 to enclose the median valley, as in *Aceratherium*. The fourth premolar resembles the molars except in the non-development of the 'anticrochet.' The true molars are characterized as follows:

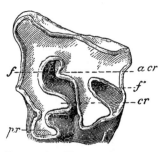

FIGURE 17.—First superior molar of *Aphelops fossiger* × ½.

by the constriction of the inner portion of the protoloph into a separate column; by the strong development of the 'crochet,' which in m^1 and m^2 unites early with the metaloph to enclose the anterior 'fossette'; by the development of the 'anticrochet' at the inner angle of the metaloph and ectoloph; by the complete enclosure of a posterior 'fossette' in the first and second molars.

The inferior molars are of the simple rhinoceros pattern, there being no trace of accessory folds. The first premolar is missing; the second is separated by a rather narrow diastema from the large lateral tooth. Between the pair of large semi-procumbent caniniform teeth are two small incisors.

The lower jaws are very massive, with a strongly arched lower border. The condyles are broad and elevated. The posterior border is broad, but not rugose.

THE SKELETON.
(Plate III.)

Vertebræ. — The atlas resembles that of *R. unicornis*, with extremely broad transverse processes. A well preserved axis has a low tuberosity representing the spine; there is some doubt whether this is the normal adult condition, although the absence of the spine would accord with the low occiput and hornless nasals. The cervicals 3–6 have deeply opisthocœlous centra, rather high and narrow in proportion, with powerful zygapophysial processes. The inferior lamellæ of the transverse processes project downwards and forwards, and · expand very slightly at the tip; the width of this lamella increases somewhat in C. 6; the superior lamellæ project opposite the vertebrarterial canal. The sixth and seventh cervicals apparently have slender elevated spines, in the remainder the spines are low or tuberous. The centrum of C. 7 is subcircular in front and broad posteriorly.

The dorsals are represented by a number of vertebræ in the mid-region. The centra are laterally compressed with distinct keels; the zygapophysial facets are very small and horizontal; the metapophyses are well developed. The length of the spines in the anterior dorsal region was apparently as in *R. javanus.* No lumbars are found in this collection.

Fore limb. — The *scapula* is very short and heavy. The general outline is triangular; the glenoid border is concave; the coracoid border is convex; the superior border rises to a point above the spine; the upper third of the spine shows a very stout recurved acromial process.

The *humerus* is remarkably short and heavy, and is distinguished by the unusually elevated position of the deltoid ridge, which is much higher upon the shaft than in the recent rhinoceroses. The tuberosities are heavy and sessile; the external condyle is unusually prominent. The *ulna* has a deep, powerful olecranon process and stout trihedral shaft, which is suddenly compressed inferiorly for the cuneiform articulation. The proximal and distal faces of the *radius* are subequal; the shaft is very slightly arched and closely united with that of the ulna, giving this segment a very massive appearance.

The structure of the *manus* is in keeping with the short and heavy upper segments; it is broader and more powerful than in any of the recent rhinoceroses. The three short, widely spreading digits are faced by rugose areas for the attachment of powerful muscles. Mtc. III. is much the largest; the lateral metacarpals, II. and IV., are short and directed outwards; the phalanges are short and wide, especially the distal series. As in all tridactyle forms the carpal displacement is extreme; the scaphoid covers the whole upper surface of the magnum anteriorly; the lunar is rather small, and rests anteriorly wholly upon the unciform; posteriorly the pivotal process of the magnum supports the lunar; the cuneiform is high and narrow. The trapezium is missing in both the carpal series before us, but is indicated by the usual facets upon mtc. II. and the trapezoid. The magnum is broad and quadrilateral. The unciform has an unusually wide mtc. III. facet, and is vertically compressed.

Hind limb. — There is a complete left innominate bone, which gives all the characters of the *pelvis.* The upper surface of the ilium, unlike that of *Cerato_rhinus,* is nearly flat. The supra-iliac border is evenly arched, and, as the ischial and acetabular borders are of approximately the same length, the ilium is unusually symmetrical. The ischium and pubis are in a plane perpendicular to that of the ilium; the pubic symphysis is short; the obturator foramen is an elongate oval. The tuber-ischii is not very prominent. The border extend_ing from the tuber to the symphysis is evenly rounded.

The *femur* is relatively longer and more slender than the humerus, having the form and proportions observed in *Ceratorhinus.* The great trochanter stands out widely; below this the shaft is of a broad flattened section; the lesser tro_chanter presents a long low ridge; the third trochanter is only half as promi_nent as in the recent rhinoceros, and is not recurved. The *tibia* is characterized by a marked asymmetry of the tuberosity; the internal malleolus is not promi_nent; the popliteal space is deeply excavated; the astragalar facets are shallow. The *fibula* is of the same proportions as in the recent rhinoceros.

The *tarsus* is unusually short and spreading. The astragalo-tibial facet is flattened laterally, and shows little fore and aft play; the ectal and sustentacu_lar facets are either confluent or slightly separate; the inferior is distinct and separate; the cuboidal facet is extremely broad. The cuboid is shallow, with subequal calcaneal and astragalar facets; posteriorly it articulates with both the navicular and ectocuneiform, anteriorly with the latter only; it has a very deep posterior hook. The presence of the entocuneiform is indicated by the articular facets for it. The mesocuneiform is narrow and deep. The ectocu_neiform is very broad; this bone and the navicular have the same proportions as in the rhinoceros. The middle digit is much the largest of the three, and Mts. III. has a considerable cuboidal facet.

The following measurements are made from specimens which belong to dif_ferent individuals, *a*, *b*, *c*, etc.; they therefore cannot be used in estimating the exact proportions of the different parts. The proportions have, however, been very carefully determined in the accompanying restoration of the skeleton.

MEASUREMENTS.

Skull.

		m.
Spec. *s.*	Total length, sagittal crest to end of nasals	.490
"	Breadth, outside zygomatic arches	.360
"	Depth, penultimate molar to top of cranium	.235
"	Occiput, diameter of, transverse, .268 m.; vertical	.198
"	From occiput to anterior end of orbit	.340
"	Antero-posterior, diameter molar-premolar series (pm. 115 m., m. 150 m.)	.265
"	Diameter first molar, antero-posterior .057 m., transverse . .	.070
	" second " " .068 " . .	.070
	" third " " .058 " .	.052

                                         m.   m.

Spec. *s.* Diameter fourth premolar, antero-posterior .045; transverse   .   .065

    "   "   third   "   "   .035   "   . . .050

    "   "   second   "   "   .028   ⟨ "   . . .032

              "   first   "   "   .017   . .017

    "   Lower jaw, length, angle to front of canine 470

              "   depth, tip of coronoid to inferior border 295

Vertebræ.

Spec. *h.* Atlas, greatest width, .356 m ; greatest depth 100

Spec. *pp.* Axis, greatest width, .18 m.; length of centrum 090

    "   "   "   depth, spine to base of centrum, estimated . . .140

Spec. *p.* Fifth cervical centrum, antero-posterior .074 m., vertical .068 m., transverse .076 m.

Spec. *o.* Twelfth dorsal centrum, antero-posterior .075 m, vertical .055 m., transverse .058 m.

Appendicular Skeleton.

Spec. *c.* Scapula, vertical diameter, approx., .295 m.; glenoid cavity, ant. post. 900

    "   Humerus, length of, .308 m.; breadth, head and tuberosity . . .155

Spec. *a.* Radius, length, .285 m.; breadth, proximal, .093 m.; distal   . .098

    "   Ulna, greatest length, .36 m.; sigmoid facet to cuneiform facet .295

    "   Carpus, greatest transverse diameter, .130 m.; ditto vertical   . .057

    "   Mtc. III., breadth . . .070 m.; length 116

    "   " II.   "   . . .043   " 100

    "   " I.   "   . . .040   " 092

Spec. *e.* Left innominate bone, diameter, antero-posterior 495

    "   Length of pubis, .185 m.; of ischium, .20 m.; of ilium 340

Spec. *f.* Femur, length of, .46 m.; diameter, head and great trochanter . .165

Spec. *g.* Tibia, length of, .37 m.; width, proximal 140

Spec. *q* and *r.* Tarsus, tuber calcis to distal facet of mts. III., approx. . .220

    "   "   transverse diameter 108

    "   Second metatarsal, length 088

RESTORATION.   (See Plate II.)

The restoration of *Aphelops fossiger* confirms Cope's statement that the proportions of the animal were rather those of the hippopotamus than the rhinoceros. The body was long, the chest deep, the limbs and feet short and massive, and supplied with powerful muscles. The skeleton is about 9 feet long and 4 feet 6 inches high. Thus *Aphelops* presented a wide contrast to its tall, comparatively slender predecessor, *Aceratherium*, of the lower Miocene. The increase in brain capacity shows that its nervous organization kept pace with its general muscular and skeletal development. We may infer that the extinction of *Aphelops* was due to climatic changes, rather than to any defects in its internal organization, because the brain, teeth, and feet are, in themselves, as adaptive as in any of the present persisting types.

COMPARISON WITH ACERATHERIUM AND RHINOCERUS.

There is nothing, however, which precludes the supposition that the American lower and upper Miocene Aceratheria are genetically related.

All portions of the skeleton of *A. occidentale* are now known to us, excepting the scapula, pelvis, and dorso-lumbar vertebræ; they indicate an animal in the same stage of skeletal evolution as the recent tapir; the proportions are practically similar; the displacement of the carpals and tarsals is in a corresponding stage. The mode of progression was also probably similar, for all the articular facets and protuberances for muscular attachment present innumerable points of resemblance. Cope [1] first pointed out the tapir resemblances in *Aceratherium*, especially in the separation of the foramina spheno-orbitale and rotundum ovale and foramen lacerum medium; the separation of the post-glenoid and post-tympanic; and the form of the femur. We have shown that this resemblance applies to the carpus [2] and tarsus; it is also true of the humerus and forearm, and of the atlas and axis. The remaining cervicals are widely different; it is probable, also, that the pelvis and scapula were different. This is of course simply an instance of functional and structural parallelism. It follows that an enumeration of the differences between the recent tapir and rhinoceros would also embrace the majority of the features which distinguish *Aceratherium* from *Aphelops*, for the latter is in most respects a fully developed rhinoceros.

Thus, if the descent from *Aceratherium* to *Aphelops* took place, it was accompanied by wide-spread modifications of the skeleton. In *Aphelops megalodus* we find a probable transition species. Its proportions are more intermediate. The narrow elevated occiput, the less degree of separation of the foramina of the skull, the lophodont character of the first upper premolar, the small development of the 'anticrochet' in the superior molars, — these characters all point towards *Aceratherium*.

A. fossiger is a highly modified form, with its broad occiput, simple first premolar, and confluent cranial foramina. In many respects the modifications it exhibits are simply steps towards the recent rhinoceros type; for example, its tridactylism, the extreme displacement of the podials, and the characters of the spinal column. But there are many points in which *Aphelops* differs from the recent rhinoceroses; namely, the sub-triangular shape of the scapula, the very elevated position and sessile character of the deltoid ridge of the humerus, the spreading manus, the oval obturator foramen, and the comparatively feeble development of the third trochanter. The marked peculiarity of the upper molars is the development of both the 'crochet' and 'anticrochet,' and absence of the 'crista.' This combination is very distinctive, since all the living rhinoceroses present combinations of the 'anticrochet' and 'crista.' [3] The molars of *Aphelops*

---

[1] Bull. U. S. Geol. Surv., Vol. V. No. 2, p. 235. Also, "On Extinct American Rhinoceroses and their Allies," Am. Nat., Dec., 1879, p. 771 c.

[2] Osborn, "Evolution of the Ungulate Foot," Mem. Uinta Mamm., p. 550.

[3] See Flower, "On some Cranial and Dental Characters of the Existing Species of Rhinoceroses," Proc. Zoöl. Soc., 1876.

resemble in this respect those of *R. tichorhinus*. Briefly stated, in all living forms the protoloph is simple,, and the accessory folds are developed, first from the metaloph, then from the ectoloph ; while in the known extinct American forms the ectoloph is simple, and the protoloph develops a fold to which a fold of the metaloph is sometimes superadded.

In view of these facts, together with the numerous divergences in the skeleton, there is strong corroboration for the opinion advanced [1] by Scott in 1883, that *Aphelops* should not be regarded as ancestral to any of the recent foreign species, but rather as the last known of an extinct American series. The question is still an open one whether its distribution was confined to this continent.

CHALICOTHERIOIDIA.[2]

CHALICOTHERIUM, KAUP.

Specimens of this genus are rare in American formations, and have not as yet been reported from the Loup Fork. Marsh [3] has mentioned the occurrence of it in the John Day Miocene of Oregon, and in view of the discoveries of Forsyth Major and Filhol, it is altogether probable that the foot-bones from that formation, which Marsh has referred to the Edentata under the names *Moropus distans* and *M. senex*,[4] belong to the same genus. A third species of the same genus is announced by Marsh [5] from the Loup Fork, *M. elatus*, which is probably represented in the Garman collection from the Loup Fork of Nebraska.

Chalicotherium elatum? MARSH.

(Syn. *Moropus elatus*, Marsh.)

The specimen is a portion of a right superior maxillary containing the third and fourth premolars and the first molar. The premolars have a flattened ectoloph connected by two convergent crests, with a large internal cone which is cleft at the summit ; the base of this cone is surrounded by a strong internal cingulum. The ectoloph is worn by two symmetrical incisions alternating with the transverse crests in the third premolar, but in the fourth these incisions are asymmetrical. The first molar is partly of the Titanotherium type, with its

---

[1] E. M. Museum Bulletin, No. 3, 1883, p. 17.

[2] Gill, Arr. of the Fam. of Mammals, Smithsonian Misc. Coll., No. 230, p. 271. This order was properly defined by Gill, but was erroneously placed among the Artiodactyla, owing to the reduced condition of the superior incisors. Filhol's forthcoming memoir upon the Mammals of Sansan will probably enable us to determine its phylogenetic relations.

[3] American Journal of Science and Arts, 3d Series, Vol. XIV. p. 362.

[4] Ibid., pp. 249, 250.

[5] Ibid., pp. 250, 251.

protocone isolated, but the hypocone, a: all known Chalicotherioids, is united with the metacone by a low ridge (met..; i).

FIGURE 18.   C...clar   *Chalicotherium elatum* × ⅓.

The available figures and descriptions re   imperfect that the relationships of this species to those of the Old \or!1 cannot be definitely made out. It is, however, decidedly smaller than t!: *lotherium*).

		m.		m.
Third premolar, antero-posterior diameter.		.024 ;	transverse,	.025.
Fourth " " "		.025 ;	"	.028.
First molar, "		.036 ;	"	.033.

Fold out or Map
Here

protocone isolated, but the hypocone, as in all known Chalicotherioids, is united with the metacone by a low ridge (metaloph).

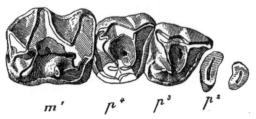

FIGURE 18. — Superior premolars and first molar of *Chalicotherium elatum* × ⅔.

The available figures and descriptions are so imperfect that the relationships of this species to those of the Old World cannot be definitely made out. It is, however, decidedly smaller than that which occurs at Pikermi (*Ancylotherium*).

MEASUREMENTS.

			m.		m.
Third premolar, antero-posterior diameter,			.024 ;	transverse,	.025.
Fourth "	"	"	.025 ;	"	.028.
First molar,	"	"	.036 ;	"	.033.

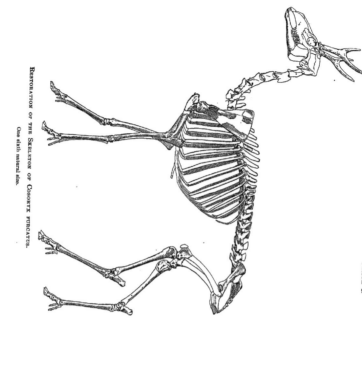

RESTORATION OF THE SKELETON OF COSORYX FURCATUS.

One sixth natural size.

PLATE I.

RESTORATION OF THE SKELETON OF APHELOPS FOSSIGER.

One twelfth natural size.

PLATE III.

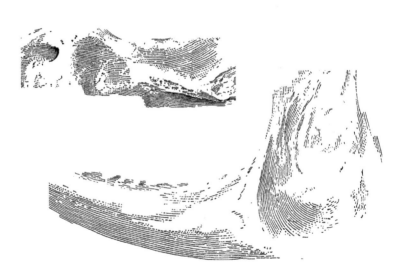

SKULL OF APHELOPS FOSSIGER.

One sixth natural size.

No. 4. — *Cristatella: the Origin and Development of the Individual in the Colony.* By C. B. DAVENPORT.[1]

I. Introduction.

AT the suggestion of Dr. E. L. Mark, I began, in the spring of 1889, the study of fresh-water Bryozoa. While at the Laboratory of the United States Fish Commission, at Woods Holl, Mass., where, through the kindness of Mr. A. Agassiz, I had the opportunity of spending the

[1] Contributions from the Zoölogical Laboratory of the Museum of Comparative Zoölogy, under the direction of E. L. Mark, No. XIX.

VOL. XX. — NO. 4.

following summer, I gathered most of the material for this study. I found an excellent place for collection in Fresh Pond, Falmouth, where Fredericella and Plumatella were also gathered. Upon my first visit to this pond (July 5th), I found at its outlet Cristatella exceedingly abundant on the leaves of the pond-lilies. A month later, the same locality yielded very few specimens; but about September 5th I found them plentiful again, and at the same time noticed the phenomenon described by Kraepelin and by Braem, — that some of the statoblasts of Plumatella had already hatched. Colonies of from five to twenty individuals were observed with the two halves of the statoblast still adhering to their bases. A few colonies of Cristatella were also gathered in the latter part of August from Trinity Lake, New York.

The material collected was killed with a variety of reagents. Cold corrosive sublimate gave the best results. In staining, I always found Czoker's cochineal the most satisfactory dye for the study of the embryonic cells of the bud.

As Haddon ('83, pp. 539-546) has reviewed the most important part of the bibliography of budding in Phylactolæmata which had been published at the time of his writing, I shall be relieved from giving here any extended historical account of the earlier researches. The contributions of Nitsche ('75) and Hatschek ('77) are well known. Reinhard has published a preliminary article ('80ª, '80ᵇ) on this subject in the Zoologischer Anzeiger; but his two more important papers ('82 and '88) I have unfortunately not seen. Braem's ('88, '89ª, and '89ᵇ) three preliminary papers concerning budding in fresh-water Bryozoa correct some erroneous statements of Nitsche, and support Hatschek's view of the origin of the polypide. The results at which I have arrived concerning this last problem are similar to those of Braem, but his work has apparently been done chiefly on Alcyonella, mine on Cristatella. Finally, I believe there will be found in this paper something new on the organogeny, which Braem does not seem to have especially studied, and which may be of general morphological importance. For these reasons, it has seemed to me desirable that I should publish my observations and conclusions, and I am the more inclined to do so because our views are not in all points the same.

In the matter of nomenclature, my studies have not led me to a final conclusion as to the homologies of the axes of the individual, and therefore I fall back by preference on non-committal terms. The individual is bilaterally symmetrical. Parts nearer the mouth end of a line joining mouth and anus (i. e. nearer the margin of the colony) will be designated "anterior" or "oral"; parts nearer the anal end, "posterior" or "anal." To parts nearer the roof of the colony will be applied the term "superior," or "tectal"; to those nearer the sole, "inferior." Parts situated at either side of the sagittal plane of the individual are "lateral," and either right or left, — the individual facing the margin of the colony. In naming organs, I have preferably used the terms employed by Kraepelin ('87). I adopt the term polypide simply because it is a convenient name for a number of organs closely united anatomically, and arising from a common source embryologically.

II. Architecture of the Colony.

The colony of Cristatella, as is well known, consists of a closed sac, which is greatly elongated in old specimens, and has a flattened base or "sole," and a convex roof. The wall of this sac is known as the wall of the colony or cystiderm (Kraepelin). Suspended from the dorsal wall, and hanging in the common cavity of the colony, which may be called the *cœnocœl*, are to be seen numerous polypides in different stages of development. A more careful observation shows that the polypides lying nearest the median plane of the colony are the largest and oldest, those nearest the margin, conversely, smallest and youngest (Plate I. Fig. 1). All young colonies of Cristatella have been derived from one of two sources, eggs or statoblasts. According to Nitsche ('72, p. 469, Fig. 1), there are two polypides of the same age first developed in the cystid, which is a product of a fertilized ovum, and regarding these he fully agrees with Metschnikoff's ('71, p. 508) statement, "Die beiden Zooiden entwickeln sich wie gewöhnliche Knospen."

Nitsche ('75, pp. 351, 352) observed that in Alcyonella the primary polypides are placed with their oral sides turned from each other, and that the younger buds arise in the prolongation of the sagittal plane of the older polypides, and from that part of the cystid lying between the œsophagus of the older buds and the margin of the colony.

As Braem ('89b, pp. 676–678) has shown, there is but one primary bud in the statoblast embryo. The younger buds formed in the statoblast arise on the oral side of the primary bud.

In Cristatella, says Braem ('88, p. 508), the newly hatched statoblast embryo already exhibits to the right and left of the adult primary polypide two nearly complete daughter individuals of unlike age, which are generally followed by two other sisters in the same relative positions, and a fifth in the median plane, — oral with respect to the mother bud. These buds may produce new ones until the whole colony has attained the size of a pea; then young buds arise analwards of the primary polypide, and as the margin of the colony is protruded on each side of this point, the colony becomes heart-shaped. The two upper lobes of the heart are regions of great reproductive activity; they separate from each other, and thus transform the heart-shaped colony into an elongated one. Through the heaping together of buds effected by this process, a misproportion between the area (Flächenraum) and the circumference of the colony results, and the buds, which lie in longitudinal rows, soon come to be crowded. After this, they each give rise to only two daughter buds, a lateral and a younger median one.

To these observations of Braem I have little to add. I have figured (Plate X. Fig. 88) a young colony of Cristatella, containing about thirty polypides. This was taken in the latter part of July, and is probably an egg colony. My reasons for thinking so are, that the statoblasts of the preceding year form colonies in the early spring; that statoblasts of any year have never been seen, like those of Alcyonella, to hatch in the fall; and that there are, occupying the centre, two polypides of very nearly equal size and development, and probably therefore of nearly equal age. Surrounding these are eight younger individuals, nearly equal to each other in size, and these are in turn followed by two generations, of thirteen and seven individuals respectively, — the last generation evidently being as yet incomplete.

As Kraepelin ('87, pp. 38, 139, 167) clearly states, the Cristatella colony is comparable with those of Pectinatella, Plumatella, etc., and may be derived from them by imagining a condensation of those branching colonies. The radial partitions seen in Figure 88, *di sep. r.*, Plate X., are thus homologous with the lateral walls of the branches of a Plumatella colony; and just as in the latter, so here young individuals arise near the tips of the branches, and the older individuals degenerate. As in Plumatella, young individuals are produced not only distad of older, but also laterad, thus founding new branches, so in Cristatella we find young buds having the same positions. These facts will be better appreciated by a reference to Figure 1, which shows a portion of the margin of a mature colony. It is here clearly seen, (1) that, as has long been known, the youngest individuals are placed nearest to the margin, and that therefore, as one passes towards the centre, one encounters successively older and older individuals; and (2) that, as Kraepelin ('87, Fig. 134) has already figured, the older individuals are arranged in a quincunx fashion.

The bit of the margin figured may be regarded as typical, not only on account of its symmetry, but also because of the fact that the youngest individuals are placed at the normal distance from the margin. Al-

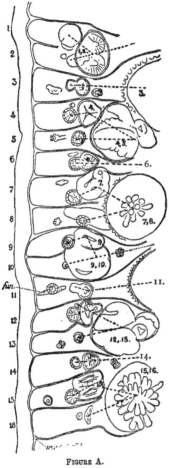

though I have seen these conditions repeated in enough instances to assure me of their normal nature, yet, owing to a crowding of polypides, both among themselves and to the margin of the colony, and also to the consequent displacement of polypides, the appearances which I am about to describe are often obscured.

First, the interrelations of the individuals included within compartments 1–8 are exactly repeated in compartments 9–16. The same repetition holds true for the remainder of this side of the colony. On the opposite side, the number varies from six to eight. At the ends of the colony, owing to crowding of individuals, it is difficult to count with accuracy. Since all individuals are derived from preceding ones, the conclusion seems reasonable that the inhabitants of these eight branches were derived from a common ancestor. It is interesting that from each of these ancestors the same number of branches and an almost equal number of individuals are produced, and that the corresponding individuals in each of these families, e. g. Figure A, 4, 5 and 12, 13, and 7, 8 and 15, 16, are similar in

FIGURE A.

position, and of the same stage of development.

Secondly, most individuals figured have given rise to two individuals; some, on the contrary, to but one. Of the two individuals produced, one (the older) passes into a second (new) compartment, and so forms a new branch. The younger, however, remains in the ancestral compartment, and thus continues the ancestral branch. See, e. g., individual

4, 5, of Figure A. The buds which give rise to new compartments may be called lateral buds, in accordance with Braem's terminology; those which prolong the ancestral branch, median buds. Where only one individual arises, it is a median bud. These conclusions regarding the relationship of buds are based solely upon the length of the radial partitions, the inner extremities of which correspond to the angle formed by two branches in branching genera like Plumatella.

Thirdly, while the lateral buds, Figure A, 4, 5, and 12, 13, give rise directly to new buds, median buds of the same or younger age, 6, 14, have moved to a considerable distance from their mother buds before giving rise to new individuals. The effect of this is, that the median bud comes to lie, not alongside of the lateral bud, but in a quincunx position relatively to it.

Fourthly, lateral buds (branches) may arise from either side of the budding individual. Although most of the branching in the part of the colony figured in the cut is to the right, yet the youngest lateral buds are being given off to the left. So in compartments 4, 6, 7, 12, the funiculus indicating the point where the median bud will arise.

To recapitulate : The descendants of common ancestors are arranged similarly in the same region of the colony; a lateral and a median bud may arise from a single individual, the first forming a new branch, the latter continuing the ancestral one; median buds migrate towards the margin before producing new buds; and new branches are formed on either side of the ancestral branches.

III. Origin of the Individual.

Two essentially different views of the origin of the polypide in the adult colony of Phylactolæmata have been maintained within recent years. The first is that advanced by Nitsche ('75, pp. 349, 352, 353), and adopted by Reinhard[1] ('80[a], p. 211, '80[b], p. 235). According to these authors, the outer of the two layers of the colony-wall gives rise, either by a typical or a potential invagination, to the inner cell layer of the bud, — the layer from which the lining of the alimentary tract and the nervous system both arise, — and pushes before it the inner layer

---

[1] Reinhard says in his preliminary article, "Meiner Meinung nach entwickelt sich die Knospe in Folge einer Verdickung des Ectoderms, in welche dann die Zellen des Entoderms eindringen," but Brandt's abstract of the paper read by Reinhard before the Zoölogical Section of the Russian Association, places entoderm for ectoderm, and *vice versa*, — a rendering more in accordance with Reinhard's statements in the context.

of the colony-wall, which thus becomes the outer layer of the bud. Hence the buds arise independently of each other.

The second view is that advanced by Hatschek ('77, pp. 538, 539, Fig. 3). He asserted that in Cristatella "Die Schichten der jüngeren Knospe stammen von denen der nächst älteren direct ab." Finally, Braem ('88, p. 505) agrees essentially with Hatschek, and believes that a typical double bud, although it does not always appear, is the fundamental condition. His preliminary account clearly shows that precisely the same condition of affairs, except in so far as modified by the less metamorphosed condition of the ectoderm, exists in Alcyonella as in Cristatella.

A. OBSERVATIONS.

1. *Origin of the Bud.* — The result of my own work has been to lead me to a conclusion differing from both of these two views, but more like the second than the first. By my view, as well as by Braem's, Nitsche's two types of single and "double" buds are united into one. I would not say, with Hatschek, that the two layers of the younger bud arise directly from those of the next older, but that each of the corresponding layers of the younger and next older buds arises from the same mass of indifferent embryonic tissue. In some cases, each of the layers of the daughter polypide does arise from the corresponding layers of the very young mother bud. In other cases each of the two layers out of which the two layers of the older bud were constructed contributes cells to form the corresponding layers of the younger bud, but the cells thus contributed have never formed any essential part of the older bud. All gradations between these two types occur. For convenience' sake, we may always call the older polypide the mother; the younger polypide, the daughter. Figure 3 (Plate I.) shows a well advanced bud (Stage VIII.) which consists of two layers of cells, an inner, *i.*, composed of a high columnar epithelium arranged about a narrow lumen; and an outer, *ex.*, of more cubical cells. In a region (I) on the bud which is near the attachment of its oral face to the body-wall there is a marked evagination of the contour, caused in part by a thickening of the outer layer, and in part by a slight increase in the diameter of the inner. This thickening o the wall is the first indication of the formation of a younger bud, which is to arise at this place. Figures 22, II., 16, VI. (Plate III.), and 11, VI. (Plate II.) show later stages of buds originating in the same manner as that of Figure 3. The mother bud has grown larger, as has also its lumen. The outline in its upper oral region has become much folded as

a result of cell proliferation, and a deep pocket has been formed lined by a layer of cells which are still a part of the inner layer of the mother bud. The outer layer of the latter has also been protruded by the activity of the inner layer, and its cells go to form the outer layer of the young bud. Still another point is to be observed. The centre of the young bud has moved away from the centre of the neck of the mother bud, and thus the former lies nearer to the margin of the colony than the latter. Figure 17, VII. (Plate III.) shows a still more advanced stage in the development of the bud, in which it is sharply separated from its parent, but its inner and its outer layers are still in direct continuity with the inner and outer layers respectively of the mother.

I have selected this series from the many which might have been chosen to show the origin of the polypide, because it is an intermediate type between two extremes, and because by it the other cases receive an easy explanation. All cases of budding, however, seem to conform to this general law : the greater the difference in age between the youngest and the next older bud, the greater the distance between the points at which they begin to develop. Thus the typical case of a "double bud" is that in which two buds appear to arise at the same time. They originate, as Nitsche observed, from a common mass of cells. A case of two buds, one only slightly younger than the other, is seen in Figure 5. By comparing with Figure 3, in which the older polypide is older than VII., Figure 5, the difference between the younger buds will be apparent. On the other hand, Figure 4 illustrates a comparatively late formation of the younger bud. The older bud had attained a stage corresponding to Figure 18 (Plate III.), but the younger bud is not older than that seen in Figure 22, II. Just as in the latter case the two layers of the older bud went respectively into those of the younger, so in the present case a direct continuity can be traced between the cells of the inner and outer layers respectively of the younger and older buds. The evidence that the cells composing the inner layer of the young bud have not arisen directly from the ectoderm is derived not only from the continuity of both cell layers of the two buds, but from the presence of the apparently unmodified ectodermic cells lying above the inner layer of the young bud, and sharply marked off from it. Figure 6 shows a later stage of this same type, in which the layers of the young bud are seen well formed, but still very sharply separated from the overlying ectodermal cells. These series afford an interpretation of the extreme type of budding shown in Figure 2, which is not uncommon. The mother polypide has reached a stage corresponding to Figure 18, Plate III. To the left of the neck of the polypide and towards the margin is the funiculus, *fun.* Between it and the neck of the polypide the cœlomic epithelium is thickened and its cell boundaries have become evident. Directly above this region, and immediately above the muscularis, is a row of cells, which stain deeply and show other evidences of being embryonic. These are directly continuous with the neck cells of the older polypide, exactly as was the case with the cells of the inner layer in Figures 4 and 6 (Plate I.). In fact, they are in every way comparable with these. Figures 8 and 9 (Plate II.) show slightly later stages. The funiculus has moved farther from the parent bud, the future outer layer (*ex.*) has become thicker, and its cells are columnar and sharply marked off from each other. The inner layer (*i.*) of the new individual is represented by a thicker, stolon-like mass of cells, which is in direct continuation with the inner layer of the mother bud, from which it was doubtless derived.

A stage which, on account of the greater distance of the funiculus from the older polypide, I believe to be slightly more advanced than Figure 8, is shown in Figure 9. In the section drawn, the inner cell mass (*i.*) exhibits few nuclei, but they are more numerous in adjacent sections. The band of protoplasm connecting this young bud with the mother is perceptibly smaller than in the preceding stage. The cells of the inner layer form a mass sharply marked off from the ectoderm; those of the outer layer are greatly thickened, as in the last stage.

A peculiar thickening of what I regard as Nitsche's "Stützlamella" takes place between the young bud and the mother polypide. This is shown in Figures 9 and 10 at *mu.* It is not stained by Czoker's cochineal, and the circular muscle fibres, here cut transversely, are very conspicuous in the midst of it. As I have noticed this appearance only in the cases of young buds which have originated like those of Figures 9 and 10, and of others of about their age, (and it is in these buds and at this age that migration from the older polypides takes place,) I believe that there is some connection between this condition of the muscularis and the disturbance which such a migration must cause.

I have already (page 106) referred to the fact that in some cases median buds are found far removed from the mother polypide, although in an early stage of development. Stage IV. is the youngest in which such buds have been found.

The cells of the mass destined to form the inner layer of the bud multiply rapidly after they have reached a proper position, and there is considerable protrusion of the cœlomic epithelium into the body cavity. The fact, that during its extensive migration the bud increased only slightly, whereas it now begins to develop rapidly, leads to the presumption that it has ceased to migrate, and has come to a state of comparative rest, relative to the surrounding ectodermal cells.

One of the first indications of further development is seen in the arrangement of the cells of the inner layer, which is such that their nuclei come to lie near the surface of a hemisphere whose convex side is turned toward the cœnocœl. The beginning of this process is seen in Figure 10 (Plate II.), and, further progressed, in Figure 11. Figure 14 exhibits a still later stage in the development of the polypide. In the lower portion of the two-layered sac of this figure a separation of the cells (*lu. gm.*) has begun. This is the first indication of the atrium.

In all cases there exists at this stage a condition of the ectoderm like that shown in Figures 5 and 14. The absence of ectodermal cells directly over the bud may be accounted for by supposing that they have come to lie upon, and form part of, the neck of the polypide. While it would be impossible to deny that they *might* migrate through the cells composing the neck of the polypide, and thus come to form the nervous elements, a careful study of the successive stages figured will not show the slightest evidence of any such migration, nor is it *a priori* probable, from what is known of the action of epithelium the world over, that such a migration would occur.[1]

According to the description of Nitsche ('75, p. 353), there is a lumen in the bud of Alcyonella (where the ectoderm is much less metamorphosed than in Cristatella), which is always in direct communication with the outer world, the bud having been formed by a typical invagination. Braem ('88, pp. 506, 507), however, states that he has never seen in Alcyonella this communication of the bud cavity with the outer world. In the much more obscure process of polypide development in Gymnolæmata the lumen first appears after the cells of that mass from which the bud is to arise have arranged themselves in two concentric layers. In Endoprocta, according to Nitsche ('75, p. 374) and Seeliger ('89, p. 179), the lumen arises by a virtual or actual invagi-

---

[1] Since writing the above paragraph I have cut some sections of Plumatella in which this process is much clearer, owing to the absence of secreted bodies in the ectoderm. Instead of a few ectodermal cells dropping down upon the upper part of the neck of the polypide, as is the case in Cristatella, there is a cup-shaped invagination of the ectoderm, which is quite deep, and thus gives rise to an elongated "neck." That none of these ectodermal cells go to form any part of the polypide proper is certain in Plumatella. But it is also true, that ectodermal cells are thus incorporated into the neck of the polypide, and probably into the stolon which proceeds from it.

nation and remains always in communication with the surrounding medium. In Cristatella the lumen is formed in the bud at the time when its diameter perpendicular to the roof of the colony slightly exceeds that parallel to it. As Figure 14 (Plate II.) shows, this cavity (*lu. gm.*) first makes its appearance in the distal part of the central mass of cells. There are always cells lying above the lumen, and thus cutting off the ectoderm from contact with it. The two layers from which, according to my view, all of the cells of the adult polypide are derived, are now completely established; and the cavity has already appeared which, by enlargement, out-pocketing, and the concrescence of its walls, gives rise to the atrium, and the lumina of the alimentary tract and supra-œsophageal ganglion.

The bud elongates, and often at this time, preparatory to giving rise to a new bud from its upper marginal angle, becomes bent or curved, the concavity always being next the daughter bud (see Figs. 3, 5, 11, and 22). By this change in form the bud becomes bilaterally symmetrical.

2. *Origin of the Alimentary Tract.* — The first organ derived from the two-layered sac is the alimentary tract. Nitsche ('75, p. 356) described the process of its formation in a very clear manner; but I believe he is in error. The original lumen of the bud represents, he says, the atrium and the lumen of the alimentary tract. The part lying nearest the attached end of the bud gives rise to the former; the latter is derived from the lower part of the lumen. These two regions become separated by the invagination of the two layers of the bud along a furrow on each side of the bud; just as though the walls of a two-layered hollow rubber ball were pressed together by a finger of each hand acting at opposite sides until the points of the fingers should be separated by the four layers of the ball only. By this process, mouth, anus, and the entire gut, would of course be formed at one time. Reinhard ('80a, p. 212) appears to agree with Nitsche as to the method of origin of the alimentary tract. Braem ('89b, pp. 677, 678) describes and figures diagrammatically this process in the statoblast of Cristatella. In the median plane of the bud there is an out-pocketing from the anal side of the atrium which involves both layers of the bud; it assumes the form of a comma, its blind end curving forward to meet the blind end of a lesser evagination of the oral part of the atrium, — the œsophagus. The blind ends of these two pockets meet, and by the breaking through of the intervening tissue their lumina freely communicate with each other, thus completing the alimentary tract. He adds: "Auch bei den Polypiden der fertigen Colonie der Darm durch Verwachsung eines analen

und eines später auftretenden oralen Schlauches, an deren Bildung freilich das äussere Knospenblatt nur secundär sich betheiligt, zu Stande kommt."

My own observations are nearly in accord with the statements of Braem, as opposed to Nitsche's. The older bud of Figure 11 (Plate II.) shows the first indication of the lumen of the posterior part of the alimentary tract near the attached portion of the bud. The cells of the inner layer are multiplying, and the lumen of the bud is broader here than elsewhere. The position of a daughter bud, VI. (on the oral side of the one under consideration), sufficiently indicates that the point marked *rt.* is in the region of the future anal opening. Figures 12 and 13 (Plate II.) show further stages in this same process. The lumen of the intestine is formed, not by constricting off a part of the original lumen of the bud, but by the rearrangement of cells at the progressing blind end of the pocket, which gradually moves towards the distal part of the larger or bud cavity. It is important to establish the fact that the alimentary tract is formed in the inner layer of the bud, and that its cells alone line the digestive cavity. Figures 20 and 21 (Plate III.) represent two successive sections out of five which pass through the inner layer; namely, the second and the third counting from the attached to the free end of the bud. The sections were cut at right angles to the plane of Figure 11 nearly along the lines 20 and 21 respectively. It will be seen that the inner layer alone is implicated in the lining of the alimentary pocket at this early age. They also show clearly the incorrectness of the statements of Nitsche on this point. Figures 24, 25, and 26 (Plate IV.) are three sections cut through a bud of about the age of that represented in Figure 13 (Plate II.), and at right angles to the plane of the latter, and in the direction of the lines 24, 25, and 26 respectively. A section cut beyond the end of the intestine, in Figure 13, is not represented. It shows that the lumen of the alimentary tract is absent at this plane. A comparison of Figures 20 (Plate III.) and 26 (Plate IV.) shows that the lumen of the bud, *lu. gm.* (which we may call the atrium from its resemblance to a space having the same relations in Entoprocta), has increased in volume owing to a growth of the lateral walls. On account of the more rapid elongation of the anal than of the oral side, the axis of the alimentary tract comes to take a horizontal position, as shown in Figures 17 and 18, Plate III. (Compare also Figs. 27–29, Plate IV.) The blind end of the digestive sac comes very close to the blind end of another pocket formed on the oral side, the œsophagus, and soon the two communicate directly. At the same time, the inner cell-layer of the middle portion of the alimentary tract has been quite cut off from that of the atrium by a constriction, the beginning of which is seen at *ex.*, Figure 24 (Plate IV.) and in a later stage at *ex.*, Figure 28. The cells of the outer layer are next pushed into the place of constriction and remove the alimentary tract at this point still further from the atrium, as is shown in Figures 18 (Plate III.) and 28, *ex.* (Plate IV.). The error of Nitsche is explainable on the ground that he believed the stage of Figure 18 to be the earliest in the development of the alimentary tract.

3. *Origin of the Central Nervous System.* — Metschnikoff ('71, p. 508) first clearly recognized that the supra-œsophageal ganglion of Phylactolæmata is derived from the inner cell-layer of the bud, — the same layer which gives rise to the inner lining of the alimentary tract. Nitsche ('75, pp. 359, 360) described and figured in an insufficient and not wholly accurate way the process of the formation of this organ. According to my observations, the central nervous system arises directly over the middle of the horizontally placed alimentary tract in the position marked *gn.* in Figure 18, Plate III. (compare also Figs. 17 and 28, *pam. gn.*). The process by which the ganglion with its internal cavity (Plate VIII. Fig. 73, *lu. gn.*) is formed will be more easily understood if the reading of the text be accompanied by reference to the following sections. Figures 17, 18, 19, Plate III., and Figure 73, Plate VIII., show successive stages in sagittal section. Figures 27–29, Plate IV., from a single individual, are vertical right-and-left sections, the positions of which are indicated by the lines 27, 28, 29 of Figure 17. Figures 30–32 are similar sections from an older individual (see lines 30, 31, and 32, numbered at the lower border of Fig. 18), and Figures 33–38 are from a still older polypide (compare lines 33–38, Fig. 19). By a study of these sections, it is seen that the cells forming the floor of the brain, *pam. gn.*, are derived from the inner layer of the bud, and indeed from the very region of the layer which furnished cells to line the alimentary tract (Plate II. Fig. 13, Plate IV. Figs. 25 and 24, *ga.*), and therefore that the layer of cells forming the floor of the ganglion is directly continuous posteriorly through the anal opening (Plate II. Fig. 13, *an.*) with the wall of the rectum, and anteriorly with the lining of the œsophagus. The first marked differentiation of this region is effected by the sinking of the centre of the floor of the neural tract (Fig. 18, *gn.*), thus forming a shallow pit, which opens directly into the atrium above.

The *closure of the walls of the ganglion* above must now be considered. Concerning this process, Nitsche says : " Die Ränder dieser Einstülpung [my 'shallow pit'] wachsen nun wie die Ränder der Medullarrinne

eines Wirbelthierembryos gegen einander, und wie in letzterem Falle
eine hohle Röhre von der dorsalen Leibeswand des Thieres abgeschnürt
wird, so wird hier eine hohle *Blase* von der Wand des Polypids abgeschnürt. In unserem Falle ist aber die Wandung an der diese *Ab-
schnürung* vor sich geht, zweischichtig." The two layers referred to were
those of the median walls of a pair of invaginations of the latero-anal
sides of the wall of the atrium, — the beginnings of the lophophoric
arms (*br. loph.*, Figs. 37, 38, Plate IV., and Figs. 61, 62, Plate VII.).

The process of closure is in reality somewhat different from Nitsche's
conception of it. The axes of the pockets which go to form the lophophoric arms are, at first, directed inward, upward, and slightly oralward (Plate I. Fig. 7, *br. loph.*). By means of these invaginations the
cell layers lining the atrium on opposite sides are brought into contact
at a point between the rectum and the ganglionic pit (Plate V. Fig. 43,
loph.'). This approximation of the walls may, perhaps, better be said
to be a continuation upward of the process by which the alimentary
tract was cut off from the atrium (after the lumen of the former was
formed), and by which cells of the outer layer of the bud came to
intrude themselves between these two regions (Plate IV. Fig. 35, *ex.*);
for the lateral furrows, by the formation of which this act is performed,
are, on each side, continuous with the lophophoric pockets, and above
end blindly in them. By the approximation and fusion of the inner
layers of the atrium several things are accomplished. The posterior
wall of the brain is formed (Plate IV. Fig. 39, *loph.'*), the anus is carried farther up (compare Plate III. Fig. 19, and Plate VIII. Fig. 73, *an.*),
and by a continuation of the constricting process the cavities of the
lophophore on opposite sides of the polypide are brought into communication between the ganglion and the rectum at a point opposite the
letters *lu. gm.* in Figure 63 (Plate VII.), whereas they formerly communicated only outside the alimentary tract.

Oralward from the lophophoric pockets there is a thickening of the
inner layer above the floor (*pam. gn.*) of the ganglion on each side
(Plate IV. Figs. 28 and 31). Later, each of these thickenings becomes
a fold involving the inner layer of the bud only (Plate IV. Fig. 35).
The upper and lower halves of this pair of folds respectively fuse in the
sagittal plane, the last point at which the union occurs being near the
œsophagus (Plate III. Fig. 19). Anteriorly the rim of the shallow brain-
pit rises up as a third fold, and the ganglion becomes a sac whose mouth
is bounded by the edges of the folds, the advance of which causes it to
become more and more constricted. These folds are the pair of folds

above the cavity of the ganglion, and the one between the cavity of the ganglion and the œsophagus. The outer layers of these three folds respectively fuse immediately behind the œsophagus; the inner layers are constricted off, but without closing the neck of the sac. Consequently the neck of the ganglionic sac, instead of opening into the atrium, now abuts upon the inner cell-layer at the angle between the floor of the atrium and the œsophagus. The lower layers of the horizontal folds thus become the upper wall of the ganglion (Fig. 35, *tct. gn.*); the upper layers form the new floor of the atrium (Fig. 73, *pam. atr.*), which lies between the lophophore arms, is continuous with its median walls, and passes over into the walls of the alimentary tract both in front and behind. The outer layer of the young bud only secondarily makes its way in between the upper and lower layers of these folds. It ultimately takes the form of a double layer embracing a space, which is the epistomic canal. (Plate VIII. Fig. 73, *lu. gn.*, Plate V. Fig. 52, *lu. gn., can. e stm.*)

4. *Origin of the Kamptoderm.* — While the alimentary tract, lophophore arms, and nervous system are being marked out in the lower portion of the bud, these organs become farther removed from the wall of the colony by an enlargement of the atrium to meet the demands of the augmenting volume of the lophophore. *Pari passu* with this enlargement of the atrium, its walls diminish in thickness (compare *kmp. drm.*, Fig. 73, Plate VIII., with Fig. 18, Plate III.). This is rather the result of a failure of the cells to multiply in proportion as the area of the wall increases, than of a decrease in the number of cells already formed. Both the inner and outer cell-layers of the bud take part in the formation of this wall, as is evident from the figures. The wall of the atrium was called "tentacular sheath" by Allman ('56, p. 12) and Nitsche, but Kraepelin ('87, p. 19) employs the name "kamptoderm" for this structure. I prefer this term to "tentacular sheath," and have employed it both on account of the reasons given by him and because it may be easily inflected, whereas "tentacular sheath" may not. The kamptoderm, then, is formed of the upper portion of the bud, and both of its cell-layers are concerned in its formation and persist in the adult.

5. *Origin of the Funiculus and Muscles.* — Nitsche ('75, pp. 353, 354) did not see the origin of the *funiculus*, but states that it suddenly occurs lying close along the oral side of the bud, to which one end is attached. Its proximal end is fastened, he says, to the inner layer of the colony-wall, and by the growth of the latter between the funiculus and the neck of the bud this end retreats from the young polypide. Braem

('88, p. 533) asserts that the funiculus arises as a longitudinal ridge on
the outer layer of the oral wall of the young polypide at the time of
the formation of the alimentary tract, and that the cells of this ridge
are cut off from the bud to form the funicular cord. Soon after this,
embryonic cells from the inner layer of the young polypide penetrate
into the midst of the cord through its proximal end, and thus lay the
foundation of the statoblast.

Concerning the origin of the *muscles*, Nitsche ('75, p. 354) states that
they are simple elements of the outer cell-layer of the bud, which were
originally situated in the angle of attachment of the bud to the inner
layer of the colony-wall, and that by the growth of this wall they become drawn out into spindle-shaped cells.

I have decided to treat of these two organs together, since their origin and development are curiously similar. According to my belief,
both arise, in part at least, from the inner cell-layer of the colony-
wall. At a stage slightly earlier than that of the first appearance of the
fully formed funiculus (Plate II. Fig. 11, *cl. fun.*), I have always found
a disturbed condition of the cœlomic epithelium. This is particularly
noticeable on that side of the young lateral bud upon which the median
bud is about to arise. In some cases I have seen the cells of this
layer taking on all the characters of wandering cells, as seen at *cl. fun.*,
Figure 22, Plate III., where some have already begun to group themselves
into a funiculus-like cord. At Figure 57, *cl. fun.*, Plate VI., the funiculus
is seen lying close to the oral wall of the polypide. That it has not
arisen in precisely the manner described by Braem is probable from this
figure alone, for the proximal end of the funiculus is not yet connected
with the wall of the colony. If my view is correct, this connection
arises only secondarily (Fig. 2, *fun.*). I am, however, inclined to believe that the distal end of the funiculus arises in a different way from
the proximal, and in the manner described by Braem. My evidence
for this is, that I have twice seen at this point cells in the act of
dividing so as to contribute daughter cells to the funiculus. Figure 53,
Plate VI., shows the condition of the distal end of the funiculus, *fun.*,
which passes, without any line of demarcation, into the outer layer of
the bud ; this layer is normally one cell thick, but in the region of
funicular formation it is two cells thick, . The proximal end of the
funiculus is, at this stage, attached to the cœlomic epithelium of the
roof of the colony, *tct.* That an attachment should occur in this manner, and become quite intimate, is not strange, considering the origin
of the funiculus from amœboid cells, and the fact that, even at a late

stage of development, this character is still retained by much of its tissue. (See, for example, Fig. 77, *fun.*, Plate IX.)

The great *retractor* and *rotator muscles* have, I believe, like the funiculus, a double origin. They arise from the outer layer of the bud, on the one hand, and from the cœlomic epithelium on the other. The first indication of the differentiation of the muscle cells consists in a disturbance in the upper lateral edge of the outer layer of the bud at about the stage of Figure 17, Plate III. This is shown in dorso-ventral sections through this region (*cl. mu.*, Figs. 24, 26, Plate IV.). Later, the disturbance becomes more marked, and cells having a semi-amœboid character appear to be proliferated (*cl. mu.*, Fig. 33, Plate IV.), and to migrate from the bud towards the cœlomic epithelium. During this process the cells of the latter layer also are active, and some of them, elongating, reach towards the young polypide, as seems to be clearly shown at *cl. mus.*, Figure 54, Plate VI. It is significant that, since each of the two upper lateral edges of the bud lies near a radial partition, the muscles also are always formed in close proximity to one (*di sep. r.*, Fig. 54, Plate VI.; Fig. 30, Plate IV.). It will thus be observed that, both in the case of the funiculus and of the muscles, the end which is attached to the wall of the colony arises at a point which is remote from that of its attachment to the adult. The migration to the later positions will be treated of farther on. (See page 141.)

6. *Origin of the Body-wall.* — As already shown (page 104), the body-wall of the individual of a Cristatella colony includes not only the endocyst of authors, — the roof and the sole, — but also the radial partitions.

Braem ('88, pp. 506, 507) concludes " dass die polypoide Knospenanlage . . . nicht allein das Polypid nebst den Tochterknospen liefert, sondern dass auch die zugehörigen Cystide aus ihr und zwar aus ihrem Halstheil entwickelt werden." I believe that a portion of the " cystid," or body-wall, is thus formed in Cristatella, but not the whole.

If one compares the relations of the polypide to its daughter bud in Figures 3 (Plate I.) and 17 (Plate III.), and reflects that later the daughter bud is to be found still farther from the mother bud, he is forced to one or the other of two conclusions: either the young bud is pushed from the mother by a proliferation of cells from the neck of the polypide without causing an increase in the length of the body-wall itself, or there is an actual increase in the length of the body-wall, produced either by the proliferation of cells already existing in it, or by the addition and subsequent proliferation of cells from the neck of the mother polypide; and this increase in length, occurring between the polypide and bud, carries the two apart. Unfortunately, I am unable to state definitely how this migration of the young bud away from the mother is effected. If the ectoderm increases in length between the two buds by the proliferation of cells already existing in it, that fact ought to be evinced by a distorted condition of the old cell-walls of the highly metamorphosed cells of the ectoderm. For, since most of the active protoplasm is at the base of the ectoderm, its area will increase faster than will the area of the surface of the ectoderm; and the latter will either rupture or stretch, or else the ectoderm will become concave on its outer side. An application of these criteria to sections of the body-wall in the budding region leads to the conclusion that the ectoderm of Cristatella increases here very slightly, if at all, by a proliferation of cells already existing in it. A search for cell division in this region has yielded the same negative results. There can be no doubt that cells are added to the ectoderm from the neck of the polypide. The process takes place, however, after the daughter bud is well established at some distance from the mother bud. The proliferation of these cells ruptures the old cell-walls of the ectoderm, and increases the area of the body-wall. I shall have occasion to speak of this process more fully in treating of the later period to which it belongs.

There remains, then, the conclusion, that the cells which go to form the inner layer of the young bud are pushed from the neck of the next older bud by a proliferation of cells in the stolon-like mass, without causing an increase in the area of the body-wall itself. Moreover, I have seen cell proliferation in the stolon-like mass. Another series of facts will lead us to this same conclusion.

Though the body-wall does not increase by cell proliferation between buds, it does so, I believe, at the margin of the colony. This, it is true, cannot be directly observed with ease, since the multiplication of cells, which tends to increase the breadth of the colony, must also occur at the margin, and one cannot be certain what dimension of the colony wall will be augmented by any given case of nuclear division. My belief rests on the following evidence. (1) In the same adult colony the distance of the youngest bud from the margin is not the same in all regions. This is not what we should expect if the distance of the youngest polypides from the margin remained unchanged during the growth of the colony. (2) There is a gradual increase in the amount of metamorphosis exhibited by the cells as one passes from the margin towards the middle of the roof. Figure 60 (Plate VI.) shows a rather marked example of a very common, although not universal, condition of the lateral margin of the colony. The epithelium of the margin is composed of columnar cells, which are higher ($54\,\mu$) than those of the roof ($48\,\mu$), and also of a less average diameter ($8.4\,\mu$) than the latter ($18.2\,\mu$). Moreover, the cells are very little metamorphosed. In passing towards the roof (*tct.*), the cells are seen to become more and more metamorphosed, the secreted bodies (*cp. sec.*) becoming relatively larger. Figure 55 represents the margin in a more metamorphosed state than Figure 60. Although this condition of things is not incompatible with the idea of a passive margin, it strongly suggests that this region is one of proliferation, by which cells are added to the roof, and thus the distance from the youngest polypide to the margin is virtually increased. This conclusion receives a very important confirmation from the study of the origin of the radial partitions, the treatment of which must be deferred for the moment. Although new cells are being added to the roof at the margin, yet the distance from the youngest polypide to the margin is not greater in old than in young colonies. How, then, is the approximate constancy of this distance maintained? Evidently it can only be by the process (which I have already shown must take place) of migration of some of the young buds at the base of the ectoderm, particularly in the case of median buds. The tendency of the migration of young buds towards the margin is to diminish the distance between the front of the budding region and the margin of the colony. The tendency of cell proliferation at the margin is to increase that distance. The actual distance is the resultant of these two opposing factors, and may be less or greater in different parts of the same colony, according as the one or the other is the more active. If we assume, further, that the cells added to the roof and sole from the margin plus those derived from the necks of the polypides are equal in amount to those lost by the degeneration of individuals in the middle of the colony, we have a sufficient explanation of the fact, observed long ago, that the adult colony of Cristatella maintains a nearly constant width.

7. *Origin of the Radial Partitions.* — I know of nothing on this subject by any previous author. The radial partitions consist of a muscularis covered on both faces by a very thin epithelial layer (Plate X. Fig. 95, 1). The muscle fibres of the muscularis arise from the already formed longitudinal muscles of the wall of the colony at the region of transition from the sole to the roof (Plate VI. Fig. 55, *mu.*). As the muscle fibres move into the cœnocœl, they carry before them the cœlo-

mic epithelium of the region from which they arise. It is owing to this method of origin that the epithelium comes to clothe both faces of the partition. The process by which the muscle fibres move into the cœnocœl appears to be this. The end of a fibre nearest the roof becomes fixed to a certain part of the muscularis of the roof, and is left behind with it when the margin is carried outward (potentially) as the result of cell proliferation. Thus from a nearly horizontal position the fibres attain a direction at first oblique, and then perpendicular to the sole. In some instances the upper ends of the fibres move through an arc of more than ninety degrees, so that they are ultimately directed upward and inward, i. e. towards the centre of the colony. (Compare *mu, mu′, mu″*, Fig. 55, Plate VI.). This process is also indicated in two horizontal sections (Plate X. Figs. 95 and 96), the former being nearer the sole than the latter. This is a region of active budding, and in consequence new compartments or *branches* are being rapidly formed. The numbers 3, 4, 5, and 6 (Figs. 95, 96) show the positions of young partitions, which are shorter above than below, owing to the oblique position assumed by the innermost muscle fibres of the partition. The oblique position is due to the fact already demonstrated (Fig. 55, Plate VI.), that the tectal end of the muscle of the partition first appears at the margin nearer the sole than the roof. At 2 (Fig. 96) there is apparently an interesting case of the formation of a new partition by the detachment of certain fibres from the muscularis of an old one. The fibres, moving away laterally, take with them a covering of cœlomic epithelium. Near the sole this process has progressed farther than it has nearer the roof, so that in Fig. 95 the detachment appears complete, whereas in Fig. 96 the union is still visible. This method of formation is intelligible when one considers that the muscularis of the partition often contains more than a single layer of muscle fibres. Thus, in Figure 87, *mu.*, there are two or three layers of fibres in the section. Figure 86 represents a section cut vertically and at right angles to a partition near its union with the marginal wall of the colony, and shows three fibres of the longitudinal or inner layer of the muscularis lying side by side in the partition.

B. — COMPARATIVE AND THEORETICAL REVIEW OF THE OBSERVATIONS ON THE ORIGIN OF THE INDIVIDUAL.

What bearing have the facts here adduced on those given for other groups of Bryozoa, and what is their probable significance in relation to the general problem of non-sexual reproduction?

1. *Origin of the Polypide.* — Lateral budding (as distinguished from linear budding, such as occurs in Turbellaria, Chætopods, &c.) may be roughly classified under two types, in one of which the young individual arises *directly* from the body-wall of the parent, as in Hydra. In the other, the young arise, one after the other, from a mass of embryonic material derived from a parent individual, — from a stolon, as in Salpa. In the group of Bryozoa both of these methods seem to be present. In such a form as Paludicella (Allman, '56, pp. 35, 36, Korotneff, '75, p. 369) we have an example of the direct type; in Pedicellina we have a stoloniferous genus. Also in the marine Ectoprocta examples of both types appear to occur (e. g. Flustra, Hypophorella). To which of these classes does budding in Cristatella belong? It seems to me that we have here an instructive example of a transitional condition. The young polypide of Figure 3 arises directly from the mother polypide, and may represent a case of the first class. Is the type of Figure 2 a representative of the stoloniferous class? It seems to me that it partakes of the essentials of that class, although, as I have shown, it may be united by intermediate stages with the first class. I understand a stolon, in its morphological sense, to signify a mass of embryonic cells derived from a parent individual, and capable of reproducing non-sexually one or more daughter individuals at some distance from that parent. The condition shown in Figure 15, Plate II., in which the embryonic cells of the two layers represent the stolon, may fairly be said to answer to this definition. The mass of cells (III.) represents, then, the distal end of the stolon. But the stolon does not end here, although its further progress towards the margin is delayed. Not all of its cells go to form the polypide which arises at this place. On the contrary, some of them remain in the " neck " of the new polypide, in an indifferent histological condition, and later give rise, either directly, or by the intervention of a typical stolon, or by both, to one or two new buds. Those cells of the neck which do not thus pass over into new buds for the most part degenerate (page 144). According to this view, the neck of the polypide is to be regarded as at first essentially a portion of the stolon. .

2. *Interrelation of the Individuals in the Colony.* — The interrelation of individuals in the colony in Cheilostomata has been most carefully investigated from a morphological standpoint by Nitsche ('71, pages 35, 36), who showed that, in opposition to Smitt's theory, each new individual arose from a single preceding one, and that the latter, in order to increase the breadth of the colony, might give rise to two individuals

instead of one. Reichert ('69, p. 311, Fig. 28, Plate VI.) has shown
that in Zoöbotryon (one of the Ctenostomata) "an der Mantelfläche,
und zwar einseitig, inseriren die Bryozoenköpfe mit Alternation in par.
allelen, wie es scheint, langgezogenen spiralig verlaufenden Reihen an.
geordnet." Nitsche ('75, p. 370) states that the buds in Loxosoma
arise from the mother alternately on opposite sides, and that the
younger the bud, the nearer it is to the foot of the parent individual.

Both Hatscheck ('77, pp. 517, 518, Fig. 33, Plate XXIX.) and Seeliger ('89, p. 176) show that in Pedicellina young individuals are devel.
oped in the plane of the older ones, and are successively formed at the
growing tip of the stolon, towards which the œsophageal side of all
individuals is turned. This relation is the same as that which we have
found in Cristatella. In Cheilostomata, however, it is apparently the
anus which is turned towards the budding margin. ·

Thus, throughout the group of Bryozoa, we find that the position which
young buds assume in relation to older individuals is very definite.

I am inclined to believe that the radial partitions of Cristatella sep.
arate the morphological equivalents of the isolated branches of such a
form as Plumatella punctata (see Kraepelin, '87, Taf. V. Figs. 124, 125).
The type of budding which gives rise to the series of median buds may,

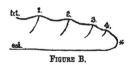

FIGURE B.

then, be represented, as seen from the side, by
Figure B. The margin (*) will then represent
that portion of the body-wall of the youngest
individual, which will give rise to a part of
the body-wall of the next younger individual.
The process by which the body-wall of the individual of Cristatella is
formed is therefore, in my opinion, different from that which Braem
describes in the case of Alcyonella, for he maintains that in Alcyonella
the proper body-wall of an individual arises later than its polypide. In
fact, the tip of the branch of Alcyonella is somewhat different from
that of Cristatella. In the former, it is occupied by the polypide of a
budding individual; in the latter, a part of the body-wall of the budding individual is pushed out beyond the polypide. In the former, the
foundations of the daughter polypide are pushed out upon the body-
wall of the mother, and begin to form their own proper body-wall; in
the latter, the young bud migrates away into the modified part of the
body-wall of the mother, which forms the extremity of the branch, and
which now becomes a part of the body-wall of the daughter polypide.
This distal part of the body-wall grows independently of the polypide
by interstitial growth, and thus differs from any part of the body-wall

of the individual of Alcyonella, for all of it, according to Braem, is derived from the neck of its own polypide.   This last method of origin of the body-wall I believe to be also present in Cristatella, as well as in Alcyonella, as I shall have occasion to show later (page 144).

In Paludicella, according to both Allman ('56, pp. 35, 36) and Korotneff ('75, p. 369), the formation of the body-wall of the new individual is begun before the appearance of the polypide.   In Cheilostomata, as both Nitsche ('71, p. 22) and Vigelius ('84, p. 75) have shown, and in Ctenostomata, as demonstrated by Ehlers ('76, pp. 91, 92), the "zoœcium" arises before the polypide takes on its definite form.

3. *Origin of the Layers.* — Although later researches have only confirmed the conclusion arrived at long ago, that in Tunicates cells from all three germinal layers of the parent pass over into the bud, the facts in Bryozoa have seemed not to favor the view of the fundamental nature of this process.   To be sure, Hatschek ('77, pp. 517–524) believed that he had found evidence of a condition in the budding of Pedicellina exactly comparable with that in the budding of Tunicates; but the more recent studies of Harmer ('86, p. 255) and Seeliger ('89) have failed to confirm his results, if they have not satisfactorily explained the source of his error.

What is the relation of the condition I have described in Cristatella to the question of the transmission of a part of each germinal layer to the bud, and in how far do the conditions here agree with our present knowledge of the budding process in other groups of Bryozoa?   Although my results accord with Hatschek's in this, that the youngest and next older buds are intimately related, that the corresponding layers in each are derived from the same cell layers, and that the inner layer of the bud is not derived directly from the overlying ectoderm, they do not strengthen the idea of the fundamental importance of his doctrine, "Die Schichten der jüngeren Knospe stammen von denen der nächst älteren direct ab."   Moreover, they afford no evidence of the accuracy of his conclusion, that the inner layer of the bud is composed of entoderm; indeed, since this inner layer does not give rise to the alimentary tract alone, as he supposed, but to the nervous system also, the facts in Cristatella tend to weaken his hypothesis.   In order to determine finally just what the origin of the stolon from which the inner layer arises is, it will be necessary to study the origin of the first-formed polypides.   This I have not yet been able to do.   Our present knowledge on the subject is still in an unsatisfactory state.

Allman ('56, pp. 33, 34) has described and figured some stages in the development of the egg, but without referring to gastrulation, or the layers involved in the first polypide.

Metschnikoff ('71, p. 508) and Nitsche ('75, p. 349) maintain that the outer layer of the embryonic "cystid" goes to form the inner layer of the primitive polypides, and that its inner layer forms the outer layer of the polypides.

Reinhard ('80ª, pp. 208–212) is more explicit concerning the early stages than preceding authors. Apparently the egg segments regularly, and undergoes embolic invagination. The blastopore closes. There is a circular groove in the anterior part of the embryo (Barrois's mantle cavity), and from the cap or "hood" which the mantle cavity sur- rounds, the wall of the "cystid" or colony-wall is subsequently formed. The embryo is already composed of three layers, "an outer, the tunica muscularis, and the entoderm." All three layers of the "hood" share in the formation of the polypides, but the fate of each layer is not clearly described.

Haddon ('83, p. 543) suggests that the gastrula is to be regarded as one in which the alimentary tract is retarded in development, and that the enlarged cœlomic diverticula, such as occur in Sagitta, etc., line nearly the whole of the so-called archenteron. From the small mass of true entoderm at the pole opposite the blastopore the alimentary tract arises. This suggestion, unfortunately, has no positive facts for its support, and could be of service only upon the assumption that the alimentary tract of the first polypide is formed from the *inner* layer of the "cystid"; but this assumption is contrary to the observation of all who have written on this subject.

Kraepelin ('86, p. 601) has also observed the "gastrulation," but he believes that it is to be interpreted as the precocious formation of an enterocœl, in which case the invagination to form the first polypide is to be regarded as the true gastrulation, the inner layer of the cystid as mesoderm, and the inner layer of the bud as entoderm.

By far the most satisfactory and complete account of the embryology of fresh-water Bryozoa is that of Korotneff, '89· The genera studied were Alcyonella and Cristatella. Since the development takes place inside of an oœcium, the use of the section method is necessary for the elucidation of the details of the embryological processes. Apparently the egg segments regularly and forms a blastula. Loose cells are given off from the inner surface at one pole of this blastula. These arrange themselves in an epithelium, lying immediately inside of the ectoderm, over a part only of its inner surface ; so that while the upper two-thirds of the embryo has two layers, the lower third is one-layered. The cavity of the lower third contains some scattered cells, which, the author hints, may be representatives of the mesoderm, while the cavity in which they lie may represent an enterocœl. The author regards the inner layer of the upper two-thirds as true entoderm. The method of its formation recalls that of the entoderm of some Cœlenterata, as demonstrated by Metschnikoff. There is no epithelial invagination, such as Kraepelin maintained, and therefore the cavity which the inner layer lines cannot be regarded, says Korotneff, as an enterocœl. Later, the entire embryo becomes two-layered by an extension of the inner layer. The two polypides arise from two distinct invaginations of the double-layered wall.

Unfortunately, Korotneff does not demonstrate by figures the method of origin of the alimentary tracts of the first polypides; but there is little reason to doubt that it is essentially like that in other buds. If it is admitted that the inner layer is entoderm, as Korotneff maintains, then the entoderm takes no part in forming the digestive epithelium; but the latter is derived solely from ectoderm.

In his discussion of the theoretical bearing of his results (p. 404), the author seems to maintain that the polypide is to be regarded neither as an individual (Nitsche's view), nor, on the other hand, as an assemblage of organs homologous with organs of the same name in other groups; but rather as a new structure, developed upon the cystid, to aid in its nutrition.

In criticism of Korotneff's view, that the loose cells given off from one pole of the blastula are entoderm, I may point out that this process bears quite as much resemblance to the process of "mesenchyme" formation (as described by Korschelt for the Echinoids), as it does to the origin of the entoderm in some Cœlenterates. Compare Figs. 13 E and 182, in Korschelt und Heider's Lehrbuch der Vergleichenden Entwicklungsgeschichte.

Braem ('89[b], pp. 676, 677) has shown that the primary polypide of the statoblast arises from the cell layers of the statoblast, exactly as the primary polypide of the egg embryo does from those of the " cystid," and the alimentary tract is formed as in buds of Cristatella.

To sum up: The outer layer of the colony-wall is ectodermal in origin; the inner layer arises by an embolic (?) invagination of the blastula, and would therefore appear to be entoderm, although the possibility of its being homologous with the mesoderm in other forms is perhaps not excluded. The first polypides so arise that their inner layers are formed by an invagination of the outer layer of the colony-wall, and their outer layer from the inner layer of that wall.

In *Endoprocta*, Seeliger ('90, pp. 176–187) has shown decisively that the inner layer of the bud is derived solely from the ectoderm, and that this inner layer gives rise to the digestive epithelium of the alimentary tract, to the nervous tissue of the brain, and to the outer layer of the tentacles. Here mesenchymatous cells, representing undoubtedly meso-dermal tissue, come secondarily to surround the polypide as a loose outer tissue. In Loxosoma the same is probably true.

The conditions of budding in *Gymnolæmata* are more difficult to un-derstand. In Paludicella the bud seems to arise as in Phylactolæmata (Allman and Korotneff). The same is probably true for Alcyonidium (Haddon, '83, p. 523, Plate XXXVIII. Fig. 23). In the Cheilostomata, however, the fact of the great development of a loose mesenchyme-like tissue obscures the process, and makes it difficult of interpretation. This tissue, which is known under three probably homologous terms, — "Funiculargewebe," Nitsche, "Parenchymgewebe" in part, Vigelius, and "Endosarc," Joliet, — is to be considered as representing the funicu-lar and cœlomic tissues of Phylactolæmata. The most careful observa-tions on the origin of this tissue are those of Joliet ('77, pp. 249, 250, and '86, pp. 39, 40) and Vigelius ('84, p. 76). Both authors assert that this tissue is derived from cells given off from an epithelium at the distal end of the budding individual. Vigelius ('84, pp. 19, 79) believes that this epithelium is ectodermal, and that it is the sole rudiment of this layer; but Ostroumoff ('85, p. 291) and Pergens ('89, p. 505) have shown that the ectoderm persists and secretes in its cells the cal-careous ectocyst. It seems more probable, however, that the "funicu-lar tissue" arises from the inner layer of the body-wall (Nitsche, '71, p. 37, Plate III. Fig. 5, c.), and is the equivalent of the cœlomic epithe-lium of Cristatella. The fact that many of these mesenchymatous cells conglomerate in the formation of the polypide sufficiently accounts for the origin of its outer layer of cells. The origin of the inner layer is problematical, if, as is asserted to be the case by several authors, the bud is not formed in the region of the body-wall.

It will be premature to speculate upon the significance of the facts of budding in the Ectoprocta until we shall have gained a more com-plete knowledge of the ontogeny of the group, and of the relationship of the Cheilostomatous to the Phylactolæmatous type through compara-tive agamogenetic studies. It may appear in the end, that, under cer-tain circumstances, undifferentiated embryonic tissue, derived from a certain germ layer, can assume the task of building organs in budded individuals similar to those derived from a different layer in the sexually produced individual.

Whatever may be the truth of the conclusions reached by Haddon ('83, pp. 548, 549, 552) and by Joliet ('86, pp. 54–56), that the nervous system and the alimentary tract arise from two distinct layers, or kinds of cells, in the species studied by them (and their evidence is certainly not conclusive even for these), their attempts (Haddon, '83, p. 540, Joliet, '86, p. 57) to apply their results to the Phylactolæmata are not justified by the observations which are here presented, nor by those which have been made upon most Gymnolæmata and Endoprocta.

4. *Origin of the Alimentary Tract.* — There is a curious difference between the Endoprocta and the Ectoprocta in the development of the organs of digestion. Seeliger ('89, pp. 182–184) has shown for Pedicellina, that the œsophagus *and stomach* arise as an evagination of the oral wall of the young bud, which secondarily becomes connected with the proctodæum. Haddon ('83, pp. 517, 518) has shown for Flustra, Barrois ('86, pp. 73–86) for Lepralia, Braem ('89[b], pp. 677, 678) for the statoblast polypides of Cristatella, and the present paper for the polypides in the adult Cristatella, that the œsophagus only is formed on the oral side, the stomach arising with the rectum on the anal side of the atrium. In all cases the œsophagus is formed first (Plate II. Fig. 13). A comparison of my Figure 18 with Figure 41, Plate XXX., of Hatschek ('77), shows a striking resemblance between the two. The form of the alimentary tract and the depression to form the ganglion are practically identical; and were the tentacles to arise directly from the immature lophophore arm (*br. loph.*, Fig. 18), and from the circumoral fold which has already appeared, it would be difficult to decide whether the anus opened outside or inside the circlet of tentacles, — whether, at this stage, the Cristatella polypide were ectoproct or endoproct.

5. *Origin of the Central Nervous System.* — The only observations on the origin of the brain in Bryozoa relate to Phylactolæmata and Endoprocta. In buds of Pedicellina, the ganglion is formed, according to Hatschek ('77, p. 520), as an invagination of the floor of the atrium, which later becomes cut off as a hollow sac. Harmer ('85, pp. 274, 275) has studied the origin of the ganglion in the bud in Loxosoma. He states that it is derived from the floor of the vestibular [atrial] cavity, and (apparently on purely theoretical grounds) that this latter is ectodermic. "In a longitudinal section through a fairly advanced bud

(Fig. 15) it is seen that a narrow slit-like diverticulum of the vestibule passes behind the epistome. This diverticulum, which remains in very much the same condition throughout life, does not give rise *in toto* to the ganglion, which is merely formed by a differentiation of some of its ectodermic cells." Harmer further doubts Hatschek's account of the formation of the ganglion in Pedicellina, and believes that the lumen of Hatschek's hollow sac is in reality the commencement of the fibrous tissue which occupies the centre of the ganglion in the adult, and which in optical sections might easily be mistaken for an empty space. "Similarly," he continues, "Nitsche has described the ganglion of *Alcyonella* as originating as a diverticulum from the tentacle sheath. I regard it as probable that the explanation which I have suggested for *Pedicellina* will hold also for *Alcyonella*." The conditions which every student of the embryology of Phylactolæmata has stated since Metschnikoff's paper in 1871, and which my own results reaffirm, do not warrant Harmer's conclusions. The nerve fibres are very evident in the adult ganglion of Cristatella, and in addition to them there is a cavity, ontogenetically derived from the atrium, which, as Saefftigen ('88, p. 96) has also shown for Phylactolæmata, contains no histological elements (Plate V. Fig. 52).

6. *Origin of the Funiculus and Muscles.* — The origin of the so-called funicular tissue in Gymnolæmata has been described already (page 126). This same tissue also gives rise, according to Vigelius ('84, pp. 34, 35) and others, to the retractor muscles of the polypide. As I have already shown (pages 115–117, Figs. 22, 54), in writing of the origin of these tissues in Cristatella, the cœlomic epithelium gives off cells, some of which take on an amœboid appearance, and, uniting together, form that end of the funiculus which is attached to the colony-wall. Other cells from the cœlomic epithelium pass directly to the adjacent outer layer of the bud, to form the nascent retractor and rotator muscles. Both of these organs are, however, formed in part from cells composing the outer layer of the bud, — itself closely related ontogenetically to the cœlomic epithelium.

These facts would seem to confirm the conclusion which the similar relation of the two layers would suggest, namely, that the cœlomic epithelium of Phylactolæmata is the homologue of the "endosarc" of Gymnolæmata.

IV. Organogeny.

1. *Development of the Ring Canal.* — Nitsche ('75, p. 358) describes the ring canal as a furrow arising from the opening of each of the lophophoric pockets, and running towards the oral side of the bud. In a later stage, both layers become deeply implicated in this furrow, and the ring canal is completed by a growing together of the edges of the furrow.

Braem ('89[b], p. 679) merely states that he cannot fully agree with Nitsche's description of the formation of the ring canal.

As a result of my own studies on this subject, I have reached the conclusion that the circumoral branch of the ring canal makes its first appearance in the median plane in the oral region at about the time that the depressions of the lophophoric pockets are first indicated. The formation of both organs is preceded by a preliminary thickening of the inner layer of the bud (Plate IV. Fig. 26, *br. loph.*, and Plate III. Fig. 17, *can. crc.*). It is only later, after the lophophoric pockets have attained considerable depth, that the groove of the incipient "ring canal" appears continuously on the side of the polypide, extending from the pre-oral region to the lophophoric pockets (Plate IV. Figs. 33, 35, 37, *can. crc.*).

As indicated in the successive stages of Figures 18 and 19, Plate III., the thickening of the inner layer anterior to the mouth is followed by a fold at this point involving both layers. The fold is deepest in the pre-oral part of the median plane, and becomes shallower as it proceeds posteriorly. Finally, the outer-layer cells of the lips of the fold approach each other and fuse, thus forming a true canal (Plate IV. Fig. 33, *can. crc.*). Kraepelin ('87, p. 57, Figs. 72, 73, *qb.*) asserts that this canal does not communicate at its neural ends with the cœnocœl, but that it is always closed by a strong "Querbrücke" connecting the "Kamptoderm" with the alimentary tract. By making sections of the colony parallel to the sole, dozens of individuals are cut through the entire length of the circumoral ring canal. Although I have examined many individuals cut in this way, I have never succeeded in finding in Cristatella this closing "Querbrücke"; but in both young and old specimens, sections nearly corresponding to Kraepelin's Figure 72 show a perfectly uninterrupted semicircular space surrounding the œsophagus, and opening freely into the cœnocœl on each side of the brain (Plate IX. Fig. 78, *can. crc.*). I must therefore conclude that in Cristatella the fluids of the cavities of the circumoral branch of the ring canal, and therefore of the tentacles also, are in free communication with the fluids of the common body cavity.   As Figure 51, Plate V., shows, the posterior ends of the ring canals open into a pair of cavities which are the bases of the lophophoric pockets, and by a comparison of Figures 61–63, *can. crc., can. crc.'*, Plate VII., it will become apparent that they each become confluent with a furrow which passes up the lophophore arm, and from which the outer lophophoric row of tentacles is developed.   Further, by a comparison of *can. crc."*, *can. crc.'''*, in Figures 61–63, Plate VII. (dextro-sinistral vertical sections), and Figure 50, Plate V. (horizontal section, compare also Fig. 52, a sagittal section), it will be seen that from the tip of the lophophoric arm a groove (*can. crc."*) passes down upon the side opposite to the ascending groove (*can. crc.'*), and, reaching the base, turns abruptly anteriorly (*can. crc.'''*, Fig. 50), and finally, in a later stage, becomes confluent with its fellow of the opposite side in the median plane just behind the epistome and above the brain. It would be quite unnecessary for me to give figures showing the course of this supraganglionic canal (cf. Fig. 52, Plate V.).   It has long been recognized, and is shown in Kraepelin's ('87) Figure 66, Taf. II.   This is probably what Verworn ('87, pp. 114, 115, Figs. 20 a, 20 b, Taf. XII.) has described as a "segmental organ."   Braem ('89b, p. 679) has given to it the name "Gabelkanal."   The "Ringkanal" of Nitsche is, then, to my mind, merely the circumoral portion of a groove which is elsewhere unclosed to form a proper canal and which lies at the base of all tentacles.   My reason for avoiding another term for the unenclosed portion of the "canal" is, that I regard the whole as morphologically equivalent to the ring canal of Gymnolæmata, which is said to be closed throughout.

2. *Development of the Lophophore.* — The early stages in the formation of this organ are well known, both from the descriptions of Nitsche ('75, pp. 357, 358) and the earlier ones of Allman and others.

I have already (page 114) shown how the cavities of the lophophoric pockets become confluent between the rectum and ganglion, and how their opposed walls, formerly passing over into each other through the floor of the brain, are now anteriorly continuous by means of the new floor of the atrium, and posteriorly are fused together.

The union of the inner layers of the two opposed walls of the lophophore arms (Plate V. Fig. 44, *loph.'*) continues, however, for some distance above the floor of the atrium, up to within a short distance of the tips of the young arms (Plate VII. Figs. 61, 62, *loph.'*).   As the arms grow longer, the relative extent of their free and fused portions remains approximately the same. The free ends of the arms are shown in Figure 99, Plate XI., just above *loph'*. The polypide figured here is only slightly older than that of Figure 77, Plate IX. The connection between the two arms is not one of contact merely, for in the region of fusion one can count roughly three layers of nuclei, whereas each of the two free portions of the same cell layer contains but one layer of nuclei (Fig. 99).

Before the atrial opening is formed, a separation of the two arms begins to take place. This process commences at the base of the arms, and proceeds upward as the tentacles of the inner row successively reach a certain stage of development. As the work of separation progresses, the cells of the connecting band lose their capacity for becoming stained and appear vacuolated. The vacuoles increase in size until the connection between the arms is reduced to a series of fine threads (Plate VIII. Fig. 75, *loph.'*), which are probably sundered when the tentacles of the inner row (*can. crc."*, Fig. 76) bend at right angles to their former position to become parallel to those of the outer row. In attempting to find an explanation of this process, it must first be ascertained how the arms of the lophophore grow in length. One is perhaps inclined to think of a terminal growth, but this does not take place. So far as I can judge from an examination of many longitudinal sections of the arms, cell proliferation goes on throughout the whole length of the arm, and with nearly equal rapidity in all parts. The distance between the centres of the terminal tentacles is about the same as in the case of the more fully developed proximal ones, but they are closer together in the young arm than in the adult one. This being the case, there ought to be as many (incipient) tentacles in the young as in the adult, and I find that to be, so far as I can determine, very nearly or exactly the case.

The horseshoe-shaped lophophore being characteristic of the Phylactolæmata, a study of its development is important, since it may be expected to throw light on the phylogeny of the group. We have in Cristatella, Plumatella, and Fredericella, a series in which the arms of the lophophore are shorter and shorter, in correspondence with other changes, by which is effected a gradual transition to the Gymnolæmata, which have a circular lophophore. In Gymnolæmata, the ring canal lies at the base of all tentacles in the adult. The anus lies outside the circle of this canal. The brain lies within the lumen of the canal.

Nitsche ('71, pp. 43–45) has given the best description extant of the development of the lophophore in Gymnolæmata. At a very early stage, the rudiments of the tentacles, he says, are seen lying in a U-shaped line, surrounding the mouth in front, but unclosed behind. The same is true for Paludicella (Korotneff, '75, p. 371). The post-oral tentacles make their appearance at the posterior free ends of the row of tentacles. They are bent slightly downward, so as to be concealed by the tentacles above. At a later stage, the tentacles lying next to the anus gradually come to lie nearer to the anal side of the mouth opening, the nearly parallel lateral rows lose their compressed appearance, and a circular basin is formed whose walls are constituted by the corona of tentacles.

In Pedicellina (Hatschek, '77, pp. 520, 521) the tentacles arise as five pairs of papilla-like processes in the upper part of the atrium. Two additional pairs are formed later nearer the anal opening. In the adult (Nitsche, '69, p. 21) the tentacles are arranged with bilateral symmetry, and so that the plane of symmetry passes through two inter-tentacular spaces, which are thus the only unpaired spaces; they are also much broader than the others.

One might be inclined to ask by what modification of the condition of the tentacles in Endoprocta we may suppose the condition in Ecto-procta to have arisen, but the question is not a fair one. I have already (page 127) shown that the young bud of Cristatella has many points of similarity to a well advanced Endoproct. This similarity leads me to the conclusion that the common ancestor of the Endo-procta and Phylactolæmata more nearly resembled the former than the latter group. But the Endoprocta are not that common ancestor; rather they are themselves more or less modified descendants of it. The proper inquiry is, To what ancestral relation between tentacles and anal opening does a comparison of the ontogeny of Endoprocta and Ectoprocta point, and by what modifications of that ancestral type may the two divergent types of the present be derived? Eliminating for a moment the evidently cœnogenetic character of the lophophore arm, an early stage of either Endoprocta or Ectoprocta reveals a U-shaped band from which tentacles are to arise. This band completely encircles the mouth, and passes posteriorly as far as the anus. This is the condition of the Endoproct bud, with only five of its seven pairs of tentacles formed; it is also the condition of the Cristatella bud of Stage XIII. (compare Figs. 19, 44). Starting from this common condition, that of the adult Endoproct, on the one hand, was attained by the addition of two pairs of tentacles posteriorly, thus nearly completing the circlet behind the anus. The condition of the adult Ectoproct, on the other hand, was reached by the curving oralwards, and the meeting of the free ends of the rows of tentacles between the mouth and anus, thus shutting the anus outside of their circle. In evidence of this latter assertion, I submit the following comparative statement.

As Nitsche has shown for Gymnolæmata, the tentacles on the ring canal are first arranged in two rows, placed bilaterally, and meeting in front, but not behind. Later the hindermost of the tentacles move forward and toward the median plane, thus completing the circlet of tentacles at a point behind the mouth, but in front of the anus. I believe the circumoral ring canal plus the early invaginations of the lophophoric arms in Phylactolæmata to be homologous with the ring canal of Gymnolæmata in its early stage; like the latter, it is closed in front, but has two free ends behind. The difference lies in the greater development of the posterior ends of the canal, which latter have become thrown into a vertical fold to afford space for more tentacles. At this stage of development it would be difficult to say whether the anus opened within or without the corona of tentacles. As in Gymnolæmata the circle is completed by a movement inward of the posterior tentacles, so in Phylactolæmata the corona of tentacles is completed in front of the anus by the two anterior processes, *can. crc.'''*, Figure 50 (cf. Fig. 44), of the lophophore arm, which come to unite just behind the epistome, Figures 52, 81, *can. crc.'''* The lumen of this process of the lophophore arm thus forms that portion of the ring canal which, as I shall show directly, is the morphological equivalent of the most posterior portion of the ring canal in Gymnolæmata. The tentacles which arise from this portion of the ring canal are ontogenetically, and therefore phylogenetically, the youngest. As in Gymnolæmata, so here the moving forward of the most posterior tentacles obliterates the basin-like floor of the atrium, such as we see in Endoprocta, and leaves the anal opening far outside the circlet of tentacles.

The answer to the question, How may the horseshoe-shaped tentacular corona of Phylactolæmata be homologized with circular ones? is involved in the answer to the preceding query. Nitsche ('75, p. 357) believed the lophophoric arms to be "primary tentacles," and the tentacles borne on them to be secondary tentacles, "Gar nicht ohne Weiteres mit den Tentakeln der Infundibulata von GERVAIS zu vergleichen." The only evidence which he offers in support of his theory is the fact that the tentacles on the lophophore arm arise later than the arm itself.

The tentacles of Phylactolæmata may be distributed into two groups. The first includes those which arise from the circumoral branch of the ring canal. The ring canal, from which they spring, begins to be formed at nearly the same time as the lophophoric arms. These tentacles are undoubtedly homologous with those of the same region in Endoprocta and Gymnolæmata. The second group of tentacles includes those which are borne upon the lophophore arms and upon the supraganglionic ring canal. Are these comparable with the posterior tentacles of Gymnolæmata? I believe they are, and for the following reasons. Nitsche's reason for supposing that they are not is unsatisfactory, since, if we regard the lophophore arms as mere upward folds of the wall of the ring canal, we should expect to have the tentacles arise later than the arms. The fact that the tentacles of the lophophore arm arise much later than those of the circumoral region is what we should expect, since the posterior tentacles arise later than the circumoral ones in both Endoprocta and Gymnolæmata, — a criticism which Hatschek ('77, p. 541) has already applied. In direct support of my belief are the facts, (1) that the ring canal is continuous along two sides of the lophophore arms, which would be the case if they were mere upward folds of the wall of the ring canal; (2) the structure of the tentacles is the same as that of the oral ones, and the relation of their intertentacular septæ to the ring canal of the arms is the same as that of the septæ of the oral tentacles to their ring canal, as Kraepelin ('87, pp. 55, 56) has shown. If both circumoral and lophophoric tentacles find their homologues in Gymnolæmata, we have only to conceive of an elongation of the postero-lateral angles of the lophophore of Gymnolæmata, after the forward movement of the posterior tentacles, to effect the condition which is found in Phylactolæmata.

The significance of the fusion of the lophophore arms is difficult to determine. I had thought it might be possible to find a phylogenetic explanation for it, by regarding the unfused tips of the arms in Cristatella as homologous with the short arms of Fredericella. In studying Plumatella, however, where the length of the lophophore arms is intermediate between that of Cristatella and Fredericella, I have been able to find no trace of this fusion. It does exist, however, in Pectinatella. I have had no material of Lophopus, upon which it is important to study this point. The evidence so far seems to indicate that this fusion of the arms during the period of their development is a secondarily acquired adaptation to some condition concerning the nature of which I am ignorant.

3. *Development of the Tentacles.* — Nitsche ('75, p. 359) observed that both layers of the bud went to form the tentacle in Phylactolæmata, and that the inner layer was derived from the outer layer of the polypide ; the outer, on the contrary, form the inner cell layer. He states, moreover, as already mentioned, that the oral tentacles arise first, then those of the outer row of the lophophore arms, of which the basal are fully formed before the terminal ones. The tentacles of the inner row, he says, are formed last, and in Alcyonella are yet lacking when the polypide is first evaginated.

My own observations confirm in general those of Nitsche. The longest tentacles in a polypide of about the age of that shown in Figure 77, Plate IX., are those arising from the region of transition from the circumoral ring canal (*can. crc.*) to the outer lophophoric ring canal (*can. crc.'*). The tentacles lying near the median plane, and in front of the mouth, are somewhat shorter than these ($75 \mu : 52 \mu$). The tentacles situated near the proximal extremity of the inner lophophoric ring canal (*can. crc.''*) are still shorter (50μ). Those situated at the tips of the lophophore arms are at this stage about 30μ in length. The tentacles behind the mouth, arising from the supraganglionic part of the ring canal (*can. crc.'''*), are shortest of all at this stage (15μ).

The two layers which, as we have seen, go to form the upper wall of the ring canal in all its parts, are the ones which give rise to the tentacles. In Figure 74, *ta.'*, Plate VIII. (compare Fig. 51, *ta.'*), young oral tentacles are cut transversely at different heights. The circumoral part of the ring canal is seen at a point (*can. crc.*) near which it opens into the cavity of the lophophore arm. The plane of the section passes obliquely upward and anteriorly from this point. The most posterior tentacle in the lower part of the figure is cut at the base. The calibre of the canal (including its walls) is evidently much enlarged at this point. The enlargements of the canal at the base of the tentacles are seen also in Figure 78, *can. crc.*, Plate IX. The more anterior tentacles in Figure 74 show the two layers well marked, but as yet enclosing no lumen. Since the tentacles arise from the ring canal at intervals only, the ring canal is a tube (or groove) whose lumen is alternately constricted and expanded laterally as well as vertically. The lumen is, indeed, often so small between the tentacles that the ring canal appears divided into separate chambers by a series of transverse septæ, which, however, are always penetrated by an opening (Fig. 78, *can. crc.*). Figures 73 and 77, *ta.'*, show, in longitudinal section, successive stages in the development of the oral tentacles. The formation of tentacles begins by a rapid cell proliferation at intervals in the upper wall of the ring canal; thus a projection is formed at each of these points, which constantly elongates to form the tentacle. Figures 70 and 69 (Plate VII.) are longitudinal sections of two later stages in the development of tentacles. The inner layer, *ex.* (Fig. 70), becomes gradually thinner as the tentacle grows older, and its cells finally become thread-like (Fig. 69, *ex.*).

Figure 81 (Plate IX.) shows the arrangement of the tentacles about the mouth and over the ganglion in a young polypide. The supraganglionic part of the ring canal is cut tangentially just behind the epistome (*can. crc.'''*). I have often noticed that, in polypides of about the age of that of Figure 77, or older, certain of the nuclei seen in a cross section of a tentacle stain more deeply than the others. These nuclei are usually two or three in number on each of the lateral surfaces of the tentacles. They are evident in Figure 81. I do not know what this difference in staining properties signifies. Vigelius ('84, p. 38, Fig. 23) describes and figures a condition of the nuclei in Flustra, as seen on cross-section, which is similar to that just described. The deeply staining nuclei in Flustra lie on the inner face of the tentacle, are larger than the others, and belong to cells which possess no cilia.

Nitsche ('71, p. 43) described the development of the tentacles in Flustra as though they were derived exclusively from the inner layers of the bud; but Repiachoff ('75ª, pp. 138, 139, '75ᵇ, p. 152) showed that in Cheilostomes both cell layers of the bud took part in their formation, and he figures an early stage which is quite similar to my Figure 70.

4. *Development of the Lophophoric Nerves.* — It has long been known that a large nerve passes along the middle of the upper wall of each lophophore arm, connecting proximally with the corresponding side of the ganglion. No observations have been made, so far as I know, upon the origin of this organ. Evidently there are, *a priori*, two possibilities. Either (1) the lophophoric nerve is formed by a direct outgrowth of the ganglion, or (2) it arises in place from the inner layer of the bud, which, since it here forms the outer layer of the lophophoric pocket, is the same as that from which the ganglion itself is constructed. By a careful study of this nerve in many stages of development, and from sections in different directions, I have come to the conclusion that it arises as an outgrowth of the walls of the ganglion, and that it penetrates between the outer and inner layers of the arm.

The facts which have led me to this conclusion are these. First, during the formation of the brain, soon after its lumen is cut off from its connection with the atrium, its cells begin to divide rapidly (Plate V.

Fig. 51, Plate VII. Figs. 63, 68) ; but that the new cells so formed do not all remain in the brain is indicated by the fact that the brain does not increase very rapidly in size. (Compare Plate III. Fig. 19, and Plate IX. Fig. 77.) This rapid cell division would be inexplicable upon the assumption of an origin *in situ*. Secondly, at an early stage the lophophoric nerve is already seen extending from the brain to the adjacent inner layer, with which it remains in contact. A longitudinal section through the middle of this nerve shows a prolongation of the lumen of the brain extending into it, so that its upper wall passes directly into the upper wall of the brain, and its lower wall into the corresponding part of the central organ (Plate VII. Fig. 68, *lu. gn., n. loph.*). The proximal part of the lophophoric nerve is thus to be regarded as a pocket of the brain. The existing condition is not what we should expect if a cord of cells derived from the outer layer of the lophophoric arm had secondarily fused with the brain. Thirdly, I have never found any good evidence that cells were being given off from the outer layer of the arm at its tip to form the nerve, where we should look for such a process, if anywhere ; on the contrary, the nerve is quite sharply marked off from the outer layer at this point, as will be seen by reference to Figures 64–67 (Plate VII.). These figures represent successive transverse sections from a young lophophore arm of about the stage of development of that shown in Figure 71. Figures 65–67 were drawn from one arm in about the position indicated by the lines 65–67 in Figure 71. Figure 64 was drawn from the opposite arm of the same individual, and in about the region of Figure 65. In Figures 64 and 65 there is a small space between the nerve (*n. loph.*) and the overlying cells of the inner layer (*i.*). This may be due to shrinkage, but in any event it indicates a complete independence between the two cell masses which it separates. Over the nerve the cells of the layer *i* are shorter than elsewhere. This might be considered as an indication that the cells had recently divided in order to give up cells to the nerve, which, on this assumption, would be formed *in situ*. Three appearances, however, indicate that the cells of the layer *i.* have been rather subjected to crowding at this point, as though by a mass of cells forcing their way between them and the layer *ex.*, and gradually increasing in volume. (*a.*) The surface of the layer *i.* is raised above the general level directly above the nerve. (*b.*) The cells of the layer *i.* are somewhat broader over the nerve than elsewhere, and the nuclei are shorter, but thicker. These are the conditions which we should expect in an epithelium subjected to pressure by the intrusion of a mass of cells at its base, for in volume the crowded cells compare fairly with their neighbors, whereas, if they had by division given rise to nerve cells, they should all be smaller. (*c.*) In Figure 67, which is a section immediately in front of the advancing tip of the nerve, the position corresponding to that opposite the nerve in the preceding sections is indicated by an asterisk (*). The nuclei are here crowded together, indicating pressure. Fourthly, there is a considerable difference in size between the nuclei of the cells of the layer *i.* of the lophophore and the nerve cells. This is not what one would expect upon the assumption of the formation of the nerve directly from the overlying cells. Fifthly, a longitudinal section through the young lophophoric nerve (Plate VII. Fig. 71) shows a more active cell division in it than in the walls of the arm (compare Fig. 64, *n. loph.*), and a crowding together of nuclei of the outer layer of the arm, *i*, at its distal end, rather than a passage of nuclei into the nerve.

The conclusion to which I have arrived from considering these facts is that the *peripheral nervous system in Phylactolæmata arises from the brain as an outgrowth of its walls.*

5. *Development of the Epistome.* — The epistome was regarded by Lankester at one time ('74, p. 80) as homologous with the foot of Mollusca, and on another occasion ('85, p. 434) as representing the preoral lobe of Annelids, — a view for which Caldwell ('83) first produced evidence from comparative embryology. In view of such divergent opinions, and of the occurrence of an organ which is possibly its homologue, in quite aberrant genera, such as Phoronis, Rhabdopleura, etc., a careful consideration of its origin and development is desirable.

After the ganglion is fully formed, its oral face remains in contact in front with the posterior wall of the œsophagus (Plate V. Fig. 52, Plate IX. Fig. 77), and on each side with the outer wall of the lophophoric pockets by means of the lophophoric nerves (Plate VII. Fig. 63, *n. loph.*). The outer layer of the bud penetrates between the ganglion and rectum, but not between the ganglion and the œsophagus (Fig. 51,*). This layer also comes to lie between the floor of the atrium above, the ganglion below, and the lophophoric nerves on either side, having made its way in from behind as a double cell-layer enclosing a flat cavity (Plate V. Fig. 52, Plate VI. Fig. 56, Plate VIII. Fig. 74, *can. e stm.*). My description of the process by which the inner layer comes to envelop the ganglion above and behind differs considerably from Nitsche's, already quoted (page 114). As the ganglion becomes farther removed from the floor of the atrium, the cavity above it (*can. e stm.*) enlarges, and the two lateral walls of this canal, each composed of two layers of cells, both belonging to the outer layer of the bud, form the " Verbindungsstrang des Ganglions mit dem Lophoderm" of Kraepelin ('87, p. 63, Taf. II. Fig. 59, *vs.*). (See Plate V. Fig. 51, Plate VI. Fig. 56, and Plate IX. Fig. 80,*.) This canal is the only one by which communication between the body cavity and the cavity of the epistome can occur. It may be called the *epistomic canal* (Plate V. Fig. 52, Plate VIII. Fig. 72, *can. e stm.*).

The epistome proper arises at the point where the epistomic canal ends blindly, above and in front of the brain (Plate VIII. Fig. 73, Plate IX. Fig. 77, *e stm.*); it is a pocket, the outer wall of which is continuous on its under surface with the œsophageal epithelium, and on its upper surface with the floor of the atrium. The growth of this organ is disproportionately great after the first evagination of the polypide. That part of its wall which is turned towards the alimentary tract is then much thicker than the remaining part; it forms the posterior wall of the pharynx (Plate VIII. Fig. 72, *e stm.*; compare Plate IX. Fig. 81). Is the epistome innervated by fibres from the brain, as maintained by Hyatt ('68, pp. 41–43)? I have not succeeded in finding such fibres, and the conditions of the formation of the epistome, cut off as it is from the brain at every point, make such a connection improbable.

Allman ('56, Fig. 8, Plate XI.) and Korotneff ('75, p. 371) have shown for Paludicella, and Nitsche ('71, p. 44) has shown for Flustra, that an epistome-like fold occurs at an early stage of development, but is absent in the adult. Such an organ has been described by Allman ('56, p. 56) and other observers in Pedicellina, and it is still more prominent in Loxosoma, in which the relation of the epistome to the body cavity is similar to that in the Phylactolæmata.

The constant occurrence of this organ in the development of Bryozoa, and its presence in so many aberrant genera which seem to be somewhat allied to this group, can only be interpreted, it seems to me, as signifying that it is an ancient and morphologically important organ. The manner of its development in Cristatella seems to throw very little light, however, upon its significance; it arises rather late, and does not become of any considerable size until the atrial opening is made.

6. *Development of the Alimentary Tract.* — The later development and histological differentiation of the alimentary tract have not been heretofore carefully studied.

At the stage at which we left the alimentary tract (Plate III. Fig. 19) only two parts were clearly differentiated, the œsophagus and the intes-

tine. In the next stage shown (Plate VIII. Fig. 73), further changes are
seen to have taken place. The most prominent is the down-folding of
the lower wall of the intestine at its middle region to form the cœcum.
Even at this early stage histological differentiation of the cells of this
region has occurred to such an extent that the lumen of the cœcum is
nearly obliterated by the great elongation of some of the cells lining it.
This condition of affairs will be understood by studying the cross section of the cœcum at a later stage, as shown in Figure 94, Plate X.
The cavity of the rectum has also enlarged, and its cells have taken
on the regular columnar appearance which exists in the adult.

At a still later stage (Plate IX. Fig. 77), the position of the cardiac
and pyloric valves, separating respectively the œsophagus (œ.) from the
stomach (ga.), and the cœcum (cœ.) from the rectum (rt.), is clearly indicated. The blind sac is still further elongated and well differentiated
from both stomach and rectum. In order to attain the adult condition
(Plate VIII. Fig. 72), the oral portion of the alimentary tract has merely
to become divided, by a difference in the character of its cells, into
pharynx (phx.) and œsophagus (œ.), the stomach (ga.) to increase in
diameter, and the blind sac (cœ.) to elongate. The anus (an.) finally
comes to lie at the apex of a small cone, or sphincter valve.

The histological changes which the cells of the different parts of the
alimentary tract undergo are considerable, and will be treated of in
order, beginning with the

Œsophagus. — At a stage a little later than Figure 77, the œsophagus,
as is shown in Figure 84, Plate X., has a small diameter relative to that
of the rest of the alimentary tract (cf. Plate VIII. Fig. 72, œ.), and its
inner lining is composed of high columnar epithelium, like that of the
oral groove. The shape of the cells is not greatly different in the adult ;
but they become vacuolated, and since these vacuoles lie near the base
of the cells, and either nearer to or farther from the lumen than the
nuclei, the latter acquire that irregular arrangement referred to by
Verworn ('87, pp. 111 and 112).

Stomach. — Figure 93 (Plate X.) represents a section across the stomach
immediately below the cardiac valve, from the same individual as that
from which Figure 84 was taken. The proximal ends of all cells stain
more deeply than the distal ends, but the cells are all alike as far as regards receptivity to stains. Already, in certain regions, the cells are
higher or lower than the average, and have even begun to group themselves as typical ridge- and furrow-cells. Figure 82 is a section through
the same region as Figure 93, but from an adult individual. The ridge-

cells are distinguishable from those of the furrows by their greater height, their weaker attraction for dyes, and their vacuolated and granular appearance. Moreover, the cell boundaries of this epithelium are gradually lost. Kraepelin ('87, p. 51) has argued that the elongated cells are the true digestive cells, and that the deeply dyed cells of the furrows are, functionally, liver cells.

Cœcum. — Figure 94 is from a cross section of the cœcum at the stage of Figures 84 and 93. The cells are more differentiated here than at any other part of the alimentary tract. They stain uniformly, however, except for a narrow light zone next to the lumen, and all reach to the muscularis. The digestive cells are swollen at their free ends; the liver cells, on the contrary, are thickest at the base. Figure 83 is from a section of the proximal part of the cœcum of an adult. The changes which the cells have undergone are of a similar character to those experienced by the gastric epithelium, only there has been an exaggeration in this region of the features shown by the stomach. Figure 85 represents a section near the blind end of the cœcum of an adult. The diameter of the tube is smaller here than in the section last described, but the inner epithelium is thrown into still higher ridges and more profound furrows. Nearly all of the cells, however, seem to extend to the muscularis. The "liver" cells do not extend so far towards the blind end of the cœcum as this region. The cytoplasm is not at all stained. Evidently, here the process of digestion reaches a maximum. The circular muscles of the muscularis are *striped*, and are developed here to an extraordinary degree, and the cœlomic epithelium is greatly thickened, another evidence, it seems to me, of the intimate relation of this layer to the muscularis. The number of ridges is not constant in different parts of the alimentary tract of the same individual, and varies somewhat for the same region in different individuals. In sections corresponding in position to Figure 83, I have, however, usually found six ridges.

7. *Development of the Funiculus and Muscles.* — It has already (page 117) been pointed out that the fixed ends of both the funiculus and muscles originate at a great distance from their position in the adult. Thus the funiculus originates upon the *oral* face of a young bud. As this bud grows older, the fixed end of its funiculus becomes gradually farther and farther removed from its neck towards the margin, until finally the funiculus is inserted upon the colony-wall at the margin, or even upon the sole. So the retractor and rotator muscles arise together on each side of the polypide and in the angle formed by the colony-wall and the radial partitions. Later (Plate V. Figs. 44, 45, *mu. ret.* + *rot.*)

they are found on the partitions immediately below the colony-wall. Still later (Plate VI. Fig. 59, *mu. rot.*, *mu. ret.*) we see them on the lower portion of the partition, and finally (Fig. 56, *mu. rot.*, *mu. ret.*) they are found attached to the sole, at some distance, it may be, from the radial partition.

The question arises at once, How do these changes of position take place? Examination shows that the union between the cœlomic epithelium and the cells of that portion of the funiculus which is attached to the roof is very slight after the funiculus has passed to some distance from the mother polypide. Although occasionally I have seen the cells of the fixed end closely applied to the cœlomic epithelium, the only connection between the two is usually effected by means of amœboid cells (Plate V. Figs. 46–48, *cl. mi.*). On cross sections of the fixed end of the funiculus these cells (Fig. 49, *cl. mi.*) are seen to surround it as a loose layer, and in longitudinal sections some of the amœboid cells are seen to be connected with the cœlomic epithelium. It is difficult to determine the origin of these cells, but they have the position and character of the cells of which the funiculus was exclusively composed before the entrance into it of the ectodermal plug described by Braem. The only explanation of the migration of the funiculus which occurs to me has been suggested by the facts given above; namely, that the " migratory cells," by which the funiculus is attached to the cœlomic epithelium, change their position, carrying with them the funiculus. Remembering that the cœnocœl is filled with a fluid in which the funiculus floats, and that by the growth of the funiculus it is elongated in proportion as the distance from its origin to the cœcum increases, this hypothesis does not seem improbable, although its truth can hardly be tested by the study of preserved material. When the funiculus has reached its permanent position its attachment to the cœlomic epithelium is more intimate. Meanwhile the end attached to the polypide has become more and more attenuated (Plate IX. Fig. 77, *fun.*), until, in the adult, I have usually been unable to discover any attachment. In any case, it must certainly be broken when the polypide begins to degenerate.

The migration downward of the ends of the muscles which are attached to the partition is even more difficult of explanation. During this migration their point of origin seems to be in the muscularis of the partition itself. The fixed point of the muscle in the adult is probably in the muscularis of the sole, since I have traced muscle fibres through the cœlomic epithelium, and to the muscularis (Plate VI. Fig. 58, *mu.*

ret.). The insertion is in the muscularis of the polypide (Fig. 56), but
I have not been able to determine the precise relation between the
muscle fibres of the great cœlomic muscles and those of the muscularis. A comparison of Figures 44, 59, and 56 shows quite plainly that
both the retractor and the rotator muscles originate from a common
mass of muscle cells, and become distinct from one another by a
separation of their points of attachment to the polypide. The retractor muscles (*mu. ret.*) are attached to the œsophagus immediately
below the ganglion (Plate IX. Fig. 78) ; the rotator muscles (*mu. rot.*),
on the contrary, to the lateral walls of the opening leading from the
cœnocœl (*cœn.*) to the cavity of the lophophore arms. These two regions are near to each other in the young polypide, but become constantly more widely separated with the growth of the lophophore.
Compare Figure 78 with Figures 74 (Plate VIII.) and 51 (Plate V.),
which are younger stages, cut somewhat above the level of Figure 78,
and more than twice as highly magnified.

I have been able to obtain in thick sections various stages in the
development of the muscle fibres, some of which are shown in Figures 89
to 92 (Plate X.). In the earlier stages, all parts of the muscle cell stain
uniformly in cochineal. Later, the cell body becomes differentiated into
two portions, easily distinguishable by their different receptivity to the
dye. The more retractile portion becomes greatly elongated, highly
refractive, and incapable of being stained. A mass of indifferent protoplasm, including the nucleus, still remains stainable (Fig. 90). The
undifferentiated portion continues to diminish relatively to the whole
mass of the cell, which has greatly increased in size, until little remains
but the nucleus, placed on one side of the muscle fibre (Figs. 91, 92).
Figure 92 is one of the retractor muscle fibres, in a partly contracted
state. The end placed uppermost in the figure was that which abutted
upon the muscularis of the œsophagus. Its more intimate relation to
the muscularis could not be traced.

8. *Origin and Development of the Parieto-vaginal Muscles.* — These
consist of two sets, the lower, or *posterior*, and the upper, or *anterior*.
The posterior arise earlier. At about the time when the neck of the
polypide begins to disintegrate in order that the polypide may become
extrusible, a disturbance is seen in the cells of the outer layer of the
kamptoderm immediately below the neck of the polypide, and in the
cœlomic epithelium opposite to them (Plate XI. Fig. 97, *mu. inf.*). As
a result, several cells of each layer become organically connected with
those of the opposite layer, and give rise to muscle cells. A later stage of

such a process is seen at Figure 98. By the time the atrial opening is established these cells have become plainly muscular (Plate IX. Fig. 79). Farther up in the angle of attachment of the kamptoderm to the roof of the colony, the cœlomic epithelium and the outer layer of the bud are both seen to be somewhat disturbed (Fig. 97, *mu. su.*). At different points, a single one of these cells reaches across, and later becomes differentiated into a genuine muscle cell (Fig. 99, *mu. su.*). Of these there may be three rows.

9. *Disintegration of the Neck of the Polypide.* — The neck of the polypide, having fulfilled its function as the most important part of the stolon, must now give way to allow of the extrusion of the nearly developed polypide. The first indication of this process is the formation within the cells of the neck of a secreted substance (*cp. sec.'*), apparently like the secreted bodies of the ectoderm. This metamorphosis first involves the outer and middle cells of the neck only (Plate XI. Fig. 97, *cev. pyd.*). Later (Plate IX. Fig. 77, *of. atr.*) a depression occurs in the ectoderm. This is due, I believe, to a cessation of cell proliferation at the centre, although it remains active at the edges of the neck. The depression gradually deepens until the atrium is closed by a thin layer of cells only (Fig. 98). The cells of the side of the neck do not disintegrate, but go to form the "Randwulst" of Kraepelin ('87, p. 40). The cells of this region remain unmetamorphosed. Only a thin layer of cells now stands between the polypide and the outside world. This ruptures, as is shown in Figure 99, and by the relaxation of the muscularis, which is thickened about the atrial opening into a sphincter (Fig. 98, *spht.*), the polypide is ready to expand itself.

10. *Development of the Body-wall.* — As already stated (page 117), Braem believes that the whole body-wall in Alcyonella is derived from the neck of the young polypide, after it has begun to give rise to daughter polypides; and I have given my reasons for believing that in Cristatella a portion of it at least is derived from the margin.

In addition to this, cells are undoubtedly added to the body-wall, as Braem states, after the time of origin of the buds. Particularly after the formation of the median bud, the neck appears to continue to furnish cells to the ectoderm. Figure 73,* Plate VIII., shows such a mass of cells. Later stages show that these cells secrete a gelatinous substance within their protoplasm (*cp. sec.'*, Plate XI. Figs. 97, 98); they gradually increase in width and height from the neck outward (Figs. 97–99), and at the same time become more and more completely metamorphosed. The result of the addition of these cells from the neck of the polypide is to carry the body-wall at the region of the atrial opening to a considerable height above the level of that portion of the roof lying between polypides. (Compare Fig. 73, Plate VIII.; Figs. 98 and 99, Plate XI.) This method of origin of the body-wall is of much less importance in Cristatella than in Alcyonella, since the extent of the proper body-wall about the atrial opening is much less in the former than in the latter case.

The development of the gelatinous bodies deserves further attention. Kraepelin ('87, p. 24) concluded, from a study of the condition in a statoblast embryo, that they are formed by a metamorphosis of the cell protoplasm, beginning at the outer end of the cylindrical cell, and finally involving, in some cases, the entire cell, together with its nucleus. Some appearances which I have noticed in the ectoderm of Cristatella lead me to conclude that the origin is not always so simple as Kraepelin describes. Figure 79, Plate IX., shows at *cp. sec.* a number of small gelatinous masses occurring at various regions in the protoplasm. Such an appearance is quite common, and must be interpreted, it seems to me, as the formation of the gelatinous balls by an intra-cellular metamorphosis of the cytoplasm. The balls, flowing together, produce the larger masses. The metamorphosed matter from several cells may also fuse into one mass (Plate VI. Fig. 55, *cp. sec.*). The final result of this process of cell metamorphosis in the ectoderm is a frame-work of old cell walls, having a thin layer of protoplasm and nuclei at its base, and inclosing the great gelatinous balls. Such a condition exists near the centre of the colony between adult polypides, and is shown in Figure 100, Plate XI.

Summary.

1. Most individuals give rise to two buds, of which one forms a new branch, the other continues the ancestral branch.

2. The median buds migrate away from the parent polypide to a considerable distance before giving rise to new buds.

3. The descendants of equal age from common ancestors are arranged similarly in the same region of the colony.

4. New branches are formed upon either side of ancestral branches.

5. The greater the difference in age between the youngest and the next older bud, the greater the distance between the points at which they begin to develop.

6. In typical "double buds," both polypides arise from a common mass of cells at the same time. From the neck of old polypides a stolon-

like process of cells is given off to form median buds. Between these two extreme types, intermediate conditions occur.

---

7. The alimentary tract is formed by two out-pocketings of the lumen of the bud in the median plane, one forming the œsophagus, the other the rectum and stomach. The blind ends of these two pockets fuse, and thus form a continuous lumen.

8. The central nervous system arises as a shallow pit in the floor of the atrium; the pit becomes closed over by a fold of the inner layer only of the polypide, which thus forms a sac, the walls of which become the ganglion.

9. The kamptoderm arises by the transformation of the columnar epithelium of the two layers of the wall of the atrium into pavement epithelium.

10. The funiculus arises from amœboid cells derived from the cœlomic epithelium.

11. The retractor and rotator muscles arise together from the cœlomic epithelium of both body-wall and bud, and in the angles formed by the radial partitions and the body-wall.

12. The wall of the colony grows by cell proliferation at its margin.

13. The radial partitions arise as follows: certain muscles of the muscularis at the margin of the colony leave the latter, and are carried into the cœnocœl, taking with them a covering of cœlomic epithelium.

---

14. Budding in Cristatella presents conditions transitional between direct and stoloniferous budding.

15. Throughout the group of Bryozoa, the youngest and next older buds are intimately related, and the place of the origin of the younger buds relatively to the older is determined by a definite law.

16. Cristatella differs from Alcyonella in possessing a region of the colony-wall, — the tip of the branch, — which grows independently of the polypides.

17. Each of the layers of the younger bud arises from a part of the same cell mass as that which gave rise to the corresponding layer of the next older bud.

18. The digestive epithelium and the nervous tissue are both derived from one and the same layer of cells, the inner layer of the bud.

19. The alimentary tract of Cristatella at an early stage is similar to that of a young Endoproct.

20. Harmer's conclusion, that the ganglion of Phylactolæmata arises exactly as in Endoprocta, is not confirmed.

---

21. The "ring canal" lies at the base of all tentacles.

22. The circumoral region of the ring canal in Cristatella is in free communication with the cœnocœl in all stages of development; and not closed, as maintained by Kraepelin.

23. The two arms of the lophophore arise independently of each other. Their adjacent surfaces undergo a secondary fusion, which persists until the inner row of tentacles is about to be formed on the lophophore. The two arms then become entirely separate.

24. The ancestor of Bryozoa probably possessed a U-shaped row of tentacles, encircling the mouth in front, and ending freely behind near the anus.

25. The tentacles near the mouth are phylogenetically the oldest.

26. Both layers of the bud are involved in the formation of the tentacles.

27. The lophophoric nerves arise as outgrowths of the central ganglion, which make their way into the lophophore arms.

28. The epistome arises as a fold continuous with the wall of the œsophagus below and the floor of the atrium above, and it communicates with the cœnocœl by means of the epistomic canal.

29. The cœcum of the alimentary tract, which occurs only in Ectoprocta, is produced relatively late in the ontogeny by an out-pocketing of the lower wall of the alimentary tract at the free end of the polypide.

30. The funiculus migrates (probably with the aid of amœboid cells) from the roof of the colony to the margin, or even to the sole.

31. The "origins" of the retractor and rotator muscles migrate along the radial partitions from roof to sole. The separation of the two muscles takes place secondarily as their points of insertion separate.

32. The parieto-vaginal muscles arise from the cœlomic epithelium of the body-wall and polypide.

33. The disintegration of the neck of the polypide is begun by a metamorphosis of the protoplasm of its cells. The metamorphosed cells break away, leaving the atrial opening.

34. The part of the body-wall lying around the atrial opening arises by proliferation of cells derived from the neck of the polypide.

35. The ectodermal cells become metamorphosed by an intercellular secretion of small "Gallertballen," which fuse to form the larger ones. Often the contents of more than one cell fuse into a single large mass.

CAMBRIDGE, June, 1890.

BIBLIOGRAPHY.

Allman, G. J.
'56· A Monograph of the Fresh-water Polyzoa, including all the known Species, both British and Foreign. London: Ray Society, 1856, pp. i.–viii., 1–119, 11 Pls.

Barrois, J.
'86· Mémoire sur la Métamorphose de quelques Bryozoaires. Ann. des Sciences Naturelles. Zoologie. 7ᵉ série, Tom. I. No. 1, pp. 1–94, Pls. I.–IV. 1886.

Braem, F.
'88· Untersuchungen über die Bryozoen des süssen Wassers. Zool. Anzeiger, XI. Jahrg., No. 288, pp. 503–509; No. 289, pp. 533–539. 17 Sept., 1 Oct , 1888.
'89ᵃ. Ueber die Statoblastenbildung bei Plumatella. Zool. Anzeiger, XII. Jahrg., No. 299, pp. 64, 65. 4 Feb., 1889.
'89ᵇ. Die Entwicklung der Bryozoencolonie im keimenden Statoblasten. (Vorläufige Mitth.) Zool. Anzeiger, XII. Jahrg., No. 324, pp. 675–679. 30 Dec., 1889.

Caldwell, W. H.
'83· Preliminary Note on the Structure, Development, and Affinities of Phoronis. Proc. Roy. Soc. Lond., Vol. XXXIV. pp. 371–383. 1883.

Ehlers, E.
'76· Hypophorella expansa. Ein Beitrag zur Kenntniss der minirenden Bryozoen. Abhand. d. königl. Gesellsch. d. Wiss. zu Göttingen, Bd. XXI. pp. 1–156, Taf. I.–V. 1876.

Haddon, A. C.
'83· On Budding in Polyzoa. Quart. Jour. of Micr. Sci., Vol. XXIII. pp. 516–555, Pls. XXXVII. and XXXVIII. Oct., 1883.

Harmer, S. F.
'85· On the Structure and Development of Loxosoma. Quart. Jour. of Micr. Sci., Vol. XXV. pp. 261–337, Pls. XIX –XXI. April, 1885.
'86· On the Life-History of Pedicellina. Quart. Jour. of Micr. Sci., Vol. XXVII. No. 101, pp. 239–264, Pls. XXI., XXII. Oct., 1886.

Hatschek, B.
'77· Embryonalentwicklung und Knospung der Pedicellina echinata. Zeitschr. f. wiss. Zool., Bd. XXIX. Heft 4, pp. 502–549, Taf. XXVIII.–XXX., u. 4 Holzsch. 18 Oct., 1877.

Hyatt, A.

'68· Observations on Polyzoa. Suborder Phylactolæmata. Salem [Mass.]. Printed separately from Proc. Essex Inst., Vols. IV. and V., 1866–68, pp. i.–iv., 1–103. 9 Pls.

Joliet, L.

'77· Contributions à l'Histoire Naturelle des Bryozoaires des Côtes de France. Arch. de Zool. Expér., Tom. VI. No. 2, pp. 193–304, Pls. VI.–XIII. 1877.

'86· Recherches sur la Blastogénèse. Arch. de Zool. Exper., 2e série, Tom. IV. No. 1, pp. 37–72, Pls. II., III. 1886.

Korotneff, A. A.

'74· Почкованіе Paludicella. Bull. Roy. Soc. Friends of Nat. Hist. Moscau, Vol. X. Pt. 2, pp. 45–50, Pls. XII., XIII. 1874. [Russian.]

'75· [Abstract of Korotneff, '74, by Hoyer, in Hofmann u. Schwalbe's Jahresber. Anat. u. Phys. f. 1874, Bd. III. Abth. 2, pp. 369–372. 1875.]

'89· Sur la Question du Développement des Bryozoaires d'Eau douce. Mémoires de la Société des Naturalistes de Kiew, Tom. X. Liv. 2, pp. 393–410, Tab. V., VI. 21 Oct., 1889. [Russian.]

Kraepelin, K.

'86· Ueber die Phylogenie und Ontogenie des Süsswasserbryozoen. Biol. Centralblatt, Bd. VI. Nr. 19, pp. 599–602. 1 Dec., 1886.

'87· Die Deutschen Süsswasser-Bryozoen. Eine Monographie. I. Anatomisch-systematischer Teil. Abhandl. der Naturwiss. Verein in Hamburg, Bd X., 168 pp., 7 Taf. 1887.

Lankester, E. R.

'74· Remarks on the Affinities of Rhabdopleura. Quart. Jour. Mic. Sci., Vol. XIV. pp. 77–81, with woodcut. 1874.

'85· [Article.] Polyzoa. Encyclopædia Britannica, Ninth Edition, Vol. XIX. pp. 429–441. 1885.

Metschnikoff, E.

'71· Beiträge zur Entwickelungsgeschichte einiger niederen Thiere. 6. Alcyonella. Bull. de l'Acad. Imp. Sci. de St. Pétersbourg, Tom. XV. pp. 507, 508. 1871.

Nitsche, H.

'69· Beiträge zur Kenntniss der Bryozoen. I. Beobachtungen über die Entwicklungsgeschichte einiger chilostomen Bryozoen. Zeitschr. f. wiss. Zool., Bd. XX. Heft 1, pp. 1–36, Taf. I.–III. 1 Dec., 1869.

'71· Beiträge zur Kenntniss der Bryozoen. III. Ueber die Anatomie und Entwicklungsgeschichte von Flustra Membranacea. IV. Ueber die Morphologie der Bryozoen. Zeitschr. f. wiss. Zool., Bd. XXI. Heft 4, pp. 416–498, Taf. XXXV.–XXXVIII. and 4 Holzschn. 20 Nov., 1871. [Also separate, pp. 1–83.]

'72· Betrachtungen über die Entwicklungsgeschichte und Morphologie der Bryozoen. Zeitschr. f. wiss. Zool., Bd. XXII. Heft 4, pp. 467–472, 2 Holzschn. 20 Sept., 1872.

'75· Beiträge zur Kenntniss der Bryozoen. V. Ueber die Knospung der Bryozoen. A. Ueber die Knospung der Polypide der phylactolæmen Susswasserbryozoen. B. Ueber den Bau und die Knospung von Loxosoma Kefersteinii Claparède. C. Allgemeine Betrachtungen. Zeitschr. f. wiss. Zool., Bd. XXV., Supplementband, Heft 3, pp. 343-402, Taf. XXIV.–XXVI. 22 Dec., 1875.

Ostroumoff, A.

'85· Remarques relatives aux Recherches de Mr. Vigelius sur des Bryozoaires. Zool. Anzeiger, VIII. Jahrg., No. 195, pp. 290, 291. 18 Mai, 1885.

'86[b]. Contributions à l'Étude zoologique et morphologique des Bryozoaires du Golfe de Sébastopol. Arch. Slaves de Biologie, Tom. II. pp. 8-25, 184-190, 329-355, 5 Pls. 1886.

Pergens, E.

'89· Untersuchungen an Seebryozoen. Zool. Anzeiger, XII. Jahrg., No. 317, pp. 504-510. 30 Sept., 1889.

Reichert, K. B.

'70· Vergleichende anatomische Untersuchungen über *Zoobotryon pellucidus* (Ehrenberg). Abhandlungen der königlichen Akademie der Wissenschaften zu Berlin, aus dem Jahre 1869, II., pp. 233-338, Taf. I.–VI., Berlin, 1870.

Reinhard, W. W.

'80[a]. Zur Kenntniss des Süsswasser-Bryozoen. Zool. Anzeiger, III. Jahrg., No. 54, pp. 208-212. 3 May, 1880.

'80[b]. Embryologische Untersuchungen an Alcyonella fungosa und Cristatella mucedo. Verhandl. d. Zool. Sect. VI. Vers. Russ. Naturf. Abstract by BRANDT, A., in Zool. Anzeiger, III. Jahrg., No. 55, pp. 234, 235. 10 May, 1880.

'82· "Skizze des Baues und der Entwickelung der Süsswasser-Bryozoen." Charkow, 1882. 7 Taf. [Russian.]

'88· "Skizze des Baues und der Entwickelung der Süsswasser Bryozoen." Arb. Naturf. Gesellsch. Charkow, Bd. XV. pp. 207-310, 7 Taf. 1888. [Russian.]

Repiachoff, W.

'75[a]. Zur Entwickelungsgeschichte der Tendra zostericola. Zeitschr. f. wiss. Zool., Bd. XXV. Heft 2, pp. 129-142, Taf. VII.–IX. 1 März, 1875.

'75[b]. Zur Naturgeschichte der Chilostomen Seebryozoen. Zeitschr. f. wiss. Zool., Bd. XXVI. pp. 139-160, Taf. VI.–IX. 8 Dec., 1875.

Saefftigen, A.

'88· Das Nervensystem der phylactolæmen Süsswasser-Bryozoen. (Vorläufige Mittheilung.) Zool. Anzeiger, XI. Jahrg., No. 272, pp. 96-99. 20 Feb., 1888.

Seeliger, O.

'89· Die ungeschlechtliche Vermehrung der endoprokten Bryozoen. Zeitschr. f. wiss. Zool., Bd. XLIX. Heft 1, pp. 168–208. Taf. IX. u. X., 6 Holzschn. 13 Dec., 1889.

Verworn, M.

'87· Beiträge zur Kenntnis der Süsswassserbryozoen· Zeitschr. f. wiss. Zool., Bd. XLVI. Heft 1, pp. 99–130, Taf. XII. u. XIII· 25 Nov., 1887.

Vigelius, W. J.

'84· Die Bryozoen, gesammelt während 3. u. 4. Polarfahrt des " Willem Barents" in den Jahren 1880 und 1881. Bijdragen tot de Dierkunde. Uitgegeven door het Genootschap Natura Artis Magistra, te Amsterdam, 11e Aflevering, 104 pp. 8 Taf. 1884.

EXPLANATION OF FIGURES.

All figures were drawn with the aid of a camera lucida from preparations of
Cristatella mucedo.

ABBREVIATIONS.

An.	Anal side of polypide.	*mu.*	Muscularis.
an.	Anus.	*mu. inf.*	Inferior parieto-vaginal muscles.
atr.	Atrium.		
br. loph.	Lophophore arm.	*mu. lg.*	Longitudinal muscle fibre of muscularis.
can. crc.	Ring canal, circumoral part.		
		mu. ret.	Retractor muscle of polypide.
can. crc.'	Ring canal, outer lophophoric part.		
		mu. rot.	Rotator muscle of polypide.
can. crc.''	Ring canal, inner lophophoric part.		
		mu. su.	Superior parieto-vaginal muscles.
can. crc.'''	Ring canal, supra-ganglionic part.		
		mu. tr.	Transverse (circular) muscle fibre of muscularis.
can. e stm.	Epistomic canal.		
cav. loph.	Cavity of lophophore arm.		
		n. loph.	Lophophoric nerve.
cev. pyd.	Neck of polypide.	*nu. ml.*	Nucleus of muscle fibre·
cl. fun.	Young cells of funiculus.	*œ.*	Œsophagus.
cl. mi.	Migratory cells.	*of. atr.*	Atrial opening.
cl. mus.	Young muscle cells.	*om.*	Ovum.
cœ.	Cœcum.	*Or.*	Oral side of polypide.
cœn.	Cœnocœl.	*or.*	Mouth.
cp. sec.	Secreted bodies of ectoderm.	*pam. atr.*	Floor of atrium.
		pam. gn.	Floor of ganglion.
cta.	Cuticula.	*phx.*	Pharynx.
di sep.	Intertentacular septum.	*pyd.* [i., ii., &c.]	Polypide.
di sep. r.	Radial septum of colony.	*pyd. fili.*	Daughter polypide.
ec.	Ectoderm.	*pyd. ma.*	Mother polypide.
e stm.	Epistome.	*rt.*	Rectum.
e t. cœl.	Cœlomic epithelium.	*sol.*	Sole.
ex.	Outer layer of bud.	*spht.*	Sphincter.
fun.	Funiculus.	*sul. or.*	Oral groove.
ga.	Stomach.	*ta.*	Tentacle.
gn.	Ganglion.	*ta.'*	Oral tentacle.
i.	Inner layer of bud.	*tct.*	Roof of colony.
kmp. drm.	Kamptoderm.	*tct. gn.*	Roof of ganglion.
loph.'	Place of union of arms of lophophore.	*vac.*	Vacuole.
		vlv. cr.	Cardiac valve.
lu. gm.	Lumen of the bud.	*vlv. py.*	Pyloric valve.
lu. gn.	Lumen of the ganglion.		

PLATE I.

Fig. 1. A portion of the lateral rim of a colony. An optical section taken just below the roof of the colony, showing the arrangement of polypides. × 72.

" 2. Origin of the stolon (I.) from the neck of a mother polypide of about Stage XII. (Fig. 18). Sagittal section of mother polypide. The margin of the colony is to the left. × 390.

" 3. Earliest stage in the origin of a bud from a young mother polypide. Sagittal section. Margin to left. × 390.

" 4. Origin of a bud from a mother polypide of about the age of that of Fig. 3. Sagittal section. The margin of the colony is to the right of figure. × 390.

" 5. Sagittal section of a double bud. Margin of colony to the left. × 390.

" 6. Later stage in bud formation of same type as Fig. 4. Sagittal section. × 390.

" 7. A part of the right side of a polypide of a stage of development intermediate between those of Figs. 19 and 73. Seen from the sagittal plane. The cut surface lies to the right of the sagittal plane, and passes through the orifice of the right lophophore arm. The alimentary tract thus lies immediately above the plane of the paper. × 150.

Pl. I

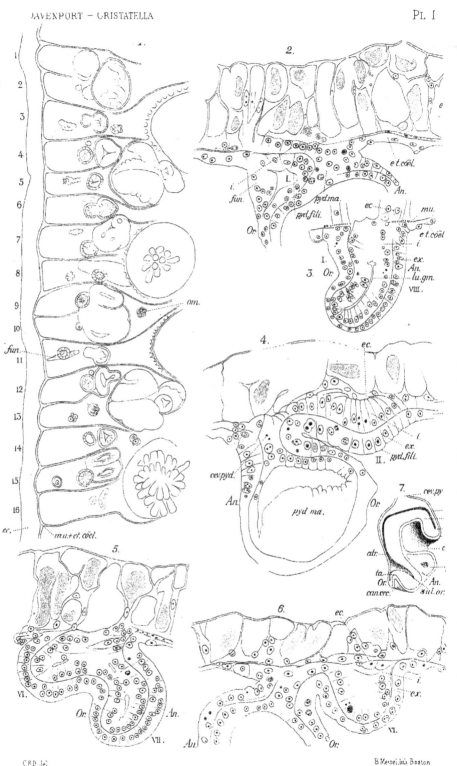

PLATE II.

All figures are magnified 390 diameters, and are from sagittal sections.

Fig. 8. Stage II. in the same series as Fig. 2. The funiculus, *fun.*, has moved farther from the mother polypide. Margin to left.

" 9. Stage IV. The inner layer, *i.*, of the bud is definitely formed, and the external layer is greatly thickened. Margin to left of figure.

" 10. Stage V. The cells, *i.*, have arranged themselves in a layer, and begin to form an invagination. Margin to right.

" 11. Stage VIII. The first indications of the alimentary tract appear as a depression in the inner layer, *rt.* The funiculus, *cl. fun.*, has begun to form, as is indicated by a disturbance of the cœlomic epithelium. Daughter bud forms Stage VI. in a series beginning with I., Fig. 3. Margin to left.

" 12, 13. Successive stages in the formation of the alimentary tract.

" 14. Stage VI. The two cell-layers are now definitely formed, and a lumen has begun to appear in the inner. Margin to right.

" 15. Stage III. in the stoloniferous type of budding. Stolon has elongated greatly, and active cell division is taking place at its distal (i. e. marginal) end.

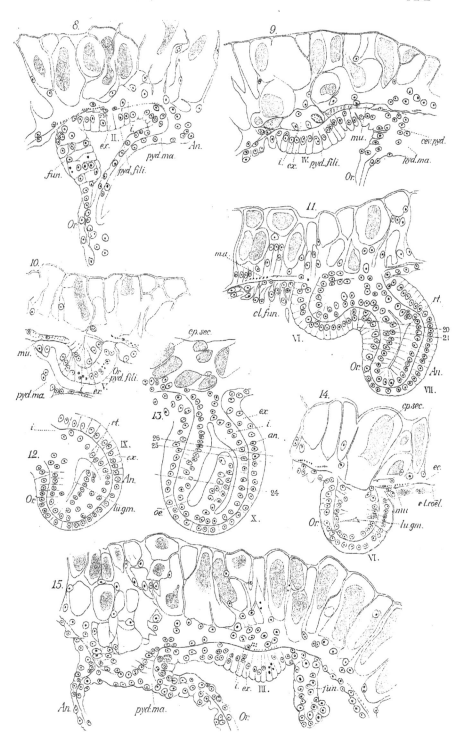

PLATE III.

All figures are magnified 390 diameters.

Fig. 16. A later stage (VI.) in the *direct* type of bud formation. The mother polypide is cut to one side of its sagittal plane, and shows the invagination of the lophophore arm (*br. loph.*). The funiculus appears as scattered cells about both buds.

" 17. A still later stage in the same series as Fig 16. The daughter bud (VII.) has a lumen. In the mother polypide (XI.) the atrium has enlarged by the inshoving of the lophophore arms. The œsophageal and rectal invaginations are not yet continuous, and the ring canal (*can. crc.*) has begun to appear oralward in the sagittal plane. Sagittal section.

" 18. Stage XII. Alimentary tract nearly complete. Beginning of the formation of the ganglion. One of the lophophore arms is cut tangentially. Sagittal section.

" 19. Stage XIII. Ganglion closing. The lophophore arm cut tangentially. Sagittal section.

" 20 and 21. The positions and directions of the planes of these sections are shown by their projections on a sagittal section (Fig. 11, lines 20, 21) of an individual of the same age. To show non-participation of the outer layer in the first stage in formation of the alimentary tract.

" 22. Early stage in *direct* bud-formation. Origin of funiculus, *cl. fun.* Sagittal. Margin to right.

" 23. The position and direction of the plane of this section are shown by its projection on a sagittal section (Fig. 19, line 23) of an individual of the same age. This figure shows the folds of the inner layer at the mouth of the ganglionic sac.

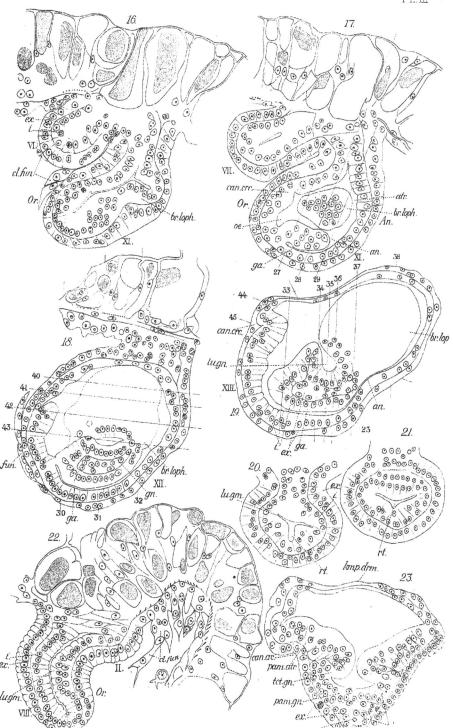

PLATE IV.

All figures, except Fig. 39, are vertical right-and-left sections, and all are magnified 390 diameters.

Figs. 24–26. Three sections from a series passing from the oral to the aboral face of a polypide of about Stage X., and cutting it in the planes indicated by the lines 24–26, Fig. 13.

" 27–29. Three sections of a series cut from a polypide of Stage XI. The planes of section are indicated in the lines 27–29, Fig. 17.

" 30–32. Three sections; whose positions are indicated by the lines 30–32, Fig. 18, cut from a polypide of Stage XII.

" 33–38. Six sections cut from a polypide of Stage XIII. in the directions indicated in Fig. 19 by the lines 33–38.

" 39. A horizontal section of a polypide somewhat older than that represented in Fig. 18, and passing nearly in the direction of the line 43.

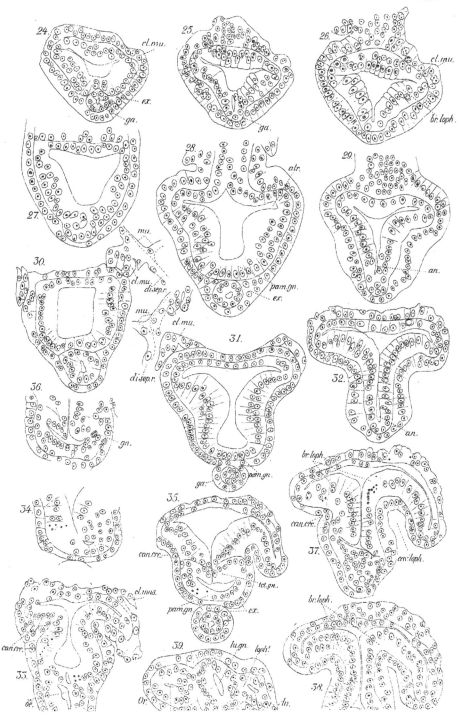

PLATE V.

Figs. 40–43 are four horizontal sections of a polypide of Stage XII. passing in the direction indicated by the lines 40–43, Fig. 18. × 390.

" 44, 45 are horizontal sections of a polypide of Stage XIII. The direction of the cutting planes is indicated by the lines 44, 45, Fig. 19. × 390.

" 46–48. Sections through the migrating end of the funiculus, showing its relation to the cœlomic epithelium of the roof. The ectoderm is not shown. The arrow indicates direction of motion. × 390.

" 49. Transverse section through the funiculus, showing the loose migratory cells. × 390.

" 50, 51. Horizontal sections of a polypide slightly younger than Stage XIV., Fig 73. Of these two sections, Fig. 50 is nearer the roof of the colony, and immediately above the ganglion. Fig. 51 is the second section below, and passes through the middle of the ganglion. × 390.

" 52. Sagittal section of the region about the brain of a polypide somewhat older than that shown in Fig. 77. This figure is reversed relatively to Fig. 77. × 600.

Pl. V.

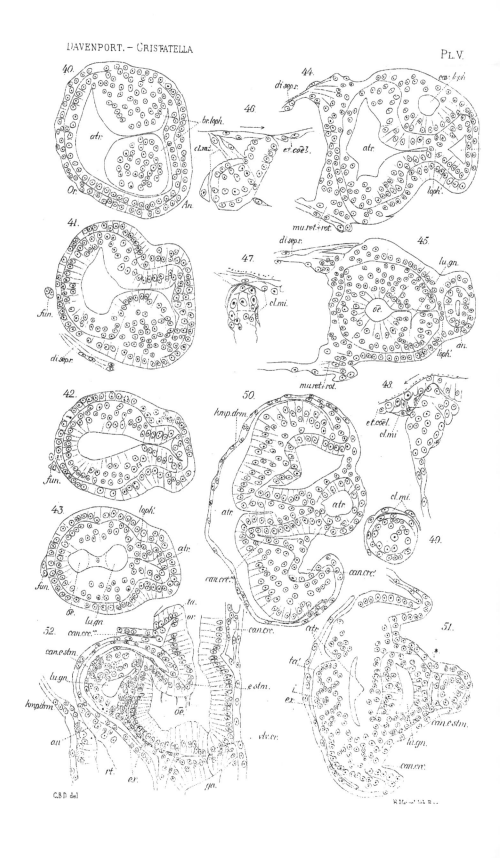

C.B.D. del

R. Mo...el Int. B...

PLATE VI.

Fig. 53. Young funiculus, showing its connection with polypide. × 390.
" 54. Origin of muscles. The section passes diagonally across a partition at the left, *di sep. r.*, and cuts the polypide tangentially at the right. × 390.
" 55. Section including a radial portion, showing the position of the muscles in the partition near the margin of the colony. × 390.
" 56. Section through the retractor and rotator muscles of a polypide of about the age of that shown in Fig. 77. × 390.
" 57. Young funiculus, whose upper end is free from the cœlomic epithelium of the roof of the colony. × 390.
" 58. Section through the sole, showing the relation between the muscle cells and the muscularis of the sole. × 600.
" 59. Section across a radial partition, and both rotator and retractor muscles which are migrating from the roof to the sole. × 390.
" 60. Section at right angles to the wall of the colony, showing the elongated and unmetamorphosed cells of the margin. × 390.

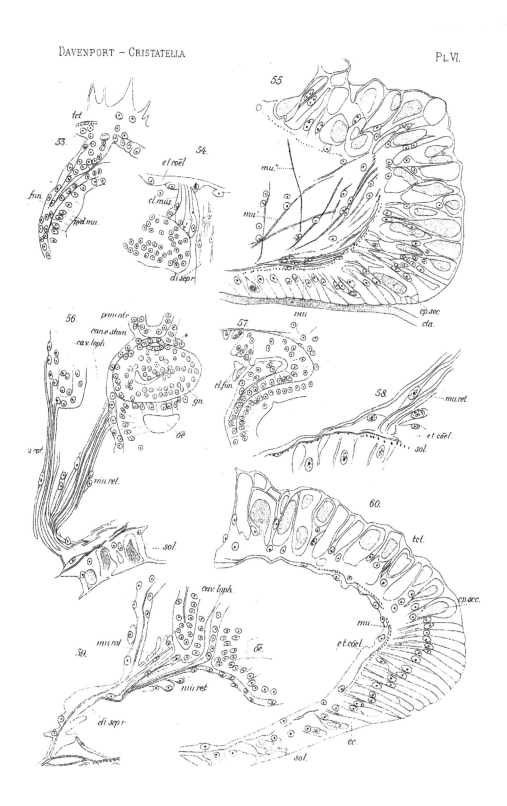

PLATE VII.

Figs. 61–63. Three vertical right-and-left sections of same polypide passing from posterior end anteriorly. About Stage XIV. (Fig. 73, Plate VIII.) × 390.

" 61. Section through lophophore arms, showing their fusion, *loph.'*, and the position of the ring canal, *can. crc.'*, *can. crc."*

" 62. Section just posterior to anal opening, showing openings of lophophoric pockets.

" 63. Section through ganglion, showing early stage in formation of lophophoric nerve, parts of the ring canal, and young tentacles.

" 64. Cross section of lophophore arm, near termination of young nerve, at place marked 64, Fig. 71. × 1000.

" 65–67. Three successive sections through end of lophophore nerve in regions marked 65, 66, and 67, Fig. 71. These figures are from the same individual as Fig. 64, but from the opposite lophophore arm. × 1000.

" 68. Vertical right-and-left section through ganglion of an individual slightly younger than Fig. 63, showing origin of cornua by outgrowth of the walls of the ganglion, with an extension of the lumen of the latter. × 600.

" 69, 70. Longitudinal sections of two stages in the development of a tentacle, Fig. 70 being the younger. × 390.

" 71. Section through ganglion and growing lophophore nerve. Stage XIV. × 490.

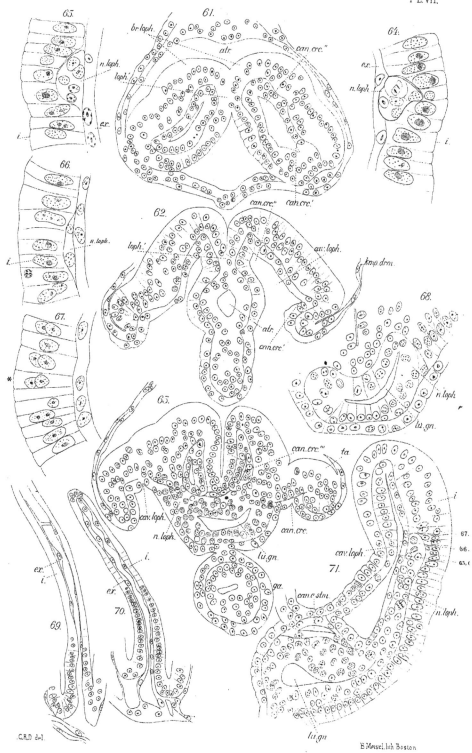

C.R.D. del.

E Meisel lith Boston

PLATE VIII.

Fig. 72. Sagittal section of an adult polypide. The lophophore has been omitted. Outlines with camera lucida. Nuclei put in free hand. × 175.

" 73. Sagittal section of bud. Stage XIV. The margin of colony to left. * Ectodermal cells derived from neck of polypide. × 390.

" 74. Nearly horizontal section of a bud a little older than that shown in Fig. 73. The plane of section passes obliquely upward and forward. The tentacles are cut at different heights. × 390.

" 75. Transverse section of lophophore arms before separation. The connecting band, *loph.'*, is reduced to threads. The polypide has already evaginated. The section figured is the seventh from the distal end of the arms, — about 40 μ distant. × 390.

" 76. Transverse section of lophophore arms immediately after separation. The tentacles arising from *can. crc.''* were previously fused. × 390.

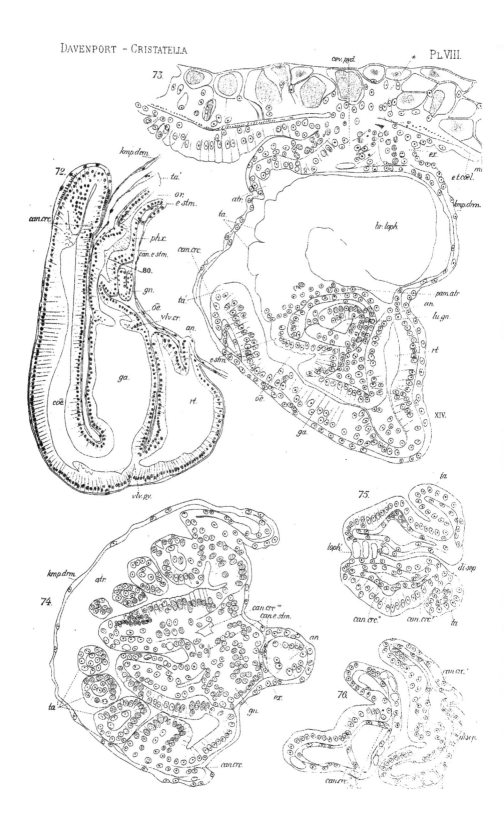

PLATE IX.

Fig. 77. Sagittal section through a polypide, of which the atrial opening (*of. atr.*) has already begun to form. × 390.

" 78. Horizontal section through the circumoral part of the ring canal, *can. crc.*, showing its free communication with the cœnocœl (*cœn.*). Adult. × 175.

" 79. Vertical section through the roof of the colony (to the left) and the kamptoderm (to the right), showing their connection by the inferior parieto-vaginal muscles (*mu. inf.*) at an early stage of their development. × 600.

" 80. Horizontal section in position marked 80, Fig. 72, Plate VIII., showing epistomic canal, *can. e stm.*, and supra-ganglionic part of ring canal, *can. crc.'''* × 390.

" 81. Section cutting lophophore at base of tentacles. The arm of the right side only is shown entire. Stage of Fig. 77. × 175.

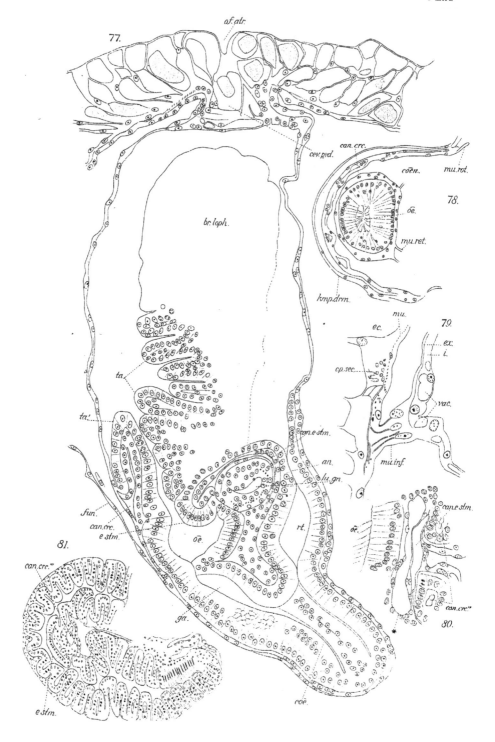

77.

of. atr.

br. loph.

can. crc.

cev. pvd.

cöen.

mu. rot.

78.

öe.

mu. ret.

kmp. drm.

mu.

ec.

79.

ex.

i.

cp. sec.

vac.

an. e. stm.

an.

lu. gn.

mu. inf.

ta.

ta'.

can. e. stm.

öe.

rt.

fun.

can. crc.

e. stm.

öe.

can. e. stm.

81.

can. crc.'''

ga.

can. crc.''

80.

coe.

e. stm.

PLATE X.

Fig. 82. Transverse section of stomach of adult polypide. × 390. Compare with Fig. 93.

" 83. Transverse section of proximal part of cœcum of same individual as that of Fig. 82. × 390. Cf. Fig. 94.

" 84. Transverse section of œsophagus of a polypide whose atrial opening is just formed. × 390.

" 85. Transverse section of the cœcum of an adult polypide near its distal extremity. × 390.

" 86. Vertical section across a radial partition at its junction with colony-wall. × 600.

" 87. Horizontal section of radial partition at its junction with colony-wall. × 600.

" 88. Small colony of Cristatella, drawn from transparent object, showing polypides in optical section at different focal planes. × *circa* 40.

" 89–92. Muscle fibres in successive stages of development. From thick sections. × 390.

" 93. Transverse section of stomach of the same polypide as that from which Fig. 84 was taken; representing, therefore, a considerably younger stage than Fig. 82. × 390.

" 94. Transverse section of cœcum of the same polypide as that from which Figs. 84 and 93 were taken, cut in a region nearly corresponding to the position of that shown in Fig. 83. × 390.

" 95, 96. Two horizontal sections of a part of the margin of a small colony in which radial partitions are being rapidly formed in correspondence with rapid budding. Fig. 95 lies near the sole; Fig. 96, near the roof. The same figures refer to the same partition. × 300.

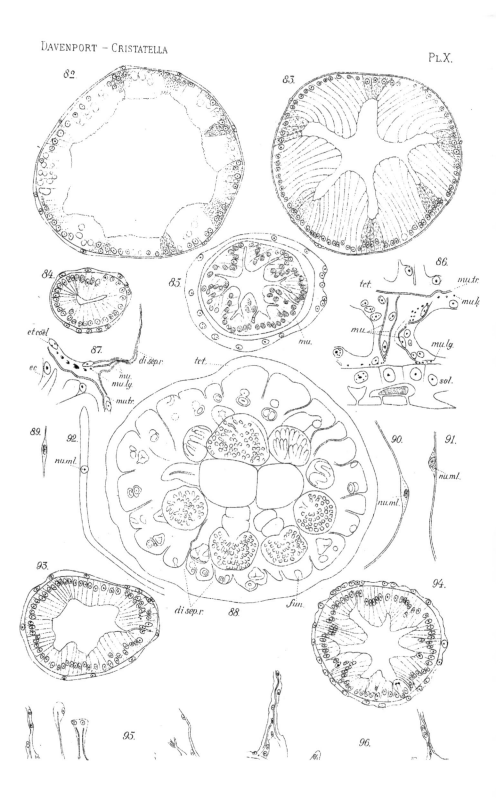

PLATE XI.

Figs. 97–99. Vertical sections, showing three successive stages in the degeneration of the roof to form the atrial opening, *of. atr.*, and development of the parieto-vaginal muscles. × 390.

" 100. Late stage in the development of the ectoderm, showing its extreme modification between adult polypides. × 390.

Pl. XI.

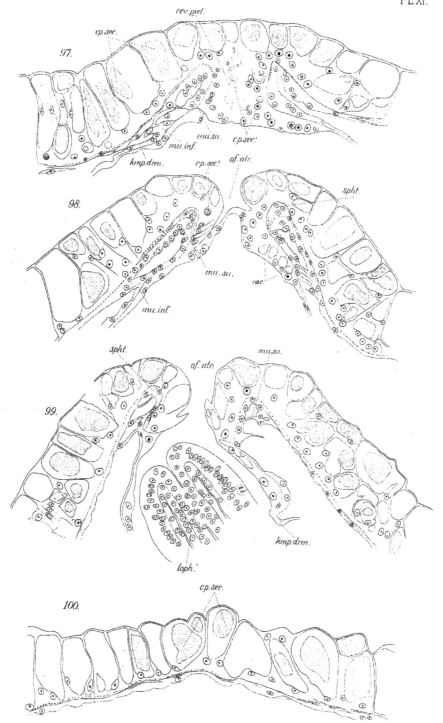

No. 5. — *The Eyes in Blind Crayfishes.* By G. H. PARKER.[1]

In the fall of 1888 Mr. Samuel Garman placed at my disposal several crayfishes [2] which had been collected by Miss Ruth Hoppin in the caves of Jasper County, Missouri. The specimens were given to me with the suggestion that I should ascertain the extent to which their eyes had degenerated, for, judging from external appearances, these organs had become as rudimentary as the eyes of the blind crayfish, Cambarus pellucidus, Tellk., from Mammoth Cave. In order to establish comparisons it was desirable to study the eyes in C. pellucidus, and for this purpose specimens of this species were kindly furnished me from the collections in the Museum of Comparative Zoölogy. These specimens, as well as those collected by Miss Hoppin, were preserved in strong alcohol. My study of this material was carried on in the Zoölogical Laboratory of the Museum, under the direction of Dr. E. L. Mark.

Notwithstanding the general interest which zoölogists have shown in the blind crayfishes there have been very few publications on the minute structure of the eyes of these animals. The earliest contribution to this subject was from Newport, who, in discussing the ocelli of Anthophorabia, incidentally described the structure of the eye in Cambarus pellucidus. According to Newport's account ('55, p. 164), the eyes in this species would seem to be only *partially* degenerated, for although the retinal region is not pigmented, the corneal cuticula is nevertheless divided into irregular facets, or "corneales," as they are termed, "and the structure [hypodermis] behind these into chambers to which a small but distinct optic nerve is given."

The second investigator who studied the eyes of blind crayfishes was Leydig ('83, pp. 36 and 37). The material which was accessible to him was unfortunately so poorly preserved that it was of little value for histological purposes. He nevertheless satisfied himself that the cuticula in the corneal region was not facetted. He also quoted from an abstract

[1] Contributions from the Zoölogical Laboratory of the Museum of Comparative Zoölogy, under the direction of E. L. Mark, No. XX.

[2] These crayfishes had previously been submitted to Dr. Walter Faxon for determination. They have since been described by him as a new species, under the name of Cambarus setosus, an account of which will be found in Mr. Garman's recent paper ('89, p. 237) on "Cave Animals from Southwestern Missouri."

of Newport's paper, to the effect that the eye is "ohne Hornhaut, Pigment und Nervenstäbe." The phrase "ohne Hornhaut" means, I believe, that a *facetted* cornea is not present; at least this seems to be the interpretation placed on it by Leydig, for the quotation is shortly followed by this sentence : "Dort wo man eine gefeldorte Cornea zu suchen hätte — am Gipfel des Kegels — zeigt sich die Haut von der gewöhnlichen Beschaffenheit." There was greater reason for Leydig's regret that he could not consult Newport's original paper than Leydig himself appreciated; for, although he probably had no reason to consider the abstract incorrect, if his quotation from it is exact, it differs at least in one respect from Newport's account. Newport described the cornea as facetted; Leydig's quotation from the abstract states that it was not facetted. I have been unable to discover where this abstract was published, but, since Leydig quotes directly from it, the probabilities are that the discrepancy between his quotation and Newport's actual statement is to be attributed to an error in 'the abstract. Aside from this difficulty, it must be borne in mind that Leydig and Newport in their observations on the cornea by no means agree; for while Newport really describes the cornea as facetted, Leydig states from his own observations that it is without facets. According to Leydig, then, the eye of C. pellucidus is more completely degenerated than the observations of Newport would lead one to suppose.

The latest account of the eyes in blind crayfishes forms a part of Packard's paper on "The Cave Fauna of North America" ('88, pp. 110 to 113). Newport and Leydig studied C. pellucidus ; Packard had the opportunity of studying not only this species, but also C. hamulatus, Cope and Packard, from Tennessee. In both species according to Packard the cornea was without facets, and the hypodermis was not thickened in the retinal region, but an optic nerve and ganglion were present. The results obtained by Packard thus confirm those given by Leydig.

From this brief historical review it will be observed that one of the principal questions concerning the eyes of blind crayfishes deals with the extent of their degeneration. This change has not only affected the finer structure of the retina, but it has also altered the shape of the optic stalk. I shall therefore begin with a description of the external form of the stalks.

The optic stalks of blind crayfishes are not only proportionally smaller than those of crayfishes which possess functional eyes, but they have in the two cases characteristically different shapes. In crayfishes with fully developed eyes the stalk is terminated distally by a hemispherical enlargement; in the blind crayfishes it ends as a blunt cone. This cone-shaped outline is especially characteristic of C. pellucidus (Fig. 2). It will be observed that in this species the optic nerve (*n. opt.*) terminates in the hypodermis immediately below the blunt apex of the cone. In C. setosus (Fig. 1) the termination of the optic nerve is also at the apex of a blunt cone. In this case, however, the axis of the cone does not coincide with the axis of the stalk, as it does in C. pellucidus, but the two axes meet each other at an angle of about forty-five degrees, and in such directions that the conical protuberance at the distal end of the stalk is directed forward and outward from the median plane of the animal. The protuberance is rather more blunt in C. setosus than in C. pellucidus (compare the regions marked *r.* in Figs. 1 and 2).

Through the kindness of Dr. Walter Faxon I was enabled to examine two specimens of C. hamulatus. In this species the stalks also terminate in blunt cones. They are not so pointed as in C. pellucidus, but approach the more rounded form of C. setosus.

The three species, C. pellucidus, C. hamulatus, and C. setosus, are the only blind crayfishes thus far known in North America, and, as they agree in having a conical termination to the optic stalks, a peculiarity not observable in crayfishes with functional eyes, it may be concluded that the conical form is characteristic of the stalks in blind crayfishes. Unquestionably, this conical shape is coupled with the degenerate condition of the retina.

In describing the finer anatomy of the eye it will be more convenient to begin with the condition found in C. setosus. Figure 1 is drawn from a longitudinal horizontal section of the optic stalk in this species. The plane of section passes through the region where the optic nerve and hypodermis are in contact. This region (Fig. 1, *r.*) corresponds to the retina of other crayfishes. The optic stalk is covered with a cuticula (Fig. 1, *ct.*), which is of *uniform* thickness and which resembles the cuticula of the rest of the body. In this respect the stalk differs from that of decapods with well developed eyes, for in these, although much of the stalk is covered with ordinary cuticula, the retinal region is provided with a thin flexible cuticula. This has been named by Patten the corneal cuticula; it cannot be said to be differentiated in C. setosus. In optic stalks with functional retinas the corneal cuticula is usually facetted, but in C. setosus no indication of facets is discoverable.

The undifferentiated condition of the cuticula leads one to anticipate a simple condition in its matrix, the hypodermis. The latter is a continuous layer of cells (Fig. 1, *hd.*) with its distal face applied to the cuticula and its proximal face bounded by a fine but distinct basement membrane (*mb.*).   The layer is throughout very nearly uniform in thick. ness ; at least it is not thicker in the region of the retina than at many other places, and the slight variations in its thickness are not in signifi. cant regions.   The only feature of the retinal hypodermis which would suggest that it was unlike the rest is the somewhat closer crowding of its cells.   This manifests itself in the arrangement of the nuclei in two or three irregular rows, instead of a single one.   In other respects the nuclei of the retinal region and the surrounding hypodermis are essen. tially similar.

The optic nerve (Fig. 1, *n. opt.*) consists of a poorly defined bundle of nerve-fibres which extend from the optic ganglion to the hypodermis. The nerve-fibres are doubtless intimately connected with the cells in the hypodermis, for the basement membrane is interrupted where the nerve and hypodermis are in contact.   It is probable that the basement mem- brane is reflected from the hypodermis to the optic nerve, although I have not been able to observe this with clearness.

Recent investigations support the conclusion that the retina in the crustacea is derived from the hypodermis.   In C. setosus that portion of the hypodermis from which the retina would be derived is scarcely distinguishable from other parts of the same layer.   The retina in this species, therefore, has so completely degenerated that it has at last returned to the condition of almost undifferentiated hypodermis.

That the optic nerve still retains its connection with the retinal area is, on the whole, not so significant a condition as one might at first sup- pose.   It is probable that the optic nerve arises in this species as it does in the lobster.   I have elsewhere (Parker, '90, p. 43) attempted to show that in the lobster it is not an outgrowth from either the optic ganglion or the retina, but that, as the ganglion was differentiated from the hypodermis, the optic nerve remained as a primitive connection be- tween these two structures.   So long, then, as an optic ganglion should be differentiated one might expect an accompanying optic nerve ; but the nerve would be present as a passive connection between hypodermis and ganglion, rather than as a structure which had retained that posi- tion by virtue of its continued functional importance.

The foregoing account of the eye in C. setosus is based upon obser- vations on three individuals of this species : Two of these measured, from the tip of the rostrum to the end of the telson, 6 cm. ; the third, 4.2 cm.   In the three individuals the eyes presented essentially the same condition. Figure 1 is taken from one of the larger individuals. In this specimen the cuticula was somewhat thinner and the hypodermis rather thicker than in the other two. This I believe was due to the fact that the animal had recently moulted.

So far, then, as the eye of C. setosus is concerned, although the optic ganglion and optic nerve are present, the retina has undergone a complete degeneration, and is now represented by a layer of undifferentiated hypodermal cells.

The eyes of Cambarus pellucidus present a somewhat different condition from that described in C. setosus. A longitudinal horizontal section of the optic stalk of C. pellucidus is shown in Figure 2. The outer surface of the stalk is covered with a cuticula (*ct.*) of uniform thickness, and there is no indication of facets. Excepting at the apex of the stalk, the hypodermis (*hd.*) is composed of a remarkably uniform layer of cells. As in C. setosus, it is bounded on its deep face by a delicate basement membrane (*mb.*). Both an optic ganglion (*gn. opt.*) and nerve (*n. opt.*) are present, the latter being connected with the hypodermis. In all these respects C. pellucidus resembles C. setosus, but when the retinal part of the hypodermis in the two species is compared a striking difference can be seen. The retinal hypodermis in C. setosus (Fig. 1, *r.*) is, as we have seen, substantially like the remaining hypodermis of the optic stalk. The retinal hypodermis in C. pellucidus (Fig. 2, *r.*) is much thicker than the hypodermis of the stalk. With this thickened region of the hypodermis the optic nerve is connected, and there is no question, therefore, that this thickening represents the rudimentary retina. Omitting minor details, the form of the thickening is that of a plano-convex lens, the curved surface of which is applied to the concave inner face of the cuticula at the distal end of the stalk. The optic nerve is attached to the central part of the flat face of the thickening.

When the retinal thickening is carefully studied by means of radial sections, one can see that it differs from the neighboring hypodermis not only in thickness, but also in the fact that it contains two kinds of substance: a protoplasmic material uniform with that of the rest of the hypodermis, and a number of relatively large granular masses (Fig. 3, *con.*). These granular masses contain two, three, four, or sometimes five nuclei, and nuclei are also to be found scattered through the undifferentiated protoplasmic substance. The nuclei in the granular masses are slightly smaller than those in the surrounding portion of the hypodermis; they are, moreover, round in outline, while the other nuclei are usually somewhat elongated. The same features can be observed in tangential sections (Fig. 6). Here, however, the outlines of the larger nuclei no longer appear oval, since these nuclei are now cut in a plane at right angles with their elongated axes. The nuclei in the hypodermis which adjoins the retinal thickening resemble the larger oval nuclei of the thickening. Nowhere in the adjoining hypodermis have the granular masses with their smaller nuclei been observed. It is therefore clear, that in C. pellucidus the retinal hypodermis is distinguished from the neighboring hypodermis, not only by its greater thickness, but also by the fact that it is composed of two kinds of substance, each with its special form of nucleus. Since the protoplasmic material of the retinal region contains nuclei which resemble those of the surrounding hypodermis, it is probable that this material represents hypodermis which has remained unmodified after the differentiation of the granular bodies. As shown in Figure 3, the granular bodies are for the most part limited to the deeper portion of the retinal thickening, and the oval nuclei occupy the more superficial part. If these oval nuclei represent undifferentiated hypodermal cells, it is only natural that they should occupy a superficial position, for it is there that the function of such cells, namely, the secretion of cuticula, could be most advantageously carried on. In tangential sections of the retinal thickening, both the nuclei of the undifferentiated hypodermis and the outlines of the cells to which they belong are distinguishable (Fig. 5). These cells when compared with those from the hypodermis of the sides of the stalk (Fig. 4) are seen to be much smaller than the latter. Like those from the sides of the stalk, however, they present no definite grouping. This accords with the fact that the cuticula presented no special markings, such as facets, etc., for such markings could of course result only from some special grouping of the secreting cells.

It is difficult to say what the granular bodies with their contained nuclei are. Doubtless they represent some element in the retina of the functional eye reduced by degeneration to this form. The ommatidium or structural unit in the retina of a crayfish consists of five kinds of cells. These are as follows: first, two cells in the corneal hypodermis, lying next the cuticula; second, four cone-cells directly below the corneal hypodermis; third, two pigment-cells, the distal retinulæ, flanking the cone-cells; fourth, seven pigment-cells, the proximal retinulæ, surrounding the rhabdome; fifth, a few yellowish accessory pigment-cells limited to the base of the retina. Excepting the accessory pigment-cells, all the cells in an ommatidium are ectodermic in origin; the accessory pigment-cells are probably derived from the mesoderm. Of these five kinds of cells, the granular bodies probably do not represent the accessory pigment-cells, for in fully developed eyes the latter lie on both the distal and proximal sides of the basement membrane, whereas the granular bodies are found only on the distal side of that structure. The granular bodies, then, more likely represent one of the four remaining elements, all of which naturally occur only on the distal side of the membrane. It is not probable that the granular bodies represent the cells of the corneal hypodermis, for these produce the cuticula of the retinal region, and if they have any representatives, those representatives must be the distal layer of unmodified hypodermal cells already indicated in the retinal thickening. The position of the granular bodies, therefore, precludes their representing corneal hypodermis. If then the granular bodies are not accessory pigment-cells nor corneal hypodermis, they must be either distal or proximal retinulæ or cone-cells. In a previous paper I have given reasons for considering the proximal and distal retinulæ as both originating from a common group of cells, the retinulæ. These are essentially sensory in function, as contrasted with the cone-cells, which are merely dioptric. The question then narrows itself to this: Are the granular masses clusters of dioptric cone-cells or sensory retinulæ?

In determining to which of these two groups of cells the granular masses belong, the relation which the latter sustain to the fibres of the optic nerve would doubtless be of great importance, for the nerve fibres in fully developed eyes are known to terminate in the retinulæ, not in the cone-cells. Unfortunately, the histological condition of my material was such as to preclude the possibility of determining this question.

The fact that each granular mass contains several nuclei clearly indicates that it consists of several cells. The number of cells in each mass, judging from the number of nuclei, varies from one to about five, the more usual number being three or four. When one compares the condition of intimate fusion which the cells of each mass present with the normal condition of the retinulæ and cone-cells, the masses must certainly be admitted to resemble more closely the cone-cells. Moreover, the number of cells in each mass, although variable, is nearer to that of the closely united cone-cells than to that of the retinulæ. Not only do the number of cells involved and the intimacy of their fusion favor the idea that each mass represents a degenerate cone, but the granular substance of the mass also closely resembles the granular material of a cone. For these reasons it seems probable that the granular nucleated masses in the retinal region of C. pellucidus are the degenerate representatives of the cones in normal eyes.

The fact that, of all the ectodermic elements of the retina, only the granular nucleated masses continue to be differentiated, throws them into strong contrast with the surrounding structures. The retention of these masses may mean that on account of their extreme differentiation they have had time to respond only incompletely to the influence of degeneration; or it may imply that phylogenetically they were among the earliest retinal structures differentiated. Admitting them to be degenerated cone-cells and merely dioptric in function, one can scarcely conceive how they could have been differentiated before the sensory cells which they serve. But even if they cannot be regarded as more primitive structures than retinulæ, their retention still may be significant, as an indication that the ommatidia of primitive crustaceans contained cone-cells as well as retinulæ.

Former studies have led me to believe that the difference in the ommatidia of various crustaceans could be explained on the assumption that the number of elements has been gradually increased from lower to higher forms by cell-division. The simplest conceivable representative of an ommatidium in the crustacea might then be a single cell. This would be of course a sensory cell; by its division, the more complicated ommatidia might subsequently be derived from it. In such an event, the cone-cells must be modified sensory cells; but the fact that these cells persist in so rudimentary a retina as that of C. pellucidus points rather to the conclusion, that they are probably almost as old, phylogenetically, as the retinulæ themselves, and that primitive ommatidia consisted of at least two kinds of cells, sensory cells or retinulæ, and cone-cells, derived not from degenerated sensory cells, but from the undifferentiated hypodermis.

As I have already shown, the results which Newport, Leydig, and Packard arrived at are not always in agreement. This might be explained by the fact that the organ under consideration is a degenerated one, and consequently subject to considerable individual variation. This supposition, however, is not supported by anything I have observed. The preceding account of the eye in C. pellucidus is based upon the examination of three individuals. These were respectively 6.5 cm., 5.6 cm., and 4.4 cm. long. Figure 2 was drawn from the optic stalk of the shortest individual. In all essential features the eyes of the two other crayfishes presented the same condition as that shown in Figure 2. In the specimen 5.6 cm. in length, the granular bodies were less distinct than in the other two, but they were nevertheless recognizable, and the retinal thickening was as pronounced in this as in either of the other specimens. The fact that these three individuals show so little variation leads me to believe that the condition of the eye in the blind crayfish is not so variable as I at first supposed it would be. The same constancy is also true of C. setosus. Hence it seems to me improbable that the differences between Newport's observation and those of the later investigators are due to individual variations in the specimens studied. The fact that Newport's work was done before the development of present methods of research offers, I believe, a more natural explanation of some of his results, than the supposition of individual variations. That the methods of his time were imperfect is evident from the fact that Newport himself seems to have overlooked the ganglion of the optic stalk, a structure readily discoverable by means of serial sections. (Compare Newport's Figure 13 ['55, p. 102] with Figure 2 in this paper.) Leydig's observations, so far as they extend, are fully confirmed by my own. Packard's account differs from mine in only one particular, but that is of considerable importance ; he states that there is *no retinal thickening* in the two species studied by him. This difference may possibly be due to individual variations in the crayfishes. Unfortunately, Packard does not state the number of specimens which he examined, and consequently one is uncertain how much weight to give to his general statements.

The conclusions to be drawn from the foregoing account may be summarized as follows. In both species of crayfishes studied, the optic ganglion and nerve are present, and the latter terminates in some way not discoverable in the hypodermis of the retinal region. In C. setosus this region is represented only by undifferentiated hypodermis, composed of somewhat crowded cells, while in C. pellucidus it has the form of a lenticular thickening of the hypodermis, in which there exist multinuclear granulated bodies. These I have endeavored to show are degenerated clusters of cone-cells. If Packard's observations are correct, the retina in C. pellucidus may be reduced in some individuals as much as it is in C. setosus, which I have studied, but my own examinations do not render this view probable.

CAMBRIDGE, February 24, 1890.

BIBLIOGRAPHY.

Leydig, F.
'83· Untersuchungen zur Anatomie und Histologie der Thiere. Bonn, Emil Strauss, 1883. 174 pp., 8 Taf.

Newport, G.
'55· On the Ocelli in the Genus Anthophorabia. Trans. Linn. Soc., London, Vol. XXI. pp. 161–165, Tab. X., Figs. 10 to 15 incl. Read, April 19, 1853.

Packard, A. S.
'88· The Cave Fauna of North America, with Remarks on the Anatomy of the Brain and Origin of the Blind Species. Mem. Nat. Acad. Sci., Vol. IV. Pt. 1, pp. 1–156, 27 Pls. Read, Nov. 9, 1886.

Parker, G. H.
'90· The Histology and Development of the Eye in the Lobster. Bull. Mus. Comp. Zoöl. at Harvard Coll., Vol. XX. No. 1, pp. 1–60, 4 Pls. 1890.

Garman, S.
'89· Cave Animals from Southwestern Missouri. Bull. Mus. Comp. Zoöl. at Harvard Coll., Vol. XVII. No. 6, pp. 225–240, 2 Pls. Dec., 1889.

EXPLANATION OF FIGURES.

ABBREVIATIONS.

con.	cone.	mb.	basement membrane.
ct.	cuticula.	nl. con.	nucleus of cone-cell.
gn. opt.	optic ganglion.	nl. hd.	nucleus of hypodermis.
hd.	hypodermis.	n. opt.	optic nerve.

r. retina.

The specimens from which the following figures were taken were killed and preserved in strong alcohol, and stained in Czocher's alum-cochineal. The crayfish from the optic stalk of which Figure 1 was drawn was 6 cm. long. That from which the remaining figures were made was 4.4 cm. long.

Fig. 1. A longitudinal horizontal section through the right optic stalk of Cambarus setosus, Faxon. The histological detail is given in the hypodermis only. The optic ganglion and the optic nerve are tinted. Between these structures and the hypodermis the space is filled with a loose connective tissue. × 65.

" 2. A longitudinal horizontal section through the right optic stalk of Cambarus pellucidus, Tellk. This drawing was made in the same manner as Figure 1. × 65.

" 3. An enlarged drawing from the distal end of the section which immediately follows that from which Figure 2 is taken. This figure shows the details in the retinal enlargement of the hypodermis. The space between this enlargement and the cuticula was artificially produced. × 275.

" 4. Tangential section of the hypodermis from the side of an optic stalk of Cambarus pellucidus. × 275.

" 5. Tangential section of the superficial portion of the retinal thickening in the eye of Cambarus pellucidus. × 275.

" 6. Tangential section of the deep portion in the retinal thickening of the eye of Cambarus pellucidus. This section is taken from the same series as the one from which Figure 5 was drawn. × 275.

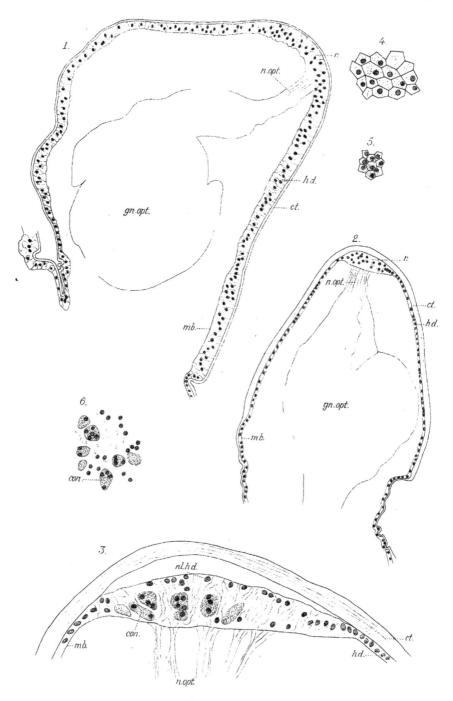

No. 6. — *Notice of Calamocrinus Diomedæ, a new Stalked Crinoid from the Galapagos, dredged by the U. S. Fish Commission Steamer "Albatross,"* LIEUT.-COMMANDER Z. L. TANNER, U. S. N., *commanding.* By ALEXANDER AGASSIZ.

[Published by Permission of MARSHALL McDONALD, U. S. Fish Commissioner.]

IN 1887, Professor G. Brown Goode, Acting U. S. Fish Commissioner, was kind enough to invite me to join the "Albatross" at Panama, and to take part in the dredging operations to be carried on between that port and the Galapagos Islands.

I always hoped to have the opportunity of comparing, at some time, the deep-water fauna of the Pacific side of the Isthmus of Panama with that of the Caribbean, and to see how far the parallelism which has been traced between the littoral fauna of the two sides was carried out with the deep-water fauna. Unfortunately, I was unable to avail myself of this exceptional opportunity, although Colonel McDonald, the U. S. Fish Commissioner, detained the "Albatross" at Panama to allow me to join her at the last moment.

To have thoroughly dredged the line from Panama to the Galapagos would have been to collect material for the solution of many an interesting problem in the geographical distribution of marine animals, to say nothing of the rich harvest likely to have been gathered, when dredging in a district so prolific as that of the Bay of Panama, in shallower waters; and if the haul made at Station No. 2818, off Indefatigable Island, is at all a measure of what we may obtain in the way of novelties, the naturalist who is the first to run that line may be prepared for remarkable discoveries.

In addition to the Stalked Crinoids collected by the "Albatross," which the Fish Commissioner has kindly placed at my disposal for study, he has also intrusted to me the Echini collected by the "Albatross" on her voyage from the east coast of the United States to San Francisco. The route she followed was about the same as that taken by the "Hassler," and the material collected differed but little from the collection made by the latter vessel. The Echini were more numerous; but with the exception of the young stages of a few species, and additional data regarding the geographical distribution of many species, there were no novelties brought to light. I shall take another occasion to publish a final report on the Echini.

The "Albatross" dredged on her voyage from New York to San Francisco, off Indefatigable Island, one of the Galapagos, at a depth of 392 fathoms, three imperfect specimens of a most interesting Stalked Crinoid. At the first glance, it might readily pass for a living representative of the fossil Apiocrinus; but on closer examination we found that it revealed some features which ally it with Millericrinus, and others with Hyocrinus and Rhizocrinus. It soon became apparent that we were dealing with a new type, combining structural features of all the genera above named. It has, like Hyocrinus and Rhizocrinus, only five arms; they are, however, not simple, but send off from the main stem of the arm three branches to one side and two to the other.

As in Hyocrinus, the first radials are high, the second radials much narrower than the first. The system of interradial plates is highly developed, as in Apiocrinus and Millericrinus, six rows of solid polygonal imperforate plates being closely joined together, and uniting the arms into a stiff calyx as far as the sixth or seventh radial, and to the third or fourth joints of the first and second pinnules. These two pinnules are on the fourth and fifth radials; the third pinnule is on the sixth radial; and they are all below the first axillary, which is the eighth radial, and which gives rise to the first branch from the main stem. The second and fifth, sixth, or seventh radials have syzygies.

The imperforated interradials are followed by smaller, somewhat thinner and perforated perisomatic plates, which extend to the prominent lateral plates of the food groove. The interradial calycinal plates extend along the arms for a considerable distance beyond the first branch.

The ventral surface extends nearly horizontally from the mouth to the level of the seventh radial, and this plane may be considered the greatest width of the cup, the interradial spaces arching very slightly toward the mouth, at the junction of the imperforate interradial plates with the perforated perisomatic plates.

The solid imperforate interradial plates extend over the prominent anal proboscis. The oral plates at the interradial angles of the food groove are small, but easily distinguished from the adjacent lateral and covering plates. They are separated from the so called calyx interradials by three or four rows of perforated perisomatic plates, except on the anal interradii. The stem was somewhat curved at the upper extremity, the terminal joints expanding slightly to form a continuation of the outline of the cup of the base of the calyx. The stem tapered very gradually, and in its general appearance recalled that of Apiocrinus, expanding again towards the base, the root of which, however, was not obtained by the "Albatross." The stem is cylindrical, without cirri. In the upper third the joints are alternately ribbed transversely, or even ornamented near the base of the calyx with more or less prominent tubercles, as in Millericrinus. The uppermost joint is convex, and in the space left vacant between it and the central part of the basal ring a small lobed delicately reticulated pentagonal disk was found resting upon the upper face of the "article basal" of De Loriol. This is probably a modified anchylosed infrabasal ring, which may or may not be resorbed in older stages of the genus.

There are five distinct basals in one of the specimens; in the second their sutures can fairly be distinguished, while in the third they were completely anchylosed, much as they so frequently are in Rhizocrinus. As in Hyocrinus, the basals are about half the height of the first radials; the second radials cut deeply into the first radials.

The stem of this crinoid must have attained a length of from 26 to 27 inches; the height of the calyx to the interradials is $\frac{7}{16}$ of an inch; its diameter at the inner base of the second radials is $\frac{11}{16}$ of an inch, at the height of the third joint of the second pinnule 1 inch, at the level of the proximal face of the radials $\frac{3}{8}$ of an inch, and at the level of the suture of the basals with the uppermost joint $\frac{1}{4}$ of an inch; and the length of the arms is probably about 8 inches.

I propose to name this crinoid Calamocrinus Diomedæ, after the vessel which discovered it. I have to thank Colonel Marshall McDonald, the U. S. Fish Commissioner, for the opportunity of studying this crinoid. With his consent, a detailed account of Calamocrinus will be published in the Museum Memoirs as soon as the plates can be prepared.

CAMBRIDGE, November 28, 1890.

No. 7. — *The Origin and Development of the Central Nervous System in Limax maximus.* By ANNIE P. HENCHMAN.[1]

FOR several years the origin of the central nervous system in Mollusks, both as to method and time of appearance, has been a matter of controversy. It has been of especial importance to determine from which of the embryonic layers its parts arise, and to ascertain if its development throws any light on the relations of Mollusks to other important groups of the animal kingdom, particularly Worms.

Since the observations of the earlier writers, down to about 1874, were carried on without the aid of sections, their conclusions do not merit that degree of confidence which is to be accorded those who have availed themselves of this means of study.

Most of the later authors agree that the central nervous system arises from the ectoderm, either by an invagination, or by a simple local thickening which later becomes detached. However, Bobretzky ('76, pp. 162–169), — the first to use sections, — while conceding that in Fusus there are invaginations of the ectoderm to form the sense organs, concludes that the supra-œsophageal and pedal ganglia arise from the mesoderm, and Bütschli ('77, pp. 227, 228) is inclined to believe that the same is true in Paludina vivipara.

Von Jhering ('74, p. 321) claims for Helix, and both Lankester ('74, pp. 382, 383) and Wolfson ('80, pp. 95, 96) for Lymnæus stagnalis, that the central nervous system arises simply from a thickening of the ectoderm.

Fol ('80, p. 664) has since pointed out, however, that Lankester's conclusions are based on an erroneous interpretation of cells ("nuchal cells"), which he believes are not at all nervous in their nature. They are the same cells which Wolfson has called the embryonic brain ; but Wolfson's opinion, previously stated, has reference to the definite nervous system, not to this so-called embryonic brain.

Haddon ('82, pp. 368–370) believes that he has seen the rudiments of the cerebral and pedal ganglia of Nudibranchs in the form of thicken-

[1] Contributions from the Zoölogical Laboratory of the Museum of Comparative Zoölogy, under the direction of E. L. Mark, No. XXI.

ings of the ectoderm, and he has made sections of Purpura lapillus and Murex erinaceus which show, as he maintains, that similar rudiments are also formed in them by proliferation from thickenings of the ectoderm.

Kowalesky ('83, pp. 23–26) shows for Chiton Polii that the lateral and pedal nerve trunks are formed simply as thickenings of the ectoderm.

Rabl ('75, pp. 206–208) maintained that the supra-œsophageal ganglia and the sense organs in Lymnæus, Physa, Ancylus, and Planorbis were formed by an invagination of the ectoderm, and that the pedal ganglia were produced by delamination from the same germinal layer. He has more recently ('83, pp. 57, 58) expressed doubt as to the manner in which the pedal ganglia arise in Bythinia tentaculata, because he has seen them so connected to the ectoderm of the dorsal wall of the foot by means of cells as to indicate that they arise by proliferation from that region.

Sarasin ('82, pp. 45–48), who has also recently studied Bythinia tentaculata, and who is the only author that has hitherto followed the development of the entire nervous system in a Gastropod, contends that the whole of it arises from ectodermic thickenings, without any invagination even for the supra-œsophageal ganglia. He also believes that the pedal ganglia arise from the dorsal wall of the foot.

Fol ('80, pp. 165–169) admits no invagination for the central nervous system in aquatic pulmonates, and he even inclines to the opinion that it may be derived from the mesoderm, which, however, has itself originated from the ectoderm. He considers it an unimportant question, and therefore one which it is useless to discuss, whether the nervous system arises from ectoderm or mesoderm. If the mesoderm were derived from the entoderm, then it would be an important question. He believes that the supra-œsophageal ganglia of land pulmonates (pp. 192–195) originate by invaginations of the ectoderm, while the pedal ganglia arise from the mesoderm of the foot.

The latest investigations are those of Salensky ('86, pp. 655–759) on the development of Vermetus, one of the Prosobranchs. He concludes that the cerebral ganglia are formed by two invaginations of the ectoderm, while the pedal ganglia arise by proliferation from the ventral and lateral walls of the foot on each side of the median depression which runs along its ventral face. These ganglia arise separately, and later become connected with each other by a commissure, and with the cerebral ganglia by connectives, both of which are outgrowths from the ganglia (pp. 694, 695).

Thus we find that, of the authors cited, Bobretzky, Bütschli, and — as far at least as regards the aquatic pulmonates — Fol consider the central nervous system as originating from the mesoderm. Rabl is a little doubtful as to its mesodermic origin in Bythinia. Rabl, Fol, and Salensky are the only investigators who consider any portion of the central nervous system as arising by invagination, and then only in certain Gastropods.[1]

The following observations were made upon embryos of Limax maximus obtained from adults kept in captivity. Under favorable circumstances, they lay abundantly during the latter part of September, and through October and November. After numerous trials, the best method found was to keep about twenty-five or thirty in a large tin pail, the cover being perforated with small holes. Instead of using moss to secure the necessary moisture, the slugs were fed upon lettuce or cabbage; the latter is the better of the two. This food affords at the same time sufficient protection against desiccation, a suitable retreat for the slugs, and a place where they may lay the eggs. It should be changed every other day, — every day if the weather is warm, — and the pail should be washed thoroughly each time. One of the advantages of using a tin vessel is the ease with which it may be kept clean. Cabbage will keep longer than lettuce, and the slugs lay more abundantly when fed upon it. The eggs were generally found in the morning, sometimes at night, in bunches of from thirty to forty. They are more abundant at first than after the slugs have been kept some time in confinement; it is therefore better to obtain at intervals fresh supplies of small numbers of slugs than to procure a larger number at one time. As soon as found, the eggs were removed to a watch-glass containing water; this was placed in a tumbler already about half filled with moss or moistened paper, having a perforated tin cover. The eggs must not be allowed to become dry. For a few days they should be carefully examined under a microscope, every twenty-four hours or oftener, and all those which fail to develop should be removed at once. In the course of a few days these can be readily detected with the naked eye by reason of the greater opacity of the eggs, and the presence of a whitish spot in them due to the disintegration of the embryo.

[1] The brothers Sarasin, in later researches in Ceylon ('87, pp. 59–69) on a species of very large Helix, find that there are *two* invaginations of the ectoderm on each side of the head to form the cerebral ganglia, and Kowalesky ('83ᵃ) had found several years before that there were in Dentalium two deep invaginations, *one* on each side.

A very large per cent of eggs kept in this way remain in good condition until hatching, which, in a moderately warm room, occurs between the twenty-second and twenty-seventh day.

The best reagents for killing embryos were found to be either chromic acid, 0.33%, or Perenyi's fluid. The chromic material when well stained with alcoholic borax-carmine shows the differentiation of nerve cells and nuclei excellently, but it is more difficult to stain sufficiently chromic material than such as has been preserved in Perenyi's fluid. The latter may be stained with alcoholic borax-carmine or picrocarminate of lithium. Good results for the study of cell division have also been obtained by staining with Czoker's cochineal. The picrocarminate of lithium is particularly valuable in the older stages, because it brings out the nerve fibres, the latter being stained yellow, while the ganglionic cells are colored red.

To obtain the embryos in an uninjured condition, it is advisable, in using the chromic-acid method, to remove only the outer envelope *before* killing. The egg may be held between the thumb and forefinger of the left hand, while with a finely pointed stick, somewhat like a wooden toothpick, the outer membrane is gently punctured; the probe should be run under the membrane a little way, to make a larger opening, and the egg carefully pressed with the thumb and forefinger, whereupon the albumen, containing the embryo and surrounded by the inner membrane, will come out in a perfect condition. This may be dropped at once into water, if several are to be treated together, for it is more convenient to put them all into the chromic acid at the same time. When all have been shelled, they should be put into 0.33% chromic acid for two or three minutes only, simply to kill the embryo without hardening the albumen. Then they should be transferred to a watch-glass of water, to which a few drops of the acid have been added. While in this fluid, the inner membrane may be removed with needles. To accomplish this, it is advisable, in the very young stages, to make as large an opening in the membrane as possible, and then with a needle gently to press the embryo out, even if the albumen adheres to it, for the albumen becomes slightly coagulated in the weak acid, and then can easily be washed off. In the stages from the tenth to the sixteenth day, the large size of the pulsating sacs of both head and foot regions makes it extremely difficult to extract the embryos uninjured; great care must therefore be taken, and *no pressure* used. While employing one of the needles to hold the membrane, the other should be forced through the membrane, which may then be ruptured and turned back over the embryo, being drawn off like the finger of a glove. In the older stages, not much care is necessary, because the embryos bear without injury considerable handling, and there is so little albumen left that their position is not readily changed while the membranes are being removed. When freed from the membranes and as much of the albumen as possible, the embryos are to be returned with a large-mouthed pipette to the chromic acid (0.33%), where they may be left for an hour or two; after washing in running water for two or three hours, they may be carried up to 70% alcohol by adding to the water, drop by drop, 35% alcohol; then 50% alcohol, etc. This dehydration must be made very carefully, to avoid shrinkage. The embryos are extremely delicate, and must be handled with great care through every step of the process.

In using Perenyi's mixture, it is best to free the embryos *while living* from the surrounding membranes and the albumen, removing the inner membrane under clear water. When set free, they should be transferred at once with a pipette into a dish of Perenyi's mixture, where they may remain from *two to three minutes*. They are then to be washed thoroughly in distilled water at least five minutes, put into a 5% aq. sol. of alum for thirty minutes, washed again in water, and finally carried through the grades of alcohol as in the chromic method. It is necessary to remove the embryo while living, because otherwise the albumen becomes in this reagent like a jelly, and cannot be removed without injury to the embryo. Material designed to be sectioned must not be left in alcohol longer than a month, since the albumen in the nutritive sac gradually becomes too hard to be cut, especially if prepared in Perenyi's mixture. The stages from the tenth to the sixteenth day can still be used, even if they have been thus overhardened, by removing the nutritive sac; but in the younger stages this is apt to destroy the embryo, and in the older ones — much of the albumen having been swallowed — its removal is still more certain to have the same effect. Attempts subsequently to soften the albumen by prolonged treatment with weak acetic acid proved to be only partially successful. If the embryos are to be kept at all, they should be left unstained; but the safest way is to carry them through to embedding as soon as possible.

They can be stained whole; but to do this successfully, they must be carried gradually through successively weaker grades of alcohol until a grade corresponding to the stain is reached. It is advisable to make the necessary steps from the stain to the parafine as quickly as possible.

Staining with picrocarminate of lithium has the advantage of saving time, since it acts rapidly,—the older specimens requiring only one or two hours, the younger from half an hour to an hour. A few grains of picric acid may be added to the dehydrating alcohols which follow the stain, in order to prevent the total extraction of the picric acid, and the consequent disappearance of the yellow color from the nerve fibres. If the object is too deeply stained, the differentiation of nerve tissue does not show well; the nerve fibres ought to be yellow, the surrounding nuclei pinkish red with a yellow tinge, and all the other tissue pinkish red. As this and Czoker's cochineal are both aqueous dyes, the chromic material is apt to macerate in them; neither does it stain so well in them as in alcoholic borax-carmine.

The chloroform method of embedding in parafine was used exclusively. When the embryo has been transferred by the well known method to a vial containing chloroform, the vial should be placed uncorked on the water-bath at 55° to 60° C. Pendent spoons in the large cups are not very serviceable, as the least jar sends the objects off, and it is almost impossible to recover them from the bottom of the cup without injury. It· is better to have ready on the bath an empty warm glass dish, — a common salt-cellar is very good; also one filled with parafine which melts at about 52° C. The embryos are to be left in the chloroform only as long a time as is necessary for them to sink, and are then to be transferred with the chloroform to the empty glass dish. The transfer is best made by means of a warm pipette, if the embryos are small. Cold soft parafine is then added, a small piece at a time, until the chloroform has so thoroughly evaporated as to leave no trace of its odor. After remaining for fifteen minutes in the soft parafine, the embryo is to be transferred to the "harder" parafine (52° C.), where it should remain from fifteen to thirty minutes. It is important to handle the object with great care, and to carry it through the period of heating as quickly as may be; the latter is necessary, because the embryos are very apt to become brittle if subjected to the heat too long. They should be embedded within an hour or an hour and a half from the time they are first put upon the bath in the chloroform. It is especially dangerous to allow the parafine to harden about the embryo before the latter is finally embedded, because upon the remelting of the parafine the object is almost certain to fall into fragments, owing to its great delicacy.

The embedding, especially for the younger stages, must be done under a lens. It is most convenient to use a dissecting microscope, the stage of which should be kept warm. I have found that parafine which melts between 50° and 52° C. is better for embedding than that which is harder, for the latter is liable in hardening to cause the embryo to crack.

Sections from 10 to 15 μ thick, and in the oldest stages even thicker, are better than very thin ones.

The central nervous system of Limax consists of four pairs of ganglia, — namely, cerebral, pedal, pleural, and visceral, — together with one abdominal ganglion. To these more central ganglia are joined in addition a pair of buccal ganglia, and one mantle or olfactory ganglion.

To summarize briefly in advance my conclusions : The ganglia arise separately. The components of three of the five pairs are joined together later by commissures. Secondarily-produced connectives [1] also serve to join the cerebral ganglia to the pedal, the pleural, and the buccal; the pleural to the pedal and the visceral; and the visceral to the abdominal. The growth of the ganglia is rapid ; they are well formed, and in their ultimate positions by the sixteenth day. The principal changes from that time until hatching, eight or nine days later, are increase in size, and modifications of the histological conditions. According to my observations, all the ganglia, with the possible exception of the pleural, are derived directly from the ectoderm, — the cerebral in part from invaginations, the others exclusively by cell proliferation without invagination. The cerebral ganglia are formed by extensive invaginations, one on each side of the head region, just below and behind the base of the ocular tentacles. During the invagination a rapid cell proliferation takes place at the deep end of the invaginated portion of the ectoderm, and also at a region of the ectoderm corresponding to the depression between the labial tentacles and the upper lips. The lateral halves of the cerebral mass arise as two separate structures, — each from a double origin, — which are only secondarily joined. This union is the result of outgrowths from each of the ganglia which, uniting, form the cerebral commissure. The invaginations begin a little later than the proliferation of cells which gives rise to the pedal ganglia, and they remain open as narrow tubes until towards the period of hatching, or even later. In one instance they have been found in this condition as late as eight days after hatching. The cerebral com-

---

[1] In accordance with the usage introduced by Lacaze-Duthiers, the term *commissure* is employed for the nerve fibres joining the components of a pair of ganglia, and *connective* for those between ganglia on the same side of the body.

missure is formed a little earlier than the commissural fibres joining the pedal ganglia. The latter are connected by *two* distinct commissures, the anterior of which is formed earlier than the posterior. The visceral ganglia precede a little in their development the pleural, abdominal, buccal, and mantle ganglia. The buccal ganglia make their appearance at about the same time as the pleural, and undergo almost no change in position.

The nervous system in Limax maximus makes its appearance on the *sixth or seventh day* after the egg is laid. At this time the foot is a conical projection, less than half as long as the diameter of the more or less spherical remaining portion of the embryo, and its pulsating sac is very small. It is a stage which is only slightly older than that represented by Fol in his Planche 17–18, Fig. 7. The ocular tentacles are now distinguishable as small elevations of the head region, near the beginning of the primitive nephridial organs, but the labial tentacles are barely to be made out. The radula sac is a nearly spherical outfolding of the floor of the oral sinus; its fundus is composed of only a single layer of cells, but the part of the sac which is continuous with the wall of the œsophagus is more than a single cell deep; the lumen of the œsophagus is traceable close up to the yolk, where it ends blindly. Both the œsophagus and the radula sac are covered with a continuous layer of somewhat flattened mesodermic cells. The shell gland has the form of a large thin-walled sac containing concretions.

When this condition has been reached, the head region (Plate I. Fig. 2) exhibits no sign of cerebral invaginations, nor have I been able to find regions of cell proliferation or thickenings in the ectoderm which were referable with certainty to the cerebral ganglia.

So far as I have been able to make out, the first contribution to the formation of the pedal ganglia occurs in the form of small clusters of cells, which are still imbedded in the ectoderm of the ventral wall of the foot (Plate I. Fig. 5), from which they are subsequently detached. Each of these clusters has a spheroidal or more ridge-like form, and contains from four to eight cells. The boundaries of the cells are not sharply marked, but the whole cluster is limited by a definite outline separating it from the rest of the ectoderm. Each cell contains a nucleus, which is large, but less deeply stained than those of the ectoderm, and each nucleus has a large nucleolus, which is very deeply stained (Plate I. Fig. 1).

The region in which this proliferation takes place is definitely located, for it lies in the same transverse plane in which the otocysts (Plate I. Figs. 3, 4) are situated, and it is found at a region in that plane intermediate between the lateral border of the foot and its middle line, but considerably nearer the former. The proliferating cells project into the cavity of the foot, and ultimately are separated from the ectoderm.

Although cells which closely resemble these are found in groups in other parts of the body wall, their nuclei do not become as large as those of the cells destined to form the ganglia. Moreover, the proliferations are constant and most abundant in the regions where the different ganglia of the nervous system take their origin. Besides, in these cases there is generally a sinking in of the surface of the ectoderm in the same region.

Somewhat later than at the stage described, usually on the *seventh day*, the external conditions still remain nearly the same, the ocular tentacles being perhaps a little more prominent, and the concretions in the shell gland more numerous.

The cells of the primitive entoderm, which surround the yolk, form a striking feature of the condition at this stage. These entoderm cells are very large. vacuolated, and only slightly stainable. They contain large ovoid nuclei, which are crowded to one margin of the cells by the nutritive contents accumulated in the cells. Each nucleus contains one large deeply stained nucleolus, and a network of chromatic substance (Plate I. Fig. 2). The ectoderm, except over the nutritive sac, consists of elongated cells, whose nuclei are so arranged as to give the appearance of two or more layers. The ganglionic cells at this time closely resemble the mesodermic cells, and this makes it difficult to distinguish between the two (Plate I. Fig. 7).

The internal ends of the primitive nephridial organs are situated one on each side of the head, immediately above and back of the ocular tentacles. These organs pass at first forwards and upwards, then in an arch backwards over the nutritive sac, and finally downward and forward. Their external openings are far back in the lateral walls of the body, behind the head region. The organs are readily distinguished in sections by their large slightly stained cells, which are arranged in a single layer around an oval lumen. The large nuclei contain each a single deeply stained nucleolus (Plate I. Figs. 2 and 6). The primitive entoderm and the nephridial organs retain this histological condition throughout the embryonic stages.

The cerebral invaginations at first appear as shallow depressions in the ectoderm at the base of the ocular tentacles, at a point immediately below the nephridial organs. At the same time that the infolding takes place, cells, whose nuclei are larger than those of the mesodermic cells, are being proliferated from the deep surface of the invaginating portion of the ectoderm (Plate I. Fig. 6).

In the region of the ventral wall of the foot referred to in the stage previously described, there are in the ectoderm of each side of the body two groups of ganglionic cells (*prf. pd.*, Plate I. Fig. 7), one behind the other. These cells project into the cavity of the foot, and reach nearly to another small group of cells situated not far from the ventral wall. The cells of the latter group (there is one group on each side of the body) have nuclei similar to those of the cells still connected with the ectoderm. Each group lies in the position subsequently occupied by the pedal ganglion of its side of the body, and is undoubtedly the beginning of that ganglion, for the cells in the ventral wall of the foot continue to be proliferated during several days, and are found in some individuals to be in direct continuity with the ganglia after the latter have attained considerable size. In the individuals shown in Figures 7 and 9 (Plate I.), the right otocyst (Fig. 7) is seen as a closed vesicle, which is not yet wholly detached from the ectoderm. The otocysts undoubtedly vary in regard to the time of their detachment, as will be seen by a glance at the left otocyst of the same individual, which has entirely lost its connection with the ectoderm (Fig. 9).

All the other ganglia, with the exception of the one near the olfactory organ and the buccal ganglia, arise by cell proliferation from ectoderm which lies between the foot and the head region, either at or a little above the posterior angle formed by the body wall with the dorsal surface of the foot, or along a depression which runs forward from this point. This angle marks the posterior limit of a furrow which passes obliquely forward and downward, partially separating the head and visceral mass from the foot. This depression will be designated as the *pleural groove*. Of the remaining ganglia, only the visceral have begun to be formed at this time. The cells destined to form these ganglia are situated immediately above the angle produced by the pleural groove (Plate I. Figs. 8 and 9). Some of those of the left ganglion are wholly detached from the ectoderm, but those of the right (Fig. 7) are still continuous with the ectoderm, though projecting into the body cavity. The cells have large, round, faintly stainable nuclei, each containing one large nucleolus, which takes a deep stain.

Twenty-four hours later, about the *eighth day*, the pulsating sac of the foot has become still larger, and the oral sinus has extended backward and downward as a very narrow tubular passage, — the œsophagus, — which follows the surface of the nutritive sac for some distance, and subsequently opens into it. The peculiar ciliated cells of great size and spongy appearance, which occupy a linear tract along the middle of the roof of the mouth and œsophagus, are at this time very prominent (Plate I. Fig. 2, *loph. cil.*). These cells form what Fol ('80, pp. 190, 191) has called the "ciliated ridge." They persist until after the completion of the nervous system. The ingrowth of the ectoderm to form the rectum is now composed of a compact group of small cells, which shows a small lumen in its central portion, but is still closed at both ends.

The cerebral ganglia remain in nearly the same condition as that last described. About twelve hours later, between the eighth and ninth days,

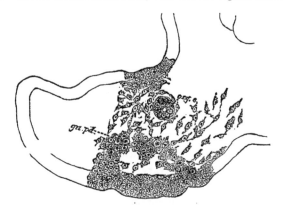

FIGURE A. — The right face of a section parallel to the sagittal plane from an embryo of the
ninth day × 220.

gn. pd. Pedal ganglion   *o cy. s.* Left otocyst.

the two cerebral invaginations have become deeper, and the two groups of cells which form the main portions of the corresponding ganglia contain a greater number of cells. (Plate II. Fig. 15.)

The pedal ganglia are also now composed of many more cells than in the previous stage. Each ganglion is usually pear-shaped, and tapers towards the posterior end of the foot. They both continue to receive accessions from the ectoderm (Figure A), and at the same time are rapidly increasing in size by division of the cells already in position. The nuclei are larger and more easily distinguished than in the previous stage from those of the mesodermic cells, the latter being more spindle-shaped than before. The cells of the mesoderm form a continuous layer along the inner surface of the ectoderm, except where cell proliferation is taking place (Figure A).

As yet nothing is to be seen of the pleural ganglia.

The visceral ganglia have increased in size (Plate I. Figs. 10, 11, 12); they are still connected with the ectoderm (Figs. 10, 12), although a few cells with large nuclei have become detached from it (Fig. 11). The ganglion and the otocyst of the same side of the body lie in nearly the same sagittal plane. Each ganglion is situated just above the angle caused by the pleural groove. The right visceral ganglion (Fig. 10) is somewhat farther forward and more dorsal than the left (Figs. 11, 12).

About in the median plane of the body, and above the angle made by the pleural groove, are the cells which form the abdominal ganglion (Plate I. Fig. 13). The greater part of them are still embedded in the ectoderm. Although in some regions they project into the body cavity, they are nowhere wholly separated from the ectoderm. The abdominal ganglion seems to be at first more intimately connected with the left visceral ganglion than with the right, but a connective is formed with both of them a little later, and the abdominal ganglion thus appears to occupy the place of a direct commissure between the two visceral ganglia. As development proceeds, the abdominal ganglion becomes closely fused with both the visceral ganglia.

Quite an advance in external conditions is made by the *ninth day*. But individuals of the same age vary so much in the degree of development attained by both their external and internal organs, that the age assigned can be taken only as an approximation to the average condition at the time indicated.

The tentacles appear as protuberances, the labial tentacles being much smaller than the ocular; the shell gland contains more concretions, the mantle is larger and bends backward over the dorsal surface of the foot. The radula sac makes its appearance and extends backward into the foot, where it ends blindly immediately back of the pedal ganglia. In transverse sections it appears flattened dorso-ventrally; its lumen is oval, and the ectoderm lining it is more than one cell deep.

The cerebral invaginations (Plate II. Figs. 15, 19, Plate III. Figs. 25, 26) are much deeper, the infolding ectoderm is greatly thickened, and the incipient ganglia receive accessions from ectodermic depressions between the rudiments of the upper lips and the labial tentacles (Plate II. Fig. 21). The cerebral commissure (Fig. 21) is also being formed, the cells of the median portion of each ganglion growing out to meet the corresponding cells from the opposite ganglion. The commissure at this stage is composed of a small number of cells, which are very much elongated. The fibres resulting from their elongation already make a continuous bridge from one ganglion to the other.

The pedal ganglia (Plate II. Fig. 20, Plate III. Fig. 27, Plate V. Fig. 60) consist of two small groups of cells, situated about midway between the sole of the foot and the posterior end of the radula sac. They are a little below and behind the pleural groove and the otocysts, and they are farther from each other than from the lateral wall of the foot. There is a slight indication of a commissure (Plate III. Fig. 27) joining their anterior portions to each other. The commissure is formed in the same manner as the cerebral commissure, the individual cells composing it being spindle-shaped, with their nuclei somewhat elongated in the direction of the fibres.

The otocysts (Plate II. Fig. 20, Plate III. Fig. 27, Plate V. Fig. 60) are on a level with the lower margin of the radula sac, and are nearer the pedal ganglia than in the preceding stage.

On each side of the body above the pleural groove is a group of a few cells, which are in all probability the first indications of the pleural ganglia (Plate II. Figs. 14 and 20). The centre of each cluster is seen on cross sections (Fig. 20) to be nearly on a level with the lumen of the radula sac. The cells at this stage are very small, and so loosely associated that it is difficult to distinguish them from mesodermic cells. I have not satisfactory evidence of their origin directly from the ectoderm, for, although I have found them at times very near to the ectoderm (Fig. 20), I have never found them at any stage continuous with it. On the other hand, I have not seen conditions which would warrant the conclusion that the ganglia were the result of outgrowths from either of the pre-existing ganglia.

A little before the ninth day the cells detached from the ectoderm to form the visceral ganglia (Plate II. Figs. 17, 18) increase rapidly in size, and the diameter of their nuclei often becomes four or five times as great as that of the ectodermic nuclei. The ganglia consist of elongated groups of such cells, still attached to the ectoderm above the pleural groove (Figs. 16, 18). The want of symmetry in the positions of the right and left ganglia is more conspicuous than in the preceding stage, the ganglion of the right side being considerably more dorsal and farther back than that of the left side (Plate II. Fig. 23, Plate V. Fig. 60). Owing to the infolding of the ectoderm on the right side of the body to form the respiratory chamber (Plate II. Figs. 16, 24), the region from which the ganglionic cells arise is now located on the ventral and median walls of the infolding. The ganglia have also grown forward, and lie between the nephridial organs and the nutritive sac (Plate II. Fig. 23). In an individual cut crosswise, the posterior portion of the ganglia is found to be two or three sections back of the otocysts. Both the ganglia may be traced through five or six sections.

A little behind the visceral ganglia, and to the left of the median plane of the body, are the prominent cells of the abdominal ganglion (Plate II. Figs. 24, *ab.*).

All of these ganglia still consist of groups of loosely associated cells. Later they become more compact, and are surrounded by connective-tissue cells.

The buccal ganglia (Plate II. Fig. 22), first seen with certainty at this stage, arise, one on each side of the radula sac, at the angle between it and the œsophagus. It is to be seen from cross sections that the cell proliferations from which they spring take place from the dorsal wall of the neck of the sac, where its lumen begins to be separated from that of the œsophagus. This is also their permanent position; they are later joined together by a commissure, which results from outgrowths of the cells composing the two ganglia.

On the *tenth day* the external appearance of the embryo remains nearly the same as before, with the exception that there is an increase in the size of the embryo, and especially of its pulsating sacs. The sac of the radula has become more elongated, and the anal opening (Plate III. Fig. 31, *an.*) is formed.

The cerebral invaginations still appear, in sections parallel to the sagittal plane (Plate III. Figs. 28 and 29), as shallow depressions. The number of cells in each ganglionic group (Plate IV. Fig. 58, Plate V. Fig. 63) has increased perceptibly. At the same time the groups have extended backward, and show indications of the cerebro-pleural connectives. In specimens cut in the sagittal plane, the cerebral commissure cut crosswise may be seen above the oral opening (Plate III. Fig. 30).

The pedal ganglia (Plate IV. Figs. 54, 58, Plate V. Fig. 63) have increased in size. Their anterior borders now reach as far forward as the plane of the pleural groove, and they extend backward into the foot much farther than before. In cross sections (Plate IV. Fig. 54) they appear as rounded groups of cells, which are far apart and not yet very compact; they still continue to receive accessions by the proliferation of ectodermic cells from the walls of the foot (Plate IV. Fig. 57, 58, *prf.*). The first decided evidence of a pedal commissure makes its appearance during this stage. It consists (Fig. 54) of a few very much elongated nerve cells, which stretch across from one ganglion to the other a little posterior to the region of the otocysts. The commissure may be traced on about half a dozen successive sections, or for a distance of some 50 or 60 μ. From its position it evidently is the beginning of the *anterior* commissure. The thickness (10 μ) of a single section contains only three or four cells, the nuclei of which have the chromatic substance so concentrated into a single nucleolus as to make the nuclei appear clearer than those of the surrounding connective-tissue cells. There is at present no trace of a posterior commissure. The otocysts are now nearer the ganglia (Plate IV. Fig. 58, Plate V. Fig. 63) than at any previous stage.

The pleural ganglia (Plate V. Fig. 63) are still inconspicuous, being composed of only a few scattered cells, which lie nearly dorsal to the otocysts, about midway between the visceral and the cerebral ganglia of the same side of the body. Many of the cells are elongated in the direction of the ganglia between which they are located, and appear to form the beginning of a connective between them.

The visceral ganglia (Plate IV. Figs. 58 and 59, *vsc.*) are still connected with the ectoderm, but project more prominently from the wall of the body, and extend forward more than before. The right (Plate IV. Fig. 59) is larger, and still lies more dorsal, than the left (Plate V. Fig. 63). The cells which compose the ganglia are numerous and large, and the nuclei of those which form the centre of the ganglion are conspicuously larger than those at the periphery. In cross sections of a stage possibly a little less developed than the one last described, the ganglia (Plate IV. Figs. 53, 56, 57, 55) lie, one on each side of the body, immediately above the pleural groove, a little below and inside the external orifices of the primitive nephridial organs. On the right side of the body the ectoderm which constitutes the anterior wall of the infolding to form the mantle chamber is seen in sagittal sections (Plate IV. Fig. 58) to be much thicker in the region adjoining the pleural groove than in that which forms the deeper portion of the infolding. The transition from the thick to the thin ectoderm is very abrupt, and is marked by a pocket-like depression. The right visceral ganglion is situated at the side and in front of this depression. Some of the cells in the anterior portion of this ganglion (Plate IV. Fig. 56) are traceable toward the median plane of the body. The left visceral ganglion

(Figs. 56, 57) is not yet as large as the right, and it consists of fewer cells.

The position of the connective between the visceral and pleural ganglia (Plate V. Fig. 63) is indicated by the presence of spindle-shaped cells with fibrous projections. The connective is at this time long, and the cells and fibres composing it are only joined to one another loosely.

As the abdominal ganglion increases in size, it extends more toward the right side of the body (Plate V. Fig. 61), and the connective be-

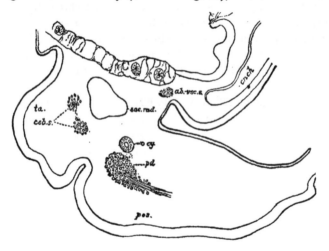

FIGURE B. — The left face of a section parallel to the sagittal plane from an embryo of the *eleventh day.* × 73.

ab.-vsc. s.	Left abdominal-visceral connective.	*pd.*	Pedal ganglion.
ceb. s.	Left cerebral ganglion.	*pes.*	Foot.
cnch.	Shell gland.	*sac. rad.*	Radula sac.
o cy.	Otocyst.	*ta.*	Ocular tentacle.

tween it and the right visceral ganglion, which is hardly perceptible at this stage, is much shorter than that to the left visceral ganglion.

The buccal ganglia remain in the same condition as in the preceding stage (Plate II. Fig. 22).

By the *eleventh day* the embryo has increased greatly in size (Figure B) ; the tentacles are prominent, and the pulsating sac of the foot is very large. A narrow slit-like infolding of the ectoderm (compare Plate VIII. Fig. 101, *gl. pd.*) has arisen in the median plane of the body at the anterior end of the foot, into which it extends backward a short distance. It is the beginning of the foot gland. The salivary glands also make their appearance during this stage as a pair of evaginations of the lateral walls of the œsophagus, immediately above its communication with the radula sac, and a little in front of the buccal ganglia (Plate VI. Figs. 77–80).

The cerebral invaginations still open broadly at the sides of the head (Plate III. Figs. 32–34, and Figure C). They are, however, quite deep, and in a series of sagittal sections the depression becomes deeper and deeper as one approaches the median plane, and at the same time the orifice which leads to the depression becomes narrower and narrower,

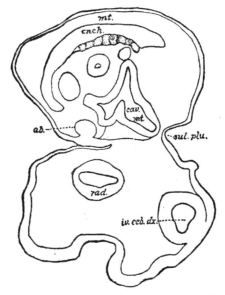

FIGURE C. — The posterior face of a transverse section from an embryo of the *eleventh day.* × 73.

ab.	Abdominal ganglion.	*mt.*	Mantle.
cav. mt.	Mantle cavity	*rad.*	Radula sac.
cnch.	Shell gland.	*sul. plu.*	Pleural groove.
iv. ceb. dx.	Right cerebral invagination.		

until it is almost slit-like (Figs. 32–40). The deep ends of the invagination are turned a little towards the median plane. These invaginated portions of the brain are composed of small, closely packed cells, whose nuclei stain deeply. The proliferated portions of the cerebral ganglia, which are deeper than the sacs (Plate V. Fig. 64, Plate VI. Figs. 70, 71), extend toward each other in the median plane, and backward and downward toward the pedal ganglia (Fig. 71). They have now become differentiated into a fibrous central part (Fig. 71), in which are lodged the larger scattered cells with their very large nuclei, and a peripheral part, where the cells are crowded together and the nuclei are smaller (Fig. 70). They are loosely enveloped by spindle-shaped, very much elongated, connective-tissue cells (Fig. 71). Immediately above the oral cavity is the cerebral commissure (Plate VI. Fig. 80ᵃ). It can be traced from one side of the brain to the other, and its cross section appears as a very small round patch of fibrous substance, surrounded on the dorsal side by a layer of flat cells.

The cerebro-pedal connectives are indicated (Fig. 71) by a few cells extending from the ventral-posterior ends of the cerebral ganglia to the anterior ends of the pedal, a little in front of the cerebro-pleural connectives (Fig. 70). The latter extend from the posterior ends of the cerebral to the anterior ends of the pleural ganglia, thus diverging somewhat from the cerebro-pedal connectives. There are found in the ganglia many cells which are in different stages of division. It is owing to this cell division that the ganglia increase rapidly in size, especially after they are wholly cut off from the ectoderm ; cells in the commissures and connectives are also found in process of dividing in planes perpendicular to the direction of their fibres.

The principal change in the pedal ganglia (Plate VI. Fig. 71) is due to an increase in size, particularly in the antero-posterior direction. The central portion of these ganglia has the same fibrous appearance as that described for the cerebral ganglia, and the pedal nerves can be traced for a considerable distance toward the tip of the foot (compare Figure E, page 191). The anterior commissure (Plate III. Fig. 44, Plate VI. Fig. 74) is now somewhat shorter than in the previous stage, and consists of a greater number of cells. Cell proliferation is still taking place from the ectoderm of the ventral wall of the foot (Plate VI. Fig. 71), and the ganglia continue to receive accessions from these sources. More highly magnified views of the regions of proliferation are given in Plate VI. Figs. 72 and 73.

The pleural ganglia (Plate VI. Fig. 70) are now easily recognized. Each ganglion is formed of a triangular group of cells, occupying a position immediately above and anterior to that part of the pleural groove which is nearest to the otocyst. The cells composing the ganglion are fewer than those of any of the other pairs of ganglia, but resemble them in their histological conditions ; they are only loosely connected, and their fibres are elongated in the directions of the three connectives. At this stage the ganglia are not closely enveloped in connective tissue.

The pleuro-visceral connectives are well developed, especially the left one (Fig. 70) ; the right one is much longer and more attenuated, since the right visceral ganglion is farther from the pleural than the left visceral. The ganglia are most distinctly seen in specimens cut in a sagittal direction.

The visceral ganglia (Plate V. Figs. 67–69, Plate VI. Fig. 70) are much larger and more elongated in the direction of the pleural ganglia — i. e. downward, forward, and outward — than they were during the previous stage. They are still connected with the ectoderm at their posterior dorsal ends, while the opposite ends are much drawn out toward the pleural ganglia (Figs. 69, 70). The right visceral ganglion (Figs. 67–69) is larger than the left, and its longest axis has a dorso-ventral direction (Fig. 68). The fibrous prolongations continue into the pleuro-visceral connectives (Fig. 71).

The abdominal ganglion (Plate III. Figs. 43, 44, 46, 47, Plate VI. Figs. 75, 76), although still connected with the ectoderm, is also larger, and projects more into the body cavity than on the tenth day. A large portion of it still lies to the left of the median plane of the body (Plate VI. Figs. 75, 76), and the connective to the left visceral is well developed (Plate III. Figs. 41, 42, Plate V. Fig. 68); that to the right is less complete (Plate III. Figs. 45, 51).

The buccal ganglia (Plate V. Fig. 62, Plate VI. Fig. 77) are now very distinct; the dorsal wall of the radula sac still contributes to their increase in size.

Cell proliferation takes place from the ectoderm bordering the entrance to the respiratory cavity. A few cells, which probably form the olfactory ganglion, are seen at this stage to be separating from the ectoderm in this region.

For the next twenty-four to thirty-six hours (*twelfth and thirteenth days*) the external appearance of the embryo remains nearly the same as on the eleventh day. In the living embryo the larval heart may be seen pulsating, and the foot gland extends somewhat farther towards the posterior extremity of the foot.

The cerebral invaginations appear simply as long narrow sacs filled with a coagulated substance; the inner ends of these sacs have grown upward as well as backward (Plate VII. Fig. 94). The proliferated portions of the cerebral ganglia (Fig. 94) are much larger, and have now assumed more nearly their ultimate positions (Plate III. Figs. 48, 49; Plate VII. Figs. 81, 82, 94). The central portion of each has become more fibrous (Fig. 81).

The connectives, both to the pedal (Plate VII. Fig. 81) and to the pleural ganglia (Plate III. Fig. 48), are well developed, and are both thicker and shorter than in the stage last described.

The pedal ganglia do not differ materially from the condition described for the eleventh day. The anterior end has increased in diameter, and has grown a little farther forward (Plate III. Fig. 50, Plate VII. Figs. 81, 91).

Both commissures are now present; the anterior (Fig. 92) is a little behind the otocysts (compare Fig. 92 with Fig. 91), and the posterior (Fig. 90) is directly above the blind end of the foot gland, and about 0.2 mm. back of the anterior commissure.

The pleural ganglia (Plate III. Fig. 48, Plate VII. Figs. 82, 83, 88) are very near the cerebral ganglia, as may readily be seen in sagittal sections (Figs. 48, 82), and the fibrous connectives to the other ganglia are plainly to be distinguished. The ganglia have become more compact and rounded, and occupy a position nearer the middle plane of the body (Figs. 86, 88).

The visceral ganglia (Plate III. Fig. 49; Plate VII. Figs. 83, 84, 86–89), although they have increased greatly in size, are still connected with the ectoderm which forms the anterior wall of the mantle chamber (Figs. 88, 89).

They have also moved inward and forward. The right ganglion (Figs. 49, 83, 87–89) is especially well developed, and much farther forward than in the previous stage. Its axis is prolonged into a nerve, which runs upward and backward, probably to the olfactory ganglion (Figs. 84, 87).

The connective from the right visceral to the abdominal ganglion passes backward and inward (Plate VII. Figs. 83, 84). Where the connective leaves the visceral ganglion (Fig. 83), the nuclei of the ganglionic cells are very large, and the fibres are very much elongated in the direction of the connective.

In specimens cut crosswise the nerve which forms the dorsal prolongation of the axis of the visceral ganglion is found far forward, in front of the anterior face of the abdominal ganglion; it passes upward and inward (Plate VII. Figs. 87, 88), and is connected with the ectoderm that forms the wall of the small infolding from the respiratory cavity (Fig. 88) referred to in the account of the tenth day. This region is at the same level as that with which the abdominal ganglion is connected farther back (Plate VII. Fig. 93). The ectodermic cells to which this nerve is distributed form the lining to an irregular infolding from the median face of the respiratory cavity, and the lumen of the infolding connects by a narrow orifice with the respiratory chamber (Fig. 88, *cav. mt.*). I believe this is the organ first described by Lacaze-Duthiers.

A little farther forward the right visceral ganglion sends to the right side of the body a nerve (Plate VII. Fig. 89 *n.*), which passes between the wall of the mantle chamber and the primitive sexual duct, probably to be distributed to the right half of the mantle.

At this time the greater portion of the abdominal ganglion (Plate VII. Figs. 81, 82, 85, 86, 93) lies on the right side of the median plane, although it is joined to the left visceral by a large and prominent connective (Plate III. Figs. 50, 52, Plate V. Figs. 65, 66, Plate VII. Fig. 93). Since the visceral ganglia have grown inward and forward, the abdominal ganglion now occupies a position considerably posterior to them (Plate VII. Figs. 83, 86); it lies above the right side of the radula sac. Its posterior dorsal margin is still continuous with the ectoderm of the wall of the respiratory cavity (Fig. 93), but farther forward it is entirely separated from the ectoderm (Fig. 85), and is surrounded by a layer of connective-tissue cells. All the other ganglia are similarly enveloped in connective tissue except where they are continuous with the ectoderm.

The connective to the left visceral ganglion (Plate VII. Fig. 93) passes downward, forward, and outward to the left side above the radula sac.

The buccal ganglia (Plate VII. Fig. 81) are larger than on the tenth day, but are closely applied, as before, to the walls of the radula sac. Their commissure (Plate V. Fig. 65) is embraced in the angle between the œsophagus and the neck of the radula sac, and in sagittal sections presents a circular outline.

On the *fourteenth day* the upper lips as well as both pairs of tentacles are very prominent, and the foot gland has grown backward still farther into the foot (Plate VIII. Fig. 102). The salivary glands have now become elongated into tubular organs with a circular lumen and thick walls consisting of a single layer of epithelial cells (Plate VIII. Fig. 106). They reach a little farther back than the buccal commissure; in passing forward they lie on either side of the œsophagus, about on a level with its lower border. They pass along the dorsal side of the buccal ganglia, and then suddenly bend downward to open into the œsophagus.

The cerebral invaginations (Plate VIII. Fig. 96) present the same general appearance as in the stage last described, but the lumen of the sacs is smaller (Plate X. Figs. 121, 126) ; in cross sections (Fig. D) it appears oval. The walls are thick, being composed of spindle-shaped cells arranged perpendicularly to the axis of the sac and so crowded that the nuclei are three or four deep.

The proliferated portion of the cerebral ganglia (Plate IX. Fig. 114) retains its pear-shaped condition, but is shorter and thicker. A ventral and

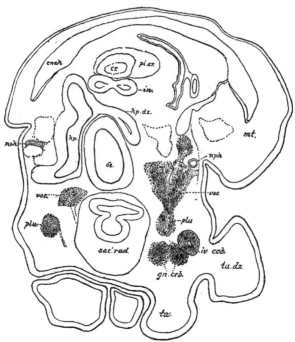

FIGURE D. — Posterior face of a transverse section from an embryo of the *fourteenth day.* ✕ 190.

cnch.	Shell gland.	*nph.*	Nephridial organ.
cr.	Heart.	*œ.*	Œsophagus.
gn. ceb.	Cerebral ganglion.	*pi. cr*	Pericardium.
hp.	Liver.	*plu.*	Pleural ganglion.
hp. dx.	Right lobe of liver.	*sac. rad.*	Radula sac.
in.	Intestine.	*ta. dx.*	Right ocular tentacle.
iv. ceb	Cerebral invagination.	*ta'.*	Labial tentacle.
mt.	Mantle.	*vsc.*	Visceral ganglion.

median portion of each ganglion forms a small rounded lobe (Figure E). These lobes are near the bases of the upper lips, and in sagittal sections appear almost completely separated from the larger part of the ganglia by ingrowths of connective tissue. It is from these lobes that the pedal connectives arise. The connectives to the pleural ganglia emerge from the larger portion of the ganglion ; they are thicker and shorter than the cerebro-pedal connectives, from which they are separated by only a narrow space.

The cerebral commissure is much shorter than before (Plate X. Fig. 126), but it has not increased much in thickness (Plate VIII. Fig. 101). In sagittal sections it is seen to be composed of a central portion made up of nerve fibres cut crosswise and a peripheral layer of nuclei ; but the nuclei are wanting on the face of the commissure which is in contact with the dorsal wall of the œsophagus.

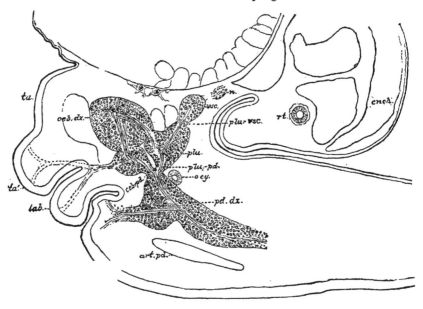

FIGURE E. — The left surface of a section parallel to the sagittal plane from an embryo of the fourteenth day. ✕ 73

art. pd	Pedal artery.	plu -pd.	Pleuro-pedal connective.
ceb. dx.	Right cerebral ganglion.	plu.-vsc	Pleuro-visceral connective.
ceb.-pd.	Cerebro-pedal connective.	pd. dx.	Right pedal ganglion.
cnch.	Shell gland.	rt.	Rectum.
lab.	Lip.	ta.	Ocular tentacle.
n.	Nerve	ta'.	Labial tentacle
o cy.	Otocyst.	vsc.	Visceral ganglion.
plu.	Pleural ganglion.		

The pedal ganglia (Plate VIII. Figs. 97–100, Plate IX. Figs. 114, 118, 119) lie between the radula sac, which is above, and the foot gland which is below them. ˙ They are nearer together than on the twelfth day, and their anterior ends are more rounded (Fig. 114). Their posterior ends are elongated and continued as two large nerves far back into the foot. In specimens cut crosswise these nerves appear as rounded patches of fibres, situated one on each side of the body, above the plane of the foot gland and about midway between it and the lateral walls of the foot. Each is surrounded by a layer of connective-tissue cells. As one approaches the pedal ganglia in passing from behind forward, the nerves increase in size and lie nearer to each other. In the region of the posterior commissure (Plate IX. Fig. 119) the ganglia are nearly as broad as in the region of the anterior commissure (Fig. 118), but they are not much more than half as thick in the dorso-ventral direction. In front of the posterior commissure they are separated by a narrow space, which is wider behind than in front, where it is terminated by the anterior commissure. The commissures are both well developed (Plate VIII. Figs. 101, 102, Plate IX. Figs. 118, 119), and owing to the approximation of the ganglia have become shorter than in the last stage. The nuclei in the region of the posterior commissure (Fig. 119) are of nearly uniform size; but in front of it each ganglion (Figs. 114, 118) contains a fibrous central portion immediately surrounded by the greatly enlarged nuclei of cells which form the most of the fibrous substance.

The pleural ganglia (Plate VIII. Fig. 106, Plate IX. Figs. 114, 116, Plate X. Figs. 123, 125, and Fig. E) have increased considerably in size, and are more compact. They have moved downward and inward; and each now lies in contact with the posterior face of the corresponding cerebral mass (Plate IX. Fig. 114), and below and in front of the ventral portion of the corresponding visceral ganglion (Figs. 106, 123, 125). They are much smaller than either the cerebral or visceral ganglia. The nuclei of their central cells are, as in the pedal ganglia, much enlarged.

The visceral ganglia (Plate VII. Fig. 95, Plate IX. Fig. 114, Plate X. Figs. 123, 125) are now entirely detached from the ectoderm, and have moved downward, forward, and inward.

The left ganglion (Plate VIII. Fig. 106, Plate X. Fig. 125) is smaller than the right, and more closely connected with the left pleural (Fig. 125) than in the previous stage. Its dorsal surface is slightly above the level of the dorsal wall of the radula sac, and its connective with the abdominal ganglion (Plate VIII. Fig. 104) is much broader than before. The right visceral ganglion (Plate VII. Fig. 95, Plate VIII. Figs. 102 and 106, Plate IX. Fig. 114, Plate X. Fig. 123) is much larger than in the last stage; it is also closely connected with the right pleural ganglion (Fig. 123). It extends dorsally much farther than the left visceral, and also nearer to the median plane (Plate VII. Fig. 95, Plate VIII. Fig. 102, Plate IX. Fig. 120). It is in contact with the lower surface of the right end of the abdominal ganglion (Plate VII. Fig. 95).

The abdominal ganglion (Plate VII. Fig. 95, Plate VIII. Figs. 101, 102, 104, Plate IX. Figs. 115–117, Plate X. Fig. 123) is entirely unconnected with the ectoderm, and has moved forward, so that there is a considerable space between it and the pleural groove, but its posterior face extends farther back than that of the right visceral ganglion (Plate VIII. Fig. 102). The greater portion of the ganglion is now situated on the right side of the body, immediately above and to the right of the radula sac (Plate VIII. Fig. 104, Plate IX. Figs. 115–117). It is elongated, and its chief axis is directed obliquely across the body, the right end being considerably higher and a little farther back than the left end. In passing downward and forward to the left side of the body, it lies between the œsophagus and the posterior part of the radula sac. Its left end is prolonged into a connective, which passes forward and outward to join the left visceral ganglion (Plate VIII. Fig. 104, Plate X. Fig. 123). A large nerve, which passes upward and backward to be distributed to the viscera, emerges from the most dorsal portion of the abdominal ganglion on the right side of the body (Plate VIII. Fig. 104, Plate IX. Fig. 117). The histological condition of the abdominal ganglion is similar to that of the previously described ganglia of this stage. The fibrous portion, as well as the enlarged cells and nuclei, are especially prominent in the portion of the ganglion which lies to the right of the median plane of the body (Plate IX. Fig 117).

The buccal ganglia (Plate VII. Fig. 95, Plate VIII. Figs. 102, 106, Plate IX. Fig. 120, Plate X. Fig. 121) have become larger, and with their commissure (Plate VIII. Fig. 101, Plate IX. Fig. 120) stretch across the dorsal wall of the neck of the radula sac, to which they are still closely united. The nuclei immediately surrounding the central fibrous portion of the ganglion are already slightly enlarged, though the cells are not so far advanced in their histological differentiation as are those of the other ganglia. A single pair of connectives passes obliquely forward, downward, and outward, to join the buccal to the cerebral ganglia (Plate X. Fig. 121).

By the *sixteenth* and *seventeenth days*, besides a general increase in size of the external organs, the foot gland extends backward much farther than the pedal ganglia (Plate VIII. Fig. 107), and the viscera lie rather more to the left side of the body (Figure G).

The central nervous system (Figure F) now consists of five well developed pairs of ganglia and an azygos ganglion (Figure G). The cerebral ganglia with their commissure form the dorsal portion of three nerve rings, the remainder of which are completed respectively, (1) by the cerebro-pedal connectives, the pedal ganglia, and their commissures; (2) by the cerebro-pleural connectives, the pleural ganglia, the pleuro-

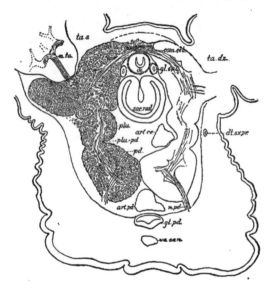

FIGURE F. — Posterior face of a transverse section from an embryo of the *sixteenth day.* × 70.

art. ce.	Cephalic artery.	œ.	Œsophagus.
art. pd.	Pedal artery.	plu.	Pleural ganglion.
com. ceb.	Cerebral commissure.	plu.-pd.	Pleuro-pedal connective.
dt. sx. pr.	Primary sexual duct.	pd	Pedal ganglion.
gl. sal.	Salivary gland	sac. rad.	Radula sac.
gl. pd.	Pedal gland.	ta. dx.	Right ocular tentacle.
n. pd.	Pedal nerve.	ta. s.	Left ocular tentacle.
n. ta.	Tentacular nerve.		

visceral connectives, the visceral ganglia, the viscero-abdominal connectives, and the abdominal ganglion; (3) by the cerebro-buccal connectives, the buccal ganglia, and their commissure. The first and second rings are further joined to each other by means of the pleuro-pedal connectives. Each of these three rings encircles the œsophagus. The posterior end of the radula sac in the earlier stages, up to the present one, is usually found to occupy a position *above* the pedal ganglia and their commissures; but with a greater concentration of the nervous ganglia toward one another, the sac is forced to occupy a position *below* the pedal ganglia and their commissures. The relations of the different ganglia to each other is even more definite than before, and can be more readily understood from transverse sections than from sagittal ones. The peripheral nerves from the cerebral, pedal, visceral, and abdominal ganglia are well developed; the principal changes from this time until hatching are histological.

The cerebral invaginations have become narrow and shorter, but are still open to the exterior (Plate X. Fig. 124, *iv.*). The deeper portion of the invagination, that in contact with the proliferated portion of the cerebral ganglion, has become a solid and rounded mass (Plate X. Fig. 122, *lob. lat.*), which is intimately connected with the ganglion by means of fibrous outgrowths from its ganglionic cells. It is composed of small deeply stained cells, which have undergone no such histological change as those which compose the proliferated portion of the brain. It forms a lobe on the antero-lateral face of each cerebral ganglion (Plate X. Figs. 122, 124, 127). From this time forward the principal change in the cerebral sacs consists in the gradual obliteration of the lumen of the invagination. This is usually completed somewhat later in the embryonic life; but, as previously stated, the sacs have in one instance at least been found open several days after hatching. Besides this, there is no other connection now remaining between the ectoderm and any of the ganglia, except such as is effected by means of the peripheral nerves.

The median proliferated portions of the cerebral ganglia now extend dorsally farther than in the last stage, and their commissure is much shorter (Plate VIII. Fig. 105, and Fig. F).

The pedal ganglia (Plate VIII. Figs. 103ᵃ, 109–113) have moved forward, and are broadly in contact with the pleural ganglia. They have become more compact, and rather more triangular in shape, than before. From the ventral portion of each ganglion emerge four or five large nerves, which terminate in the ventral wall of the foot; from the dorso-lateral region two nerves are given off to the lateral walls, and the antero-ventral part of each ganglion tapers off into a stout nerve running forward to the anterior wall. The connectives with the cerebral ganglia are well developed (Plate VIII. Figs. 103, 107).

The pleural ganglia (Plate VIII. Figs. 103ᵃ, 111–113) are nearer to the median plane than previously. The ventral posterior face of each is closely joined to the corresponding pedal ganglion (Figs. 103ᵃ, 112), the dorsal median face to the visceral ganglion (Figs. 103ª, 112, 113), and the anterior face to the cerebral ganglia (Fig. 107). No nerves arise from the pleural ganglia.

The visceral ganglia (Plate VIII. Figs. 103ª, 110–113) have also moved nearer to the median plane. The left ganglion is directly below

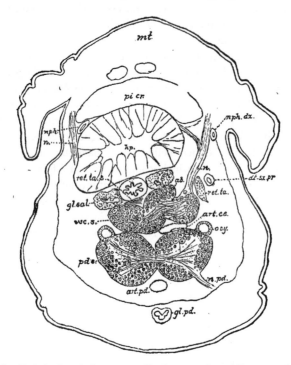

FIGURE G. — Posterior face of a transverse section from an embryo of the *seventeenth day*. × 60.

ab.	Abdominal ganglion.	n. pd.	Pedal nerve.
art. ce.	Cephalic artery.	nph.	Nephridial organ.
art. pd.	Pedal artery.	nph. dx	Right nephridial organ.
dt. sx. pr.	Primary sexual duct.	o cy.	Otocyst.
gl. sal.	Salivary gland.	pd. s.	Left pedal ganglion.
gl pd.	Pedal gland.	pi. cr.	Pericardium.
hp.	Liver.	ret. ta.	Retractor muscle of right ocular tentacle.
mt.	Mantle.	ret. ta. s.	Retractor muscle of left ocular tentacle.
n.	Nerve.	vsc. s.	Left visceral ganglion.

the œsophagus (Fig. 111 and Figure G), since the latter occupies a position more to the left side of the body than before. The right visceral ganglion still remains larger, and extends farther dorsally than the left (Figs. 103ª, 111, 112). It is nearer the median plane than in the stage last described (Figure D., page 190) ; it lies in front and only a little to the right of the abdominal ganglion (Fig. 111).

The abdominal ganglion (Plate VIII. Figs. 109–111) is less elongated than in the last stage (Fig. 104). It is wedge-shaped, and appears as though crowded in between the two visceral ganglia from behind and above. It is so intimately connected with these ganglia that it almost appears to form a part of them (Fig. 111). But the presence, between the ganglionic masses, of connective-tissue cells, which reach nearly to the connectives, enables one to make out with some certainty the extent of each of the three ganglia. Since the planes which separate them are oblique to the transverse planes of the body, these boundaries are not always readily seen in cross sections. The right and left visceral ganglia have no *direct* commissural nerve fibres uniting them ; they are joined only by such fibres as pass through the abdominal ganglion.

The buccal ganglia (Plate VIII. Fig. 108, Plate X. Fig. 124) are now entirely separated from the dorsal wall of the radula sac, from which they arose, and are surrounded by a layer of connective-tissue cells. The differentiation of their ganglionic cells is well advanced.

Summary.

1. In Limax maximus the whole of the central nervous system arises directly from the ectoderm.

2. The cerebral ganglia originate in part as a pair of true invaginations, one on each side of the body in front of the pleural groove and behind and below the bases of the ocular tentacles. In the course of their development, the neck of each invagination becomes a long, narrow tube-like structure, which remains open throughout the period of embryonic life. The main part of the cerebral ganglia is formed from cells which are detached at an early period from the deep ends of these cerebral invaginations, or from neighboring ectoderm ; the portions which persist as the walls of the infoldings finally form distinct lateral lobes of the brain.

3. All the other ganglia originate by cell proliferation from the ectoderm without invagination.

4. The ganglia arise separately, and, with the exception of the abdominal and mantle ganglia, in pairs, one on each side of the body. Their connection with each other is the result of a secondary process in the development, — the outgrowth of nerve fibres.

In advanced stages, the central nervous system consists of five pairs of ganglia and an azygos ganglion. Together these form three complete rings surrounding the œsophagus.

The relative positions of the ganglia are best appreciated from cross sections. In passing from behind forward, they are encountered in the following order: (1) the pair of pedal ganglia, which lie under the radula sac, and are joined to each other by an anterior and a posterior commissure; (2) one abdominal ganglion a little to the right of the median plane; (3) a pair of visceral ganglia occupying the posterior angle formed by the outgrowth of the radula sac from the œsophagus. They are separated by the abdominal ganglion, from which connectives pass to them; (4) a pair of pleural ganglia, not joined by a commissure, and not giving off nerves. They are united by means of connectives to the pedal, visceral, and cerebral ganglia of the same side; (5) a pair of cerebral ganglia, with their supra-œsophageal commissure and connectives to the pleural, pedal, and buccal ganglia; (6) a pair of buccal ganglia, with a commissure under the œsophagus posterior to its connection with the sac of the radula.

The mantle ganglion lies far back, and is joined to the abdominal ganglion by a large nerve.

It seems as if there could be no doubt that the infolding of the ectoderm of the anterior wall of the respiratory cavity on the right side of the body gives rise to the special-sense organ discovered by Lacaze-Duthiers ('72, pp. 483–494). It corresponds in its position and its connection with the right visceral ganglion to his description of the adult, and also to Fol's description ('80, pp. 166–168) of the origin and position of that organ in the aquatic pulmonates.

As is well known, Limax belongs to that group of Gastropods in which all the nerve centres, except the cerebral and buccal, lie on the ventral side of the intestinal tube; not to the group in which the connection between the right pleural and right visceral ganglia passes above the œsophagus, and in which that of the left lies below it. Limax, therefore, is not directly referable to Von Jhering's group of Chiastoneura, although the want of symmetry in the position of its ganglia does not allow one to say that it is orthoneuric.

The Gastropod in which the details of the origin and fate of the nervous centres have been most carefully studied is Bithynia, a chiastoneuric form, in which Sarasin has found that the abdominal ganglion is joined to the right visceral ganglion *only*, and is located at the fundus of the gill cavity. The relation is different from that found in Limax maximus, where the abdominal ganglion is intimately fused with the right visceral, and is *also* in close connection with the left visceral ganglion.

As was to have been anticipated, the abdominal ganglion of Limax corresponds more nearly in position to that in Lymneus and other fresh-water pulmonates, as described for the adult by Lacaze-Duthiers ('72, pp. 437–500).

Of the authors who have studied the origin and development of the *cerebral ganglia* in Mollusks, Fol ('80, pp. 168, 169, 193–195) is the only one who has pursued his investigations on Limax maximus. He says ('80, p. 193): " Vers l'époque de la fermeture de la vésicule oculaire, se montrent deux autres enfoncements de l'ectoderme. L'un des deux, assez vaste et situé à la base du tentacule, à son bord intérieur, est l'origine du ganglion cérébroïde ; je le décriai plus loin. L'autre enfoncement, plus petit, est situé au-dessous de ce dernier, à la base du pied, et mène à la constitution de la vésicle auditive."

As to the method by which the cerebral ganglia originate, this agrees in part with that which I have found ; but as to the time of origin, my investigations lead me to a different conclusion. The otocysts are present as small groups of cells (Plate I. Fig. 4), and the cellular elements which go to form the beginning of the pedal ganglia are also being proliferated (Fig. 3), before there is a trace of the invaginations which go to form the cerebral ganglia (Fig. 2).

A little later the otocysts assume the form of closed vesicles, unconnected with the ectoderm (Plate I. Fig. 9), while the cerebral invaginations are now seen as shallow pits (Fig. 6). Therefore, in Limax maximus the formation of both pedal ganglia and otocysts precedes that of the cerebral invaginations.

Sarasin ('82, pp. 1–68) maintains that in Bythinia tentaculata there are no invaginations to form the cerebral ganglia. They arise as thickenings of the ectoderm, one on each side of the body, which he calls *die Sinnesplatte*.

In the recent researches of the Sarasin brothers ('87, pp. 600–602, '88, pp. 59–69) on Helix Waltoni, of Ceylon, it is asserted that each of the cerebral ganglia is at first represented by a group of cells derived from the part of the ectoderm called " Sinnesplatte " before there is any invagination. There are two groups of these cells, one on each side of the body. Somewhat later two infoldings arise from each Sinnesplatte, one above the other. These infoldings become long, narrow "cerebral tubes," the deep ends of which are enlarged ('88, Fig. 24). From their inner ends a rapid cell proliferation takes place, the products of which join the cerebral cells already in position. The invaginated portions later form the "accessory lobes" of the brain. At a late stage only one pair of tubes remains open to the exterior, and the openings to these are closed before the end of embryonic life. The Sarasins ('88, p. 61) do not know the precise time at which they are closed, but are certain that the openings do not persist. They express their belief that the cerebral tubes are homologous with the organs of smell in Annelids, which, according to Kleinenberg's studies on Lopadorhynchus, also originate as invaginations of the Sinnesplatte, and by cell proliferation furnish a part of the material for the brain.

Prior to any knowledge of the investigations on Helix by the Sarasins, I found very similar conditions in Limax maximus. In this case, however, there is but one invagination of the ectoderm on each side of the body. It corresponds in position to those described in Helix, being perhaps the equivalent of the upper or larger invagination in that species.

The invaginations in Limax have the form of shallow pits before any other ganglionic cells are to be seen. The cell proliferation, which results in the production of the main portion of the ganglia, takes place during their ingrowth. Possibly the proliferation from the depression between labial tentacle and upper lip represents what was originally a true invagination, and corresponds to the lower of the two invaginations described by the brothers Sarasin. In Limax maximus the external openings persist until a late stage, and occasionally even after hatching. Here, also, the invaginations form a lobe of the brain, exactly as in the case of Helix (Sarasin, '87, p. 601).

Two well developed "Seitenorgane" were found by the Sarasins ('88, p. 54) in Helix Waltoni, situated near each other in the "senseplate"; and they think (p. 60) that these may correspond in position to the cerebral tubes of later stages.

The groups of cells embedded in the ectoderm, from which, in my opinion, the greater part of the nervous system in Limax maximus takes its origin, resemble both in the arrangement of the cells and their histological condition the "Seitenorgane" described by the brothers Sarasin ('88, pp. 53–57). But I have never observed bristles, or other terminal structures, projecting toward the outer world. Moreover, in Limax unmodified ectodermic cells usually lie between these groups of large cells and the outer surface of the body.

The Sarasins ('88, p. 57) consider these clusters of cells homologous with the "taste-buds" and "lateral organs" of vertebrates, and say that they are to be found in and at the margin of the Sinnesplatten, and along the sole of the foot, — more rarely on the sides of the foot. I think these organs are probably the same as those which I have seen in Limax, and to which I attribute simply the function of contributing to the formation of the ganglia.

Salensky ('86, pp. 685–690) describes the cerebral ganglia of Vermetus as arising from a pair of ectodermic thickenings, which early show pocket-like invaginations, and become deeper and narrower. From the inner ends of these invaginations are formed the main portion of the ganglia. The latter are united to each other by a very small commissure, composed of fibrous prolongations of the ganglionic cells surrounded by other nerve cells.

The principal difference between the method of development in Vermetus and that in Limax maximus consists in the fact that the detachment of the deep portion of the invaginations to form the ganglia in Vermetus is not effected until the invaginations have reached their ultimate size, whereas in Limax the detachment of cells from the invaginated area begins as early as does the invagination, and accompanies it during the whole of its formation.

Kowalesky ('83[a], pp. 1–54) found in Dentalium two deep invaginations, which he calls the "sincipital tubes," one on each side of the head region, a little ventral to the middle of the velar area. From the posterior deep ends of these sacs the cerebral ganglia are subsequently formed ; but he is uncertain whether all the cells concerned in the involution share in the formation of the ganglia. If his Figure 65 is compared with Figures 27 and 33 A in Salensky's paper, the close resemblance in the method of origin of the cerebral ganglia in the two types becomes apparent.

Fol ('80, pp. 169, 170) asserts that the *pedal ganglia* of the aquatic pulmonates appear as condensations in an already formed mesoderm, and that they are nearer the pharynx than the ectoderm when they begin to be discernible. "One may therefore say," he adds, "that these ganglia arise from the mesoderm without prejudging the unsettled question, viz. from which of the primordial layers arises the mesoderm which forms them." Of the pedal ganglia of the terrestrial pulmonates, he says that they are diffentiated *en lieu et place* in the midst of the mesodermic tissues of the foot.

With this I cannot agree, although I admit that at the time when the groups of cells which form the ganglia begin to be proliferated from the ectoderm, it is extremely difficult to distinguish them from the mesodermic elements (Plate I. Fig. 5). It is to be observed, however, that Fol considers it an unimportant distinction, whether the ganglia are formed from groups of mesodermic cells which have themselves recently originated from the ectoderm, or by a proliferation of cells directly from the ectoderm.

I am unable to reconcile the account of the development of the pedal ganglia in Bithynia given by P. Sarasin ('82, pp. 47–49), with the conditions seen in Limax; nor can I think it probable that any considerable difference exists between nearly related mollusks in regard to the *place* whence the ganglionic cells arise. Sarasin maintains that in Bithynia the pedal ganglia arise from a *single median thickening of the ectoderm of the dorsal wall of the foot,* in the region where that wall bends over to become continuous with the posterior wall of the visceral sac. Anteriorly, in the region of the oral invagination, this median band of cells forks, and each branch becomes joined to the corresponding cerebro-pleural cell mass by a slender cord of cells. Subsequently, the posterior unique portion of the proliferated cell mass is completely divided into lateral branches by a separation which progresses from in front backward. It seems to me that, according to this account, both the pedal ganglia must be regarded as arising from a common mass of cells, and that they are not from the beginning wholly separate, as I maintain for Limax.

The relative positions of pedal ganglia and otocysts present, to my mind, a serious objection to Sarasin's view, which may not have seemed so important to him on account of his uncertainty about the origin of the otocysts. I believe it is sufficiently evident that the otocysts do not arise, as Sarasin thinks probable, from the cerebro-pleural proliferations, but independently, and from the dorso-lateral wall of the foot in the region of the "pleural groove." They ultimately lie immediately dorsal to the corresponding pedal ganglia. If Sarasin's view as to the origin of the pedal ganglia as a *median* dorsal proliferation were correct, the ganglia would have to migrate to a lower plane than that occupied by the otocysts. But there is no evidence either in Limax or the figures given of Bithynia which would confirm such a supposition. As further corroboration of my opinion that the pedal ganglia arise from the *ventral* and lateral walls of the foot, I would cite the conclusion reached by Salensky ('86, pp. 691, 692) for Vermetus. He has shown that the pedal ganglia originate from the ventral wall of the foot, in a region and by a method corresponding to that seen in Limax maximus, as will be seen by comparing his Figures 21 C to 23 with my Plate I. Figs. 5 and 7, and Plate IV. Fig. 57. The only important difference between Vermetus and Limax lies in the fact that, in the case of the former, the cells forming the ganglion remain from the beginning a more compact mass than they do in the latter.

No one except Lacaze-Duthiers ('72, pp. 456, 457) has mentioned the existence of more than a single pedal commissure. He maintains that there are in Lymnæus as many as *three*. After speaking of the cerebral ganglia as being connected by one commissure, he goes on to say (p. 456), " Au contraire les ganglions pédieux ont trois commissures réelles." He seems, however, uncertain as to whether the most *posterior* ought to be considered a true commissure : " La troisième commissure mérite-t-elle bien ce nom? elle est constante dans les Pulmonés et se présente sous la forme d'un petit nerf grêle transversal naissant à peu près à la hauteur du troisième nerf pédieux inférieur ; elle donne vers son milieu naissance à un filet nerveux trés-délié, impair médian que l'on suit dans les tissus de la fosse pédieuse sans trop pouvoir définir et limiter exáctment son rôle." (p. 457.) His investigations were made exclusively upon the adult.

In Limax maximus two commissures are certainly distinguishable during a greater part of the embryonic life ; no trace of a third has been seen. The adult has not been studied.

None of these authors, with the exception of Sarasin, say anything conclusive concerning the origin of the *remaining ganglia*, although Salensky (86, p. 697) speaks as if the pleural ganglia of Vermetus originated in the cerebro-visceral connectives, which are shown in his Figures 31 B to 31 F.

Sarasin asserts ('82, pp. 46, 47) that in Bithynia the pleural ganglia originate as part of the " Sinnesplatte," from which the cerebral ganglia arise, and that these ganglia, cerebral and pleural, are so closely fused with each other in the later stages of development as to form on either side of the body a single mass.

I believe that they arise in Limax maximus by cell proliferations from the lateral walls of the body, behind the cerebral ganglia, and just above the pleural groove ; they are closely connected (not fused) with the cerebral ganglia only in late stages.

Sarasin ('82, pp. 50–52) says that the visceral ganglia in Bithynia arise by cell proliferation from the dorsal margin of the ventral wall of the head or trunk region, above that which I have called the pleural groove. Further, that the right visceral (or supra-intestinal) ganglion is connected by a nerve fibre to the olfactory ganglion under the gill cavity. Farther back than the visceral ganglia he finds a median proliferation of cells lying at the ventral margin of the gill cavity, from which the abdominal ganglia arise. He asserts that there are two abdominal ganglia, — one connected with the supra-intestinal ganglion, the other with the sub-intestinal ganglion.

In Limax maximus the visceral ganglia and the abdominal ganglion arise by the same method as that described by Sarasin; but the former are produced from the lateral walls of the head region, above the pleural groove, one on each side of the body. The right ganglion in later stages is more dorsal than the left. It appears to be formed in part from the inner wall of the respiratory cavity, to which it remains connected by a nerve. It is in this region that is developed an organ which I believe to be the olfactory organ of Lacaze-Duthiers.

There is only one abdominal ganglion; this takes its origin a little to the left of the median line of the body, from the anterior margin of the body wall immediately above the pleural groove.

Sarasin ('82) is the only author who gives attention to the origin of the buccal ganglia. He describes them as arising in exactly the same manner, and in the same situation in relation to the walls of the radula sac and the œsophagus, that they do in the case of Limax maximus.

CAMBRIDGE, November, 1889.

BIBLIOGRAPHY.

Bobretsky, N.
'76· Studien über die embryonale Entwicklung der Gasteropoden. Arch. f. mikr. Anat., Bd. XIII. pp. 95-169, Taf. VIII.-XIII. 1876.

Bütschli, O.
'77· Entwicklungsgeschichtliche Beiträge. Zeitschr. f. wiss. Zool., Bd. XXIX. pp. 216-254, Taf. XV.-XVIII. 1877.

Fol, H.
'80· Développement des Gasteropods Pulmonés. Arch. de Zool. exp. et gén., Tom. VIII. pp. 103-232, Pls. IX.-XVIII. 1877.

Jhering, H. von.
'75· Ueber die Entwicklungsgeschichte von Helix. Zugleich ein Beitrag zur vergleichenden Anatomie und Phylogenie der Pulmonaten. Jena. Zeitschr., Bd. IX. Heft 3, pp. 299-388, Taf. XVIII. 1875.
'77· Vergleichende Anatomie des Nervensystems und Phylogenie der Mollusken. Folio, x + 290 pp., 8 Taf., 16 Holzschnitte. Leipzig, 1877.

Jourdain, S.
'85. Sur le Système nerveux des Embryons des Limaciens et sur les Relations de l'Otocyste avec ce Système. Compt. Rend. Acad. Sci. Paris, Tom. C. pp. 383-385.

Kowalesky, A.
'83· Embryogénie du Chiton Polii (Phillippi), avec quelques Remarques sur le Développement des autres Chitons. (Odessa, Dec., 1882.) Ann. Musée Hist. Nat. Marseille, Zool., Tom. I., Mém. No. 5, 46 pp., 8 Pls. 1883.
'83ᵃ· Étude sur l'Embryogénie du Dentale. Ann. Musée Hist. Nat. Marseille, Zool., Tom. I., Mém. No. 7, 54 pp., 8 Pls. 1883.

Lacaze-Duthiers, H. de.
'60· Mémoire sur l'Anatomie et l'Embryogénie des Vermets. 2ᵉ Partie. Ann. Sci. Nat., 4 série, Tom. XIII. pp. 266-296, Pls. VII.-IX. 1860.
'72· Du Système nerveux des Mollusques Gastéropodes pulmonés aquatiques et d'un nouvel Organ d'Innervation. Arch. de Zool. exp. et gén., Tom. I. pp. 437-500, Pls. XVII.-XX. 1872.
'85· Le Système nerveux et les Formes embryonales du Gardinia Garnotii. Comp. Rend. Acad. Sci. Paris, Tom. C. pp. 146-151.

Langerhans, P.
'73· Zur Entwickelung der Gastropoda Opisthobranchia. Zeitschr. f. wiss. Zool., Bd. XXIII. pp. 171-179, Taf. VIII. 1873.

Lankester, E. R.
'73· Summary of Zoölogical Observations, etc.   Ann. Mag. Nat. Hist.,
       4th series, Vol. XI. pp. 85–87. 1873.
'74· Observations on the Development of the Pond Snail (Lymnæus stag-
       nalis), and on the early Stages of other Mollusca.   Quart. Jour. Micr. Sci.,
       Vol. XIV. pp. 365–391, Pls. XVI., XVII. 1874.
Rabl, C.
'75· Die Ontogenie der Süsswasser-Pulmonaten.   Jena. Zeitschr., Bd. IX.
       Heft 2, pp. 195–240, Taf. VII.–IX.   July, 1875.
'79· Ueber die Entwicklung der Tellerschnecke.   (Wien. Mitte Feb., 1879.)
       Morph. Jahrbuch, Bd. V. Heft 4, pp. 562–655, Taf. XXXII.–XXXVIII.,
       7 Holzschnitte. 1879.
'83· Beiträge zur Entwickelungsgeschichte der Prosobranchier.   Sitzungsb.
       der K. Acad. der Wissensch. Wien, Math.-naturw. Cl., Bd. LXXXVII.
       Abtheil. III. Heft 1, pp. 45–61, Taf. I., II.   1883.
Salensky, W.
'72· Beiträge zur Entwicklungsgeschichte der Prosobranchier.   Zeitschr. f.
       wiss. Zool., Bd. XXII. pp. 428–454, Pls. XXXV.–XXXVIII. 1872.
'86· Études sur le Développement du Vermet.   Archives de Biologie,
       Tom. VI. pp. 655–759, Pls. XXV.–XXXII. 1886.
Sarasin, P. B.
'82· Entwicklungsgeschichte der Bithynia tentaculata.   Arbeit. a. d. Zool.-
       Zoot. Inst. Würzburg, Bd. VI. Heft 1, pp. 1–68, Taf. I.–VII. 1882.
Sarasin, P. und F.
'87· Aus der Entwicklungsgeschichte der Helix Waltoni Reeve.   Zoolo-
       gischer Anzeiger, Jahrg. X., No. 265, pp. 599–602.   21 Nov., 1887.
'88· Aus der Entwicklungsgeschichte der Helix Waltoni Reeve.   Ergeb-
       nisse naturwissenschaftlicher Forschungen auf Ceylon in den Jahren
       1884–1886, Bd. I. Heft 2, pp. 35–69, Taf. VI.–VIII. 1888.
Wolfson, W.
'80· Die embryonale Entwickelung des Lymnæus stagnalis.   Bull. Acad.
       Sci. St. Pétersbourg, Tom. XXVI. No. 1, pp. 79–99, 10 Figs. 12 Mars,
       1880.

EXPLANATION OF FIGURES.

All the figures were drawn with the aid of the camera lucida, and were made from preparations of Limax maximus.

INDEX TO STAGES.

The Roman numerals indicate Plates. The Arabic numerals, Figures; those which are enclosed in a parenthesis belong to the same specimen. Skeleton numbers on the plates refer to the number of the section in its series.

6th day. $\left(\frac{\text{I.}}{1,\,5}\right)$, $\left(\frac{\text{I.}}{2-4}\right)$.

7th " $\left(\frac{\text{I.}}{6-9}\right)$, $\left(\frac{\text{I.}}{10-13}\right)$, $\left(\frac{\text{II.}}{15,\,17,\,18}\right)$, $\left(\frac{\text{IV.}}{53,\,55-57}\right)$, $\left(\frac{\text{IV.}}{4}\right)$.

8th " $\left(\frac{\text{II.}}{14,\,16,\,19}\right)$.

9th " $\left(\frac{\text{II.}}{20-24},\,\frac{\text{III.}}{25-27}\right)$, $\left(\frac{\text{V.}}{60}\right)$.

10th " $\left(\frac{\text{III.}}{28-31},\,\frac{\text{IV.}}{58,\,59},\,\frac{\text{V.}}{61,\,63}\right)$.

11th " $\left(\frac{\text{III.}}{32-47,\,51},\,\frac{\text{V.}}{62,\,64,\,67},\,\frac{\text{VI.}}{70-73}\right)$, $\left(\frac{\text{V.}}{68,\,69},\,\frac{\text{VI.}}{77-80}\right)$.

12th " $\left(\frac{\text{III.}}{48,\,49},\,\frac{\text{V.}}{65,\,66},\,\frac{\text{VII.}}{81,\,82}\right)$, $\left(\frac{\text{III.}}{50,\,52},\,\frac{\text{VII.}}{85-94}\right)$, $\left(\frac{\text{VI.}}{74-76,\,80^{a}}\right)$, $\left(\frac{\text{VII.}}{83-84}\right)$.

14th " $\left(\frac{\text{VII.}}{95},\,\frac{\text{VIII.}}{96,\,101,\,102},\,\frac{\text{IX.}}{114}\right)$, $\left(\frac{\text{VIII.}}{97-100,\,104,\,106},\,\frac{\text{IX.}}{115-120},\,\frac{\text{X.}}{121,\,123,\,125,\,126}\right)$.

16th " $\left(\frac{\text{VIII.}}{107.}\right)$, $\left(\frac{\text{X.}}{124}\right)$.

17th " $\left(\frac{\text{VIII.}}{103,\,103^{a},\,105,\,108-113},\,\frac{\text{X.}}{122,\,127}\right)$.

ABBREVIATIONS.

The right side of the animal is indicated by the letters dx., the left side by s. These letters are usually affixed to one or more of the abbreviations used to designate organs. The skeleton figures immediately under the number of a figure on the plate indicate the number of the section in the series to which the figure belongs. Consult also " Index to Stages " (p. 207).

ab.	Abdominal ganglion.	*lab.*	Upper lip.
ab.-vsc.	Abdomino-visceral connective.	*lns.*	Lens.
an.	Anus.	*lob. lat.*	Lateral lobe of brain.
buc.	Buccal ganglion.	*loph. cil.*	Ciliated ridge.
cav. mt.	Mantle cavity.	*mt.*	Mantle.
ceb.-buc.	Cerebro-buccal connective.	*n.*	Nerve.
ceb. dx.	Right cerebral ganglion.	*nph-*	Nephridial organ (primitive kid-
ceb. s.	Left cerebral ganglion.	*oc.*	Eye. [ney.)
ceb.-pd.	Cerebro-pedal connective.	*o cy.*	Otocyst.
ceb.-plu.	Cerebro-pleural connective.	*œ.*	Œsophagus.
cnch.	Shell gland.	*pd.*	Pedal ganglion.
com. a.	Anterior pedal commissure.	*pes.*	Foot.
com. buc.	Buccal commissure.	*plu.*	Pleural ganglion.
com. ceb.	Cerebral commissure.	*plu.-pd.*	Pleuro-pedal connective.
com. pd.	Pedal commissure.	*plu.-vsc.*	Pleuro-visceral connective.
com. pd. a.	Anterior pedal commissure.	*pr f.*	Cell proliferation.
com. pd. p.	Posterior pedal commissure.	*rad.*	Radula sac.
dt. sx. pr.	Primary sexual duct.	*ret. ta.*	Retractor muscle of tentacle.
dx.	Right.	*s.*	Left.
en.	Entoderm.	*sul. plu.*	Pleural groove.
gl. pd.	Pedal gland.	*ta.*	Ocular tentacle.
gl. sal.	Salivary gland.	*ta.'*	Labial tentacle.
gn.	Ganglion.	*vsc.*	Visceral ganglion.
iv. ceb.	Cerebral invagination.	*vsc.-plu.*	Viscero-pleural connective.

PLATE I.

All the figures of this plate were made from material killed in Perenyi's fluid, and all except Fig. 1 are magnified 250 diameters.

Fig. 1. A small portion of Fig. 5 more highly magnified to show the cell proliferation for the right pedal ganglion.

" 2. Posterior face of a transverse section from an individual about *six days* old. The section passes anterior to the "pleural groove," and through the region where the cerebral invaginations subsequently arise; the left side is cut a little anterior to the right. Stained in alcoholic borax-carmine.

" 3. A section from the same individual posterior to the pleural groove in the region of the cell proliferation for the pedal ganglia.

" 4. A section from the same, still farther back.

" 5. Transverse section from an embryo a few hours older than the preceding, in the region of the proliferation to form the pedal ganglion. Stained in alcoholic borax-carmine.

" 6-9. The left surface of sections parallel to the sagittal plane from an embryo of the *seventh day*. Figs. 6, 8, and 9 represent respectively the 11th, 16th, and 18th sections of the series, and are from the left half of the embryo. Fig. 7 is from the right half, and passes through the right otocyst. Stained in Czoker's cochineal.

" 10-13 exhibit the right surface of sections from another individual (between the *seventh and eighth days*) cut parallel to the sagittal plane, the anterior portion a little in advance of the posterior. Fig. 10 is a section passing through the proliferation forming the right visceral ganglion. Figs 11 and 12 are two successive sections passing through the left visceral ganglion; the latter also passes through the left otocyst. Fig. 13 shows the region of the forming abdominal ganglion. Stained in picro-carminate of lithium. In both these individuals the left ganglia and otocysts are more developed than the right.

PLATE II.

All the figures of this plate are magnified 250 diameters.

Fig. 14. A section parallel to the sagittal plane from an individual of the *eighth day*. It passes through the cerebral and pleural ganglia of the left side of the body, and also shows four cells of the left otocyst posterior to the pleural groove. The material was killed in 0.33% chromic acid, and stained in alcoholic borax-carmine.

" 15. The left surface of a section cut parallel to the sagittal plane from an embryo of the *seventh day* (but more advanced than in Figs. 6–9), passing through the cerebral invagination and a group of cells belonging to the proliferated portion of the cerebral ganglion of the right side. The material was treated as in that of Fig. 14.

" 16. A section from the same individual as Fig. 14, passing through the cell proliferation to form the visceral ganglion of the right side.

" 17 and 18 are from the same individual as Fig. 15.

" 17. A section passing through the visceral ganglion and the external opening of the nephridial organ of the right side.

" 18. The second section nearer the median plane than Fig. 17, showing the cell proliferation to form the right visceral ganglion.

" 19. From the same individual as Fig. 14, showing the cerebral invagination and proliferation of the right side, and also a cross section of the primitive kidney.

" 20–24. The anterior surfaces of transverse sections from an embryo of the *ninth day*. Material killed in Perenyi's fluid, and stained in alcoholic borax-carmine.

" 20. Portion of a section which passes through the proliferation of cells forming the pleural ganglion (dorsal to the pleural groove), and through the pedal ganglion and the otocyst of the left side.

" 21. The 37th section, which passes through the cerebral commissure and shows the proliferation of cerebral cells on the left side.

" 22. The 51st section, which passes through the buccal ganglion of the right side.

" 23. The 69th section, showing the unsymmetrical position of the visceral ganglia and a cross section of the right nephridial organ.

" 24. The 75th section of the series, passing through the abdominal ganglion and the invagination to form the mantle cavity. It is in the region where the anterior portion of the embryo is bent backward over the foot by the nutritive sac; the foot is not represented in the figure.

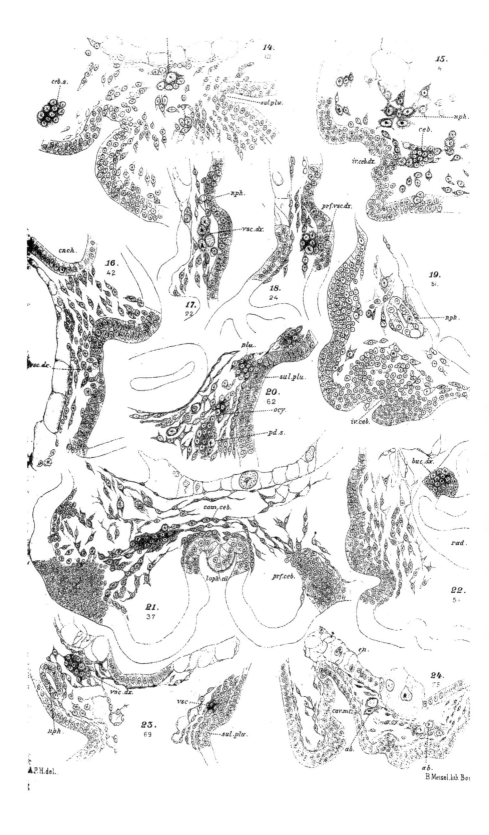

PLATE III.

Figs. 25–27 are from the same individual (*ninth day*) as Figs. 20–24 on Plate II.
They are magnified 83 diameters.

" 25. This section passes through the left cerebral ganglion, and the cerebral
invagination of the right side.

" 26 shows, in addition to the cerebral invagination, that of the right eye (at
the left of the figure).

" 27. Section passing through the pedal ganglia and their anterior commissure.
It shows a cross section of the œsophagus, the radula sac, and the right
otocyst.

" 28–31.. The left surface of sections parallel to the sagittal plane from an embryo of the *tenth day*. (Figs. 58, 59, Plate IV., and Figs. 61, 63, Plate
V., also belong to this series.) This individual was killed in Perenyi's
fluid, and stained in picro-carminate of lithium. × 100.

" 28 and 29 are successive sections passing through the invaginations of the cerebral ganglion and the eye of the left side.

" 30 shows a cross section of the cerebral commissure.

" 31. The second section nearer the median plane than Fig. 30, showing the
cerebral commissure, the position of the right visceral ganglion, and
the anus already open to the exterior.

" 32–47, 51. The left surface of sections cut parallel to the sagittal plane, from
an embryo of the *eleventh day*, magnified 100 diameters. Killed in
Perenyi's fluid, stained in Czoker's cochineál. (Sections shown in
Plate V. Figs. 62, 64, 67, and Plate VI. Figs. 70–73, also belong to
this series.)

" 32–38. Seven successive sections passing through the invagination for the
cerebral ganglia of the left side.

" 39. The second section nearer the median plane than Fig. 38, showing the
inner sac-like end of the invagination.

" 40. The next section, showing the blind end of the invagination and the
proliferated portion of the ganglion.

" 41 and 42. Successive sections (32d and 33d of the series) passing through the
left pedal ganglion and the connective between the abdominal and
left visceral ganglia.

" 43. Section cutting the abdominal ganglion crosswise.

" 44. The 35th section shows, in addition to the abdominal ganglion, a cross
section of the anterior pedal commissure. (The 36th, 37th, 39th, and
40th sections of the series are shown in Figs. 47, 46, 45, and 51,
respectively.)

" 45. Section passing through the connective between the abdominal ganglion
and the right visceral ganglion.

" 46 shows the abdominal ganglion still connected with the ectoderm.

" 47. The next section to that shown in Fig. 44, passing through the abdominal
ganglion.

" 48, 49. The left surface of sections cut parallel to the sagittal plane from an
individual of about the *twelfth day*, magnified 100 diameters. Killed in

(See obverse)

PLATE III. (*continued.*) .

0.33% chromic acid, and stained in alcoholic borax-carmine. (Sec.
tions shown in Plate V. Figs. 65, 66, and Plate VII. Figs. 81, 82, also
belong to this series.)

Fig. 48. A section passing through the cerebral and pleural ganglia of the right
side, and also through the abdominal ganglion.

" 49 shows the proliferated portion of the cerebral ganglion and the visceral
ganglion of the right side.

" 51. (See explanation of Figs. 32–47.) The section following that shown in
Fig. 45. It passes through the connective between the abdominal
ganglion and the visceral ganglion of the right side.

" 50, 52. The posterior surface of transverse sections of an embryo at the same
stage of development (*twelfth day*) as that represented in Figs. 48 and
49. The ventral portion and the right side cut a little in advance of
the dorsal portion and the left side, magnified 100 diameters. Killed
in 0.33% chromic acid, stained in alcoholic borax-carmine.

Figs. 85–94, Plate VII., continue this series. The following shows
the sequence of the sections : —

Section 77, 101, 103, 109, 110, 112, 113, 117, 125, 126' 128, 146.
Figure 90, 92, 91, 93, 50, 52, 85, 86, 88, 87, 89, 94.

" 50. This section passes through the left pedal ganglion and the abdominal
ganglion where the latter is still attached to the ectoderm. The
mantle cavity open to the exterior.

" 52. The second section in front of Fig. 50, showing that in this region the abdominal ganglion is free from the ectoderm.

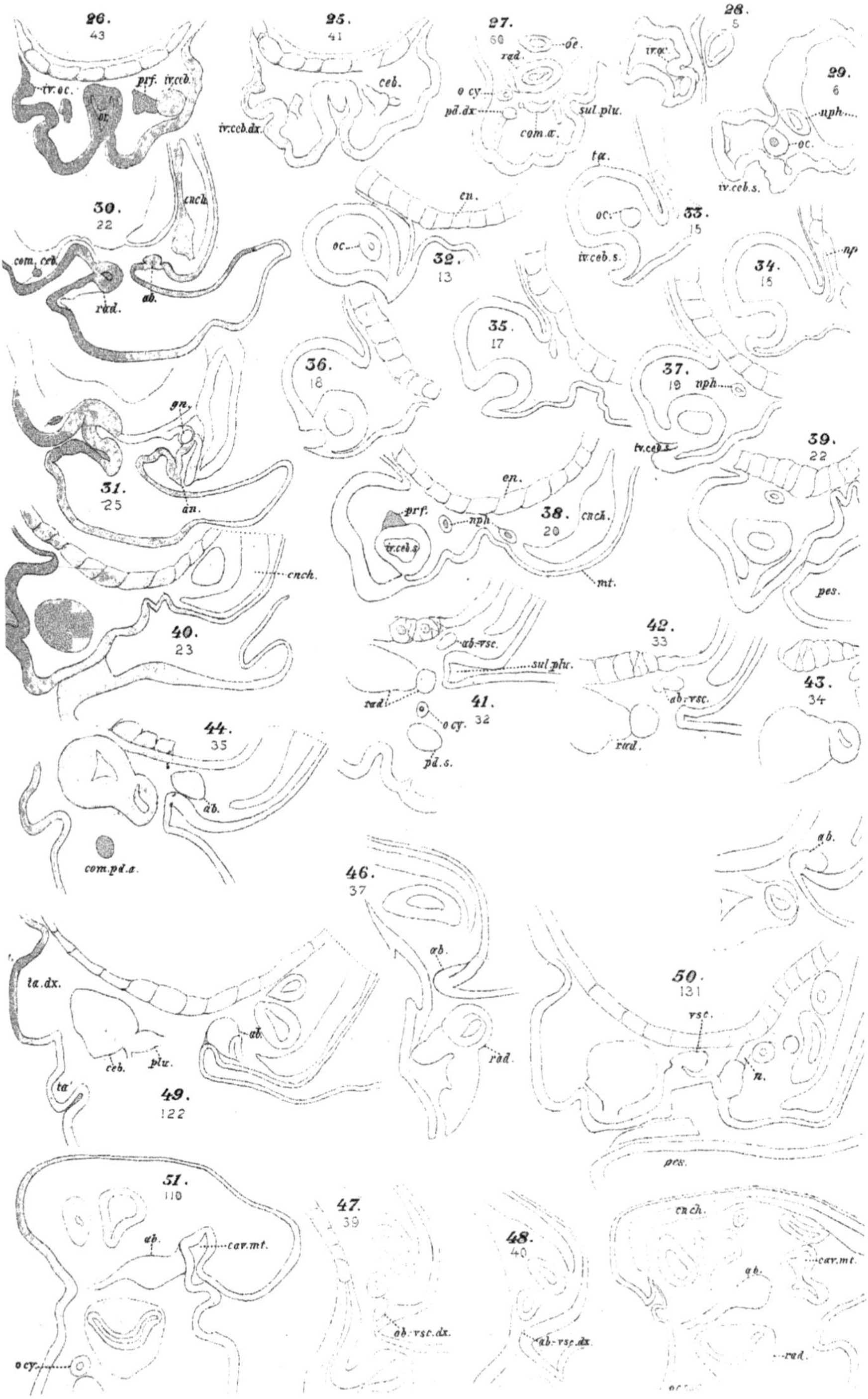
26.
43
iv.oc.
prf.
iv.ceb.
oc.
25.
41
ceb.
iv.ceb.dx.
27.
60
rad.
oe.
o cy.
pd.dx
sul.plu.
com.æ.
28.
5
iv.oc.
29.
6
nph.
oc.
iv.ceb.s.
30.
22
cnch.
com. ceb.
rad.
ab.
en.
oc.
32.
13
ta.
oc.
33.
15
iv.ceb.s.
34.
15
np
35.
17
36.
18
37.
19
nph.
iv.ceb.s.
39.
22
gn.
31.
25
an.
en.
prf.
nph.
38.
20
cnch.
iv.ceb.s.
mt.
pes.
cnch.
40.
23
ab.vsc.
sul.plu.
41.
32
rad.
o cy.
pd.s.
42.
33
ab.vsc.
rad.
43.
34
44.
35
ab.
com.pd.a.
46.
37
ab.
ab.
rad.
ta.dx.
ceb.
plu.
ab.
ta'
49.
122
50.
131
vsc.
n.
pes.
51.
110
ab.
cav.mt.
o cy.
47.
39
ab.vsc.dx.
48.
40
ab.vsc.dx.
cnch.
cav.mt.
ab.
rad.

PLATE IV.

All figures of this plate magnified 250 diameters.

Figs. 53, 56, 57, and 55 are four successive sections from the same embryo. *This was seven days old, but much more advanced than the embryos represented in Figs. 6–13, 15, 17, and 18.* Killed in 0.33% chromic acid, stained in alcoholic borax-carmine.

" 53. Posterior face of a transverse section passing through the visceral ganglia, the external opening of the nephridial organ of the right side, and the opening into the mantle chamber or respiratory cavity. (Compare Figs. 56 and 55.)

" 54. The posterior surface of a transverse section passing through the anterior pedal commissure. Embryo of about the same stage of development as the preceding, and prepared in the same way as that.

" 55. Portion of the third section following that shown in Fig. 53; it passes through the pedal and visceral ganglia of the right side.

" 56 shows both the visceral ganglia.

" 57. Portion of section showing the left pedal ganglion and cell proliferation from the ventral wall of the foot; the visceral ganglion and the external opening of the nephridial organ of the left side are also seen.

" 58, 59. Consult explanation of Figs. 28–31, Plate III.

" 58. Sagittal section passing through the proliferated portion of the right cerebral, pedal, and visceral ganglia and right otocyst. Shows cell proliferation from the ventral wall of the foot, and that the visceral ganglion is attached to the ectoderm at the margin of the mantle cavity.

" 59. A small portion of the section following Fig. 58 to show the right visceral ganglion and its attachment to the ectoderm.

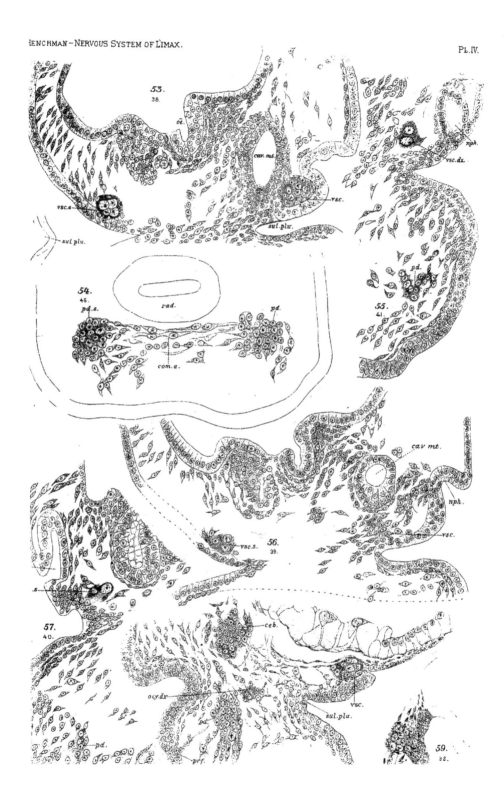

PLATE V.

All figures magnified 250 diameters.

Fig. 60. The left surface of a section cut parallel to the sagittal plane from an embryo of the *ninth day*. The section passes through a few cells of the cerebral, the pedal, and the visceral ganglia of the right side of the body, and also shows a section of the right otocyst. Killed in Perenyi's fluid, and stained in alcoholic borax-carmine.

" 61. (Consult explanation of Figs. 28–31, Plate III.) Sagittal section (the 22d) to show the abdominal ganglion, which lies embedded in the ectoderm anterior to the pleural groove.

" 62. (Consult explanation of Figs. 32–47, Plate III.) The 32d section of the series, showing a transverse section of the cerebral commissure and a portion of the left buccal ganglion.

" 63. (Consult explanation of Figs. 28–31, Plate III) The 15th section of the series; it passes through the left cerebral, the pedal, the pleural, and the visceral ganglia, and the wall of the left otocyst.

" 64. (Consult explanation of Fig. 62) The 24th section; it shows the internal end of the cerebral invagination and the cell proliferation to form the larger part of the brain of the left side.

" 65, 66. Compare explanation of Figs. 48, 49, Plate III.

" 65. The 91st section of the series, showing transverse sections of the buccal commissure and the connective between the abdominal and the left visceral ganglia.

" 66. The 102d section of the series; it passes through the abdominal ganglion.

" 67. (Consult explanation of Fig. 62.) A section through the right visceral ganglion.

" 68. The posterior surface of a transverse section from an embryo of the *eleventh day*. It shows the right visceral and the abdominal ganglia. Killed in Perenyi's fluid, stained in Czoker's cochineal.

" 69. The second section anterior to Fig. 68, passing through the visceral ganglion of the right side.

(Additional sections from this specimen are shown in Figs. 77–80, Plate VI.)

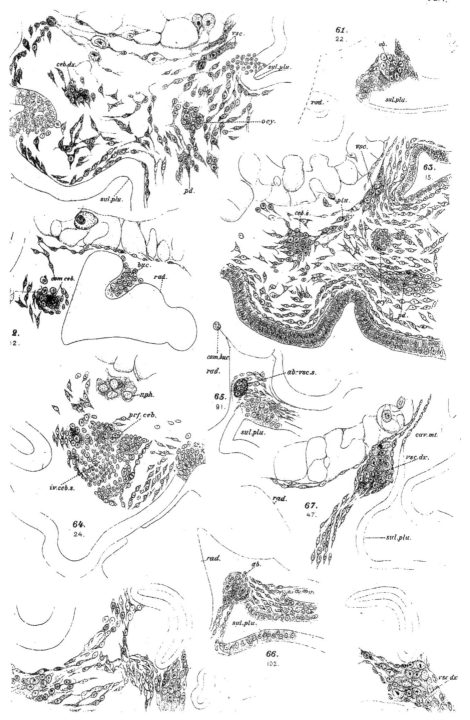

PLATE VI.

All the figures of this plate were made from material killed in Perenyi's fluid, and all except Figs. 72, 73, and 77 are magnified 250 diameters.

Figs. 70–73. Consult explanation of Figs. 32–47, Plate III.

" 70 shows the left cerebral, pleural, and visceral ganglia in the region of the cerebro-pleural and pleuro-visceral connectives.

" 71. Section passing through the otocyst and the cerebral, pedal, and visceral ganglia of the left side. The visceral ganglion is connected with the ectoderm, and the pleuro-visceral connective is much more elongated than on the *ninth day*.

" 72. A portion of the ventral wall of the foot from the 32d section, to show the cell proliferation for the pedal ganglion. × 665.

" 73. Same as Fig. 72, but from the right side of the body.

" 74–76, 80ᵃ. The posterior surface of transverse sections from an individual of the *twelfth day*. The right side cut slightly in advance of the left. Stained in picro-carminate of lithium.

" 74. The 81st section, which passes through the otocysts and both pedal ganglia in the region of their anterior commissure.

" 75 and 76. The 59th and 60th sections of the same series, showing a part of the left side of the embryo in the region of the abdominal ganglion.

" 77–80. (See explanation of Fig. 68, Plate V.) Posterior faces of four successive transverse sections through the mouth of the left salivary duct where it connects with the œsophagus. Fig. 77 shows also the buccal ganglia.

" 80ᵃ. (See explanation of Figs. 74–76.) The 107th section of the series. It shows the cerebral commissure, a few cells of the right cerebral ganglion, and the right eye.

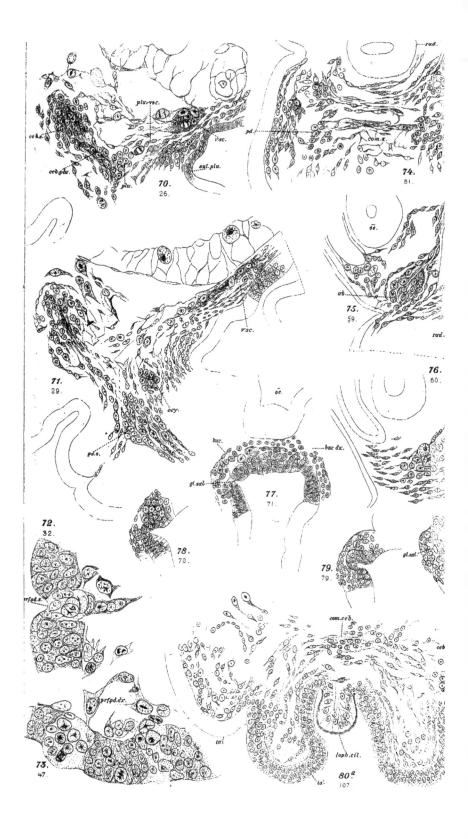

70.
26.

74.
81.

71.
29.

75.
59.

76.
60.

72.
32.

77.
71.

78.
73.

79.
79.

73.
47.

80ᵃ
107.

PLATE VII.

All the figures of this plate, except Fig. 95, were made from material killed in 0.33% chromic acid, and stained in alcoholic borax-carmine. All figures are magnified 250 diameters.

Figs. 81–82. (See explanation of Figs. 48, 49, Plate III.)
" 81. The 114th section of the series; it shows the right cerebral and pedal ganglia, together with the cerebro-pedal connective, the right buccal ganglion, and the abdominal ganglion.
" 82. The 126th section of the series. It passes through a portion of the right cerebral ganglion, the right pleural and the abdominal ganglia, from the last of which a nerve runs dorsalward.
" 83, 84. The left surface of two sections (the 96th and 99th) parallel to the sagittal plane from an embryo of the *twelfth day*.
" 83. The section shows a very small portion of the cerebral ganglion, the pleural and visceral ganglia of the right side in the region of the pleuro-visceral connective, and the abdominal ganglion together with its connective with the right visceral ganglion.
" 84 shows the right visceral and the abdominal ganglia; also, a portion of the connective between the abdominal and the visceral ganglia, and a large nerve extending dorsalward from the latter.
" 85–94. (See explanation of Figs. 50, 52, Plate III.) Transverse sections, *twelfth day*.
" 85. This section (113th) shows a portion of the abdominal ganglion separated from the ectoderm.
" 86 (117th section) shows a small portion of the abdominal ganglion, as well as the pleural and visceral ganglia of the right side, together with their connective.
" 87. (126th section; compare also 125th section, Fig. 88.) A portion of the right visceral ganglion and the large nerve running dorsalward from its left dorsal margin are shown.
" 88 (125th section) shows both visceral ganglia — the left one still connected with the ectoderm — and the left pleural ganglion, together with the pleuro-visceral connective.
" 89. The 128th section, which touches the right visceral ganglion, and a large nerve running dorsalward between mantle cavity and sexual duct from the right dorsal margin of the ganglion.
" 90. The 77th section; it passes through both pedal ganglia and their posterior commissure, which is directly above the blind end of the pedal gland, the tip of which is cut.
" 91. The 103d section, which shows the right pedal ganglion and otocyst.
" 92. The 101st section, which passes through the pedal ganglia and their anterior commissure, above which is the radula sac, and below which a blood-vessel and the pedal gland are to be seen.

(See obverse)

PLATE VII. (*continued.*)

Fig. 93. The 109th section; this passes through the abdominal ganglion at the place of its connection with the ectoderm lining the median wall of the mantle cavity. It also shows a portion of the connective from the abdominal ganglion to the left visceral ganglion.

" 94. The 146th section of the series; it shows a cross section of the neck of the cerebral invagination, and a small portion of the cerebral ganglion of the left side.

" 95. The left surface of the 54th section, from an individual of the *fourteenth day*. The section passes through the right visceral and the abdominal ganglia, showing their close connection with each other at this time, and it also cuts the right buccal ganglion. The material was killed in Perenyi's fluid, and stained in picro-carminate of lithium.

The following figures are drawn from sections of the same series as Fig. 95 : Plate VIII. Figs. 96, 101, 102; Plate IX. Fig. 114. The sequence of sections is this : —

Section 52, 54, 56, 66, 82.
Figure 101, 95, 102, 114, 96.

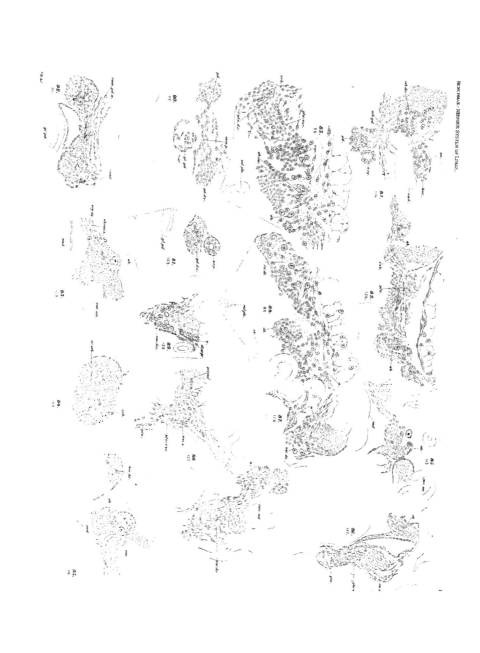

PLATE VIII.

Fig. 96. (See explanation of Fig. 95, Plate VII.) The 82d section; it passes
through the ocular tentacle, showing in section the cerebral invagination and ganglion of the right side. × 100.

" 97-100. Posterior faces of transverse sections, the right side a little in advance of the left, from an embryo of the *fourteenth day.* Killed in
Perenyi's fluid, stained in picro-carminate of lithium. — Figs. 104,
106; also Plate IX. Figs. 115-120, and Plate X. Figs. 121, 123, 125,
126, belong to the same series as Figs. 97-100. The sequence of
sections is : —

Section 81, 87, 92, 98, 102, 104, 109, 111, 113, 113, 115, 121,
Figure 97, 119, 98, 99, 118, 100, 115, 116, 104, 117, 123, 120,
122, 122, 127, 130.
106, 125, 121, 126.

" 97. The 81st section ; it passes through the pedal ganglia, a few sections behind the posterior commissure. × 100.

" 98 and 99. The 92d and 98th sections; they pass through the pedal ganglia
between the two commissures. × 100.

" 100. The 104th section, two sections in front of the anterior commissure. It
shows the right otocyst in addition to the radula sac and pedal gland.
× 100.

" 101, 102. (See explanation of Fig. 95, Plate VII.)

" 101. The 52d section of the series, showing the abdominal ganglion, and a
cross section of the cerebral, buccal, and both pedal commissures.
× 100.

" 102. The 56th section ; it passes through the visceral and buccal ganglia of
the right side, and a portion of the abdominal ganglion. It shows
the cerebral and pedal commissures, as well as a sagittal section of the
foot gland. × 100.

" 103, 103ᵃ. The posterior surfaces of transverse sections from an embryo of
the *seventeenth day.* 0.33% chromic acid ; alcoholic borax-carmine. —
Additional sections from this series are shown in Figs. 105, 108-113,
and Plate X. Figs. 122, 127. The sequence of sections is indicated by
the following : —

Section 111, 176, 180, 186, 187, 194, 211, 212, 217, 225, 225.
Figure 103ᵃ, 109, 110, 111, 112, 113, 103, 105, 122, 108, 127.

" 103. The section passes through the cerebral ganglia in the region of the
cerebro-pedal connective. The wall of the body is not represented,
but merely the ganglia, together with the œsophagus, the ducts of the
salivary glands, the radula sac, and the right ocular tentacle. × 140.

" 103ᵃ. This section (111th) passes through the pedal, pleural, and visceral
ganglia in the region of the pleuro-pedal connective. A large nerve
passes from the dorsal margin of each pedal ganglion to the lateral
wall of the body. × 83.

(See obverse.)

PLATE VIII. (*continued.*)

Fig. 104. (See explanation of Figs. 97–100.) The 113th section; it passes through the pleural ganglia and the abdominal ganglion. A large nerve connects the abdominal ganglion with a pocket-like infolding from the wall of the mantle cavity. × 100. Figure 117 (Plate IX.) shows a portion of this section more highly magnified.

" 105. (See explanation of Figs. 103, 103ᵃ.) The 212th section; it passes through the cerebral commissure. × 100.

" 106. (See explanation of Figs. 97–100.) This section, the 122d, passes through the right cerebral invagination, the right cerebral and buccal ganglia with their connective, and the right visceral ganglion. It also shows the left pleural and visceral ganglia with their connective. × 100.

" 107. Is a combination of two successive sections cut parallel to the sagittal plane from an embryo of the *sixteenth day*. It shows the cerebral, pedal, and pleural ganglia, with their connectives, and the otocyst of the left side. The position of the pedal gland is shown by dotted lines. Perenyi's fluid; picro-carminate of lithium. × 100.

" 108–113. See explanation of Figs. 103, 103ᵃ.

" 108. The 225th section of the series shows the cerebral invaginations. The right one (an enlarged view of which is seen in Fig. 127) is cut lengthwise, it being still open to the exterior; the left one transversely. The cerebral ganglia and their lateral lobes, and the buccal ganglia with their commissure crossing between the radula sac below and the œsophagus above, are also shown. × 83.

" 109. The 176th section, which passes through the pedal ganglia and the abdominal ganglion. × 83.

" 110. The 180th section; it shows, in addition to the organs seen in Fig. 109, a small portion of the left visceral ganglion and the otocysts.

" 111. The 186th section; it passes through the pedal ganglia, a portion of the pleural ganglion of the right side, both visceral ganglia, and the abdominal ganglion. It also shows the connective from the abdominal to the right visceral ganglion, and a stout nerve arising from the latter. × 83.

" 112. The 187th section shows the pedal, pleural, and visceral ganglia. × 83.

" 113. The 194th section. This shows, in addition to the ganglia, the nerve which arises from the left visceral ganglion. × 83.

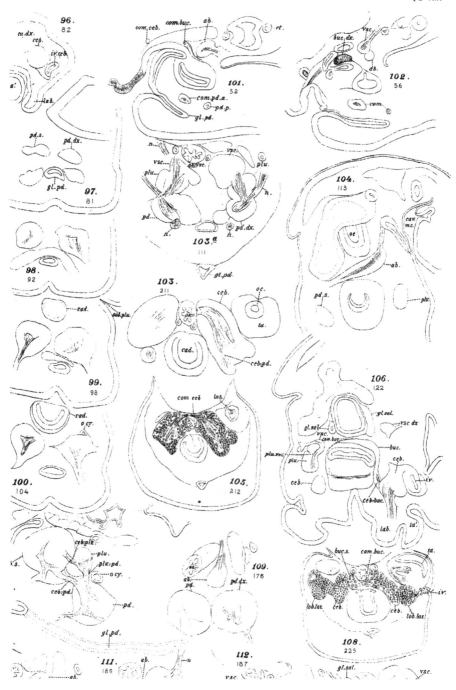

PLATE IX.

All the figures of this plate are magnified 250 diameters, and were made from material killed in Perenyi's fluid, and stained in picro-carminate of lithium.

Fig. 114. (See explanation of Fig. 95, Plate VII.) The 66th section; it passes through the cerebral, the pedal, the pleural, and the visceral ganglia of the right side in the plane of the cerebro-pedal and cerebro-pleural connectives. It also shows the right otocyst.

" 115–120. See explanation of Figs. 97–100, Plate VIII.

" 115. The 109th section; it shows a portion of the abdominal ganglion at the right of the radula sac.

" 116. The 111th section; it passes through the abdominal and right pleural ganglia.

" 117. The 113th section; it shows the abdominal ganglion where it passes above the radula sac, and a portion of the right pleural ganglion. (Compare Fig. 104, Plate VIII.)

" 118. The 102d section; it passes through the pedal ganglia in the plane of their anterior commissure. It also shows the right otocyst.

" 119. The 87th section. The pedal ganglia in the plane of their posterior commissure.

" 120. The 121st section, which passes through the visceral and buccal ganglia of the right side, and shows a portion of the buccal commissure.

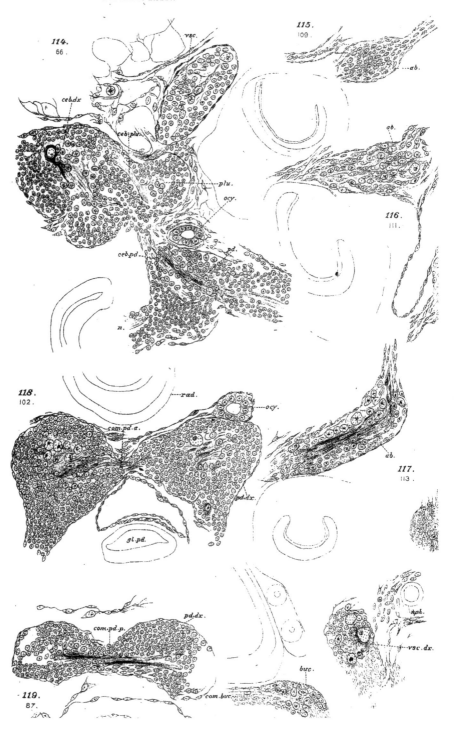

PLATE X.

Fig. 121. (See explanation of Figs 97–100, Plate VIII.) The 127th section; it passes through the left cerebral invagination, also through the left cerebral and buccal ganglia, and their connective. × 237.

" 122. (See explanation of Figs. 103, 103ᵃ, Plate VIII.) The 217th section of the series. It passes through the right cerebral ganglion and its lateral lobe; it also shows the ocular tentacle and the wall of the right eye, transverse sections of the œsophagus, salivary glands, radula sac, and primary sexual duct. × 237.

" 123. (See explanation of Figs. 97–109, Plate VIII) This section passes through the right pleural and visceral ganglia, a small portion of the left pleural and visceral ganglia, and the abdominal ganglion, which lies between the œsophagus and radula sac. × 250.

" 124. The posterior face of a transverse section from an embryo of the *sixteenth day*. The left side is cut a little in advance of the right. The section passes through the cerebral invagination, — still open to the exterior, — the cerebral ganglion of the left side and its lateral lobe, and the left buccal ganglion It also shows in cross section the duct of the salivary gland, and a small portion of the wall of the radula sac. × 237. Perenyi's fluid; picro-carminate of lithium.

" 125, 126. See explanation of Figs. 97–100, Plate VIII.

" 125. The 122d section; it shows the left pleural and visceral ganglia, with the pleuro-visceral connective, and a small portion of the left cerebral ganglion. × 250.

" 126. The 139th section; it shows in section the left cerebral invagination, the cerebral ganglion, and the cerebral commissure. × 250.

" 127. (See explanation of Figs. 103, 103ᵃ, Plate VIII.) The 225th section; it shows the cerebral invagination, and the right lateral lobe of the brain. (Compare Plate VIII. Fig. 108.) × 237.

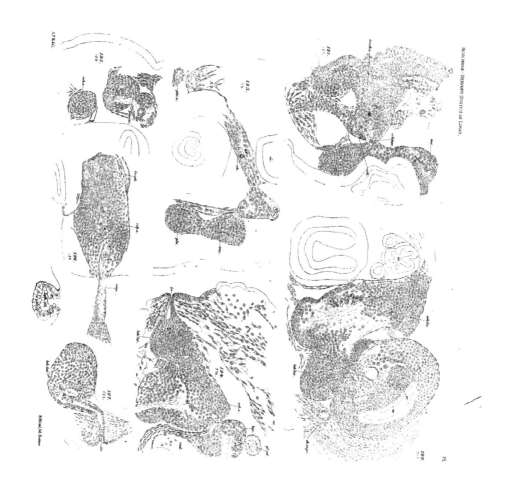

No. 8. — *The Parietal Eye in some Lizards from the Western United States.* By W. E. RITTER.[1]

WITH a single though notable exception, the numerous authors who have written on the parietal organ in vertebrates since the papers of de Graaf ('86ª and '86ᵇ) and Spencer ('86 and '87) appeared, have agreed that the structure is, or at least was in ancestrâl vertebrates, an eye. This belief is based entirely on the structure of the organ, no physiological experiments or observations on the habits of the animals possessing it having yet been produced in proof of its function.

Leydig ('89) alone, in a recent preliminary paper on the subject, has denied its optical nature, and has assigned to it an entirely different function; though in a second preliminary, still more recent ('90), he expresses his denial with considerably less confidence. He rejects the eye hypothesis, however, on the same grounds that have led others to adopt it; namely, on the grounds of its structure, and especially of its relation to the brain.

He believes that what is generally held to be an optic nerve is in fact merely a string of connective tissue.

Among those who believe the organ is or has been an eye, there are important differences of opinion as to its present value. By Ahlborn ('84), de Graaf, Spencer, and several other more recent writers, it is believed to be degenerate and entirely functionless in all living vertebrates. Rabl-Rückard ('86) has expressed the opinion that the organ may still be of use in furnishing its possessors with a more delicate means of detecting differences of temperature than exists elsewhere on the body. Béraneck ('87) believes that, while the structure is probably of an optical nature in some vertebrates, it has become so secondarily; that the primitive function of the epiphysis, common to the brains of all vertebrates, was something entirely unknown to us now, though not concerned with vision; but that in the Cyclostomes, the Amphibians, and the Reptiles it has taken on, secondarily, the function and form of an eye.

[1] Contributions from the Zoölogical Laboratory of the Museum of Comparative Zoölogy, under the direction of E. L. Mark, No. XXII.

But even were it established beyond question that the organ is a degenerate eye, there would still remain several quite distinct and very interesting problems to be solved. The most fundamental of these is probably that of its homology. Much has been written on this question by the various recent authors, but even less unanimity of opinion has been reached here than on the question of its structure and function. The question why the organ has remained so well developed in a few systematically widely separated groups of vertebrates, while in all others the process of degeneration has gone so far as to leave but a mere trace of the proximal portion of the epiphysis, has not been much discussed. It is not my purpose in the present paper to enter upon a discussion of the theoretical questions involved, and they are here adverted to merely to point out the need — as indicated by their importance and the discordance of the opinions now held with regard to them — of a larger body of facts on the subject than we yet possess. For the present, I confine myself to a presentation of the facts observed, and my interpretation of them as bearing upon some of the minor conclusions reached by other writers, hoping to be able to pursue the subject further in the near future, when situated in a region where an abundance and a variety of material, adult and embryonic, can be obtained, and where observations on the habits of the animals can be made.

The present work was undertaken at the suggestion of Prof. E. L. Mark. I wish here to acknowledge my indebtedness to Mr. G. H. Parker, of the Museum of Comparative Zoölogy; to Mr. J. J. Rivers, Curator of the Museum at the University of California; and to Mr. T. C. Palmer, of the United States Department of Agriculture, Washington, for material used; and also to Mr. S. Garman, of this Museum, for assistance in determining the species studied.

A word as to technique. For studying the structure of the retina it is very desirable to remove the great quantity of pigment that invariably obscures the histological elements in this region. Neither nitric nor hydrochloric acid, nor the alkalies, have any visible effect on this pigment, but the desired result was reached by the use of chlorine gas. The mounted, unstained sections were covered by a film of ninety per cent alcohol, and placed in a tight glass chamber, in which was also confined a small vessel containing a mixture of potassium chlorate and hydrochloric acid for generating the gas. By being careful that the slide on which the sections were mounted occupied a perfectly horizontal position, and was so placed that the film of alcohol could not be drawn off by capillary attraction, the film soon became saturated with the gas, and did not need renewing. From forty-five minutes to an hour, depending on the quantity of pigment, was sufficient time in which to accomplish the work. Considerable difficulty was found in removing the chlorine from the sections. As it had thoroughly penetrated the tissue, simple washing, even though prolonged, did not wholly remove it ; but by washing carefully, and then leaving the whole slide immersed in ninety per cent alcohol for twelve or fourteen hours, the gas was entirely removed. A good quality of Schällibaum's fixative held the sections perfectly through all this and the subsequent staining.

For decalcifying and hardening the tissues I have found Perenyi's fluid more satisfactory than anything else tried, the two processes being accomplished at the same time by this reagent.

Of the several species of lizards which I have studied I shall describe the structure in only three, namely, Phrynosoma Douglassii, P. coronata, and Uta Stansburiana, these being the only ones that have presented anything new or of special interest.

Phrynosoma Douglassii.

1. *External Appearance.* — Concerning the external appearance of the organ little need be said, since it differs in no essential particular from what has been amply described and illustrated in numerous other lizards. The scale marking the position of the eye is quite conspicuous, especially in very young individuals, where it is of a rather lighter color and larger size, relatively, than in the adult. In old individuals the great development of the surrounding scales and tubercles renders it somewhat less noticeable than it otherwise would be, but it is always readily distinguished, not only by its median position, but also by the absence of pigment and by its translucent appearance.

2. *The Parietal Vesicle.* — Figure 1, drawn from a sagittal section through the dorsal wall of the head, shows the form of the vesicle and its position within the parietal foramen and with reference to the external and internal surfaces of the wall. It lies within the parietal foramen, though extending somewhat above the dorsal surface of the parietal bone, firmly embedded in connective tissue, so that when the wall of the head is separated from the brain the vesicle always goes with the former. The tissues composing the dorsal wall of the head are, excepting the corneous layer of the skin, quite different immediately over the vesicle from those of the surrounding regions. The epidermal layer of the skin

elsewhere sends down irregular cone-shaped masses, which penetrate
and become lost in the underlying connective tissue, thus firmly uniting
the two layers. Over the vesicle, however, these processes are wholly
wanting, the under surface of the epithelial layer being even, and
sharply limited from the connective tissue. These processes are especially well developed immediately beyond the margin of the disk of
the vesicle, where they carry the cells of the epidermal layer (*e'drm.'*)
considerably deeper than their general level. The connective tissue
between the vesicle and the epidermal layer is composed of fibres considerably finer and looser than those found in other places, and, furthermore, the fibres are here disposed at various angles to the surface of the
skin, whereas elsewhere they are approximately parallel to this surface
(*con't. tis.'*). Pigment, which is found in great abundance in the skin in
all other regions of the body, is always entirely absent here. It will
thus be noticed that each of the tissues over the vesicle is considerably
more penetrable to light than are the corresponding ones elsewhere.
The connective-tissue fibres immediately around the vesicle are arranged
concentrically to its surface, and are, especially in the proximal two-
thirds of their extent, considerably finer and closer than elsewhere.
A kind of capsule for the vesicle is thus formed, and it is this alone
which separates it from the cranial cavity. The fibres of a string of
tissue extending from the distal end of the epiphysis can be traced,
though with some uncertainty, to this capsule, but I find no indication
of their passing through it, or even entering it, though I have given
special attention to this point.

The internal surface of the cranial wall in the region of the vesicle
presents a depression, which is much less marked, however, than a corresponding one in P. coronata, to be referred to hereafter. Running
through the connective tissue at the bottom of this depression, and
hence near the deep surface of the vesicle, are found a number of blood-
vessels of considerable size and well filled with blood corpuscles (*va. sng.*).
The vesicle itself is elliptical in sagittal section, the major axis, 258 μ
long in the specimen figured, having the direction of the long axis of
the head. In transverse section it is slightly elongated dorso-ventrally,
and measures in this axis 171 μ.

The cavity in sagittal section shows a triangular outline, the base of
the triangle being on the dorsal or lens side. From this outline in the
sagittal section the form gradually changes to that of an ellipse in
the last sections on each side that cut the cavity; so that the form of
the cavity is approximately that of a broad, flat cone, the base directed

outward and the apex inward. The base of the cone is slightly concave, corresponding to the convexity of the inner surface of the lens.

The wall of the vesicle is very distinctly differentiated into lens (*lns.*) and retinal (*rtn.*) portions, the latter forming about two thirds of the whole. The lens is slightly biconvex, the two convexities being very nearly equal. The line of demarcation between the lens and the retina is a sharp one, though the two portions are plainly continuous. The cells composing the lens are large and distinct in outline, each one extending entirely through its thickness (Plate II. Fig. 5, *cl. lns.*). Their nuclei are large, easily stainable, and somewhat granular ; they are uniformly situated near the internal ends of the cells. The lens is entirely without pigment.

Figure 5 represents a highly magnified portion of a longitudinal vertical section of the vesicle taken from near the median plane. In the retinal portion six regions or zones may be distinguished. Passing from the external surface toward the cavity, we find (1) a basement membrane (*mb. ba. ex.*). This is very thin, but uniform in thickness, and is of a structureless nature. From many points on this membrane fine processes radiate into the connective tissue enveloping the vesicle (Plate I. Fig. 3, *prc. r.*). These processes do not appear to be of a muscular nature, but rather the same in structure as the basement membrane from which they arise. (2) A zone containing a few scattered nuclei (*nl.'*), and fine-grained sparsely but evenly distributed pigment (*pig.*). No cell boundaries can be made out in this zone. The nuclei, few in number, form a single layer, and are situated near the basement membrane. They are very nearly round, exhibiting no tendency to elongate in the radii of the vesicle. Areas in their centres, which are somewhat more deeply stained than the rest of the nuclei, and which are probably nucleoli, are to be seen. (3) A zone (*z.''*) in which are distinguishable neither cells, nuclei, nor pigment ; only a uniform, fine-granular, slightly stainable substance, of much the same nature, apparently, as the cell substance in those regions of the retinal portion in which cell boundaries can be distinguished. Whether or not this zone represents the centrally directed ends of a layer of cells, the nuclei of which are the ones found in zone 2, I am unable to say, but it probably does. (4, 5) The next two zones are distinguished from each other only by the difference in the elements composing them, no distinguishable line of separation existing between the two. The most obvious difference between the constituent elements of these two regions is in the shape of the nuclei, those in zone 4 being approximately spherical

(*nl.''*), while those in zone 5 are much elongated in the radii of the vesicle (*nl.'''*). The suggestion at once comes that this difference is due
solely to the crowding together of the cells nearest the internal surface
of the retina, and hence that the two zones should in reality be regarded as but one. If, however, the difference in shape of the nuclei
were the result solely of such crowding, we should find a complete
gradation from the spherical to the elongated form in passing from
without inward ; but such a gradation is not found in fact. Furthermore, on close examination with high powers, it is found that the nuclei
differ in structure as well as in form. An irregular stellated area can
be detected in the centres of some of the spherical ones which does not
exist in the elongated ones ; also, the entire substance of the former
is slightly more granular than that of the latter. In the fifth zone
cell boundaries (though not well shown in the figure) can be quite distinctly traced to the internal basement membrane ; but how the cells
of the fourth and fifth zones are related I have been unable to determine, since cell boundaries in the fourth zone cannot be traced. (6) The
last layer may be designated as an internal basement membrane (*mb.
ba. i.*), though it differs somewhat in structure from the external basement membrane, being of a granular nature. It extends over the
surface of the lens, as well as over the retina, and is rather more compact in the former than in the latter region. Projecting into the cavity
of the vesicle from the retinal portion are found certain structures concerning the nature of which I am not quite sure, but believe them to
be secretions from the cells of the fifth zone. They are in general
elongated, and pointed at their free ends, though their outlines are
ragged and indefinite. They always stain most deeply at their internal
free ends. In many cases, as at *, they are seen to be continuous with
the cells of the fifth zone through the internal basement membrane.
These structures may correspond to what de Graaf has described and
figured as existing on the internal surface of the retina of Anguis, and
has called "Staafjeslaag," but which Spencer and others believe to
be merely a coagulum from the fluid that probably filled the cavity in
the recent state. It is, however, scarcely possible to account for the
structures here under consideration in this way, as is to be seen from
my description and figures of them ; furthermore, a coagulum (*cog.*)
does exist in addition to these.

Within the substance of the retina (Fig. 5, *va. rtn.*) are found a number of cavities varying in diameter, as measured in the plane of the
sections, from $5.5\,\mu$ to $22\,\mu$. The sections of these cavities are never

quite circular, but are never much elongated. In many, though not in all, an exceedingly thin endothelial lining can be seen, and in a few instances blood corpuscles are found in the cavities (Plate I. Fig. 4, *en'th. va.* and *cp. sng.*). Although none of these cavities were found to extend through more than four or five sections, each 7.5 μ in thickness, and although in no instance was it possible satisfactorily to trace a connection between them and the blood-vessels lying outside the vesicle, it still seems quite certain that they form a network of fine blood-vessels ramifying through the substance of the retina. Owing to the fact that in some instances no lining membrane to these cavities can be found, and that their outlines are not sharply marked, the possibility of their having been artificially produced by the removal of pigment masses suggests itself; but the definiteness of the outline of many others and their endothelial lining membranes, in which much-flattened nuclei are found, strips this conjecture of its plausibility. If these are really blood-vessels, it might appear that some of them would be seen cut longitudinally; and while it is true that in many cases focusing shows the cut walls to be very oblique to the plane of the section, still no sure instance of a vessel cut lengthwise has been seen. When, however, one considers the exceeding delicacy of the endothelial lining, and the fact that no differential staining takes place, it does not seem impossible that such sections may exist, and yet escape detection. These cavities have no regularity of arrangement, but are for the most part confined to zones 2, 3, and 4. In no instance has one been seen confluent with the cavity of the vesicle.

These may possibly correspond to what Owsjannikow mentions as having been seen by him in Chamæleon vulgaris. He says: "Am hintern Rande der Retina findet sich an einigen Schnitten das Lumen eines Rohrs, von dem nicht mit Bestimmtheit gesagt werden kann, ob es einem Blutgefässe oder einem anderen Gewebe angehört." (Owsjannikow, '88, p. 16.)

3. *The Epiphysis.* — Figure 9 (Plate III.) represents a sagittal section of the epiphysis, and so much of the brain as is necessary to show the relation of the former to the latter. The entire structure, or, more properly, the combination of structures that must be considered at this time, presents the form of a curved cylinder, one end of which is produced into a cone, while the other end has a hopper-shaped excavation. In keeping with the usual method of designation, I shall call the whole structure the epiphysis, though, as the sequel will show, it is doubtful if this is justifiable. The excavated end is proximal, the

excavation being the continuation of the cavity of the third ventricle into the epiphysis. The conical end, then, is distal, and rises somewhat above the level of the cerebral hemispheres. The curved axis forms very nearly a segment of the circumference of a circle, and is directed upward and forward from its point of origin from the brain. Continuing anteriorly from the apex of the cone is a string of connective tissue (*con't. tis.*), which passes to the region of the parietal vesicle, and in the distal portion of its course comes close in contact with the dura mater of the brain. The axis of the cylinder, if we consider it as continued to the anterior termination of this connective-tissue string, describes very nearly a semicircumference. The most anterior point in the connection of the epiphysis with the brain is at the junction of the cerebrum with the optic thalamus, somewhat anterior and dorsal to the superior commissure (*com. su.*). For a short distance above its connection with the brain in this anterior part, the epithelial nature of the epiphysial wall is less distinct than at a higher level, where the wall becomes thicker, and is composed of a single layer of more or less cuboid nucleated cells, which stain readily in borax carmine or hæmatoxylin (Plate III. Figs. 8, 9, *e'th.*). Also at this level the wall becomes thrown into a highly complicated system of folds; and it is this folded epithelium, containing within its folds great quantities of blood corpuscles, that forms a large bulk of the whole epiphysis (Figs. 8 and 9, *e'th.* and *cp. sng.*).

In the section represented in Figure 9 no connection exists between the epithelium of the posterior portion of the epiphysis and the brain, and it is doubtful if such connection exists here in any of the sections of this specimen; at any rate, if it does exist, it is exceedingly thin and limited in extent. There is, however, an undoubted connection in this region in P. coronata, which will be described later; but even in this latter species the posterior wall of the epiphysis is much less developed than the anterior wall. The exceedingly thin epithelium that forms the posterior wall in P. Douglassii would, as is evident from its position and from comparison with P. coronata (Plate IV. Figs. 11 and 12), form a connection with the brain roof had not a separation taken place, either artificially or as a result of degeneration. This wall is closely applied to the anterior, concave side of the blood sinus to be presently described, and at a considerable distance above the brain is continuous with the anterior wall of the epiphysis. The space included by these walls is the hopper-shaped excavation in the proximal end of the cylinder already mentioned, — an extension of the cavity of the third ventricle (*vnt.*[3]) into the epiphysis. Intimately connected with the distal end of the portion of the epiphysis thus far described is found a vesicle (*eph. vs.*), the thick walls of which are composed of columnar epithelium, and thus differ markedly from the folded epithelium of the anterior wall previously described. This vesicle is much flattened antero-posteriorly, its longest axis lying very nearly in the axis of the cylinder to which the epiphysis as a whole has been compared. That the structure here described is a separate vesicle, and that its cavity is not continuous with the cavity already described as a continuation of the third ventricle, admit of easy and satisfactory demonstration, not only in this particular instance, but also in all other individuals both of this species and of P. coronata of which sections have been made. In passing through the entire series of sections, it is easily seen not only that the two cavities nowhere approach more nearly to confluence than in the one represented in the figure, but also that the walls of the vesicle and those of the more proximal part of the epiphysis with which they are in relation are clearly distinct. The separateness of these two structures will appear more clearly when we come to consider the same parts in P. coronata. Passing upward and forward from the distal end of this vesicle is to be seen a bundle of connective-tissue fibres which becomes blended with the string of connective tissue already described as running from the apex of the cone to the region of the parietal vesicle. There is no indication that the epithelial wall of the epiphysial vesicle, as it may be called, passes into this string.

Covering the whole postero-dorsal convex side of the portion of the epiphysis thus far described, and even extending considerably beyond its distal extremity, is an immense blood sinus fully distended with blood corpuscles (Fig. 9, *sn. sng.*, and Fig. 8, *cp. sng.*).

Phrynosoma coronata.

1. *General Description.* — Figure 2 (Plate I.) represents a transverse section of the dorsal wall of the head, passing through the middle of the parietal eye of P. coronata. The description of the external appearance and of the vesicle and its surrounding structures given for P. Douglassii requires modification in only a few points to become applicable to this species. The depression mentioned as existing on the internal surface of the wall of the brain-case immediately under the vesicle in P. Douglassii becomes in this species a deep pit. To correspond with this pit the external surface of the wall immediately over the vesicle forms

a low, broad cone, a condition which gives quite a different general appearance to the sections in the two species.   In P. coronata the vesicle
is situated somewhat nearer the external surface of the cranial wall than
in P. Douglassii; and the intervening connective tissue differs less,
both as regards the fineness and direction of its fibres, from the adjacent
tissues, than in the case of P. Douglassii.   The vesicle, with its connective-tissue capsule, protrudes into the bottom of the pit considerably.
The pit is bridged over by the dura mater of the brain, and thus a
chamber is formed in which a great quantity of blood corpuscles is
found (*cp. sng.*).   It will be remembered that no such blood sinus in
this region exists in P. Douglassii, but that numerous blood-vessels do
occur here.   In P. coronata, however, the sinus replaces the vessels.

2. *The Parietal Vesicle.* — With regard to the vesicle itself, the only
points in which it differs very essentially from that found in P. Douglassii
are the absence of the cavities in the retina regarded as blood-vessels, and
the far less perfect development of the structures projecting from the
internal surface of the retina into the cavity of the vesicle.   The latter
difference I am inclined to think due to the probably somewhat greater
degree of degeneration of the retinal cells which secrete these structures.   That this portion of the retina is more degenerated in P. coronata may be supposed from the fact that we find here considerably
more pigment than in the corresponding region in P. Douglassii.   However, too much stress must not be laid on the greater or less quantity
of pigment, since the quantity is quite variable even within the same
species.   In one individual of this species pigment was found, though
in small quantity, in the lens.

3. *The Epiphysis.* — Although this structure does not differ in any
essential particular from what we have already seen in the preceding
species, the fact that several of the points which go to make the study
of the epiphysis of much interest are here well brought out, has made it
seem best to describe and illustrate the organ in detail.   Figures 10, 11,
and 12 (Plate IV.) present vertical longitudinal sections from the same
animal at different planes to the left of the median plane, Figure 12
being very nearly median, and Figure 10 farthest removed from it.   It
should here be said, however, that the sections are not quite vertical;
so that, while the epiphysial vesicle is situated more to the left than to
the right side of the sagittal plane, yet it is less so than would be
inferred from the way in which it appears in the figures.   The form
of the epiphysis, as a whole, is nearly the same as that found in P.
Douglassii, and it is composed of the same parts ; — namely, a proximal

part with an anterior much-folded epithelial wall, and a posterior not folded and thinner epithelial wall; an epiphysial vesicle; a blood sinus; and a string of connective tissue extending from the distal end of the vesicle and blood sinus to the region of the parietal vesicle. In the anterior wall of the proximal portion the folding extends down somewhat nearer to the brain than is the case in P. Douglassii, and just at its junction with the brain a large blood-vessel is found filled with blood corpuscles (Fig. 12, *cp. sng.*). As already said in describing the posterior wall in P. Douglassii, the connection (opposite the letters *vnt.*[3]) with the brain is here complete and very evident, though the roof of the third ventricle (*tct. thl. opt.*) appears in the section to constitute a part of this wall.

The cells composing the walls of the proximal part are about two or three deep, but not arranged in layers. They are small, distinctly nucleated, and the nuclei are apparently perfectly round. They stain readily. On the outer surface of this wall is found, throughout most of its extent, a very thin layer of tissue, the cells of which are much flattened. This layer becomes continued from the apex of the epiphysis as the connective-tissue string (*con't. tis.*) already mentioned as passing to the region of the eye; another portion of it also becomes continuous with the pia mater of the brain.

Figure 10 represents a section through the longest portion of the epiphysial vesicle. In this plane the proximal portion of the epiphysis has not yet appeared, and is not found till we pass to a section in which the long axis of the vesicle has become considerably shortened. In the wall of the vesicle three zones or layers are found. The external one is similar to — in fact, on the posterior surface is continuous with — the thin external layer mentioned in the proximal portion. The second zone, comprising more than half of the entire thickness of the wall, is composed of cells apparently of the same nature as those described as forming the chief portion of the wall of the proximal part; but the layer is considerably thicker here than there, and on the whole rather more compact (*e'th.*, Figs. 10 and 11). The third and most internal zone is a deeply pigmented one (*pig.*). This pigment is so dense that when destroyed no distinguishable structure remains. In the presence of this pigment the species now under consideration differs entirely from P. Douglassii, where no pigment in this region is found. Again, however, attention is called to the fact that great importance cannot be attached to the presence or absence of pigment. Figure 11 shows the relation between this vesicle and the proximal portion of the epiphysis. In this section it will be seen that a distinct line of demarcation exists between the true epithelial portions of the two walls where they come in contact. This distinctness is maintained throughout the entire series of sections. When the median section is reached, the vesicle has entirely disappeared. From the distal end of the vesicle the connective-tissue string extends forward to the region of the eye, as in the case of the proximal portion (*con't. tis.*). The blood sinus (Fig. 12) does not, in this species, come in contact with the epiphysial vesicle, but occupies the same position on the proximal part as in the case of P. Douglassii. It is much smaller in P. coronata, but in other respects is of the same nature. Whether or not this epiphysial vesicle may be homologized with the secondary vesicle in Petromyzon (Ahlborn, '83, Beard, '89, Owsjannikow, '88, Wiedersheim, '80) can be profitably discussed only after its development has been studied. So far as the condition in the adult is concerned, there is little to indicate such a homology.

I mention here an observation which may be of significance in connection with this complicated structure of the epiphysis. In both species and in all the individuals of Phrynosoma of which I have made sections favorable for exhibiting the entire dorsal surface of the brain, I have noticed that the pia mater appears to form a junction with the connective-tissue string described as passing from the distal extremity of the epiphysis to the region of the parietal eye, and also that it is thrown into several folds on the dorsal surface of the cerebellum. The membrane where folded is considerably thicker than elsewhere, contains within its folds numerous blood-vessels, and is composed of a single layer of cells very regular and distinct in outline and of a decidedly epitheloid appearance. The condition reminds one strongly of the folded portion of the wall of the epiphysis.

Uta Stansburiana.

As I have had but two specimens of this species, both preserved in alcohol, and hence not in the best histological condition, my study of it has been less satisfactory than that of the species of Phrynosoma. A few points, however, have been observed which are of some interest; but these can be presented without entering into a detailed description of the structure. Figure 6 (Plate II.) represents a portion of a sagittal section through the dorsal wall of the head and the parietal vesicle. The parietal foramen, too broad to be embraced in the figure, is much larger here than in Phrynosoma, and the vesicle can scarcely be said to be embedded in the connective tissue of the brain roof, as in the case of Phrynosoma, but rather is suspended from the under side of the wall in a connective-tissue capsule.

The most striking features about this vesicle, as seen in the section, are its dorso-ventral flattening, and the entire separation of the lens from the retina. The lens, a well defined structure, composed of much elongated, almost fibrous, non-stainable cells, has its margins widely separated from the retina, and the intervening space is occupied by a uniformly fine granular substance (cog. ?), which also occupies the narrow space corresponding to what would be the cavity of the vesicle, were the lens and retina continuous at the margins of the former. The retina shows no structure beyond two deeply pigmented layers, corresponding to its external and internal surfaces, connected at short but irregular intervals by pillars of pigment, between which are seen a few scattered nuclei. This distinct separation of the margins of the lens from the retina is the only undoubted case of the kind, so far as I know, that has been seen, and if normal may be of significance in connection with the theory of the origin of the eye recently advanced by Beard ('89). I am, however, inclined to believe, notwithstanding the fact that the condition here found is apparently confirmed by the sections of my second specimen of this species, that the separation is in reality due to the extreme differentiation of the two structures, by means of which the connection between them was weakened, and then to artificial rupture by the flattening of the vesicle. The point certainly needs confirmation in more carefully preserved specimens.

I was unable to study the epiphysis in the material which I had, but no trace of anything like a nerve or even like a connective-tissue string extending from the parietal vesicle could be detected, nor were there any indications of blood-vessels or sinuses corresponding with those existing in Phrynosoma found here.

Conclusions.

The general bearing of the facts here presented I discuss at present only in connection with the question of the function, past and present, of the parietal organ. I concur in the opinion held by most of the persons who have written on the subject, that the organ is a degenerate eye, although my observations furnish, perhaps, no evidence in addition to what has been presented by former writers, in support of the belief. From the morphologist's point of view, the evidence that would remove all doubt as to the correctness of this opinion would be that the vesicle regarded as the eyeball should be composed of elements essentially similar to elements found *somewhere* in organs known to perform the mechanical part in the act of vision ; and, second, that this vesicle should be connected with the brain by a nerve comparable with the optic nerve of *some known* functional eye. I think no one familiar with the structure of the vesicle as it exists in many Lacertilia and in Petromyzon, will refuse to accept as satisfactory the evidence on the first point. The evidence on the second point is less conclusive. In many cases where the vesicle is well developed, as in Phrynosoma, it is certain that nothing which can be justly compared to an optic nerve exists. Spencer ('86 and '87) and several succeeding writers have held it as beyond doubt that in several species, notably of the genera Lacerta, Hatteria, and Varanus, there is a nervous connection between the brain and vesicle. Leydig ('89), however, in his preliminary, based on his study of Lacerta ocellata, Varanus elegans, and other forms, says " der von Spencer beschriebene Nerv ist kein Nerv sondern das strangartig ausgehende Ende der Zirbel." Lacerta ocellata is one of the forms in which Spencer ascribes, with least question, a nervous nature to the structure under consideration ; but apparently Leydig has not examined either of the species of Varanus, viz. gigantea and Bengalensis, which Spencer studied ; while, on the other hand, V. elegans, Leydig's species, is not mentioned by Spencer as having been studied by him. This denial *in toto* of the existence of the nerve as described by Spencer, Leydig practically repeats in his most recent contribution to the subject (Leydig, '90), and adds, as further confirmation of his opinion, that he has studied Hatteria (he does not tell us what species) and finds that here also the so-called nerve is of the nature of connective tissue. He also comes to the conclusion in this communication, that, while from the structure of the vesicle alone the organ must at least be put among the sense organs, it is yet " as good as impossible to do so while it is recognized that in the parietal structure of all the animals investigated by me not one contains a nerve, for we must hold fast to the proposition that for the equipment of a sense organ the peripheral end of a nerve is necessary." It appears to me, however, that we are not compelled to relinquish the belief that the organ was originally an eye, even though we accept Leydig's statement, as against Spencer's and others, regarding the nature of the supposed nerve in the cases which both have examined ; or even should it appear that in no case does the nervous connection *now* exist.

It seems to me that Leydig has not given sufficient prominence to the possibility, not to say great probability, that the nervous connection has been lost by the modification and degeneration which the whole structure has certainly undergone; and especially must we hesitate in rejecting this explanation, when we remember that by so doing we are compelled to seek another. To be obliged to ascribe a function other than that of vision to a structure entirely like an organ of vision in most of its essential parts, and differing widely from one in no essential point, is requiring us to accept a conclusion that would throw suspicion on all our morphological reasoning. Should it be shown conclusively that the vesicle never has, in any vertebrate, *either in the adult* or *during its ontogeny*, nervous connection with the brain, then we should be obliged to abandon the optical explanation of its origin, and turn to the exceedingly difficult task of finding another. But until such knowledge is at hand, it seems to me we must suppose that the organ was produced as an eye, that in some way entirely unknown to us it lost its optical function, and that, in the consequent modification and degeneration, the optic nerve degenerated more rapidly in some cases than did the optic vesicle; and that in this way the separation which we now find took place.[1]

In previous discussions of the nature and function of the parietal organ, I believe sufficient attention has not been given to the structure and development of the epiphysis and its relation to the parietal vesicle, and especially its relation to the so-called choroid plexus. I have designated the entire structure found in connection with the roof of the thalamencephalon as the epiphysis; but, as already said, I have considerable doubt as to the wisdom of so doing. For the sake of precision it would seem best that the term epiphysis should be limited to the structure which arises as an evagination from this portion of the brain. Certain it is that the large blood sinus which I have described as a part of the epiphysis in Phrynosoma cannot be regarded as forming an essential portion of the structure, and I think it quite possible that what I have called the epiphysial vesicle is not a portion of the epiphysis, should

---

[1] Concerning the nervous connection between the eye and the epiphysis in Anguis fragilis, Strahl and Martin say ('88, p. 154), " Der Nerv der nach hinten am Vorderrand der Epiphyse scheinbar verschwindet, tritt von unten her in das Auge ein." Francotte ('88, p. 782) also describes essentially the same condition in this species. But such a condition would be so anomalous that C. K. Hoffmann ('88, p. 1991), notwithstanding the agreement of these independent statements, has, it seems to me with reason, expressed doubt as to the trustworthiness of the observations.

the term be limited as I have suggested that it ought to be. The distinctness of the epiphysial vesicle from the proximal portion of the epiphysis in the adult Phrynosoma is without exception, so far as my observations have gone ; and if it is regarded as having been derived from the epiphysis, then we have two vesicles instead of one that have arisen in this way, and the difficulty of explaining the nature and function of the whole structure is correspondingly increased.

In his recent paper, Leydig ('90) has expressed the belief that there are two forms of parietal organs. He says : " From the posterior portion of the embryonic thalamencephalon (Zwischenhirn), especially in Lacerta agilis, two thick-walled vesicles (Blasen) bud out just in the middle line, lying one behind the other and springing from a common root (einem Wurzelpunkte). The anterior vesicle gives rise to the parietal organ, and the posterior one constitutes the epiphysis (Zirbel)." It is only, he says, from the anterior of these two vesicles (Blasen) that a vesicle (Blase) becomes cut off, and attains an eye-like character; the posterior one ends in the expanded blind terminal portion of the epiphysial thread (Zirbelfaden). But Selenka ('90) informs us, in a still more recent communication, that, after studying the development of the brain in a large number of reptiles and other vertebrates, he is unable to confirm Leydig's statement as to the origin of the parietal eye. He does find, however, in all cases, an evagination from the dorsal wall of the *fore* brain very similar to the one that forms the epiphysis from the roof of the thalamencephalon ; also that the two structures elongate *pari passu*, the epiphysis becoming directed upward and forward, while the anterior evagination, which he calls the " paraphysis," becomes directed upward and backward. After the parietal vesicle is cut off from the epiphysis, the distal end of the paraphysis grows in between the vesicle and the end of the epiphysis from which it was detached, and the vesicle comes to lie on the paraphysis as on a pillow.

The relation of the two structures in the adult he does not know. C. K. Hoffmann ('85) has also described an evagination from the roof of the brain at the place of transition from the fore brain to the thalamus, which he calls the ependyma, — the beginning of the choroid plexus, — and he says that in the grown animal "it comes to take a not inconsiderable part in the formation of the epiphysis." Although there is nothing in the brief papers of either Leydig or Selenka to indicate whether or not the additional more anterior evagination seen by them is the same as that described by Hoffmann, yet, since all have studied the same forms, viz. of the genus Lacerta, it seems quite probable that they have all observed the same structure. Whether or not any portion of the epiphysis as I have found it in Phrynosoma corresponds to the paraphysis of Selenka, or the ependyma of Hoffmann, can of course be determined only by studying the development of this portion of the brain.

Bearing in mind the highly vascular condition of all parts of the parietal organ, the numerous large blood-vessels surrounding the vesicle in P. Douglassii, and the great sinus in the same region in P. coronata, the sinuses of the epiphysis in both species, as well as the great quantity of blood contained in the much folded anterior wall of the epiphysis, it seems to me impossible to escape the belief that, in this genus at least, the organ must have some physiological significance. Leydig ('89) has expressed the opinion that it belongs primarily to the lymph system. From what has already been said, it is evident that I cannot accept this conclusion; but it does appear to me highly probable that the structure has become secondarily of such a character. From the numerous instances of change of function in the animal organism to which attention has been directed by Dohrn ('75), Kleinenberg ('86), Lankester ('80), Weismann ('86), and others, there are certainly no *a priori* objections to such a view, and it seems to afford more nearly a satisfactory explanation of the present condition of the organ than does any other.

CAMBRIDGE, August 15, 1890.

BIBLIOGRAPHY.

Ahlborn, F.
'83· Untersuchungen über das Gehirn der Petromyzonten. Zeitschr. f. wiss. Zool., Bd. XXXIX. pp. 191–294, Taf. XIII.–XVII.
'84· Ueber die Bedeutung der Zirbeldrüse. Zeitschr. f. wiss. Zool., Bd. XL. pp. 331–337, 1 Taf.

Beard, J.
'87· The Parietal Eye in Fishes. Nature, Vol. XXXVI. pp. 246–248 and pp. 340, 341.
'89· Morphological Studies. — I. The Parietal Eye of the Cyclostome Fishes. Quart. Jour. of Micr. Sci., Vol. XXIX. pp. 55–73, Pls. VI. and VII.

Béraneck, Ed.
'87· Ueber das Parietalauge der Reptilien. Jenaische Zeitschr., Bd. XXI. pp. 374–410, Taf. XXII., XXIII.

Carrière, Justus.
'85· Die Sehorgane der Thiere vergleichend-anatomisch dargestellt. München. 205 pp.
'89· Neuere Untersuchungen über das Parietalorgan. Biolog. Centralbl., Bd. IX. pp. 136–149.

Cattie, J. Th.
'82· Recherches sur la glande pinéale (epiphysis cerebri) des Plagiostomes, des Ganoïdes et des Téléostéens. Arch. de Biologie, Tom. III. pp. 101–194, Pls. IV.–VI.

Dohrn, A.
'75· Der Ursprung der Wirbelthiere und das Princip des Functionswechsels. Leipzig. 87 pp.

Ehlers, E.
'78· Die Epiphyse am Gehirn der Plagiostomen. Zeitschr. f. wiss. Zool., Bd. XXX. (Supplement), pp. 607–684, Taf. XXV., XXVI.

Francotte, P.
'88· Recherches sur le développement de l'épiphyse. Arch. de Biologie, Tom VIII. pp. 757–821.

Götte, A.
'75· Die Entwicklungsgeschichte der Unke. Leipzig. 956 pp., 22 Taf.

Graaf, H. W. de
'86ª. Zur Anatomie und Entwickelungsgeschichte der Epiphyse bei Amphibien und Reptilien. Zool. Anzeiger, Bd. IX. pp. 191–194.

'86ᵇ. Bijdrage tot de Kennis van den Bouw en de Ontwikkeling der Epiphyse bij Amphibien en Reptilien. Leyden. 4to, 61 pp., 4 pls.

Hoffmann, C. K.

'85· Weitere Untersuchungen zur Entwickelung der Reptilien. Morph. Jahrb., Bd. XI. pp. 176–218, Taf. X.–XII.

'88· Epiphyse und Parietalauge. Bronn's Klassen und Ordnungen des Thier-Reichs, Bd. VI. Abth. III., Reptilien, pp. 1981–1993.

Hanitsch, P.

'88· On the Pineal Eye of the Young and Adult of *Anguis fragilis*. Proc. Liverpool Biolog. Soc., Vol. III. pp. 87–95, Pl. I.

Herdman, W. A.

'86· Recent Discoveries in Connection with the Pineal and Pituitary Bodies of the Brain. Proc. Liverpool Biolog. Soc., Vol. I. pp. 18–25, Pls. I. and II.

Julin, Ch.

'87· De la signification morphologique de l'épiphyse (glande pinéale) des vertébrés (avec trois planches). Bull. Scientifique du Nord de la France et de la Belgique, 2ᵉ série, 10ᵉ année, pp. 81–142, Pl. I.–III.

Kleinenberg, N.

'86· Die Entstehung des Annelids aus der Larve von Lopadorhynchus. Nebst Bemerkungen über die Entwicklung anderer Polychaeten. Zeitschr. f. wiss. Zool., Bd. XLIV. pp. 1–227, Taf. I.–XVI. (Ueber die Entwickelung durch Substitution von Organen.)

Kölliker, A. von.

'87· Ueber das Zirbel- oder Scheitelauge. Sitzungsb. d. Würtzburger phys.-med. Gesellsch., Sitzung vom 5. März, pp. 51, 52.

Lankester, E. Ray.

'80· Degeneration. A Chapter in Darwinism. London. Nature Series, 76 pp.

Leydig, Fr. von.

'72· Die in Deutschland lebenden Arten der Saurier. Tübingen. 4to, 262 pp., 12 Taf.

'89· Das Parietalorgan der Reptilien und Amphibien kein Sinneswerkzeug. Biolog. Centralbl., Bd. VIII. No. 23, pp. 707–718

'90· Das Parietalorgan. Biolog. Centralbl., Bd. X. No. 9, pp. 278–285.

McKay, W. J.

'89· The Development and Structure of the Pineal Eye in *Hinulia* and *Grammatophera*. Proc. Linn. Soc. N. S. Wales, 2d ser., Vol. III. Pt. 2, pp. 876–889, Pls. XXII.–XXIV.

Owsjannikow, Ph.

'88· Ueber das dritte Auge bei Petromyzon fluviatilis, nebst einigen Bemerkungen über dasselbe Organ bei anderen Thieren. Mém. de l'Acad. imper. de St.-Pétersbourg, 7ᵉ série, Tom. XXXVI. No. 9, 21 pp., 1 Taf.

Ostroumoff, A. von.

'87· On the Question concerning the third Eye of Vertebrates. Beilage zu den Protocollen der Naturforsch. an der kaiserl. Universität zu Kasan. 13 pp. (Russian.)

Peytoureau, S. A.

'87· Le glande pinéale et le troisième œil des vertébrés. Thèse, avec 42 figs. Paris. Doin, 68 pp.

Rabl-Rückard.

'82· Zur Deutung und Entwickelung des Gehirns der Knochenfische. Arch. f. Anat. u. Phys., Jahrg. 1882, Anat. Abth., pp. 207–320.

'86· Zur Deutung der Zirbeldrüse (Epiphysis). Zool. Anzeiger, Bd. IX. pp. 405–407.

Selenka, E.

'90· Das Stirnorgan der Wirbelthiere. Biolog. Centralbl., Bd. X. Nr. 11, pp. 323–326.

Spencer, W. Baldwin.

'86· The Parietal Eye of Hatteria. Nature, Vol· XXXIV. pp. 33–35.

'87· On the Presence and Structure of the Pineal Eye in Lacertilia. Quart. Jour. of Micr. Sci., Vol. XXVII. pp. 165–238, Pls. XIV.–XX.

Stieda, L.

'65· Ueber den Bau des Haut des Frosches (Rana temporaria). Arch. f. Anat. Phys. u. wiss. Med., Jahrg. 1865, pp. 52–66, 1 Taf.

'75· Ueber den Bau des Centralnervensystems des Axolotl und der Schildkröte. Zeitschr. f. wiss. Zool., Bd. XXV. pp. 361–406, Taf. XXV., XXVI.

Strahl, H.

'84· Das Leydig'sche Organ bei Eidechsen. Sitzungsb. d. Gesellsch. zur Beförd. d. gesam. Naturwiss. zu Marburg, Nr. 3, 1884, pp. 81–83.

Strahl, H., und Martin, E.

'88· Die Entwickelungsgeschichte des Parietalauges bei Anguis fragilis und Lacerta vivipara. Arch. f. Anat. u. Phys., Jahrg. 1888, Anat. Abth., pp. 146–163, Taf. X.

Weismann, A.

'87· Ueber den Rückschritt in der Natur. Ber. d. Gesellsch. zu Freiburg, Bd. II. pp. 1–30.

Wiedersheim, R. von.

'80· Das Gehirn von Ammocœtes und Petromyzon Planeri. Jenaische Zeitschr., Bd. XIV. pp. 1–23, Taf. I.

'86· Ueber das Parietalauge der Saurier. Anat. Anzeiger, Bd. I., pp. 148, 149.

EXPLANATION OF FIGURES.

All the figures are camera drawings excepting where otherwise indicated in the explanations.

ABBREVIATIONS.

cav. e'phy. Cavity of the epiphysis.
cbl. Cerebellum.
ceb. Cerebrum.
chs. opt. Optic chiasm.
cl. i. Cells of zone 5 of the retina.
cl. lns. Cells of the lens.
cog. Coagulum.
com. a. Anterior commissure.
com. p. Posterior commissure.
com. su. Superior commissure.
con't. tis. Connective tissue.
cp. sng. Blood corpuscles.
e'drm. Ectoderm.
en'th. va. Endothelium of retinal blood-vessels.
eph. vs. Epithelium of the epiphysial vesicle.
e'th. Epithelium.
la. trm. Lamina terminalis.
lns. Lens.
lob. opt. Optic lobes.
m. scu. Scale of the parietal eye.

mac. opt. Spot marking the position of the parietal organ.
mb. ba. ex. External basement membrane. [brane.
mb. ba. i. Internal basement membrane.
nl. Nucleus.
nl'. Nuclei of zone 2 of retina.
nl''. Nuclei of zone 4 of retina.
nl'''. Nuclei of zone 5 of retina.
os par. Parietal bone.
pig. Pigment.
prc. r. Processes radiating from the external basement membrane.
rtn. Retina.
sn. snq. Blood sinus.
tct. thl. opt. Roof of the optic thalamus.
thl. opt. Optic thalamus.
va. rtn. Retinal blood-vessels.
vnt.[3] Third ventricle of brain.
vs. Epiphysial vesicle.
z''. Second zone of retina.

PLATE I.

Fig. 1. Left face of a section through the dorsal wall of the head of *Phrynosoma Douglassii* in the sagittal plane, and consequently passing through the middle of the parietal organ. Diagrammatic in unimportant details. × 140.

" 2. Transverse section through the dorsal wall of the head and middle of the parietal organ of *P. coronata.* Diagrammatic in unimportant details. × 140.

" 3. Section of a small portion of the deep wall of the parietal organ and the enveloping connective-tissue capsule, to show the processes radiating from the external basement membrane. × 1060.

" 4. A transverse section of one of the retinal vessels, in which a blood corpuscle is seen. × 1060.

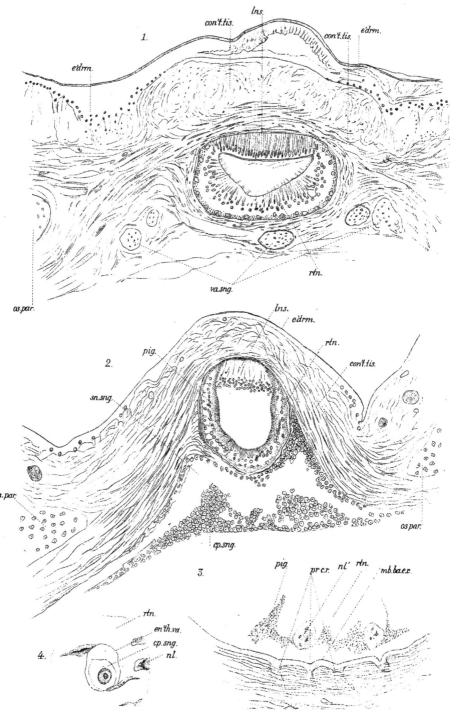

ABBREVIATIONS.

cav e'phy.	Cavity of the epiphysis.	*mac. opt.*	Spot marking the position of the parietal organ.
cbl.	Cerebellum.		
ceb.	Cerebrum.	*mb. ba. ex.*	External basement membrane. [brane.
chs. opt.	Optic chiasm.		
cl. ı.	Cells of zone 5 of the retina.	*mb. ba. ı.*	Internal basement membrane.
cl. lns.	Cells of the lens.	*nl.*	Nucleus.
cog.	Coagulum.	*nl'.*	Nuclei of zone 2 of retina.
com. a.	Anterior commissure.	*nl''.*	Nuclei of zone 4 of retina.
com. p.	Posterior commissure.	*nl'''.*	Nuclei of zone 5 of retina.
com. su.	Superior commissure.	*os par.*	Parietal bone.
con't. tis.	Connective tissue.	*pıg.*	Pigment.
cp. sng.	Blood corpuscles.	*prc. r.*	Processes radiating from the external basement membrane.
e'drm.	Ectoderm.		
en'th. va.	Endothelium of retinal blood-vessels.		
		rtn.	Retina.
eph- vs.	Epithelium of the epiphysial vesicle.	*sn. sng.*	Blood sinus.
		tct. thl. opt.	Roof of the optic thalamus.
e'th.	Epithelium.	*thl. opt.*	Optic thalamus.
la. trm.	Lamina terminalis.	*va. rtn.*	Retinal blood-vessels.
lns.	Lens.	*vnt.*[3]	Third ventricle of brain.'
lob. opt.	Optic lobes.	*vs.*	Epiphysial vesicle.
m. scu.	Scale of the parietal eye.	*z''.*	Second zone of retina.

* Processes secreted from the inner surface of the retina.

PLATE II.

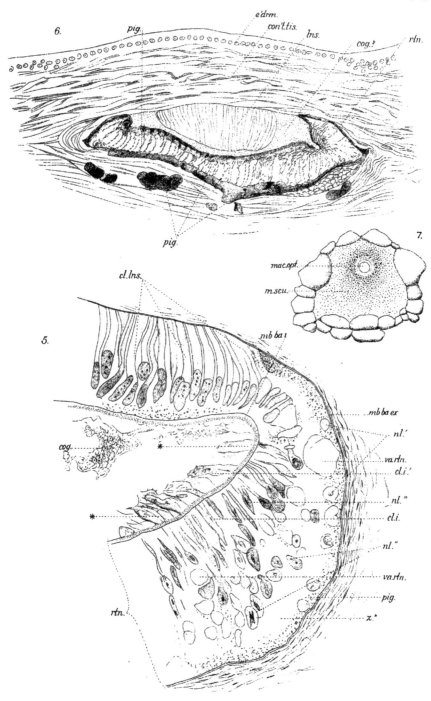

ABBREVIATIONS.

cav. e'phy.	Cavity of the epiphysis.	mac. opt.	Spot marking the position of the parietal organ.
cbl.	Cerebellum.		
ceb.	Cerebrum.	mb. ba. ex.	External basement mem-brane. [brane.
chs. opt.	Optic chiasm.		
cl. i.	Cells of zone 5 of the retina.	mb. ba. i.	Internal basement mem-
cl. lns.	Cells of the lens.	nl.	Nucleus.
cog.	Coagulum.	nl'.	Nuclei of zone 2 of retina.
com. a.	Anterior commissure.	nl''.	Nuclei of zone 4 of retina.
com. p.	Posterior commissure.	nl'''.	Nuclei of zone 5 of retina.
com. su.	Superior commissure.	os par.	Parietal bone.
con't. tis.	Connective tissue.	pig.	Pigment.
cp. sng.	Blood corpuscles.	prc. r.	Processes radiating from the external basement mem-brane.
e'drm.	Ectoderm.		
en'th. va.	Endothelium of retinal blood-vessels.		
		rtn.	Retina.
eph. vs.	Epithelium of the epiphysial vesicle.	sn. sng.	Blood sinus.
		tct. thl. opt.	Roof of the optic thalamus.
e'th.	Epithelium.	thl. opt.	Optic thalamus.
la. trm.	Lamina terminalis.	va. rtn.	Retinal blood-vessels.
lns.	Lens.	vnt.[3]	Third ventricle of brain.
lob. opt.	Optic lobes.	vs.	Epiphysial vesicle.
m. scu.	Scale of the parietal eye.	z''.	Second zone of retina.

PLATE III.

Fig. 8. Left face of a sagittal section through a portion of the epiphysis, a short distance above its connection with the brain in *P. Douglassii*. It is in part diagrammatic, though the outlines of the figure as a whole, and of most of the foldings of the epithelium, were drawn with the camera. From the same individual as Figure 1. × 312.

"   9. Similar view of a sagittal section from the same individual, to show the relation of the epiphysis to the brain and the blood sinus. × 30.

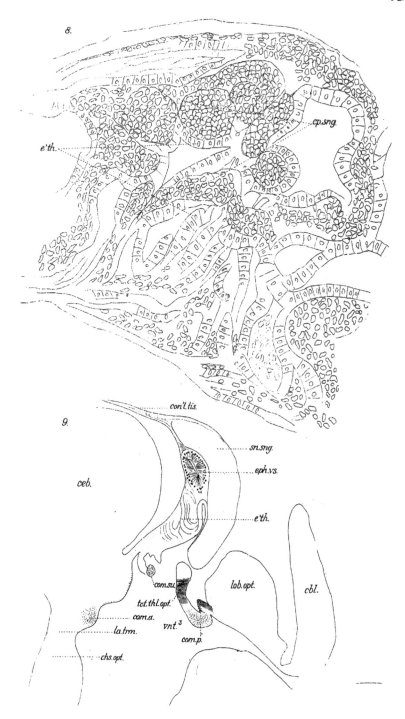

ABBREVIATIONS.

cav. e'phy.	Cavity of the epiphysis.	*mac. opt.*	Spot marking the position of the parietal organ.
cbl.	Cerebellum.		
ceb.	Cerebrum.	*mb. ba. ex.*	External basement membrane. [brane.
chs. opt.	Optic chiasm.		
cl. ı.	Cells of zone 5 of the retina.	*mb. ba. i.*	Internal basement membrane.
cl. lns.	Cells of the lens.	*nl.*	Nucleus.
coj.	Coagulum.	*nl'.*	Nuclei of zone 2 of retina.
com. a.	Anterior commissure.	*nl''.*	Nuclei of zone 4 of retina.
com. p.	Posterior commissure.	*nl'''.*	Nuclei of zone 5 of retina.
com. su.	Superior commissure.	*os par.*	Parietal bone.
con't. tis.	Connective tissue.	*pig.*	Pigment.
cp. sng.	Blood corpuscles.	*prc. r.*	Processes radiating from the external basement membrane.
e'drm.	Ectoderm.		
en'th. va.	Endothelium of retinal blood-vessels.		
		rtn.	Retina.
eph. vs.	Epithelium of the epiphysial vesicle.	*sn. sng.*	Blood sinus.
		tct. thl. opt.	Roof of the optic thalamus.
e'th.	Epithelium.	*thl. opt.*	Optic thalamus.
la. trm.	Lamina terminalis.	*va.rtn.*	Retinal blood-vessels.
lns	Lens	*vnt.[3]*	Third ventricle of brain.
lob. opt.	Optic lobes.	*vs.*	Epiphysial vesicle.
m. scu.	Scale of the parietal eye.	*z''.*	Second zone of retina.

PLATE IV.

Figs. 10, 11, 12. The left faces of three sections of *P. coronata*, parallel with the sagittal plane, — Figure 12 nearly median, Figures 10 and 11 to the left of it. Figure 10, farther to the left of the median plane than Figure 11, passes through the longest part of the epiphysial vesicle. Figure 12 is more highly magnified, to show the histological structure. Figs. 10 and 11 × 40. Fig. 12 × 90.

Lightning Source UK Ltd.
Milton Keynes UK
UKHW012209030119
334668UK00006B/771/P